Chemistry and Technology of Oils and Fats

Chemistry and Technology of Oils and Fats

M.M. Chakrabarty

ALLIED PUBLISHERS PVT. LIMITED

New Delhi • Mumbai • Kolkata • Lucknow • Chennai
Nagpur • Bangalore • Hyderabad • Ahmedabad

ALLIED PUBLISHERS PRIVATE LIMITED
1/13-14 Asaf Ali Road, **New Delhi**–110002
Ph.: 011-23239001 • E-mail: delhi.books@alliedpublishers.com
47/9 Prag Narain Road, Near Kalyan Bhawan, **Lucknow**–226001
Ph.: 0522-2209942 • E-mail: lko.books@alliedpublishers.com
17 Chittaranjan Avenue, **Kolkata**–700072
Ph.: 033-22129618 • E-mail: cal.books@alliedpublishers.com
15 J.N. Heredia Marg, Ballard Estate, **Mumbai**–400001
Ph.: 022-42126969 • E-mail: mumbai.books@alliedpublishers.com
60 Shiv Sunder Apartments (Ground Floor), Central Bazar Road,
Bajaj Nagar, **Nagpur**–440010
Ph.: 0712-2234210 • E-mail: ngp.books@alliedpublishers.com
F-1 Sun House (First Floor), C.G. Road, Navrangpura,
Ellisbridge P.O., **Ahmedabad**–380006
Ph.: 079-26465916 • E-mail: ahmbd.books@alliedpublishers.com
751 Anna Salai, **Chennai**–600002
Ph.: 044-28523938 • E-mail: chennai.books@alliedpublishers.com
5th Main Road, Gandhinagar, **Bangalore**–560009
Ph.: 080-22262081 • E-mail: bngl.books@alliedpublishers.com
3-2-844/6 & 7 Kachiguda Station Road, **Hyderabad**–500027
Ph.: 040-24619079 • E-mail: hyd.books@alliedpublishers.com

Website: www.alliedpublishers.com

Reprinted 2010, 2012

ISBN : 81-7764-495-5

Published by Sunil Sachdev and printed by Ravi Sachdev at Allied Publishers Private Limited, Printing Division, A-104 Mayapuri Phase II, New Delhi - 110064

Preface

The chemistry and technology of oils and fats deals with a very broad group of natural raw materials which have an important place in the overall economy of India. During the publication in 1927 of his monograph on *Industrial Chemistry of Fats and Waxes,* Professor T.P. Hilditch lamented at the time that 'The chemistry of glycerides has been a neglected chapter of organic chemistry.' Since then, quite a number of books have appeared dealing with both fundamental and applied aspects of the subject. Among them one of the most authoritative books have been the series *'The Industrial Oil and Fat Products'* authored originally by A.E. Bailey and later edited by Daniel Sweren in the US of which the 5th edition 1994-95 is currently in print. The purpose of writing this monograph was to write a textbook for Indian students and also for those who are working in industries based on there important raw materials. It has been, therefore, the aim of the author along with the collaborators, Professor D.K. Bhattacharyya of Calcutta University and Professor D.N. Bhowmick of Bombay University, to try to produce a book of a size and priced so that it may be used by Indian students.

The author wishes to acknowledge with gratitude the financial assistance given for the purpose by the Department of Science and Technology, Government of India, which made this task possible.

The author also wishes to thank the Asiatic Society of Calcutta, in general, and the General Secretary, (Late) Dr. Chandan Roychaudhuri and the Treasurer, Professor Anil Sarkar, in particular, who also came forward and acted as an agency for making this book a reality.

It is but natural that well known books and literature on this subject have been taken advantage of and also the published materials in journals like the *Journal of the American Oil Chemists' Society* and the published monographs on the subject from time to time by the American Oil Chemists' Society and other organisations. Wherever assistance has been taken from any publication, it has been acknowledged. The author acknowledges in particular the following: Academic Press Limited, London, and John Wiley & Sons, New York, for using some of the information and illustrations from the books on the subject published by them. The author also would like to acknowledge the assistance of Shri. O.P. Narula, Industrial Consultants (India), New Delhi, for permission to use some of the materials published in their monograph, e.g. *'Treatise on Fats, Fatty Acids and Oleochemicals.'*

The author would like to acknowledge the help in preparing the manuscripts from the collaborators, Professor D.K. Bhattacharyya, Rashbehari Ghosh Professor, Department of Chemical Technology, University of Calcutta, Calcutta - 700009; and Professor D.N. Bhowmick, University Department of Chemical Technology, University of Bombay, Matunga, Bombay - 400 019.

The author wishes to put on record in this connection the encouragement received from Professor Mrs. Asima Chatterjee, Coordinator, Centre for Advanced Studies in Natural Products, University of Calcutta, Science College, Calcutta - 700 009.

The author wants to further thank Ms. Sonali Sengupta, Technical Assistant who helped in preparing the manuscript.

The author would like to record with gratitude the assistance received from Shri Anil Chatterjee for his stenographic assistance and for help in compiling all materials in an appropriate format.

The author also would like to acknowledge the assistance received from Shri Shanti Nath Ghosh of the Department of Chemical Technology, University of Calcutta.

Finally, it is a pleasure to acknowledge the help, sympathy, patience and understanding of my wife, Sreemati Bharati Chakrabarty in the absence of which the task would have been impossible.

Contents

CHAPTER 1

Introduction

INTRODUCTION

The use of fats and oils for edible and industrial purposes has as long as history as mankind itself (1). Their use in cooking and other purposes in everyday life in ancient cultures like the Chinese, the Indian, the Egyptian and the Greco-Roman was just as important as it is in present time. The ancient Aryans in India are known to have used butter fat or ghee in offerings to gods as also for cooking purposes apart from its use in massage oils. The old Sanskrit texts of the Hindus give specific instructions of how to anoint the body with oils before taking a bath. In fact, the Sanskrit word for oil 'taila' is derived from 'til' i.e. sesame seed which was known in India from very ancient time. Some of the oldest known fats are those found in pots of the Pharaohs of Egypt. There is a description, as old as 1400 BC, regarding the probable use of animal fats for moving large stones or wooden logs in ancient Egypt. In the ruins of Pompcii, residue of soap was found in a soap boiler which was used to produce soap from potash and fats. The Romans knew how to make candles from tallow and made the better sorts with a mixture of tallow and bees' wax. Regarding the sources of oils, olives of the Mediterranean, rapeseeds of Europe, sesame seeds of India and soyabeans of China are amongst the oldest described by the ancients.

The technology for the recovery of oils was also discovered by the ancient people. The Chinese knew the use of wedges and the Romans knew the use of the double-edged running grinding stone rotating in a stone plant for crushing olives. The crushed mass of seeds were heated before being extracted by pressure. Stones were used to put pressure in pounding the masses. The use of animal energy and water energy instead of human labour for grinding and stamping or pounding the masses were the mediaeval developments. In India the 'ghani' crushing system consisting of a stationary mortar and a rotating pestle has a very ancient history. It was almost the only production system till about the turn of the present century. While some 400,000 units were functioning all over the country, their numbers are now estimated to have dwindled to between 1.0 and 1.5 lakhs, but there have been upgraded designs developed by the Khadi and Village Industries Commission (KVIC). However, in the western world, the development of the roller type grinder introduced in 1752 and hydraulic presses discovered between 1795 and 1800 led to the application of plate presses by Montgolfier in 1890 along with the hydraulic press. Screw presses were developed later. The first patent for solvent extraction was issued between 1855 and 1858 using carbon disulphide as the solvent. Later, in 1864, a patent was issued for the use of hydrocarbon solvation, the basis of present-day technology. Although the proposal for the use of screw presses was made as early as 1877 by Marticke, the first true expeller or the modern screw press was introduced by Anderson in 1906. Mege Mourie obtained the first patent on the production of margarine in 1869 thus providing the fundamental basis for the production of a cheap butter-like high calorie spread. The invention of catalytic fat hardening by Normann in 1902 gave the oil-based industry independence from hard animal fats like lard and tallow. This led to the use of cheap marine oils such as fish and whale oil after hardening by hydrogenation as a hard fat. Other extractors have been developed like the continuous chain and basket type extractors.

The sources of fats and oils were then, as they are now, plants and animals, the latter being primarily the land animals, the use of marine animals being secondary. Mainly because of growing world population, new fatty raw material resources are now being discovered. Further improvements in the technology of agriculture by chemical, biochemical and genetic engineering research have led to high-yielding, climate and disease-resistant crop varieties. In times of serious shortages (viz., war crisis) both fatty acids and glycerols, two main building blocks could be synthesized or prepared by fermentation using micro-organisms and it has been shown that the cost of production of these oils could be almost at par with those produced from plants. If we recall some major points of scientific development regarding the cxtraction of oils and fats historically, they are as follows.

In 1741 Geoffroy observed that soap acidified by mineral acids is decomposed to a substance soluble in alcohol which was not the property of the original raw material. Scheele in 1779 saponified fats with lead oxide and produced a water-miscible product, which he called 'sweet oil', that we call glycerol today. Between 1813 and 1825 the French chemist Chevreul identified the acid character of the substances which we know today as fatty acids and established the alcohol character of glycerol and the role of water in saponifying fats. Because of this work Chevreul is considered to be the founder of modern fat chemistry. Berthelot established the triglyceride (triacyl-glycerol) character of fats and hydrolyzed and esterified them. The introduction of many useful group characteristics known as 'values' have been due to a number of scientists. In about 1879 Merz described acid values and Koettstafar first told about saponification value. Iodine value in saturation was first proposed by Hueble in 1879. The modern elucidation of triglyceride composition was taken up by T.P. Hilditch in England and his co-workers between 1927 and 1951. He developed the ideas concerning the triglyceride composition of oils and fats as we know today. The introduction of chromatography executed originally on columns by Tswett in 1906 and developed in gas phase by James and Martin in 1952 opened a previously unknown chapter in establishing the chemical composition of fats and oils. Chromatography in combination with other methods of instrumental, physical analysis and the application of computers have been very useful, not only for elucidating structure of oils and fats, but also for the control of production processes.

CHEMISTRY OF FATS

Oils and fats both for edible and industrial purposes may be of either vegetable or animal origin. Among the fats of animal origin there are three categories, viz. (1) land animal body fats, (2) land animal milk fats, and (3) fats from marine animals (fish, whale etc.).

(a) Among the fats of vegetable origin there are those which are from annual plants, namely the seed fats like groundnuts, sunflower, linseed, i.e. mainly those which are cultivated and harvested seasonally every year.

(b) There are also fats of perennial plants or trees which once planted may bear fruits, the seeds or kernels of which are obtained every year for quite a number of years. Examples are coconut kernel, palm, palm kernel, mowrh seed, sal seed etc. There are also those obtained primarily as by-product, for example, animal fats may be obtained as by-product of the meat or fishery industry. In many cases there are fats, for example soyabean oil where the non-oily product, the protein part, is the main product and oil is a by-product. In the case of cotton seed, cotton is the main product and cotton seed oil is a by-product. The residues, either from vegetable or from animal origin, are also considerably important from the point of view of overall economy. For example, oil cakes left after the extraction of oil from seeds or nuts are used as animal-feed, poultry-feed, fish-feed and also for human food. The cakes of soyabean, groundnut, mustard and other seeds are very valuable by-products. Both source material and the oil can lead to other product or products. For example, although many oils after extraction are taken without further refining, not only in India but also in other countries of the world, the products obtained as refinery by-products may be of considerable commercial value. Phosphatides or lecithin is an important by-product of the oil refining industry and used as an emulsifier for both edible and inedible products like surface coatings. It should be mentioned that both terms, oil and fat, are often used to indicate liquid and solid respectively. In chemical terms the difference is only in their melting points. There are some oils which are liquid in summer and semi-solid or solid in winter. The butter fat or ghee which is a milk product also belongs to this category. So, the term 'fat' will be used to indicate both oil and fat hereafter.

Chemically, oils and fats are esters of glycerol and fatty acids. The fatty acids are usually long-chain aliphatic mono-carboxylic acids. Chemically, the oils and fats are the reaction products of one molecule of glycerol with three molecules of fatty acids which give three molecules of water and one molecule of triglyceride. This means that three fatty acids are incorporated into one molecule of glycerol since glycerol is a tri-hydroxylic alcohol. The ester formed is known as triglyceride or triacyl-glycerol. Acyl glycerols and many other compounds which by origin are either fat incorporated or fat-soluble substances, are called lipids. Lipids which contain the glycerol residue are called glycero-lipids. Lipids which after hydrolysis give only two products such as glycerol and fatty acid or fatty alcohols and fatty acids (the latter are called waxes) are known as simple lipids. Fats are not composed solely of triacylglycerol. Fats in which only two of the three available triacyl groups are substituted by acyl groups are called diacylglycerol (diglycerides) and those with just one acyl substituent, are called monoacylglycerol (monoglycerides). Mono and diacylglycerol (glycerides) are also called partial acylglycerols or partial glycerides according to the traditional denomination.

When separating (rendering) fats from their natural surroundings the processes used rupture and destroy the cell structure of housing tissues. For this reason fats in the crude stage are always mixed physically, chemically or colloid-chemically as micelle solutions or colloid dispersions with smaller or larger amounts of the constituents of their parent source (the so-called accompanying components) some of which are also lipids. The parent sources are quite characteristic of the nature and sources of the fats and therefore may give very important analytical clues regarding their origin and quality. Among the component substances we should first mention free (non-esterified) fatty acids (FFA). These, if present in higher amounts than is accepted for the given source, are the first and most important (necessary but insufficient) measure for testing quickly the quality of a given sample.

Besides the full and partially substituted acyl glycerol and free fatty acids, it is necessary also to consider the other component lipids, viz. the phosphoacyl glycerols (phosphoglycerol, glycerol phospholipids), the main building elements of these compounds is an ester of glycerol with or through orthophosphoric acids, the common basis for fat and phospholipids synthesis. This ester is further esterified with a glyceride residue with fatty acids and on the phosphoric acid residue in a particular manner, for example, the mono alcohol or glycerol like alcohol (inositol) esters of orthophosphoric acid can be formed. This can be substantiated by analysis upon which more than two types of products are formed. They may also be called complex lipids. All the lipids mentioned above occur in crude fats in quite distinct amounts and comprise upto 99 per cent of the fat contents. These lipids form the so called major lipid components.

COMPOSITION OF FATS

Natural fats and oils vary widely in their physical properties even though they are made up of the same or similar fatty acids (2). This is because (a) individual fats and oils vary over relatively large ranges in the proportions of the component fatty acids, and (b) the structures of the individual component fatty acids vary although these two factors are interdependent. There are aspects of the overall effects that can be more or less attributed to one or the other. There are a number of factors which affect the composition of fats and oils in the vegetable kingdom, for example climatic condition (hot or cold, wet or dry), soil type in which the parent plant was grown, geography of the location (groundnut in Gujarat and groundnut in Karnataka may vary in composition), maturity of the plant, health of the plant, numerous environmental conditions (presence of certain moulds, bacteria and enzymes), specific location of the oil seeds within the flower itself (3,4) and, most important, the genetic variation in the plant (rapeseeds which generally contain 40–50 per cent erucic acid was bred to produce a low percentage of erucic acid e.g., 0–5 per cent).

In land animals fat and oil composition varies according to animal species. In addition, it varies in a very complex fashion: (a) with the type of the animal (for example, increased linoleic acid fat with supplemented diet), and (b) with location of the fat within the animal (bovine milk fat is much different from bovine body fat). Also, liver fat in many animals differs from the adipose tissue fat in the proportion of palmitic acid in the two positions (5), and there is much difference in composition in pork, beef and lamb fats (6). In addition, composition varies in relation to the health and well-being of the animal in the environment in which it matures.

Marine animal fats and oils are not only characteristically different in many ways from those of land animals or vegetable origin, but they show differences depending on whether the fish is an oceanic or fresh-water species, and within a particular species whether it is the cold or warm water derived and the season when it is caught, winter or summer.

The great number and variety of fats and oils in nature and the effect of slight and imperceptible changes in the environment on their composition make it unlikely that the structures of fats and oils (triglyceride) are governed by a set of simple rules. However, well known chemists from time to time have tried to formulate on the basis of their experimental evidences a number of guidelines to the composition and structure of fats. Before 1900 AD, sufficient qualitative evidences have been obtained which tended to show that although fats and oils were mainly mixtures of mixed triglycerides and not of simple triglycerides. The first full and systematic investigation of the quantitative glyceride composition of fats were initiated in about 1927 by Hilditch and co-workers and a large volume of work has been reported during the period 1927–1964. These gave rise to the well known theory of 'Even Distribution', first proposed by Hilditch and thereafter expanded by others. According to this theory the component fatty acids in natural fat tend to be distributed as broadly as possible among all the triglyceride molecules described in some specific terms (7). The theory is as follows:

A) When a given fatty acid A forms about 35 mole per cent or more of the total fatty acid (A + X) in a fat, it will occur at least G (AX_2) in practically all the triglyceride molecules of the fat in question.

B) If it forms about 35–65 per cent of the total fatty acid (A + X) it will occur twice G (A_2X) in any given triglyceride molecules in some instances and more frequently the higher the proportion of this acid in the total fatty acids.

C) If it forms 70 mole per cent or more of the total fatty acid, the remaining fatty acid X can at most form mixed glyceride $G(A_2X)$. All the excess of A thus will only appear as simple triglyceride $G(A_3)$.

D) A minor component acid that forms less than about 1/3 of the total fatty acid (for example, 15 mole per cent or less) will not occur more than once in any triglyceride molecule and not at all in many of the triglyceride molecules.

Shortcomings of the 'Even Distribution' theory appeared quickly and many oils such as linseed, safflower, soyabean etc. gave a more random pattern of distribution (8). Some fruit coat fats deviated and animal body fats generally exhibited marked deviations especially those rich in stearic acid which appear to concentrate together, for example, fully saturated triglyceride appearing in fats where the saturated fatty acid content does not exceed 60 per cent. Although there is absolutely no reason why vegetable oil and animal body fats would be governed by same rules of structure, opinions were advanced that the animal fats were originally synthesized in the animal body by a process of even distribution which were then partially hydrogenated to produce characteristic saturated glycerides of the stearic-rich animal body fats.

In 1952 it was discovered that pancreatic lipase selectively hydrolyses triglycerides attached at the primary hydroxyl groups (1 and 3 or α and α') of the glyceryl molecules long before those at the central position (secondary hydroxyl) or 2 or β are removed. Within a few years many investigators employed this technique on oil seed fats and it was evident that a very high proportion (usually 96–99 per cent) of the acyl groups attached at the 2 position consisted of oleic and/or linoleic acid. Saturated fatty acids and any other unsaturated acids (for example erucic acid) with any excess oleic or linoleic acid necessary to combine with 2 glyceryl position were almost wholly attached to the α, α' (1 and 3 glyceryl groups). In the animal fats the position was less positive; the proportions of saturated acid (mainly palmitic acid) combined at the central glyceryl position were greater than in seed fats, ranging in a number of species from 20–35 per cent of the total fatty acids in this position and in one group of pig fats forming 70–80 per cent (9). So, the distinct limitation of the 'even distribution' of fatty acids in triglyceride was appreciated.

Some investigators (10) have suggested that animal fats conform to a pattern of 'random' distribution. They assumed that in the biochemical synthesis of fats each set is governed by the laws of chance. The chemical preponderence of the α position in glyceryl for any initial esterification notwithstanding, in any biosynthesis everything is simple, completely statistical and subject to no directional or steric effects whatsoever. It is ofcourse entirely possible that there should be a basic difference between activities of plant and animal enzymes so that the fatty acids of animal fats are distributed at random, whereas those in most vegetables are distributed otherwise, but the mechanism does not explain those vegetable fats which appear to conform to a random distribution.

Again, evidence has been presented (11) that some glycerides, for example that of corn oil, conform to neither the 'even' nor the 'random' distribution pattern, but a pattern designated as 'partial random'. An interesting proposal concerning the distribution of fatty acids in a glyceride by the Indian chemist Kartha is called 'Restricted Random Distribution' and it was supported by many of the facts available from about 1951 and 1964. But after 1964 exceptions to this theory appeared again. Kartha also reported some. However, according to this theory (a) fully saturated triglyceride (GS_3) can be present in each substituent of animal or plant fat only to the extent they can remain fluid (soluble) *in vivo*; (b) *in vivo* the fatty acids are distributed at random among the four glyceryl types GS_3, GS_2U, GSU_2 and GU_3 provided the GS_3 thus produced can exist in the liquid state; and (c) when the proportion of GS_3 which can exist in a liquid state *in vivo* is less than that which can be produced by unrestricted random distribution, when fatty acids are nevertheless distributed according to chance among the glyceryl types. Kartha has formulated a rule for calculating glyceride type distribution of any natural fat from S - GS_3 contents. He has determined the proportion of the four glyceryl types in innumerable fats by an important oxidation isolation procedure and he has also compared the glyceride composition of fats by fractional crystallization. The composition was calculated according to 'Restricted Random Distribution' theory. Bailey has given the detailed calculation of composition of fats determined by fractional distribution. However, later the experimental basis for Kartha's restricted random distribution theory has been reported to be unsound. Later workers have tended to elucidate the complexity of structure, the limitations of methodology employed and some of the remaining unsolved problems.

Youngs (12) using a gas liquid chromatography (GLC) and oxidation method has concluded that some seed oils like sunflower and linseed oil and animal fat like tallow are restricted random. He along with his co-workers analysed some vegetable and animal fats with the use of GLC Lipase Hydrolysis method. They found some oils followed the random distribution pattern, but there were also others which did not conform to this pattern. Various workers have studied the specific distribution of fatty acids and glycerides of animal and vegetable fats by means of pancreatic lipase hydrolysis, counter-current distribution techniques and other newly developed methods. However, almost in every case deviations are found. Gunstone (13) has proposed a theory of acyl group distribution in vegetable fats based on enzymatic hydrolysis and has related this theory with biosynthesis of triglycerides. According to this view the secondary hydroxyl group of glyceryl is preferentially acylated by C_{18} unsaturated acids. The primary triacyl of glyceride are acylated subsequently by all remaining fatty acids and by any excess of C_{18} unsaturated acids not required at the secondary position. Within these limits the distribution of acyl groups at each position is statistical. Other workers have also tried by using instrumental methods and lipase hydrolysis method. While the problem has been under critical examination for a long time (studies are being continued even now) Carter Litchfield (14) tried to summarize on the basis of evidence collected upto 1972 the various distribution hypotheses as follows:

(a) 1,3 random - 2 - random hypothesis has proven fully accurate for predicting the triglyceride composition of seed fats containing only common fatty acids provided enantiomorphic triglycerides are not considered; (b) the 1,2,3 random hypotheses may possibly describe the triacyl position of adipose tissue fats from higher animals maintained on a constant diet for 4–6 months before sacrifice; and (c) with every natural fat triglyceride mixture the two theories mentioned before give only a rough guess of the triglyceride present. Such calculation provided 'best guess answers' which can be useful in planning and interpretation of experiment, but they do not approach the accuracy of direct analysis procedure.

In recent years advance techniques like high performance liquid chromatography, thin layer chromatography, ion-exchange chromatography, radiation chromatography along with gas chromatography have provided further insight into the analysis and composition of fats.

FATTY ACIDS

The molecular weight of the glyceryl portion (C_3H_5) of the triglyceride molecules is 41. The combined molecular weight of the fatty acids RCOO—comprising the remainder of the molecule will vary with different fats from 650 to 970. Thus the fatty acids contribute 94–96 per cent of the total weight of the molecule because of their preponderance in glyceride molecule and also because they contribute to both the chemical and physical properties of the glyceride. The fatty acids influence the physical and chemical character of the glycerides.

$$
\text{Molecular weight 41} \quad
\begin{array}{c}
\text{H} \\
| \\
\text{H—C} \quad - \quad \text{OOCR}_1 \\
| \\
\text{H—C} \quad - \quad \text{OOCR}_2 \\
| \\
\text{H—C} \quad - \quad \text{OOCR}_2 \\
| \\
\text{H}
\end{array}
\quad \text{Molecular weight 650–970}
$$

(a) In nature the fatty acids occur with an even number of carbon atom with few exceptions. The fatty acids may be saturated, unsaturated or may belong to other types. In nature some species of seed fats may consist of an oxygenated derivative, such as hydroxyl group, epoxy group, keto group. Although most fatty acids in nature are straight chained, there are also branch chained fatty acids of certain origin. As the molecular weights of the branched fatty acids are the same as those of the corresponding non-branched ones, they are also called *iso* or isomeric fatty acids. Some of the important fatty acids with their systematic name, trivial name, chain length, melting point and melting point of the corresponding triacylglycerol are given in Table 1.1.

(b) *Physical properties of fatty acids*: Going along with the series of fatty acids from those of short chain length to those of long chain length the physical state changes from liquid to solid consistency. The members of the longest chain length even under ambient temperature in moderate climate have a wax like appearance. The melting point (MP), boiling point (BP) and the refractive index (nD) all increase with increase in chain length while the density (D) decreases with increasing chain length. The shorter chain members are easily distilled at atmospheric pressure, whereas the longer chain ones which have very low vapour pressures should preferably be distilled under vacuum because of the danger of decomposition. The shortest chain fatty acids with upto 12 carbon atoms (lauric acid) can be distilled easily by steam distillation. Fatty acids with more than 12 carbon atoms should preferably be distilled under reduced pressure using special columns. The solubility in water decreases as the carbon number increases. Decanoic acid or capric acid is only very slightly soluble even in boiling water. The solubility or miscibility of fatty acids with different organic solvents in the liquid state (above their melting point) depends on the affinity of the acid to the solvent, different solvents being more or less polar in character. However, in practical terms nearly all fatty acids are fairly soluble or readily miscible with polar (alcohols, ketones) or apolar (for example, aliphatic, aromatic branched or cyclic hydrocarbons) solvents.

(c) *Chemical properties*: In general fatty acids are easily transformed by alkalis or metallic oxides to their corresponding salts called soaps. This neutralization reaction is easily performed even in heterogeneous phase (just-mixed aqueous alkalis or metallic oxides heated with fatty acids). Inspite of their saturated nature the saturated fatty acids are not quite inert. Free radicals initiate oxidation and halogenation processes which become very enhanced at higher pressures and elevated temperatures and ionic mechanisms easily cause substitution of one or more hydrogen atoms in the chain. The cleavage of the chain (rupture) and the formation of lower molecular weight scission products at exaggerated reaction conditions (for example, temperature above 300°C) are quite possible. Heating to high temperatures in an inert atmosphere may cause decarboxylation of fatty acids by loss of carbondioxide with the formation of hydrocarbons.

Unsaturated Fatty Acids

Fatty acids having hydrogen deficient carbon atoms bonded by double or triple valencies are called unsaturated. According to the number of double bond present they are called monoethenoid (monoenoic) or di- or tri-, tetra-, penta enoic acid or in general polyenoic fatty acid. Ethylene is the simplest ethene monoene. The less common triple body ones are called ethynoic acid, acetylene being the simplest ethyne.

(a) *Nomenclature*: Nomenclature of unsaturated fatty acid has been assigned by modifying the saturated fatty acid of corresponding chain length; the 'a' in the last syllable of the Greek name is changed to 'e' for the ethynic type 'y' to ethynoic type (15). Repetitive presence of the same unsaturated fatty acid is built into the name by writing di-, tri-, tetra-, etc. according to the occurrence. The viable part of the name is 'enoic acid', e.g. octadecadienoic acid is double ethynic unsaturated 18 carbon chain (C_{18}) fatty acid. The place, i.e. location of any change to or substituent on the carbon chain, is given by an Arabic numeral describing the first substituted carbon just counter from the carboxyl group as carbon no. 1 (C_1). The name of the substituent group (oxo) according to hyroxy, epoxy, methyl, ethyl, methoxy, cyclopropane, cyclopentane, chloro-, bromo- etc. is written as a prefix, the locant or a location as an Arabic numeral before the prefix. The unsaturation or locant is incorporated between the prefix and the modified Greek name by an Arabic numeral, e.g. 12-hydroxy 9-octadecenoic acid is C_{18} fatty acid at the 9 position, mono unsaturated at the 12 position substituted by a hydroxyl group (trivial name: Ricinoleic acid).

(b) *Isomerism*: Unsaturation can be found at different locations which give rise to the phenomenon of positional isomerism. Isomers are compounds which have the same molecular formula and molecular weight but are different in structure. Another type of isomerism is Geometric Isomerism. For fatty acid this concerns the relative position of the two hydrogen atoms attached to the double-bonded carbon atoms against the symmetric plane, parallel to the longitudinal axis of the fatty acid molecule. If the two hydrogen atoms are located on the same side of the bisecting plane, then the double bond is called a *cis* (c) bond or as agreed by IUPAC a Z (German Zusammen) configuration. If the hydrogen atoms are on opposite side of the plane, the bond formed is called a *trans* (t) bond or an 'E' (German entgegen) configuration. The geometric isomerism is expressed by writing the type of configuration and the double or the triple bond as an Arabic numeral separately with solvent that prefix ethylene and the modified generic name. No sign at all could mean unsaturation or the natural c or Z configuration, e.g. *cis*-9, *cis*-12 octadeca dienoic acid ≡ Z–9, Z–12 octadeca dienoic acid ≡ trivially 9c 12c linoleic acid.

(c) *Shorthand notations*: There are two shorthand notation systems for unsaturated fatty acids e.g. one used by organic and analytical chemists and the other used by biochemists. The first system abbreviates fatty acids in the form of $x : y$, zc (or z t) where, x stands for number of carbon atoms in fatty acid chain, y for number of double bonds present and z for the position of the double bonds. All are numbered from the carboxyl group as C_1 plus the geometric configuration of the double bonds: *cis* or *trans* (Z or E). In the second system which is the ω or the end of the carbon chain (ECC) (or more up-to-date n–m system, ($x : y$, ω, m; or $x : y$, n–m), the meaning of x and y is the same as before. The terms ω and n stand for the fact that the terminal methyl group of the fatty acid chain is taken as C_1 and m is now the position of the locant of the first double bond observed from the methyl end.

(d) *Physical and chemical properties*: Double bond of the *cis* configuration seen sterically give a bending in the straight chain profile of the fatty acid molecule. Actually because of the valence increase (about 109 grades for a single bond) the profile is of a zigzag type.

Table 1.1: Shorthand Notation for Fatty Acids

Trivial name	*Analytical shorthand*
α-Linolenic acid	18:3, 9c, 12c, 15c

$$\overset{C_{15}\downarrow}{} \quad \overset{C_{12}\downarrow}{}$$

$$CH_3\text{–}CH_2\text{–}CH = CH\text{–}CH_2 - CH = CH\text{–}CH_2\text{–}CH{=}CH\text{–}CH_2\ (CH_2)_6 - COOH$$

ECC Shorthand	*Systematic name*
18:3, n-3 18:3, ω 3(6,9)	Z-9, Z-12, Z-15-Octadeca-trienoic acid

Double bonds in the *trans* configuration fits sterically more smoothly into the zigzag line of the rest of the molecules which calls our attention to the distinct differences between the physical properties of isomeric states, e.g. *trans* acids melt at higher temperature than do *cis* ones which explains the liquid state of the latter at room temperature. The *cis* isomers occur more frequently in nature than the *trans* ones.

The presence of double (or triple) bonds between vicinal carbon atoms provides special reactivity to the neighbouring so called methylenic carbon atoms. The bonds exert a more strong attraction on the electrons of the valency, binding its hydrogen atoms. This interaction causes acids containing double bonds to be very reactive or sensitive to certain chemical origins and/or external physical conditions.

Unsaturated fatty acids may undergo reaction when (even unevenly) confronted with oxygen, hydrogen, halogen, inorganic or organic acids, sulphur compounds, heavy metals, heat, radiation or generally free radical or ion induced processes. Fatty acids containing more than two so called isolated (at least one methylene group interrupted or ejected non-methyl interrupted) double bonds are very easily auto-oxidized, isomerized or polymerized. These reactions can occur even in the cold (e.g. drying oils). Fatty acids can also accept the above reagent by addition or substitution reactions.

Most of the poly-unsaturated fatty acids are to be found in the triacylglycerol, all oils of vegetable and in the oils of aquatic animals. The latter contain fatty acids with four, five or even six mainly isolated double bonds. Because of this high accumulation of unsaturation (many 1,4 pentadecanoic group in series) these fatty acids are even more prone to the alterations mentioned above.

The average degree of unsaturation of fatty acids can be expressed analytically by the iodine value. Differences in stereo configurations can easily be detected spectroscopically in the infra-red (IR) region by measuring the specific absorption of light (by the out of plane vibrating CH linkages) contained in the *trans* configuration. The refractive index (nD) increases with the number of double bonds present and with increasing conjugated character of the double bonds. The density decreases with decreasing unsaturation. Therefore, changes in these two criteria may serve as quick controls for saturation during the catalytic hydrogenation of fats.

(e) *Other types of fatty acids*: Besides saturated and unsaturated fatty acids, nature also produces oxygenated species of which ricinoleic acid is a hydroxylated member, vernolic acid an epoxydised member and licanic acid of the keto group containing C_{18} representative of all fatty acids of vegetable origin.

Exceptionally, some fatty acids, e.g. those present in micro-organism and some synthetic ones are not built up from straight aliphatic chain, but may be substituted on one or more carbon atoms of the chain by one or more methyl (ethyl) groups. These fatty acids are called branched. Certain bacteria in the gastric tract of ruminant animals can decompose food to C_3 unit, more specifically to propionic acid. A resynthesis can then incorporate some propionic coenzyme A building blocks into the chain, the coupling being performed again on the β carbon atom of the coenzyme compound propionic group. As the molecular weight of the branched fatty acids are the same as those of the corresponding non-branched straight chained ones, they are called *iso* or isomeric fatty acids. Branched chain fatty acids are to be found in small amounts in the milk and fat (tallows) of ruminants and can be detected by using a suitable technique such as gas chromatography (GC). Ingested linoleic, linolenic fatty acids, e.g. in grass, can be partially hydrogenated bacterially in the rumen leading to the formation of 5–10 per cent natural *trans* isomers and some positional isomers of linoleic or oleic acids again in the depot fat of ruminants.

Some fatty acids in plants contain at the end of their chain a cyclopentanoic unsaturated five-member ring. The existence of cyclo-propane three-member rings in the middle of the fatty acid chain is exemplified by the low level presence of sterculic, malvalic and dihydrosterculic acids in cotton seeds and hemp oils.

(f) *Bio-synthesis of fatty acids*: Both plant and animal cells are surrounded by non-rigid semi-permeable membranes which allow the diffusion of small molecules into and out of the cells. Plant cells have more definite shape because of cellulose content of their more rigid cell wall surrounding the membranes. Cells have nucleus which contains chromosomes. The chromosomes are built up with nucleic acids and carry the genetic information required for the maintenance of the cells. The cytoplasm inside the cell is an aqueous colloidal solution of proteins and other constituents. The endoplasmic reticulum is a complete network of filaments which together with ribosomes are the sites of protein built up from amino acids. The green matter called chlorophyll is responsible for the photosynthesis of organic material in the green plants. More specifically chlorophyll catalyses the transformation of the sun's radiation energy to chemical energy by which the photolysis (cleavage) of water to its elements and the binding of atmospheric carbon dioxide outside is made possible and last but not the least, the mitochondria, small rod like structures in the cytoplasma are 'power houses' of the cell releasing the chemical energy required by the cell for its metabolic processes. The enzymes needed for fat synthesis are housed either inside or outside the mitochondria.

One of the most important pathways for producing fatty acids is step-wise two carbon atoms 'denovo' synthesis of a saturated carbon chain up to hexadecanoic (palmitic) acid. Fatty acid synthesis is executed by two mechanisms: chain elongation and desaturation. Elongation is accomplished by coupling (polycondensation) of malonyl-coenzyme A molecules with a 'starter' acetyl coenzyme A molecule to a butyryl group. All these acetyl and malonyl building blocks are attached to so called 'acyl carrier protein (ACP)'. Although the malonyl coenzyme A constitute C3 unit, it ends up in C2 unit in each step losing its acquired carboxyl group by decarboxylation (the carbon dioxide is originally delivered by the enzyme biotin fixing carbon dioxide from the atmosphere in the 'active form'). This synthesis via further malonyl coenzyme A molecule is terminated with the hexadecanoic acid. The necessary condensation, reduction, dehydration and hydrogenation steps (four in total) are executed by suitable interaction of different plant enzymes and coenzymes all of which are proteinaceous and/or nucleic acid in nature. This so-called multi-enzyme complex is not present in the mitochondria that are found in the cytoplasm of the cell.

Further elongation to longer chain length by the two carbon (acetyl coenzyme A) units is mediated by the enzymes present in the mitochondria leading primarily to stearic acid. The elongation of some unsaturated fatty acids is accomplished in the plant cell by this mitochondrial mechanism as for example, the formation of 22-1 n-9 (erucic acid) cruciferae rape or mustard seed oils from 18-1 n-9 (oleic acid).

Unsaturation of the fatty acids in plants occurs by 'oxidative desaturation'. For example, by the enzymic process the 18:0 saturated fatty acid (stearic acid) already synthesised in the plant cell is first transformed to monounsaturated 9 octa-decenoic acid (oleic acid).

Animal organisms can also desaturate stearic acid to oleic acid but cannot desaturate the latter to linoleic or linolenic acid. These two acids belong to the group of the so called polyunsaturated fatty acids (PUFA). There are some polyunsaturated fatty acids which are very important for animal health but cannot be synthesized by the animal organism. These fatty acids must then be introduced ready-made by ingesting fat containing them. These unsaturated fatty acids are known as the essential fatty acids (EFA).

The starting point of synthesis is again stearic acid. Suitable plant enzymes introduce the first double bond on the n-9 carbon atoms, oleic acid having been formed. The second double bond is placed at the n-6 carbon thus between the existing double bond and the terminal methyl group. This pattern is the n-6, 9 (1C-4C-pentadienoic or skipped) one:

$$\overset{cis}{-CH = CH} - CH_2-\overset{cis}{CH = CH} -$$

The n-6 (9) pattern or configuration has been proven as the common denominator for essentially (physiological activity), independent of the chain length of the fatty acid.

Linoleic acid (18:2-9*cis* 12*cis* or n-6, 9) is the most important essential fatty acid, its absence in fat causes growth retardation in living animals, skin permeation problems (darmatosis, other deficiencies also relevant to human beings). Linolenic acid is well converted by the enzymes of animal organism via chain elongation and desaturation to dihomo-γ-linolenic acid C_{18} (20:3 8C, 11C, 14C or n-6, 9, 12) and consequently to arachidonic (20:4 5c 8c 11c 14c or n-6, 9, 13, 15) acid, all of which are essential fatty acids. γ-linolenic acid (C18:3, 6c, 9c, 12c or n-6,9,12) present in the evening-primrose (Oenothera) seeds is also considered as an essential fatty acid). On the other hand, the polyunsaturated α-linolenic acid (18:3, 9c, 12c, 15c or n-3, 6, 9) is not absolutely required by animal organisms.

In animals the conversion of linolenic acid to arachidonic acid occurs mainly via ω-linolenic acid and its C_{20} homologue, i.e. dihomo ω-linolenic acid (20:3 8c, 11c, 14c or n-6, 9, 12) by similar but more extensive series of reactions furnishing mainly the 20:5 and 22:6 acids (eicosopentaenoic and decosohexaenoic acids) thus: (15)

18:3 (9,12,15) ⟶ 18:4 (6,9,12,15)

↓

20:3:.......... (11,14,17)   20:4 (8,11,14,17) ⟶ 20:5 (5,8,11,14,17)

↓

22:4 (10,13,16,19)   22:5 (7,10,13,16,19) ⟶ 22:6 (4,7,10,13,16,19)

Two C_{16} polyethynoic acids (4,7,10,13) also produce 20:5 and 22:6 acids when fed to animals. Some essential fatty acids form the building blocks of the physiologically important substances called prostaglandins (1). These are short-lived smooth muscle and vasodepressor regulators formed in the animal organism by the enzymic cyclization of the above mentioned polyunsaturated fatty acids with the n-6 configuration but with a chain length extended to C_{20}.

Bio- and Agro-techniques have provided the industries with plant hybrids which have great uniformity in the quality and quantity of their oil production. This and their resistance to disease were mainly achieved by breeding techniques based on selection and crossing or tissue culture methods, the latter being successfully applied in the case of palm oil trees. Similarly, the very significant changes in the unsaturated fatty acid pattern of sunflower, safflower and rapeseed oils have been achieved mainly by single intact seed selection and/or tissue culture methods. On the other hand more detailed studies of the biochemical pathways of elongation and even more so of the desaturation of fatty acids are also of great commercial interest. Therefore, there are enormous changes in modifying the chain length and unsaturation of fatty acids by using the new genetic engineering techniques.

(g) *Arachidonic acid, essential fatty acids and prostaglandins*: Arachidonic acid (20:4 5c, 8c, 11c, 14c) is the best known of the higher polyunsaturated fatty acids because of its widespread occurrence in animals, as a trace component in some glycerides and a minor or major component in phosphoglycerides (e.g. ox adrenal phosphoglycerides containing about 10 per cent of this acid). It contains four methylene-interrupted double bonds and belongs to the linolenic series (ω–6).

Interest in essential fatty acids has developed since Burr and Burr (1929) reported that linoleic acid is essential for the growth of rats. The term, often abbreviated to EFA, is reserved for fatty acids required for the maintenance of normal growth, reproduction and skin permeability. These properties are characteristic of acids of the linoleic series which have a common terminal diethenoid structure.

$$\overset{c}{}\qquad\overset{c}{}$$
$$CH_3{\cdot}[CH_2]_4{\cdot}CH = CH{\cdot}CH_2{\cdot}CH = CH-$$

It is likely that some essential fatty acids are converted to other acids of this type before use for their specific purpose. Arachidonic acid is the most important among the fatty acids and linoleic acid acts as a precursor of this acid through the sequence:

$$18{:}2(9,12) \longrightarrow 18{:}3(6,9,12) \longrightarrow 20{:}3(8,11,14) \longrightarrow 20{:}4(5,8,11,14).$$

Exciting development in this field have followed the recent discovery of a series of compounds (Prostagladins) which have smooth muscle and vaso-depressor activity. These compounds are present particularly in seminal plasma and in accessory genetical glands and also in lung tissue. The structure of one such compound is given below and often differs in the presence or absence of double bonds 5 and 17 and in the degree of oxidation (oxy or hydroxy) at c(9). These compounds are related to the C_{20} polyethynoid acid and there is evidence for the biological conversion of such acids to prostaglandins. Archidonic acid gives PGE_2.

Formula of Prostaglandin

$${}^{20}CH_3[CH_2]_4CH(OH)CH = {}^{13}CH-\underset{\text{ring: 12, 8, 11, 9, 10; H at 12 and 8; HO at 11, OH at 9}}{\boxed{}}-CH_2CH = {}^{5}CH[CH_2]_3\,{}^{1}COOH$$

H H

12 8

11 9

HO 10 OH

Acyl Glycerols (glycerides)

(a) *Nomenclature*: In (tri) acyl glycerols the component fatty acids esterified to the glycerols are named by adding post-fix -oly or -yl to the trivial or systematic name of the substituting fatty acids and attaching it to the word glyceryl, e.g., 1-stearoyl-, 2-oleyl-, 3-linoleyl glycerol or 1-octadecanoyl, 2-octadecenoyl, 3-octadecadienoyl glycerol. The numbers indicate the place of the substitution in the glycerol molecule.

(b) *Physical and chemical properties*: Triaclyglycerols are synthesized in living cells by the interaction of an intermediate of carbohydrate metabolism called glycerol phosphate formed by the reduction of dihydroxy acetone phosphate and the complete fatty acid residue coupled with coenzyme A. Direct coupling delivers first a monoacyl glyceride phosphate which after release of phosphoric acid residue by a phosphate enzyme is substituted by a third coenzyme A coupled with fatty acid residue.

Because of the tri- functionality of glycerol tri-, di- and mono acyl glycerols can be composed of a wide variety of fatty acid (acyl) substituents. As the triacyl glycerols in natural fats may contain more than one type of fatty acids, fats are composed of the so called mixed triacyl glycerol. Basically these fatty acids are those principal constituents which parallelly determine the physical and chemical properties of a given fat. These properties manifest this in values which are often more close to the average value of the component fatty acids, if allowance is made for the higher molecular weight of the triacyl glycerol molecule as compared with those of the component fatty acids.

Di- and mono-acyl glycerols possess one or two free hydroxyl groups and can be considered as tracers of a not fully. completed biologically triacyl glycerol synthesis (unripe fruits, seeds etc.) or as the sign of an early hydrolysis caused by enzyme activities in deteriorating fatty commodities. Depending on the fatty acid composition, they may be liquid or solid-at ambient temperatures. Their main importance (besides being important from quality criteria) is their strong affinity for both water and apolar substances such as other lipids, hydrocarbons etc. which explains their decreasing use as emulsifiers (many foods, e.g. margarine).

The low vapour pressure of triacyl glycerols makes it impossible to distil them under conventional conditions without decomposition primarily to glyceryl and secondly to the fatty acid residues. The volatilities of di- and even more of the monoacyl glycerols are increasingly somewhat higher due to their lower molecular weights.

As some of the physical properties of triacyl glycerols have a profound influence on their technical quality and applicability, they are discussed elsewhere. Briefly speaking, the physical properties influence characteristics such as fractional crystallization and winterization, dewaxing, consistency and phenomenon of polymorphism. The density of acyl glycerols is higher in solid state than in the liquid state. For this reason they show specific volume expansion, when being transformed from solid to the liquid phase by heating. As fats are varying mixtures of solid and liquid acyl glycerols at a given temperature, the measurement of this melting dilatation can be used to determine the solid fat index (SFI) of the given fat. Although this is practised in USA and elsewhere the SFI is proportional but not equal to the percentage of solid fat content (SFC) of fats, and SFI must be converted to a percentage by using suitable formula. The SFC is a very important characteristic in edible fat processing.

The differences in the structure related to the position of the fatty acids in the triacyl glycerol molecule are the source of tendencies of this substituent in the natural fat to attain a fully random distribution during catalytic intra- and intermolecular ester interchange.

Other chemical reactions of triacyl glycerol are generally comparable in effect, although not always in rate, with those carried out on mixtures of component fatty acids and glycerides. One of the most technically important chemical reactions of glycerides is addition of hydrogen to the double bonds of fats containing many liquid unsaturated fatty acids, i.e. the hydrogenation of fats.

In the absence of the separate techniques required for a more individual analysis, fats in the past had very often been characterized on the basis of their average fatty acid composition. The similar reactivity of given samples of any fat towards different functional group reagents has resulted in the so-called 'values' or 'numbers' which give the mass average of certain properties, e.g. the presence of free fatty acids give rise to the so-called acid value, the determination of which is regularly required to neutralize the free fatty acids.

"Iodine value" gives the average unsaturation in terms of amount of iodine added to the double bonds. The "saponification value" gives an indication of the average chain length in terms of the amount of alkali required to totally saponify the sample, and "titre value" is the average solidification temperature of the mass obtained after fully hydrolysing, i.e. free fatty acids. These 'values' still have some commercial importance as the tests do not require complicated equipment of techniques and the values give a quick clue of the identity and quality of the fat. Modern analytical techniques tend to individualize the determinations, the more or less individual acyl-glycerol separation and analysis becoming as feasible as individual fatty acid (and accompanying components) analysis.

Phosphoacyl Glycerols (Phosphatides)

The cell wall of living animal and plant material and some structural elements of certain animal organs contain numerous derivatives of orthophosphoric acid. In case of fats the most important ones are called glycerol esterified derivatives, also called glycerophospholipids. Others like the sphingosine based phospholipids of the sphingophospholipids of the nerves and the carbohydrates contain the glucosphongolipids such as ceramides, cerebrosides of plant parts and animal organs, blood, brains etc. which are of great biological importance.

Of the phosphoacyl glycerols at least one and generally both hydroxyl groups of the glycerol are esterified with long chain fatty acids. Actually at this stage of the biosynthesis it has already been decided whether a triacyl glycerol or phosphoacyl glycerol will be formed. Sometimes the glycerol is etherified with a fatty alcohol, either saturated or unsaturated. On the other hand, the phosphoric acid is esterified e.g. an aminoalcohol inositol (a hexahydroxycyclohexene). All these substances, trivially called phosphatides are generally oils insoluble in the molecular sense, but they are co-extracted with the triacylglycerol during oil milling as colloidal aggregates called micelles. They are present in crude vegetable oils in amounts 0.1 to 3 per cent and are generally easily removed during refining as a sludge because of their high affinity to water. The calcium (Ca) and magnesium (Mg) salts of phosphatidic acid and the free acids themselves being less hydrophilic are less easily removed and so they are called nonhydratable phosphatides. Glycerophospholipids are classified into five main groups:

i) Phosphatidic acids (no substituents), PA: their dimers are called cardiolipins, CL;

ii) Phosphatidyl ethanolamine (cephalin), PE

iii) Phosphatidyl choline (lecithin), PC

iv) Phosphatidyl serine, PS

v) Phosphatidyl inositol, PI

If only one fatty acid (O-acyl residue) is present in the molecule of the above compounds they are considered to be partially hydrolysed members. So, the prefix lyso- is put in front of the above given names.

The presence or diminished presence of phosphoacylglycerols (commercially called lecithin or often as gums) is an important clue when assessing the quality of fats. The amount of these compounds can be measured by phosphorus or phosphate determination. Commercially, lecithin is used as a dispersion stabiliser; an anti-spattering agent and a viscosity reducer in many foods and in technical, cosmetic and pharmaceutical preparations.

MINOR ACCOMPANYING (LIPID) COMPONENTS

(a) *Alcohols*: Glycerols or 1, 2, 3-trioxy propane, although the basic component of triacyl glycerols giving them their name is not found in its free form in fats but is produced if hydrolysed by aqueous alkalis, acids or enzyme type catalysts of plant, animal or microbial origins. Free glycerol is water soluble in contrast to its state when esterified with fatty acids.

Primarily long carbon chain alcohols can be derived from the fatty acids by converting the carboxyl group to a hydroxymethylenic ($-CH_2-OH$) one. Alcohols of the 22 carbon atoms are called fatty alcohols and those with longer chain are called wax alcohols. The C_{14}–18 and 18:1, 9c members are mainly present in the tissue fat of aquatic animals (mammals and fish) as palmitic, stearic, oleic and linoleic monoesters. They have a wax like appearance. The C_{16} (cetyl) alcohol esterified with oleic acid is more liquid and is present in large amounts in the skull fat of sperm whale and is called spermaceti or sperm oil. In the past the spermaceti oil was hydrolysed and the alcohol used for the cosmetic preparation and detergents (fatty alcohol sulphates). The discovery of fatty alcohols in the Jojoba *simmondsia* shrub fat and catalytic reduction of fatty acids to the corresponding alcohols has at least relieved the need for whaling of sperm whales. C_{26}–30 alcohols can be found as esters of fatty acids in minerals, plant and animal and insect waxes.

(b) *Sterols*: The basic sterol structure is the so-called cyclopenta-phenanthrene 4-ring system with 17 carbon atoms (Fig. 1.1). Sterols are alcohols. The substitution at this skeleton by one hydroxyl and by one or more methyl and side chain alkyl groups, together with some unsaturation causes the main differences between the different classes of these compounds. Powerful techniques are available in all areas of chromatography which if sensibly combined can differentiate between them. After saponification of the fat to be analysed, the compounds are recovered in the so called 'unsaponifiable matters'. Therefore, the previously routine measurement for detecting the mineral oil contamination has become of increasing importance in modern quality assessment. Sterol level vary in the range 0.05 to 0.6 per cent, the extreme range being 0.6 to 1.7 per cent in wheat and maize germ oils.

Sterol esters and sterol glucosides (steroids) are more or less fat soluble and completely water insoluble. Sterol dissolves easily in ether, pyridine, chloroform, acetone and also in warm alcohol, alcohol or glacial acetic acid. Predominantly free sterols and sterol glucosides having polar hydroxyl groups and apolar condensed ring elements have a strong affinity both to water and to fats. Because of this they have the ability to reduce the interfacial tension of water-oil mixtures. This makes them powerful emulsifiers for fats when occurring e.g. in the blood plasma or in medicinal or cosmetic preparations. Sterols can be classified by their origin as animal sterols (zoo sterols) or as plant sterols which can be sub-divided into higher plant sterols (phyto sterols) and sterols present in the body of yeast and fungi (mycosterols). Inspite of this classification of origin which remains important for the fat analysis when dealing with unknown fatty mixtures, they can be classified in the more upto-date way:

i) 4-dimethyl sterols
ii) 4-methyl sterols and
    4,4-dimethyl sterols

FREE STEROLS

(a) *4-dimethyl sterols*: Cholesterol with a C_{27} carbon skeleton is a typical animal sterol to be found in large concentration in organs such as the brain, liver and the spleen (as sediment), in the adrenal gland, in the white nerve tissue, in blood plasma and in the egg yolk. Cholesterol is very useful as an analytical characteristic in the identification of animal fats and their mixtures adding extra clues in addition to the common GC determination of the fatty acids composition. Palm (kernel) oil contains a small percentage of cholesterol among their plant type sterols and this fact must be considered when analysing mixtures containing palm or palm kernel oil.

Ergosterol: A C_{28} sterol present in yeast is one representative of lower plant sterol. It is the starting material in the formation of vitamin-D_2.

Stigmasterol and β-Sitosterol are typical plant sterols with a C_{29} carbon skeleton and they can be found in the unsaponifiable part of nearly all vegetable oils. The same situation exists in Campesterol of C_{28} sterol. Brassica sterol is also a C_{28} sterol. C_{28} sterol is characteristic of cruciferae oils such as rapeseed and mustard seed oils.

(b) *4-methyl and 4,4-dimethyl sterol*: As 4-methyl sterols are only minor constituents of the unsaponified parts, so vegetable oils can be better discriminated (additionally to the 4-dimethyl sterols) by analysing the 4-dimethyl sterols, which are also called triterpenic alcohols. The most abundantly found representative alcohols are cycloartenol (C_{30}) and 24-methylene cycloartenol (C_{31}) accompanied by small amounts of α and β amyrins, lupenol, butyrospermol and cycloartenol, all being C_{30} 4,4-dimethyl sterols. The varying amounts of these compounds present in different oils can help the analyst to arrive at a better assessment of the origin and composition of an unknown mixture of fats.

Sterol esters: In some vegetable oils a part of the sterols is esterified with fatty acids because of which their fat solubility is increased and their emulsifier activity is decreased. Particularly in the germ oils, such as maize and wheat-germ oils, a quite significant percentage of the sterols present is in the esterified form.

TOCOPHEROLS

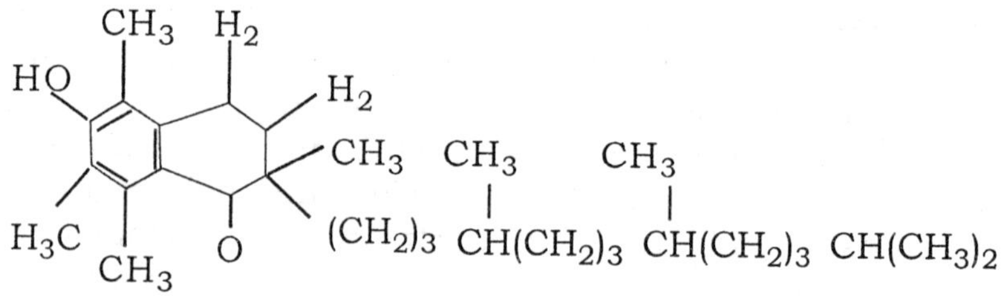

CH_3 H_2

HO H_2

CH_3 CH_3 CH_3

H_3C CH_3 O $(CH_2)_3$ $CH(CH_2)_3$ $CH(CH_2)_3$ $CH(CH_3)_2$

α-Tocopherol

The tocopherols are widely distributed in nearly all the vegetable oils but not in animal oils and are the most important natural anti-oxidants of phenolic nature. All tocopherols are fat soluble and generally occur in their free form, although some fats contain them in a partial esterified form, the phenolic hydroxyl being esterified by a fatty acid. Therefore, in esterified form their solubility infact increases but their anti-oxidant activity is totally removed. Their basic structure is a hypothetic tocol molecule a chromanol structure containing a farnesyl side chain, substituted by one, two or three methyl groups. A homologous series of tocopherols is also known (e.g. in palm oil) which differs only by the presence of three double bonds in the farnesyl side chain. The standard (100 per cent) Vitamin E activity is ascribed by agreement with α-tocopherol. The amount of tocopherols varies between 30 and 700 mg/kg, extremes being 1000–2800 mg/kg as measured in germ, linseed, soyabean and cotton seed oils. The other tocopherol (β-, γ- and δ) thus have less Vitamin E activity than that of the α compound. The different amounts, and even more importantly, the different types and proportions of tocopherols present in different oils can additionally help the analyst when trying to determine the origin of an unknown mixture of fats. Because of their oxidizability, the proportions can change during storage and processing to such an extent that the clue may be rather uncertain. Tocopherols (and Tocotrienols) while protecting the fat are generally those which oxidize to quinones and tocopherol dimers. Some oxidation products have distinct colours like toco-red (5, 6 chromanoquinone) in heavily oxidized soyabean oils, this *in vitro* antioxidant activity of tocopherols in oils and fats is different according to the temperature and possible contaminants (heavy metals) present in the system. It is generally accepted that the least substituted ones which more easily form stabilized free radicals and thus lose their mobile hydrogen atoms in the hydroxyl group are the more potent ones thus:

$$\delta > \beta = \lambda > \alpha$$

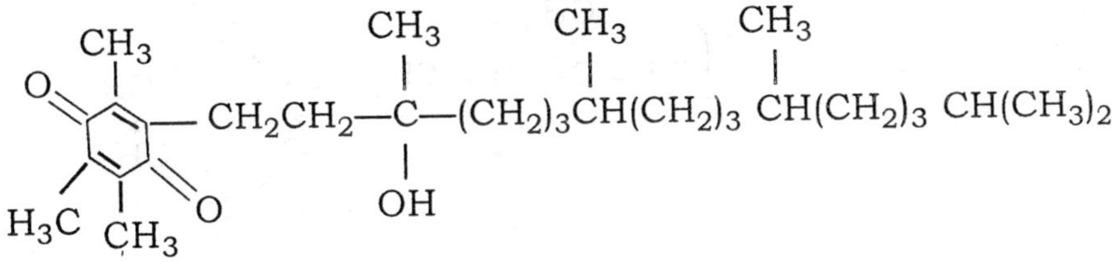

$$CH_2CH_2-C(CH_3)(OH)-(CH_2)_3CH(CH_3)(CH_2)_3CH(CH_3)(CH_2)_3CH(CH_3)_2$$

α-Tocoquinone

WAXES

Waxes by definition are esters of long chain aliphatic alcohols and fatty acids, partly of plant and partly of animal origin. The so-called mineral waxes are ultimately also of plant origin representing the wax remnants of past vegetation in a progressively carbonified state. The waxes of vegetable oils as minor substituents are mere contaminations originating from seed covers such as brans and husks like wax of rice-bran oil or sunflower seed oil. Waxes derived from leaves like candelilla and carnauba waxes or from animal secretions like beeswax or woolswax of sheeps also contain among many others aliphatic alcohols, free fatty acids, sterols, triterpenic alcohols and aliphatic hydrocarbons which together form the substances popularly known as waxes. Some marine animals contain acyl glyceride etherified fatty alcohols which are wax-like materials.

PIGMENTS

It has been seen that the colour of different oils and fats is highly influenced by the presence in traces of some oil-soluble pigments like the carotenoids, the chlorophylls and gossypols—the characteristic colouring pigments e.g. red palm oil, green olive oil and yellow cotton seed oil respectively.

```
H3C   CH3
   \ /
    C              CH3
   / \              |
H2C   C — (CH = C H - C = CH)2  - CH = CH -
 |    ||
H2C   C
   \ / \
    C   CH3
    H2
                                      H3C   CH3
                  CH3                    \ /
                   |                      C
          (CH = C- CH = CH)2  — C       /   \   CH2
                                |           |
                                C          CH2
                              /  \        /
                           H3C     C
                                   |
                                   H2
```

β-Carotene

Carotenoids: The carotenoids are oil-soluble polyisoprenoids of hydrocarbon alcohol or carboxylic acid nature. They contain a considerable amount of conjugated double bond which explains their pronounced colour in the yellow, orange or red origin of the visible spectrum. Thus about 70 different carotenoids have been isolated, the most important being the α-, β- and γ-carotenes and lycopene which all have the molecular formula of $C_{48}H_{56}$. Carotenoids are widely distributed in the plant and animal kingdoms, although their concentration in fats (except in palm oils) is generally low. They are highly oxidizable, yet this generally provides a protective action against radiation (light) damage of the medium in which they are dissolved. The oxidation, a free radical process, causes their ultimate destruction and in this way a bleaching of the colour. Carotenoids are not sensitive towards dilute alkalis, but they suffer considerable damage under acidic conditions and even more by heating above 150°C. They can be removed by the adsorbents like bleaching earth and activated carbon. Hydrogenation of the double bonds, if done selectively and without saturating too much of the unsaturated fatty acids of the medium, also causes a chemical type of hydro-bleaching. Bleaching of oils and fats by chemicals in the oxidative way, use of peroxides and generally all reagents producing active oxygen and other free radicals are out of the question for use in edible oil refining.

Some carotenoids are closely related to vitamin A. Thus α- and β-carotenes of vitamin A are used as such (pro-vitamin A).

Carotenoids are produced from roots and seeds (carrots, anatto seeds etc.) and also synthetically. They are used in large amounts as safe colouring agents in foods and drinks.

Chlorophylls: The green plant pigments chlorophyll α and β are found mainly in crude olive and grape seed oils and also in vegetable oils harvested from unripe seeds like rapeseed and soyabean. The later two both contain the so called porphyrin ring centred around a magnesium atom. The magnesium centred and strongly conjugated structure of the ring are responsible for the green colour. If the ring contains iron as an anchoring metal ion in the so called 'chelates' the colour is brownish and the compound is haemin. When chemically bound to proteins like haemoglobin in blood, myoglobin in meat and also in different enzymes and catalysts like cytochromes, peroxidase, catalases, it becomes of the highest importance for animal metabolism as an oxygen transporter or transformer. Some animal fats, mainly tallows, may have greenish tinge, most probably because of decomposition of the brown haemoglobin and myoglobin of animal blood and meat tissues, which have been in contact with e.g. copper appliances like tubes and valves. In plants the chlorophylls were the earliest discovered catalysts of photosynthesis, the radiation-mediated assimilation of water and carbondioxide in the plant material. Chlorophylls are not really oil soluble and yet, if combined with oxidized fatty acids, it is very difficult to remove them by either chemical (acids, alkalis) or adsorptive means from the greenish coloured oils. The loss of magnesium turns chlorophylls into a brownish colour and leads in the first instance to pheophytins.

Gossypols: Gossypols, an aldehydic polyphenol, is the precursor of the brownish discolouration of cotton seed oils of less than average quality. It is originally of yellowish colour and only by oxidation due to bad conditions during harvesting, delintering and seed pre-treatment before storage, it is transformed to the brownish red colour quinones or dimers of quinoid structure. Once dissolved in oil and probably combined with the oxidized fatty acids its chemical or absorptive removal during refinement becomes a very trouble-some operation. There are methods which can preferentially accumulate the pigment either in the extracted meal or in the oil. Gossypol causes male sterility if orally administered and has been used for this purpose.

HYDROCARBON

The biosynthesis of fatty acids like fatty alcohols, tocopherols, carotenoids and sterols is based on the action of coenzyme A, the sulphhydril groups containing protein which has been mentioned before. When producing methyl group substituted repetitive linear or ring structures, the enzyme uses acid as the main building block in order to synthesize them in consecutive anaerobic steps. One of the most important intermediates is squalene, a polyisoprenoid C_{30} hydrocarbon present in higher amounts in the fatty tissues of certain species of shark living in the Pacific Ocean around Japan. However, in many vegetable oils (olive oils, 0.5 per cent) and animal fats it can be found as evidence of the intermediate of the biosynthesis of other compounds formed. Squalene contains six non-conjugated double bonds in the molecule. The rubber like mass of about 10 per cent karitene in the vegetable butter Shea fat, pristane in the liver oils of fishes and other saturated and unsaturated, branched and unbranched hydrocarbons are examples of the existence of these unsaponifiable substances in plant and animal fats.

$$(CH_3)_2C = CHCH_2CH_2\overset{\displaystyle CH_3 \atop |}{C} = CHCH_2CH_2\overset{\displaystyle CH_3 \atop |}{C} = CHCH_2CH_2CH$$

$$= \overset{\displaystyle CH_3 \atop |}{C}CH_2CH_2CH = \overset{\displaystyle CH_3 \atop |}{C}CH_2CH_2CH = C(CH_3)_2$$

Squalene

Glucosides: Glucosides are ethers of mostly monosaccharide, tetrosid, pentosid and hexosid sugars with a non-sugar component. The non-carbohydrate part which can be liberated e.g. by acids or enzymes from the other linked compounds is called aglucone. This is generally the biologically active or technically important part providing the specific character of the total oil. Many oil bearing seeds and fat parts contain glucosides, the best one being amygdalin, the glucoside of bitter almond. The aglucon is the benzaldehyde cyanohydrin which after further decomposition liberates benzaldehyde having the characteristic bitter almond flavour. Sesamin has been used as a margarine tracer for detection of adulteration of butter. It gives a characteristic red colour when reacted with concentrated hydrochloric acid and furfural (Villavecchia-Baudouin test).

Thioglucosinolates in cruciferae oils are also important glucosides e.g. Sinigrin in mustard seed oil, their aglucone originally being a sulphate coupled mustard oil, allylisothiocyanate, which may form (goitrogenic) cyclized sulphur containing derivatives such as vinylthio-oxazolidons. Non-sterol or diacylglycerol glucosides are generally oil insoluble but if extracted with more or less colloidally soluble micelle forming lipid substances, they can be traced back in small amounts in some crude oils and fats vegetable origin.

Sterol and diacylglycerol glucosides: Small amounts of mainly β-Sitosterol etherified with tetrosid or pentosid sugars in the cyclofuranose or pyranose form can be found in some vegetable oils. Besides these compounds, diacylglycerols are also often etherified with monosaccharides. Together with phospholipids, the above constitute the main building blocks of the cell membrane bilayer material.

Saponins are sterol glucosides which possess surface active soil removing and haemolytic properties. The richest source of saponines are saponaria roots and horse chestnuts but they are also found in vegetable butter, illipe fat and in smaller amounts in soyabeans. The sapogenins are mainly 4,4-dimethyl sterols like the amyrins and lupenol etherified with monosaccharide sugars. Some of these glucosides are also esterified by fatty acids via the hydroxyl groups which explains their colloidal solubility in the originally apolar triacyl glycerols. All these glycolipids can often be found together with the phosphatides in the so-called 'lecithin sludges' of the oil degumming process.

VITAMINS

Vitamins are organic substances which must be acquired by the human organism from outside sources i.e. with its food. They regulate certain metabolic processes being autonomic in their action or being involved in some enzymatic systems as co-enzymes. . Some forms are fat soluble and others are water soluble. The most important fat soluble vitamins which occur in different parts of plant or animal organisms are briefly described.

(i) *Vitamin A or Retinol*: The principal sources of vitamin A (Retinol) are fish liver oils, egg, liver, butter, cheese and milk, and it has been totally synthesized. It is a polyisoprene primary alcohol and it represents the biologically hydroxylated half of α- and β- carotenes both of which function as pro-vitamin for retinol. A shortage of vitamin A in humans may cause night blindness and growth deficiency. Generally Vanaspati and Margarine are vitaminized with a synthetic preparation which is either acetic or palmitic acid ester which increases the fat solubility.

(ii) *Vitamin D or Calciferol*: The principal source of vitamin D (Calciferol) is fish liver oil (codliver or shark liver oil) and it is present in lesser amount in eggs and liver and dairy products. It can be totally synthesized. The two most important variants are vitamin D_2 and D_3 which regulate phosphate adsorption in the animal organism. Deficiency causes diseases of the bones like rachitis in children and ostemalaccae in the aged. Vitamin D_2 and D_3 are sterol derivatives formed by irradiation of ergosterol (vitamin D_2) or of 7-dehydrocholesterol (vitamin D_3) present in skin tissue. Margarines are vitaminized with the synthetic preparation of these vitamins. Vitamin E or α-tocopherol is present in liquid vegetable oils, tomatoes and leafy vegetables. Its deficiency may cause some muscular trouble. Vitamin E is medically used for preventing the oxidation of cholesterol in human body and also it generally supports good health. It is produced either by isolates from vitamin E rich oils like palm oil or totally synthesized. It is used as an anti-oxidant of fats.

(iii) *Vitamin K (or phytomenadion)*: is an important factor in the clotting of blood. If vitamin K is not present in blood, the persons injured may bleed to death. Spinach and other green vegetables together with tomatoes and some vegetable oils contain vitamin K.

FAT MODIFYING REACTIONS

Lipid transformation products are derived from lipid compounds which are probably either not present in only small amounts in the oils and fats included in the cells. The products are thus formed mainly during or after oil production as a result of external influences (reagents, conditions, processes). They are thus not necessarily the result of any maltreatment of the fats, although they are inevitably due to chemical changes. As lipid transformation products are different in structure from the parent lipids, they may give important analytical clues for monitoring quality differences. Their more important reactions which cause alterations are discussed.

Hydrolysis: Hydrolysis can affect all acyl glycerols, sterol esters, sterol glucosides, ester waxes, tocopherol esters and esterified chlorophylls. The reaction requires an aqueous medium and usually a catalyst. The enzyme lipases act as a biochemical catalyst whereas mainly mineral acids or bleaching earth act as chemical catalyst for the hyrolysis. If short chain hydroxyl fatty acids are cleaved in acid or neutral medium as in the case during deodorization of coconut and palm kernel oils, these acids can be secondarily cyclized to lactones, some of which are important flavour carriers in butter and margarine.

Auto-oxidation and anti-oxidants: Air consisting of 1/5 of its volume of oxygen is freely admitted to oils and fats. Unsaturated fatty acid chains are therefore more or less easily oxidized. It is to be noted that by its very 'negative' character a double bond exerts an attractive force on the bonding electrons of the valence electrons of one of the hydrogen atoms of the methylenic group next to it, also called the active α-methylenic group. This hydrogen atom is easily eliminated by radical formation or inducing conditions or agents such as heat, radiation (light) and oxygen. After cleavage of the hydrogen bond a free radical is formed on the fatty acid chain. The removal of this hydrogen atom is easier if the α-methylenic group is situated between two double bonds (1,4-pentadienoic or 'skipped' methylenic structure) as is generally the case in linoleic and linolenic acid.

(i) Primary Oxidation Products

A. Different steps are shown in the figure below:

$$RH \xrightarrow{\text{light-heat}} R. \qquad \text{A}$$

$$\left.\begin{array}{l} R. + O_2 \longrightarrow ROO. \\ ROO. + RH \longrightarrow ROOH + R. \end{array}\right\} \qquad \text{B}$$

$$\left.\begin{array}{l} 2\,ROO. \longrightarrow \\ 2\,R. \longrightarrow \\ ROO. + R. \rightarrow \end{array}\right\} \qquad \text{C}$$

Fig. 1.1: Auto-oxidation reaction: A. initiation; B. propagation; C. termination. RH: fatty acid molecule; R: alkyl radical; ROO: peroxy free radical

The initiation step A cannot be stopped by additives. Only the complete exclusion of the radical formers can inhibit the free radical formation.

B. The next so called propagation step B consists of two parts. During the second part a number of protective measures can be taken. In the first propagation step, oxygen addition leads to peroxide free radical which is the second rate determining step which reacts with an unattached unsaturated fatty acid. This yields a highly reactive alkyl free radical and a so-called fat hydroperoxide. The alkyl free radical reacts with oxygen molecules in a new cycle. The first stable product of the auto-oxidation reaction is the fat hydroperoxide. This compound is tasteless and odourless. The steps are illustrated below:

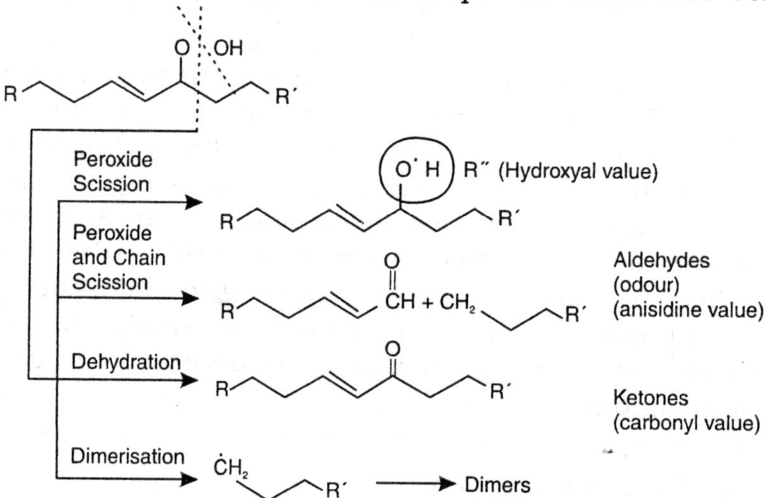

Fig. 1.2: Secondary oxidation products: R′: rest of triacylglycerol molecule, R: alkyl R″H: another triacylglycerol molecule.

As a rule the free radical reactions do not continue until total consumption of the fat substrate. There is usually a depletion of oxygen supply. Oxygen transfer is inhibited by a (polymeric) film formed in the course of the reaction at the structure of the bulk of the oil or the fat. In the absence of oxygen the reaction ceases in the so called termination step (C), when leads to different end products. The hydroperoxides, depending on whether oleic, linoleic and linolenic acids are attached, differ as regards their stability. Oleic hydroperoxides are relatively the most stable ones.

(ii) Auto-oxidation Reaction

Conjugation by Auto-oxidation

Auto-oxidation is initiated by radical formation at the α-methylenic carbon atoms. This situation causes mobility of all the bonding electrons in the whole group composed of three vicinal carbon (–CH– methyne) atoms. This mobility can lead to a shift of the double bonds within this three carbon unit, the so called 'allylic group'. The initial attack on a 1,4-pentadienoic system leads to free radical formation on the methylene group between the two double bonds. This unstable situation leads, by shifting, to the conjugation of two double bonds which represents a more stable state. The *Z* double bond which was actually shifted is transformed to an *E* one.

Isolated double bonds with an absorption maximum around 190 nm cannot be detected with a commercial uv/visible spectrophotometer. On the other hand, if for any reason a conjugated dienoic group is present, this can be detected in the uv region of the spectrum by an absorption maximum around 232 nm.

The degree of auto-oxidation of fats and oils containing polyunsaturated fatty acids can also be followed by the increase in the dienoic conjugation. This increase results from the rearrangement caused by a radical formation and temporary stabilization in the form of hydroperoxides. If the hydroperoxide is then decomposed during the course of secondary reactions, a large part of the dienoic conjugation may still persist, thus giving some clue to the past oxidative history of a given fat sample. Conjugated trienoic systems, in which the shifted double bonds are generally in the *E* configuration, absorb with a triplet of absorption peaks having maxima around 262, 268 and 274 nm.

Acid catalyzed dehydration of hydroperoxides may give rise to the formation of trienoic conjugation, the dehydration extending the dienoic conjugation already present to a trienoic one. This fact can be used, for example, as a clue to previous bleaching of an oil which has been more or less attacked by oxygen, or by acid activated bleached earth.

(iii) Secondary Oxidation Products

Hydroperoxides decompose to secondary oxidation products as a result of heating, radiation or the presence of heavy metals (copper, iron) and other radical initiating agents. This decomposition is usually also radical generating. Secondary lipid oxidation products can be formed either by peroxide scission alone or by simultaneous peroxide and chain scission. The latter leads to short chain volatiles and new radicals (chain branching or to dimerization of the scission products).

OFF-FLAVOUR FORMATION

One of the most important decomposition reactions of hydroperoxides is that involving chain scission which often leads to the short chain aldehydes. Depending on their structures and the amounts formed, they can cause very typical odours, some of them characterized as off-flavour like soyabean oil reversion odours, the green, beany, painty, fishy, tallowy, epithets or just plain rancidity. If present in substantially high concentration, aldehydes (carbonyls) formed can be determined in the bulk by specific colour forming reagents. The individual aldehydes may be different but sometimes have extremely low flavour threshold levels (1 mg ≡ 1 $\mu g\ kg^{-1}$), which readily explains the off-flavour formation in fats without measurable auto-oxidation. The threshold level or value is the minimum concentration of a flavour carried, in which 50 per cent of the observers can perceive it positively. There is distinct difference between the sensitivity of the human nose/tongue and the sensibility of the carbonyl group reagents towards the carrier aldehydes of different off flavours. Inspite of the confirmed organoleptic assessment of a certain off-flavour aldehyde, the bulk carbonyl reagent may fail to detect its presence. The threshold levels and detection limits can differ by a factor of up to 1000, a discrepancy which should be remembered when assessing fat quality by different means.

(iv) Retardation of Auto-oxidation

(a) *Anti-oxidants*: Radical quenchers like carotenoids act by protecting crude fats and oils once they have been produced only against light induced photo-oxidation by transforming the absorbed radiation energy into heat. Storage in the dark is, therefore, a good proposition for extending the shelf life of oils and fats. Exclusion of air by using vacuum or inert covering gases and low temperatures during storage are also excellent, but not always feasible or practical. For this reason suitable measures must be taken to reduce damage in the most critical steps of the auto-oxidation reaction.

Most vegetable oils and fats are protected by their natural anti-oxidants which are quite effective in slowing down the rate of oxygen absorption by reacting with the fatty acid peroxy free radical in the second part of propagation step. These peroxy free radicals react preferably with the phenolic hydrogen atoms of the tocopherol molecule. This reaction leads to a fat hydroperoxide as in the case of normal, uninhibited oxidation reaction. However, as the tocopherol free radical which is produced simultaneously has much less affinity to react to the oxygen molecules present, the reaction chain is stopped. The rate of oxidation is thus effectively reduced and the stability of the fat is increased. The formation of the same, sometimes very unstable, fat hydroperoxides, explains why, despite the presence of anti-oxidants, a strong and early off-flavour formation can still very often be perceived. Decomposition of the hydroperoxide due to the simultaneous presence of heavy metal traces may lead to the formation of the same aldehydes as mentioned previously.

Animal fats (except marine oils) although unprotected by natural anti-oxidants are mainly composed of fatty acids which are less unsaturated than the liquid vegetable oils. This characteristic renders them more stability to auto-oxidation. If they do become oxidized, the resulting decomposition aldehydes generally have higher threshold value. The addition of synthetic or natural anti-oxidant is therefore the feasible protective measure. Examples of such anti-oxidants are phenols and polyphenols; tertiary butylated hydroxy anisole (TBHA), tertiary butylated hydroxy quinone (TBHQ), synthetic tocopherols, gallates and flavonoids, e.g. quercitine.

(b) *Anti-oxidant synergists (metal scavengers)*: Retardation of auto-oxidation can also be achieved by removal of the free radicals initiating impurities, generally admixed with the fats during their extraction, processing and handling. Many of these impurities are of metallic character and in either ionized or complexed form (such as the ions of myoglobin and haemoglobin) as trace contamination in animal fats. The decomposing action of transition metals on hydroperoxides as shown in Fig. 1.3 leads to chain initiating free radical formation and off-flavours.

$$\text{A} \quad \text{ROO.} + \text{AH} \longrightarrow \text{RO:OH} + \text{A.}$$

$$\text{B}_1 \quad \begin{cases} \text{ROO:H} + \text{M}^{3+} \longrightarrow \text{ROO.} + \text{M}^{2+} + \text{H}^{+} \\ \text{RO:OH} + \text{M}^{2+} \longrightarrow \text{RO.} + \text{M}^{3+} + \text{OH}^{-} \end{cases}$$

$$\text{B}_2 \quad \begin{cases} \text{ROO:H} + \overset{\text{Ch}}{\underset{\text{Ch}}{(\text{M}^{3+})}} \longrightarrow \text{no reaction} \\ \text{RO:OH} + \overset{\text{Ch}}{\underset{\text{Ch}}{(\text{M}^{2+})}} \longrightarrow \text{no reaction} \end{cases}$$

Fig. 1.3: Retardation of auto-oxidation. (A) Chain stopping by auto-oxidants (A. < R. AH. antioxidant with mobile H-atom). (B_1) Scission by metal ions: M. metal ion in different valence states (B_2) Inactivation by metal scavengers: Ch. chelators.

Inorganic or organic scavenging agents, which inactivate transition and other metal ions in the form of inclusion compounds (chelates), also give some relief (Fig. 1.3). Removal of metal compounds still seems to be the best proposition for protecting final quality. Citric, ascorbic and lactic acids and (disodium) ethylene diamine tetra acetate (EDTA) are effective metal scavenging or sequestering agents. These compounds are also the so-called 'synergists' as used in combination with the true anti-oxidants, strengthening their primary radical catching action. The strength of an acid is expressed by its dissociation constant and in this light the following series has the relevance:

oxalic > phosphoric > lactic > citric > formic > ascorbic > acetic.

The strength of synergist action can be expressed by salt formation constants with a metallic (e.g. a trivalent ferric) cation. In this respect the following series can be established:

oxalic > citric > phosphoric > acetic > formic.

ISOMERIZATION

Heating oils which contain poly-unsaturated fatty acids with different catalysts may cause, at temperature above 100°C, the positional and geometrical isomerization of 1-*cis*, 4-*cis* pentadienoic double and *cis*-single bonds. The process is essentially the same as the free-radical based double bond shifting or recombination as described for the auto-oxidation-caused conjugations. Some isolated dienoic double bonds may be transformed into conjugated 1,3-tetradienoic isomers, whereby some single bonds may undergo *cis*/*trans* isomerization. Catalysts which induce these transformations are nickel, copper, selenium, sulphur, sulphur dioxide and nitrogen oxides, and also bleaching earth residues which contain iron traces.

Oils containing 1,4-pentadienoic unsaturation in their fatty acids if heated with strong alkali above 200°C for a long time, may undergo extensive conjugation of these dienoic acids. If heated more excessively the conjugated bonds, e.g. linoleic acid, may form cyclic monomers like cyclohexadiene derivatives.

POLYMERIZATION

Heating oils at higher temperatures (240–280°C) may cause shifting and *cis*/*trans* isomerization of double bonds in unsaturated fatty acid molecules. Besides the use of metal catalysts the heat induced mobility of double bonds is also assisted by a limited access of air, both of which act as free radical initiators. During shifting, the alkyl free radicals formed temporarily can combine with those present in adjacent fatty acid molecules, leading first to dimeric, then to trimeric and finally to polymeric acids. The viscosity increases as three dimensional cross links are formed. The resulting products are insoluble in the normal oil solvents.

The inclusion of air during heating as in deep frying or shallow frying and baking can lead partly to dimers and later to trimers (polymers) of the above-mentioned three-dimensional cross linked structures. The dimers formed in household use are found to be completely harmless not being resolved by the animal organism. The oxidation of highly unsaturated oils in thin layers as in paints etc. in the cold, leads to polymerization which is called 'drying'.

PYROLYSIS

Pyrolysis of acyl-glycerols: Heating of fats above 300°C for longer times leads first to cleavage of the ester function; the glycerol residue is dehydrated to acrolein and unsaturated lacrimatory C_3 aldehyde. The fatty acid moiety generally first loses carbondioxide from the carboxyl group (decarboxylation), the residual part forming straight chain hydrocarbons.

Pyrolysis of carotenoids (*heat bleaching*): Palm oil contains quite high amounts of carotenoids (400–2000 ppm) which are responsible for its deep orange colour. The removal of this colour can be achieved by adsorption of carotenoids on adsorbents like bleaching earth or activated carbon. Another way to remove carotenoids is by pyrolytic cleavage in the so called 'heat bleaching' operation.

SPECIALITY FATS

The speciality fats are those which are used for chocolate or butter based confectionery products.

Chocolate is the mixture of roasted fat free cocoa powder, sugar, milk or other proteins, emulsifiers, flow regulators like lecithin, added flavours and fat which, due to legal requirements, is generally genuine cocoabutter. Some legislation permits the admixture of (5 per cent on fat, 15 per cent on chocolate) so called cocoabutter equivalents (CBE) or cocoabutter extenders. Cocoabutter is composed of large amounts of symmetrical triacyl glycerols of POP and POST types, resulting in very steep and uniform melting at its melting point which is close to, but below body temperature. It has a strong tendency to recrystallize after solidification into its highest melting β-modification, which causes some brittleness and after some repeated and non-deliberate temperature changes during storage, results in the well known 'bloom' of chocolate products. This can be reduced by the use of crystallization retarders. This brittleness, which causes this 'snapping' sound when breaking a chocolate bar, is the welcome quality aspect of a good quality product.

Although cocoabutter is produced in substantial amounts because of its price (6–12 times that of common fats and oils), it cannot be incorporated in very popular but generally much cheaper cocoapowder based confectionery products like candy bars etc. For reasons of economy the cocoabutter component is partially or totally replaced by new fats and fat fractions which are produced by a specialized branch of the fat industry as speciality fats.

Speciality fats which are used to replace part (or all) of the cocoabutter fraction in the butter are those (1) compatible for blending in any proportion with cocoabutter—the cocoabutter equivalents (CBE), (2) compatible for blending in nearly all proportions with cocoabutter—the cocoabutter extenders (CBE); and (3) similar, but not compatible for blending with cocoabutter, but can replace it in the above mentioned cheaper products—cocoabutter replacers (CBR).

The most commonly used raw materials are natural vegetable butters like shea, illipe, kokum, sal fats and palm oil which contain high levels of SUS fraction (symmetrical triglyceride fraction). Also used are coconut and palm kernel oils, high palmitic level cotton seed oil and groundnut oils or other vegetable oils like fractions of mahua or mango kernel oils. These materials can be processed by hydrogenation, interesterification, fractionation or a combination of some or all these techniques. The processing can be done on single oils or on blends. All the above choices depend on the final quality characteristics required, the availability and price of the raw materials and the technological possibilities.

Partial hydrogenation of palmolein with no trisaturates by selective (two stage or high *trans* promoting) hydrogenation techniques and/or catalysts can lead to useful margarine bases or after fractionation, to non-lauric based replacers. The nearly full or full hydrogenation of lauric fats (mostly palm kernel and to a lesser extent coconut oil) will lead to products with drastic solid fat content changes over a narrow temperature range, the steepness of the solid fat content/temperature (melting point) being the necessary minimum quality requirement for a good cocoabutter replacer. The desired level of the combination of hydrogenation and interesterification leads to useful replacers.

Other possibilities are the partial/multi-step fractionation of hard vegetable butters, high palmitic fats, the laurics and partially hydrogenated fats, which is carried out by mechanical pressing and dry, wet or solvent aided fractionation techniques. The stepwise execution of solvent-aided processes on palm oil, or mixtures of it with hard vegetable butters, leads to fractions which are equivalent to cocoabutter. The equivalence is valid not only in terms of their physical behaviour but also their chemical constitutions as regards the high levels of symmetrical SUS type triacyl-glycerols. These fractions are thus real cocoabutter and can be blended without any (except legal) restrictions, with the much more expensive natural product (cocoabutter).

The fraction of selectively hardened vegetable oils, by any of the above means can also deliver fat fractions which can be used as replacers in confectionery products.

Of course cocoabutter equivalents may also be produced chemically by re-esterification of suitable monoglycerides with the relevant acids protecting the free hydroxyl groups with suitable (hydrolysable) constituents. An alternative method which is showing promise is the enzymatic interesterification of separated POP fractions from palm oils with stearic acid or tristearin by 1, 3 position specific microbial lipases (e.g. rhizopus and candida strains). Lipase enzymes are in the presence of only a little amount of water, typical esterases and therefore under equilibrium conditions give the ester interchange on fats instead of hydrolysing them in one variant (Macrae 1983). The reaction is done continuously at 40°C with e.g. 0.05–5 per cent of immobilized enzymes on the carrier (kieselguhr) impregnated with eight per cent water. Other methods involve immobilizing the enzyme on macroporous ion-exchange resins with particle sizes 200 μm, facilitating its use in solvent free continuous systems (Novo Industry, Denmark). Another two-step process (Asahi Denko Kogyo, Japan, 1984) involves 90 per cent interesterification in aqueous medium and the other 10 per cent interesterification in dry medium. After fractionation the industrially produced product is significantly enriched in POS—the genuine symmetric triacyl-glycerol of cocoabutter.

The reinteresterification of medium chain (C_8–C_{10}) fatty acid fraction from (hydrogenated) lauric fats with glycerols gives the necessary fat basis in a special 'diet margarine' for children and older people (MCT triglycerides).

REFERENCES

1. The Chemistry and Technology of Edible Oils and Fats and their High Fat Products by G. Hoffmann, Academic Press Limited, 1989.
2. Bailey's Industrial Oil and Fat Products, Volume 1, Fourth edition, edited by Daniel Swern, 1979.
3. E. Fedell, G. Pavini, F. Camurati and G. Jacini, *J. Am. Oil Chemists' Soc.*, **45**, 676–679 (1968).
4. G.N. Fick and D.C. Zimmerman, *J. Am. Oil Chemists' Soc.*, **50**, 529–531 (1973).
5. R.A. Barford, F.E. Luddy, S.F. Herb, P. Magidman, and R.W. Riemenschneider, *J. Am. Oil Chemists' Soc.*, **42** 446-448 (1965).
6. G.K. Chacko and E.G. Perkins, *J. Am. Oil Chemists' Soc.*, **42**, 1121–1124 (1965).
7. T.P. Hilditch, *J. Am. Oil Chemists' Soc.*, **26**, 41–45 (1949).
8. H.J. Dutton and J.A. Cannon, *J. Am. Oil Chemists' Soc.*, **33**, 46–49 (1956), C.R. Scholfield and M.A. Hicks, *J. Am. Oil Chemists' Soc.*, **34**, 77-80 (1957). C.R. Scholfield and H.J. Dutton, *J. Am. Oil Chemists' Soc.* **35**, 493-496 (1958). C.R.Scholfield, J. Nowakowska, and H.J. Dutton, *J. Am. Oil Chemists' Soc.*, **38**, 175–177 (1961).
9. T.P. Hilditch, *J. Am. Oil Chemists' Soc.*, **42**, 745–747 (1965).
10. H.E. Longenecker, *Chem. Rev.*, *29*, 201–224 (1941). R.W. Riemenscheider, F.E. Luddy, M.L. Swain, and W.C. Ault, *Oil Soap*, **23**, 276-282 (1947), K.F. Mattil and F.A. Norris, *Science*, **105**, 257–259 (1947); *J.*

Am. Oil Chemists' Soc., **24**, 274–275 (1947). H.K. Dean, Utilisation of Fats, Chemical Publishing Co., New York, 1938.

11. A.P. Doerschuck and B.F. Daubert, *J. Am. Oil Chemists' Soc.*, **25**, 425–433 (1948).
12. C.G. Youngs, *J. Am. Oil Chemists' Soc.*, **36**, 664–667 (1959). C.G. Youngs and H.R. Sallans, *J. Am. Oil Chemists' Soc.*, **35**, 388-393 (1958), C.G. Youngs, *J. Am. Oil Chemists' Soc.*, 62–67 (1961).
13. F.D. Gunstone, in Progress in the Chemistry of Fats and Other Lipids, Vol. IV, Chapter 1, Pergamon, New York, 1958; also in Progress in Organic Chemistry, Vol. IV, Chapter 1, Academic Press, New York, 1958, E.M. Meade, in Progress in the Chemistry of Fats and Other Lipids, Vol. IV, Chapter 2, Pergamon, New York, 1958. F.R. Earle, E.H. Melvin, L.H. Mason, C.H. Van Etten, T.A. Wolff and Q. Jones, *J. Am. Oil Chemists' Soc.*, **36**, 304–307 (1959).
14. C. Litchfield, *J. Am. Oil Chemists' Soc.*, **48**, 467–472 (1971).
15. An introduction to the chemistry and biochemistry of fatty acids and their glycerides by F.D. Gunstone, Chapman and Hall Limited, 1967.

CHAPTER 2

Physical Properties of Fats and Fatty Acids

PHYSICAL PROPERTIES OF FATS AND FATTY ACIDS

The material which is solid at ordinary temperature is termed as fat and that which is liquid is termed as oil (1). The physical properties of fats have many technical applications and practical importance in a variety of fields as in the edible fat industries in some physical treatments for the manufacture of a number of commercial fat products or in testing and analysis of physical properties of fat products. The melting and solidification are the most important physical properties of fats and oils.

OILINESS AND VISCOSITY

Fats and oils have the property 'oiliness' which is used for lubrication. Nowadays fats and fatty materials have replaced the petroleum products as lubricating agents in various applictions. In some edible products also this lubricating action is important.

Viscosity is a physical property which is a measure of internal friction in molecules. The coefficient of viscosity is expressed in poise. Kinematic viscosity is expressed in stokes which is poise divided by the density of the sample. Viscosity is measured by

Saybolt viscometer in case of lubricating oils and Ford cup is used for paints and lacquers. In these instruments efflux time is measured which is the number of seconds required for discharging a fixed volume of sample.

The viscosities of heated or blown oils are higher than normal oils. Generally, with increase in unsaturation viscosity of oils decreases. So the viscosities of oils having low molecular weight fatty acids are more or less equal to the viscosities of oils having an equivalent degree of unsaturation but containing only high molecular weight acids. Castor oil is a class by itself because it is much more viscous than ordinary oils. The high value of viscosity of castor oil is due to the presence of ricinoleic acid which forms intermolecular hydrogen bonding. Ordinary oils have more or less the same range of viscosities and they show less change in viscosity with temperature than do the mineral oils. Table 2.1 lists the viscosities of some common oils.

Table 2.1: Viscosities of Fats and Oils (2)

Oil	*Specific gravity (20°/4°C)*	*Kinematic Viscosity*		*Saybolt Viscosity*	
		100°F	*210°F*	*100°F*	*210°F*
Olive	0.9158	46.68	9.09	216	55.2
Rapeseed	0.9114	50.64	10.32	234	59.4
Mustard	0.9237	45.13	9.46	209	56.9
Soyabean	0.9228	28.49	7.60	134	50.1
Castor	0.9619	293.4	20.08	1368	97.7
Coconut	0.9226	29.79	6.06	140	45.2
Palm Kernel	0.9190	30.92	6.50	145	46.5

Viscosities of fatty acids are higher than their methyl or ethyl esters because of the intermolecular hydrogen bonding between the carboxyl groups of the fatty acids.

The viscosities of mixtures of binary, ternary and quarternary fatty acids of C_{12}- C_{18} depends on the average chain length. This is called the 'Principle of Congruence'. Fatty oils behave as true Newtonian liquids but they may exhibit thixotropic properties at very high rates of shear. Under this condition the viscosities of polymerized oils decrease markedly. This happens because at high rate of shear the aggregates or micelles try to orient themselves under high shearing stresses.

SURFACE AND INTERFACIAL TENSION

Surface tension and interfacial tension of fats and their derivatives are important physical properties. A molecule in the bulk of a liquid is subjected to a force of attraction from all direction by the surrounding molecules, but when the molecule is in the surface of the liquid it is attracted more by the bulk molecules than by the molecules of the vapour over the liquid surface. Hence, the surface layer seems to behave as a stretched layer and is in a state of tension. So an attempt to make a penetration along any line in the surface will require an application of force to hold the separate portions of the surface together. This force is called surface tension. Surface tension increases with increase in chain-length and decreases linearly with increase in temperature. With increase in temperature interfacial tension also decreases. Table 2.2 lists the surface tension (dynes/cm) of some oils at different temperatures.

***Table 2.2:* Surface Tension of some Oils at Different Temperatures (dynes/cm)**

Temperature °C	*Cottonseed oil*	*Castor oil*	*Coconut oil*
20	35.4	39.0	33.4
80	31.3	35.2	28.4
130	27.5	33.0	24.0

Interfacial tension of an oil or fat is decreased by the presence of mono-glycerides or phospholipids and is lowered by free fatty acids or soaps.

Neutral glycerides have hardly any surface activity and since glycerides, fatty acids and methyl esters differ in surface activity they can be easily separated by chromatographic methods.

Phospholipids act as emulsifiers because they can decrease the interfacial tension between fats and water and for this reason they are used in many food products and paints.

Surface tension and interfacial tension of soaps are important properties which are studied elaborately.

DENSITY AND EXPANSIBILITY

(a) *Density in liquid state*: In case of fats and oils relative density, i.e. the ratio of mass of the substance to the mass of an equal volume of water at 4°C, is used more than the absolute density (gm/c.c). Densities of fatty acids and glycerides increase with increase in their unsaturation and decrease in molecular weight. An empirical relationship is given below according to (3) for the determination of specific gravity (15°/15°C):

$$\text{Specific gravity} = 0.8475 + 0.00030\ (\text{saponification value}) + 0.00014\ (\text{iodine value})$$

During the processing of fats and oils the density decreases linearly with temperature; at lower temperature range the change is greater:

Densities of some liquid fatty acids, methyl and ethyl esters and triglycerides are listed in table 2.3.

The density change per unit change in temperature is approximately 0.0005 – 0.001 depending upon the unsaturation, chain length and temperature range. Molar volume (Vm) is obtained by dividing the molecular weight by density.

(b) *Density in the solid state*: Fats and their derivatives have higher density in their solid form than in liquid state. When they are subjected to solidification or melting, they show shrinkage or expansion much greater than the thermal expansion of solid or liquid phase. Density relationship for mixtures of mixed triglycerides are more complex than the density of pure single acid glyceride.

(c) *Density and volume of plastic fats — Dilatometry*: Dilatometry is defined as the study to observe the changes in density or volume with varying temperature and time and the instrument used for this purpose is known as dilatometer. To characterize and control the consistency of commercial fat mixtures, like margarines and shortenings, dilatometry data is useful. It also gives information about the melting point of a fat.

Table 2.3: Densities of Some Fatty Acids, Methyl and Ethyl Esters and Triglycerides

Compound	Density (gm/ml) Acid	Methyl ester	(Temperature °C) Ethyl ester	Triglycerides
Butyric	0.9292 (50 0.9043 (75)	0.8359(75)	0.8205(75)	—
Caproic	0.8796 (75) 0.8751 (80)	0.8313(75) 0.8270(80)	0.8198(75) 0.8141(80)	— —
Caprylic	0.8662 (75) 0.8615 (80)	0.8287(75) 0.8240(80)	0.8191(75) 0.8148(80)	— —
Capric	0.8583 (75) 0.8531 (80)	0.8271(75) 0.823 (80)	0.8196(75) 0.8151(80)	— 0.8913 (80)
Lauric	0.8516 (75) 0.8477 (80)	0.8259(75) 0.8220(80)	0.8195(75) 0.8153(80)	0.8801 (80) —
Myristic	0.8481 (75) 0.8439 (80)	0.8252(75) 0.8210(80)	0.8197(75) 0.8156(80)	— 0.8722 (80)
Palmitic	0.8446 (75) 0.8414 (80)	0.8249(75) 0.820 (80)	0.8198(75) 0.8158(80)	— 0.8663 (80)
Stearic	0.8431 (95) 0.8390 (80)	0.8244(75) 0.821 (80)	0.8218(75) 0.8161(80)	— 0.8652 (80)
Oleic	0.863 (60) 0.850 (60)	0.875 (25) 0.860 (45)	— —	0.9078 (25) —

The density of a solidified fat cannot be compared with the density of a liquid oil or fat because solidified fat is a mixture of solids and liquids and its density depends on the proportions of these two phases. Dilatometric curve shows the change in specific volume of a fat with increase in temperature. The proportion of solid material in the fat indicates the consistency of the fat.

(d) *Density of oil-solvent mixtures*: Density of oil-solvent mixtures are important for the solvent extraction processes. Commercial solvent used for extraction processes is mainly hexane. A general equation is given to obtain the density of oil-solvent mixtures.

The equation is:

$$\frac{1}{D} = a - (a - b + 4K)x + 2Kx^2$$

where '*D*' is the density of the oil-solvent mixture, '*a*' and '*b*' are the reciprocals of the densities of the solvent and oil respectively, '*x*' is the weight fraction of the oil in the mixture and '*K*' is a constant depending on the solvent and temperature.

MELTING POINTS

Melting point of fatty materials is a very important physical property as melting point identifies the material, and it is also used for many technological applications. Melting points of fatty acids increase with increasing chain length and decrease with unsaturation.

Pure fatty acids, esters and glycerides often exhibit distinct melting points as they are polymorphic. Even and odd numbered carbon fatty acids and glycerides show a regular alternation of melting points. Larsson explained this behaviour as the differences in packing of chains in crystal (4). Figure 2.1 shows this behaviour of melting points of fatty acids.

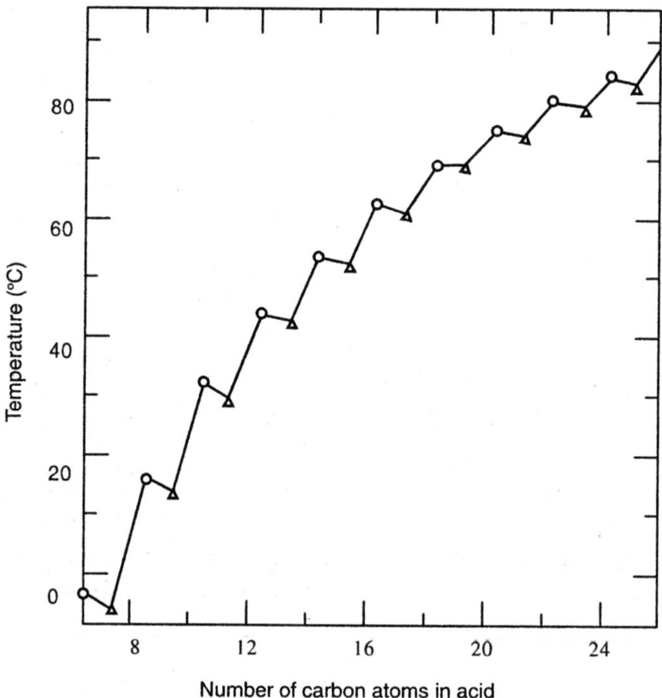

Fig. 2.1: Melting point alternation of straight chain saturated acids. O, Even; Δ, Odd,

THERMAL PROPERTIES

(a) *Heat of combustion*: Heat evolved during combustion is called heat of combustion. Heat of combustion of saturated fatty acids depends on the chain length; it increases with increase in the number of carbon atoms in the straight chain. Heats of combustion of unsaturated fatty acids are slightly lower than saturated acids of the same chain length. For example stearic acid has heat of combustion 9600 cal/gm but oleic acid and linoleic acid have 9450 cal/gm and 9350 cal/gm respectively. Triglycerides have more or less the same heats of combustion as their parent acids. An empirical relation is given to calculate the heat of combustion of variety of fatty oils.

$$\text{Heat of combustion} = 11380 - (\text{iodine value}) - 9.15\ (\text{saponification value}).$$

(b) *Specific heats-heats of fusion or crystallization*: Specific heat of a substance is defined as the number of calories required to raise 1 gm of the substance by 1°C. Specific heats of some common saturated fatty acids are listed in table 2.4.

***Table 2.4*: Specific heats of some saturated fatty acids (5,6,7,8).**

Acid	*Specific heat (cal/gm)*	*Temperature (°C)*
	Solid	*Liquid*
Caproic	0.4495 (10-33)	0.5105 (0-23)
Caprylic	0.4650 (0-12)	0.5050 (18-46)
Capric	0.5009 (0-24)	0.4989 (35-65)
Lauric	0.5116 (19-39)	0.5146 (48-78)
Myristic	0.5209 (24-43)	0.5157 (to 84)
Palmitic	0.4920 (22-53)	0.5416 (to 68)
Behenic	0.4854 (18-71)	0.5556 (to 109)

Specific heats of triglycerides are different in their different forms, e.g., the specific heat of liquid tristearin at temperature 79°C is 0.530 whereas in its β-form (solid) the specific heat is 0.160 at –177.3°C and in β-form (solid) it is 0.273 at –81°C. It is seen that in the solid state the specific heats of saturated fatty acids or their simple triglycerides do not change much with the chain length, but as the unsaturation increases the specific heat also increases steadily. However, the specific heat of solid is less than that of liquid at the same temperature.

In the case of liquid state, the specific heats of fatty acids increase with increase in chain length and decrease in unsaturation.

Heats of fusion of fatty acids are generally high and depend on chain length and unsaturation. Mixed triglycerides have lower heats of fusion than simple triglycerides. Table 2.5 shows heats of fusion and transformation of some simple saturated triglycerides.

Table 2.5: Heats of Fusion and Transition of Simple Saturated Triglycerides (9)

Triglyceride	*Transformation*	*Temperature °C*	*Heat (cal/gm)*
Trilaurin	β-form to liquid	46.3	46.2
Trimyristin	β-form to liquid	57.0	50.3
	β-form to liquid	32.3	34.6
	β-form to β-form	32.3	−12.6
Tripalmitin	β-form to liquid	65.7	53.1
	β-form to liquid	44.7	37.4
	β-form to β-form	44.7	−13.3
Tristearin	β-form to liquid	72.5	54.5
	α-form to liquid	54.0	38.9
	α-form to β-form	54.0	−13.7

(c) *Vapour pressure and boiling point, heat of vaporization*: Vapour pressure, boiling point and heat of vaporization are the important physical properties of fats which have specific applications in the practical field.

Monoglycerides have generally high vapour pressures and they can be distilled using proper temperature and pressure conditions. But if interesterification reaction is to be avoided (10,11) monoglycerides should be distilled under high vacuum. Triglycerides on the other hand have extremely low vapour pressures and for this reason they are distilled by molecular distillation. Fatty acids are more volatile than their mono- and triglycerides. They are distilled by vacuum distillation or fractional distillation under reduced pressure. Fatty acid mixtures behave as non-ideal mixtures and thus they show deviation from Raoult's Law. For this reason, the fatty acids are transformed to their methyl or ethyl esters which behave more or less ideally and hence fractional distillation is employed to the ester mixtures. Methyl oleate and methyl stearate can be separated by using fractional distillation column owing to the difference in their vapour pressures.

Boiling point data of some saturated fatty acids and fatty alcohols are given in the tables 2.6 and 2.7 respectively.

***Table 2.6*: Boiling Points of Saturated Acids at Atmospheric Pressure (12)**

Acids	*Boiling points (°C)*
Caproic	205.8
Caprylic	239.7
Capric	270.0
Lauric	298.9
Myristic	326.2
Palmitic	351.5
Stearic	376.1

***Table 2.7*: Boiling Points of Fatty Alcohols**

Compound	*Boiling points °C (mm)*
Hexanol	157 (760)
Octanol	194.5 (760)
Nonanol	215 (760)
Decanol	231 (760), 42 (0.1)
Dodecanol	255 (760), 64 (0.1)
Tetradecanol	171.5 (20), 85 (0.1)
Octadecanol	210 (15), 121 (0.1)

Heat of vaporisation of fatty materials is obtained by using Clausius-Clapeyron equation which is stated as

$$\frac{dP}{dT} = \frac{L}{(V - v)T}$$

where, P = pressure, T = absolute temperature,

V = molar volume of vapour, v = molar volume of liquid

L = latent molar heat of vaporisation.

The volume of liquid at low pressures can be neglected and the equation is reduced and can be presented as,

$$\frac{-d \ln P}{d(1/T)} = \frac{L}{R}$$

'R' is a constant. If we plot log (vapour-pressure) against $\frac{L}{T}$, the slope at every point of the curve gives latent heat L.

(d) *Heats of mixing*: Heat associated with mixing of two substances is called heat of mixing. Mixing of an organic solvent with an oil is usually an endothermic process but when a chlorinated solvent is added to an oil, heat is evolved.

(e) *Thermal conductivity*: Thermal conductivity is defined as the heat flow required across unit area of an isothermal surface to make unit change in temperature gradient. This can be expressed by an equation

$$\frac{q}{A} = -K\frac{dT}{dn}$$

where q = rate of heat flow across the surface in the direction normal to the surface,

A = area of isothermal surface,

T = temperature,

n = distance measured normally to surface

K = proportionality constant called thermal conductivity.

In C.G.S. unit K is expressed is Cal/cm^2-hr. (^{o}C/cm).

Fats are poor conductors of heat. The thermal conductivity data cannot be determined accurately. However, some thermal conductivity data are available such as thermal conductivity of olive oil is 0.00040 Cal/cm^2. sec. ^{o}C/cm at 19°C and for castor oil it is 0.104 Btu/hr. ft^2 ^{o}F ft at 68°F etc.

SOLUBILITY AND MISCIBILITY

(a) *Miscibility with organic solvents*: Fats and fatty acids show complete miscibility with organic solvents at temperatures above their melting points but they are completely immiscible with water.

Since castor oil contains hydroxyl group in its constituent ricinoleic acid it is miscible with aromatic solvents and alcohols but it cannot mix with straight chain hydrocarbons easily. Organic solvents near their critical temperatures and pressures show anomalous behaviour being incompletely soluble with fats and this nature of solvents is used is liquid-liquid extaction.

(b) *Solubility in organic solvents*: Fats and fatty materials are soluble in organic solvents and solubility decreases with decrease in temperature. They form non-ideal solution with organic solvents and show positive deviation from ideality. Solubility data of pure compounds cannot be applied to the mixture due to the inter solubilizing effect of one component on the other within the mixture. Separation techniques are employed depending on the solubility data of pure compounds in various solvents over a range of temperatures.

Solubility of fatty materials in organic solvents decreases with increase in chain length and decreases in unsaturation. So, a marked difference in solubilities of saturated and unsaturated fatty acids is seen which is an essential requirement of these acids for fractional crystallization. All trisaturated, disaturated, monosaturated and poly-unsaturated glycerides have different solubilities in the same solvent at a definite temperature and hence they can be separated by using fractional crystallization method. Pure compounds are prepared by fractional crystallization.

(c) *Mutual solubility of fats and fatty acids in water*: Small chain fatty acids are more or less soluble in water but solubility decreases with increase in chain length and with decrease in temperature. The solubility of $C_6 - C_{18}$ saturated fatty acids in water has been obtained in the temperature range of -8° to $+92^{o}C$ (13). At different temperatures the solubility of fatty acids is represented as percent solubility, e.g., at $46^{o}C$, the solubility of caproic acid is 10 per cent by weight of water. In fat splitting high temperature, miscibility of fats and water is utilized. Long chain fatty acids do not dissolve in water.

(d) *Solubility of gases in fats*: Gases are soluble in fats. Air, oxygen, nitrogen, hydrogen and carbon dioxide have been reported to have solubility in lard, cottonseed oil, butter fat etc. at $23^{o} - 60^{o}C$ temperature range. Besides carbon-dioxide, all gases dissolve to a higher extent with increase in temperature.

OPTICAL PROPERTIES

(a) Refractive Index: Refractive index is a very important physical property of fats and fatty acids as it is a unique method for their identification. It is useful as it can be performed easily and quickly with a small amount of sample. Moreover, refractive index is used as a tool for observing the progress of reactions like catalytic hydrogenation and isomerization.

Relationships of the refractive indices of methyl esters of saturated acids having six or more carbon atoms presented by Gonw and Vlugter (14) are stated as:

$$n_{D}20 = 1.47706 - \frac{0.72436}{4.0969 + n}$$

$$n_{D}20 = 1.46977 - \frac{0.72922}{3.9252 + n}$$

The values of refractive index corresponds to the wavelength 5890 Å, the wavelength of the D line of sodium.

Refractive index of fats and fatty acids depends on the chain length and unsaturation. With increase in chain length, double bonds and also conjugation, refractive index of the material increases. Glycerides have higher refractive indices than their corresponding fatty acids. Refractive indices of monoglycerides are higher than those of corresponding triglycerides. Refractive index is an additive property i.e., refractive index of a mixture of simple glycerides is more or less close to the sum of refractive indices of its components.

Conjugation increases the refractive index of a compound to a marked degree and hydroxyl group also has the same effect.

Normally, refractive index is measured at a wavelength 5890 Å corresponding to the D line of sodium. When refractive index is measured at two wavelengths using monochromatic radiation, the difference between the two refractive indices is called dispersion. In case of fats and fatty acids the measure of dispersion is a valuable source for distinguishing a sample. For this purpose, besides using D line of sodium, another wavelength 4358 Å (G line of mercury) is used for determining dispersion. Dispersion data are usually utilized for non-conjugated and conjugated double bonds but it has been less used in the case of saturated fatty acids, glycerides and other esters.

(b) Absorption spectra: Absorption spectra like ultraviolet, infrared and more recently nuclear magnetic resonance spectra have been studied extensively. UV and IR spectroscopies are useful for the study of analysis of composition and identification of fats and fatty materials. A very important application of spectral study is in the field of observing and following chemical reactions of fats and fatty materials, specially *cis*, *trans* and position isomerization, oxidation and polymerization.

Normally, crude oils and fats are dark coloured due to some pigments which are removed upto a certain extent in refining process. Usually Lovibond tintometer is used for the determination of colour of different fats and oils. Now special methods have been developed (15) which can be used as replacement of visible Lovibond method in controlling, refining and bleaching operation.

Double bonded fatty materials exhibit absorption in the ultraviolet region between 200– 400 mμ. Saturated and non-conjugated materials have weak absorption band and so UV spectral data are not employed for their analysis. For the use of conjugated compounds which exhibit strong absorption band in uv region, spectral data are used for quantitative determination of a sample. For example 9, 11 octadecadienoic acid shows an intense absorption band at 233 mμ.

Ultra-violet absorption spectra have unique applications in the determination and analysis of the mixtures of non-conjugated unsaturated fatty acids. For example, linoleic and linolenic acids undergo isomerization reaction to produce conjugated dienoic and trienoic acids. By observing the uv spectra of the isomerized acids quantitative determination of the Linoleic and Linolenic acid content has been possible. But this method has a limitation, for the *cis* forms of acids this method is accurate but it is not satisfactory for *trans* forms.

Infra-red spectroscopy is another important tool for determination and identification in the fat chemistry. Infra-red spectrum region is from about 0.75 to 1000m but 2–15m region has been extensively studied.

Infra-red absorption spectroscopy has been used widely for the quantitative determination of *trans*-monounsaturated acids, esters, alcohols and glycerides in a mixture of *cis*-isomers and saturated compounds. The *trans* materials absorbs at about 10.3μ and this important characteristic is used in this method of quantitative determination. This method is more rapid and accurate than chemical methods.

Infra-red spectrum has other important applications which are extensively used. These include the determination of oxygen containing groups formed during auto-oxidation reactions like epoxy, hydroxy etc., identification of monoglycerides, determination of the number of methylene groups (finger-print region of IR is used here) in a chain etc.

X-ray diffraction is used for the crystal study and polymorphism in the fats field. X-rays are electromagnetic vibrations of atomic dimensions (~ 1°A) which diffract by a crystal and so crystalline materials give a sharp x-ray diffraction pattern but non-crystalline compounds give only diffuse pattern.

Polymorphism is defined as the ability of a compound to exist in the more than one crystalline form. Polymorphism is of great value from the scientific and technological points of view in the fats and oils field.

Polymorphic forms of a solid compound differ only in their crystal structure, free energy and other physical and chemical properties but are of the same chemical composition (16). They yield identical liquid or gaseous phase upon fusion and evaporation. According to the phase rule, each polymorphic form is considered as a separate phase and hence the number of solid phases for a single component is the number of existing polymorphic forms.

These definitions are applied to the equilibrium conditions and it is possible to form a meta-stable crystalline solid from a meta-stable liquid phase. This meta-stable solid melts and resolidifies as the temperature is raised to a more stable or high melting form.

There are two types of polymorphism: one is enantiotropic and the other is monotropic. In case of enantiotropism, different forms possess different vapour pressure curves which approach each other with varying temperature and cut at a point which is called transition point. At this point the vapour pressures of the two forms are identical. Only at this transition temperature both the solid phases are stable and an equilibrium exists between them such that an isothermal transition is possible from one phase to another. The phase which is stable at higher temperature can be obtained from the other phase by supplying heat until the other phase has disappeared. The reverse is also true when heat is subtracted. These changes occur at constant temperature and this point is called invariant or triple point.

In the case of monotropism, the vapour pressures of the two forms do not meet and hence no transition point is obtained. In monotropism, one form is always stable and the other form is metastable under the condition where both forms can exist. Metastable form always has higher vapour pressure than stable form, and it can be changed to the stable form by some processes like distillation. Enantiotropism is completely reversible in the solid state but monotropism is not.

Polymorphism is commonly found in long chain aliphatic compounds like long chain fatty acids, glycerides, alcohols, hydrocarbons etc. Polymorphism is important mainly in some food products and fatty derivatives like lard, butter, margarine, shortenings etc. because the consistency, plasticity and other physical properties are dependent on polymorphic forms which are produced during the processing of these products. During crystallization of long chain aliphatic compounds, polymorphic forms are obtained and the particular polymorphic form which is assumed to form depends on various factors like the solvent used, purity of the compound, the presence or absence of particular crystalline nuclei, temperature, rate of cooling, degree of supercooling etc. Each polymorphic form has its characteristic melting point, resolidification point, heat of crystallization, specific volume, x-ray spacing etc. and every form differs in these respects from other forms of the same compound. The polymorphic forms have different looks, e.g., they may be dense, opaque crystals or transparent crystals. The derivatives of fatty acids exhibit both enantiotropism and monotropism but acids themselves exhibit only monotropism.

During the crystallization of fat phase for the manufacture of different food products, e.g. chocolate, it is seen that a number of different melting points are shown by fats which depend on the temperature condition and the type of polymorphic crystalline form produced. It has been found from x-ray diffraction studies that each triglyceride molecule exhibits a tuning fork type of structure in which the fatty acid at the 2-position is projected away from the other two acids (17). The crystal structure and, therefore, the melting point depend on the way by which the triglyceride molecules are packed with each other. There are three main types of crystal structure found in fats:

i) α-form: This form has the lowest packing density, the lowest melting point, the fastest growth rate and is the least stable.
ii) β′-form: Its packing is denser than that of the α-form; it has higher melting point and is more stable.
iii) β-form: This form has the highest packing density and melting point and is the most stable.

ELECTRICAL PROPERTIES

(a) *Resistance*: Fats and fatty acids may be considered as non-conductors of electricity. So, they have a high value of specific resistance. For example, oleic acid has a specific resistance 2×10^{11} Ohms at 100°C.

(b) *Dielectric Constant*: Dielectric constant of a molecule is a property by which it can alter the attraction between oppositely charged ions once they are solvated.

Dielectric constants of most oils are within the range of 3 – 3.2. However, oils containing oxygenated fatty acids (castor oil etc.) have higher values. Oxygen being a polar group increases the dielectric constant.

For emulsions like butter the dielectric constant depends on emulsion structure and moisture content and is useful to control the manufacturing of these products.

REFERENCES

1. Bailey's Industrial Oil and Fat Products, Volume 1, Fourth edition, edited by Daniel Swern, 1979.
2. A.R. Rescoria and F.L. Carnahan, *Ind. Eng. Chem.*, **28**, 1212 (1936).
3. J. Lund, Z. Untersuch Nahr. *Genussm.*, **44**, 113 (1922).
4. K. Larsson, *J. Am. Oil Chemists' Soc.*, **43**, 559 (1966).
5. W.E. Garner, and F.C. Randall, *J. Chem. Soc.*, **1924**, 881,
6. W.E. Garner, F.C. Madden and J.F. Rushbrook, *J. Chem. Soc.*, **1926**, 2491.
7. W.E. Garner and A. M. King, *J. Chem. Soc.*, **1929**, 1829,
8. D.Q. Kern and W. Van Nostrand, *Ind. Eng., Chem.*, **41**, 2209 (1949).
9. G.H. Charbonnet and W. S. Singleton, *J. Am. Oil Chemists' Soc.*, **27**, 143 (1950).
10. N.H. Kuhrt, E.A. Welch and F.J. Koranik, *J. Am. Oil Chemists' Soc.*, **27**, 310 (1950).
11. J. Ross, A.C. Bell, C.J. Arrowsmith and A.T. Gebhart, *Oil Soap*, **23**, 257 (1946).
12. W.O. Pool and A.W. Ralston, *Ind. Eng. Chem.*, **34**, 1104 (1942).
13. C.W. Hoerr, W.O. Pool and A.W. Ralson, *Oil Soap*, **19**, 126 (1942).
14. T.H. Gonw and J.C. Vlugter, *J. Am. Oil Chemists' Soc.*, **41**, 426 (1964).
15. American Oil Chemists' Society, Official and Tentative Methods, 3rd ed. 1975, Method C_c, 13C–50.
16. Fatty acids — Their Chemistry, Properties, Production and Uses edited by **Klare. S. Markley**, 2nd edition, Part – I, Chapter 7, Inter-Science Publishing Co., New York, 1960, p. 609.
17. Fats and Oils, Chemistry and Technology edited by R.J. Hamilton and A. Bhati, Applied Science Publishers Ltd. England, p. 173, 1980.

CHAPTER 3

Reactions of Fats and Fatty Acids

REACTIONS OF FATS AND FATTY ACIDS

This chapter mainly covers the manufacture and processing of several commercially important fat-derived products. Fatty acids having eight or more carbon atoms undergo most of the reactions of the simpler members of the series, e.g. acetic acid and propionic acid (1). The higher members do not react as the lower ones, but this difficulty can be overcome by altering the conditions of the reaction.

Fats and oils are likely to be spoiled to some extent by bacterial action but toxic bodies called aflatoxin develop as result of the action of certain molds. Main damage of fats and oils is caused by the atmospheric oxidation during storage. To prevent this, special attention is given to minimize the reaction between oxygen and the oils or fats under storage.

HYDROLYSIS, ESTERIFICATION AND RELATED REACTIONS

(1) *Hydrolysis*: The triglycerides of fats and oils are hydrolyzed to give fatty acids and glycerol under proper fat and water miscibility condition.

$$C_3H_5\ (OOCR)_3 + 3HOH \longrightarrow C_3H_5(OH)_3 + 3HOOCR$$

The reaction occurs in steps and is reversible. For the reaction to proceed in the forward direction a large excess of water is taken to ensure full hydrolysis and the glycerine-rich phase is continuously withdrawn from the reaction medium.

High temperature and high pressure favour the reaction, although the selection of temperature depends on the content of polyunsaturated, particularly conjugated polyunsaturated acids in the fats, because at high temperature, polymerization reaction may be a competing reaction with the hydrolysis. Hydrolysis is catalyzed by some catalysts like acids, certain emulsifiers etc. Use of lipolytic enzyme permits the reaction to occur at ordinary temperature.

(2) *Esterification*: Esterification reaction occurs when fatty acids react with alcohols, glycols, glycerols or other polyhydroxy compounds like sugars and starches. Esterification can be considered as the reverse of hydrolysis. The reaction is reversible and goes to completion when water is removed.

Triglycerides of any desired composition can be made by hydrolysing fats, purifying the resulting fatty acids by fractional distillation and then reacting the fatty acids in the desired proportions with glycerol by the process of esterification.

Mono- and di-glycerides can also be prepared by properly controlling the reactants. The products obtained in these esterification reactions are the mixtures of mono- and di-glycerides and unreacted glycerol. These products are used in various fields, mainly as food emulsifiers. Another important reaction of esterification is the reaction of fatty acids with acetylene or vinyl acetate which yields high proportion of fatty acids (2).

Interesterification: Methyl esters of fatty acids are prepared by displacing the glycerol from fats, generally with an alkaline catalyst. The reaction is called methanolysis.

$$C_3H_5\,(OOCR)_3 + 3CH_3OH \longrightarrow 3CH_3OCOR + C_3H_5(OH)_3.$$

A mixture of mono- and di-glycerides are obtained by the reaction of triglycerides with glycerol and a catalyst (glycerolsis). If a small quantity of glycerol is used, the desired partial glycerides cannot be obtained but with increasing proportion of glycerol the formation of monoglycerides increases.

Saponification with alkalis: When fat is reacted with an alkali it yields glycerol and a salt (called 'soap') of the alkali metal and the reaction is termed as 'saponification' reaction.

$$C_3H_5(OOCR)_3 + 3NaOH \longrightarrow C_3H_5(OH)_3 + 3NaOOCR.$$

Soaps may also be formed by reacting free acids with alkali,

$$RCOOH + NaOH \longrightarrow RCOONa + H_2O$$

This reaction between an acid and an alkali is the basis of two important analytical determinations. The acid number or neutralization number is the measure of the hydrolysis of a fat and is expressed as the number of milligrams of potassium hydroxide required to saponify the free fatty acids (FFA) in 1 gm of fat or to neutralize 1gm of fatty acids. The acidity of the fats is expressed in terms of FFA and the molecular weight of acids are always assumed to be equivalent to oleic acid for easy calculation. The saponification number of fatty materials is defined as the number of milligrams of potassium hydroxide required to saponify one gram of fat. The saponification equivalent of fat or ester is the number of milligrams of fat or ester saponified by one mole (56.104 gm) of potassium hydroxide. A relationship holds between saponification value and saponification equivalent: saponification value × saponification equivalent = 56.104.

OTHER REACTIONS INVOLVING THE CARBOXYL GROUP

(1) *Formation of metal soaps*: Soaps are produced when fatty acids react with sodium or potassium hydroxides, oxides and carbonates:

$$RCOOH + NaOH \longrightarrow RCOONa + H_2O$$

$$2RCOOH + K_2CO_3 \longrightarrow 2RCOOK + H_2O + CO_2$$

The soaps of heavy metals like zinc, lead, manganese, cobalt or tin may usually be formed by double decomposition reaction of a sodium or potassium soap and a salt of the metal:

$$2NaOOCR + Pb(OOCCH_3)_2 \longrightarrow Pb(OOCR)_2 + 2NaOOCCH_3$$

These reactions are utilized in the manufacture of lubricating greases, metallic dryers for paints and other protective coatings and in the preparation of vinyl resin stabilizers.

(2) *Hydrogenation of the carboxyl group*: Fatty acids are converted to the fatty alcohol by the addition of hydrogen to the carboxyl group under high pressure and metal catalysts.

$$RCOOH + 2H_2 \longrightarrow RCH_2OH + H_2O$$

There are many other methods of preparation of fatty alcohols, such as the hydrogenation of glycerides or methyl esters ($RCOOCH_3$):

$$RCOOR^* + 2H_2 \longrightarrow RCH_2OH + R'\,OH$$

or, sodium reduction of esters of fatty acids:

$$RCOOR' + 4Na + 4ROH \longrightarrow RCH_2OH + 4NaOR' + R'\,OH$$

It has been shown (3) that homogeneous catalytic hydrogenation in a normal high pressure autoclave of oleic acid, methyl oleate and olive oil in the presence of copper or cadmium oxides or soaps gives almost selective hydrogenation of the carboxyl group without double bond hydrogenation to a substantial degree.

Fatty alcohol reacts with ethylene oxide to produce a series of polymeric polyoxethylene alcohols:

$$ROH + n\overset{\frown O \frown}{CH_2 - CH_2} \longrightarrow RO(CH_2CH_2O)_nH$$

They also react with acetylene under pressure to produce vinyl ethers which, when polymerized, give products useful as adhesives, surface coating materials and water proofing agents.

(3) *Formation of Nitrogen Derivatives*: Reaction of amine or ammonia with fatty acids yields amides at elevated temperatures:

$$RCOOH + NH_3 \longrightarrow RCONH_2 + H_2O$$

$$RCOOH + R'NH_2 \longrightarrow RCONHR' + H_2O.$$

Nitriles can be obtained when amides are dehydrated at higher temperature:

$$RCONH_2 \longrightarrow RCN + H_2O$$

Nitriles can be transformed into primary or secondary amines by catalytic hydrogenation:

$$RCN + 2H_2 \longrightarrow RCH_2NH_2$$

$$2RCN + 4H_2 \longrightarrow (RCH_2)_2\ NH + NH_3$$

Triglycerides as esters undergo ammonolysis to give mixed amides. The reaction is reversible and is catalyzed in non-aqueous medium by ammonium chloride and several other ammonium salts. Tertiary amines and quarternary ammonium compounds are also important nitrogen compounds of fats which have large commercial importance.

(4) *Formation of Acid Chlorides*: Fatty acid chlorides are prepared by a number of ways, but the route employing phosphorous trichloride has industrial significance:

$$2RCOOH + PCl_3 \longrightarrow 2RCOCl + H_3PO_3$$

$$RCOOH + PCl_5 \longrightarrow RCOCl + POCl_3 + HCl$$

$$RCOOH + SOCl_2 \longrightarrow RCOCl + SO_2 + HCl$$

$$RCOOH + (COCl)_2 \longrightarrow RCOCl + HCl + CO_2 + CO$$

Fatty acid chlorides are primarily used as the reactive intermediates for acylation reactions. A comprehensive review of the extensive chemical reactions of fatty acid chlorides has been published (3).

(5) Dehydration of anhydrides: Fatty acids when thermally dehydrated give acid anhydride, but by heat alone satisfactory yield cannot be obtained. Lower molecular weight acids such as myristic acid etc. can be dehydrated at elevated temperatures (250° * $600^{\circ}C$) but for efficient dehydration, dehydrating catalysts such as phosphoric acid and its salts are used.

$$2\ RCOOH + heat \rightleftharpoons (RCO)_2O + H_2O$$

It is seen that the higher members of fatty acid give hydrocarbons or ketones when the temperature is raised or the time of exposure is increased:

$$RCOOH \longrightarrow RH + CO_2$$

$$2RCOOH \longrightarrow R_2CO + CO_2 + H_2O$$

The best choice for dehydration are the acyl compounds and acetic anhydride:

$$2RCOOH + CH_3COCl \longrightarrow (RCO)_2O + CH_3COOH + HCl$$

$$2RCOOH + (CH_3CO)_2O \longrightarrow (RCO)_2O + 2CH_3COOH$$

Another efficient method of preparation of acid anhydrides is the reaction between corresponding acid chloride and acetic anhydride:

$$RCOCl + (CH_3CO)_2O \longrightarrow 2CH_3COCl + (RCO)_2O$$

Decanoic, lauric, myristic, palmitic, stearic and oleic anhydrides have been prepared in high purity (90 per cent or more) by this method when high purity acid chlorides were used (4).

(6) *Alkoxylation*: There are two main modes of alkoxylation reaction between alkylene oxide and fatty acids. One is uncatalyzed reaction, which is slow and involved esterification and alkylene oxide hydrolysis along with alkoxylation:

$$RCOOH + \underset{\text{(epoxide, bridged by O)}}{CH_2 - CH_2} \xrightarrow{\text{slow}} RCOOC_2H_4OH$$

$$RCOOH + RCOOC_2H_4OH \xrightarrow{\text{slow}} RCOOC_2H_4OCOR + H_2O$$

$$\underset{\text{(epoxide, bridged by O)}}{CH_2 - CH_2} + H_2O \xrightarrow{\text{Very slow}} HOC_2H_4OH$$

$$RCOOC_2H_4OH + \underset{\text{(epoxide, bridged by O)}}{CH_2 - CH_2} \longrightarrow RCOOC_2H_4O(CH_2CH_2O)H$$

The other mode is alkali-catalyzed ethoxylation which is more complex and involves interesterification:

$$RCOOH + HO^- \xrightarrow{\text{Very fast}} RCOO^- + H_2O$$

$$RCOO^- + \underset{\text{(epoxide, bridged by O)}}{CH_2 - CH_2} \xrightarrow{\text{Slow}} RCOOC_2H_4O^-$$

$$RCOOH + RCOOC_2H_4O^- \xrightarrow{\text{Very fast}} RCOOC_2H_4OH + RCOO^-$$

$$RCOOC_2H_4O^- + \underset{\diagdown O \diagup}{CH_2 - CH_2} \xrightarrow{\text{Fast}} RCOOC_2H_4OCH_2CH_2O^-$$

$$HOH + RCOOC_2H_4O^- \xrightarrow{\text{Very fast}} RCOOC_2H_4OH + HO^-$$

$$RCOO^- + RCOOC_2H_4OC_2H_4OH \longrightarrow RCOOC_2H_4OC_2OCOR + HO^-$$

$$HO^- + \overset{\diagup O \diagdown}{CH_2 - CH_2} \xrightarrow{\text{Fast}} HOC_2H_4O^-$$

Ethoxylation and propoxylation reactions of common fatty acids such as stearic acid and oleic acid are important as they are used industrially to produce valuable polyoxyethylene ester emulsifiers and surfactants.

(7) *Pyrolysis*: Pyrolysis is the conversion reaction of one substance to another by heat in the absence of air or oxygen and with or without the presence of a catalyst.

Dialkyl ketones are prepared by the pyrolysis of calcium or barium salts of fatty acids:

$$2(RCO_2)_2Ca \longrightarrow RCOR + 2CaCO_3$$

Decarboxylation of fatty acids is a pyrolysis reaction, and by this reaction saturated fatty acids give mainly ketones and unsaturated acids, yield olefins:

$$RCH = CHCOOH \longrightarrow RCH = CH_2 + CO_2$$

Methyl esters are more thermally stable than ethyl or other simple esters. Ethyl esters undergo pyrolysis reaction at about 400°C to give ethylene and the corresponding fatty acid:

$$RCOOC_2H_5 \longrightarrow RCOOH + CH_2 = CH_2.$$

Methyl esters do not undergo pyrolysis. For this reason as well as for economic reasons, methyl esters are chosen when fractional distillation of simple esters is required.

REACTIONS IN THE FATTY ACID CHAIN

Hydrogenation and reduction

Unsaturated fatty acids form saturated or less unsaturated fatty acids by the reaction of gaseous hydrogen in the presence of a suitable catalyst like nickel, platinum, palladium etc. Catalytic hydrogenation is a very important industrial process.

Theoretically as the hydrogenation reaction is an addition reaction, it can be used for the determination of unsaturation:

$$-CH = CH^- + H_2 \longrightarrow -CH_2-CH_2^-$$

The partial reduction of double bonds may be obtained by the use of hydrazine and oxygen and this reaction, unlike partial hydrogenation, does not involve migration of bonds:

$$-CH{=}CH^- + NH_2NH_2 + 1/2\ O_2 \longrightarrow -CH_2-CH_2- + N_2 + H_2O$$

It is seen that the partial reduction with hydrazine retains the double bond position and the geometrical configuration about the non-hydrogenated bonds.

Dehydrogenation

Dehydrogenation reaction when accomplished under a suitable condition increases the unsaturation of fats. But due to its non-specificity and random nature this reaction does not have a wide industrial application.

Halogenation

Analytical aspects: Halogenation is an addition reaction and chlorine, bromine, iodine monochloride, iodine monobromide etc. add to the double bonds of unsaturated acids and their derivatives readily. Addition of halogen may be quantitative under special condition:

$$\begin{array}{cccc} H & H & H & H \\ | & | & | & | \\ -C- & C= & C- & C- \\ | & & & | \\ H & & & H \end{array} + Br_2 \longrightarrow \begin{array}{cccc} H & H & H & H \\ | & | & | & | \\ -C- & C- & C- & C- \\ | & | & | & | \\ H & Br & Br & H \end{array}$$

The quantitative addition of iodine monochloride or monobromides gives the basis of the important characteristic of fat, 'iodine number' (or iodine value). It is defined as the number of grams of iodine absorbed under standard conditions by 100 gms of fat. Iodine number is the representative of the total unsaturation of fats or fatty acids unless their double bonds are unconjugated and the structure is not severely hindered. Wiji's method for the determination of iodine number is one of the standard methods of the American Oil Chemists' Society. Another standard method is Hanus method. For the halogen addition analysis of α-β unsaturated fatty acids, Kaufmann's method is used as standard.

Preparative aspects: Halogenated fatty derivatives have some uses:

(1) The brominated fatty acid derivatives give anti-inflammatory property to textile additives.

(2) Halogenated fatty compounds are used as reactive intermediates for the preparation of other compounds by the replacement of the halo groups with hydroxyl, cyano, sulfonic acid or phosphonic acid groups.

(3) Those prepared by substitution offer the possibility of yielding chains of increased unsaturation by dehydro-halogenation process. Halogen derivatives may be prepared by addition to double bonds or by substitution of the hydrogen atoms in the chain.

Addition of sulfur dichloride: Addition of sulfur dichloride SCl_2 to the double bond of various mono-ethenoid fatty materials (methyl oleate, methyl elaidate, oleonitrile etc.) and ethyl linoleate has been studied and it is found that the products, β, β′-dichlorosulfides (a) can be easily oxidized by peracetic acid to the corresponding β, β′-dichlorosulfoxides (b) and then to β–β′ dichlorosulfones (c):

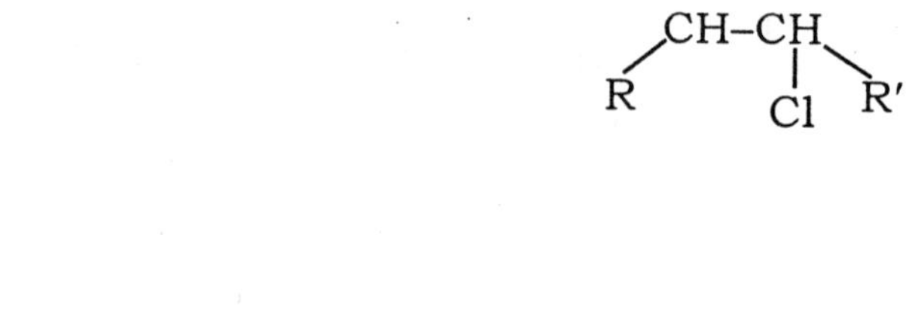

$$2RCH = CHR' + SCl_2 \longrightarrow \begin{array}{c} R\diagdown \quad Cl \;\; R' \\ CH\text{—}CH \\ | \\ S \\ CH\text{—}CH \\ R\diagup \quad | \;\; \diagdown R' \\ \quad Cl \end{array} \qquad (a)$$

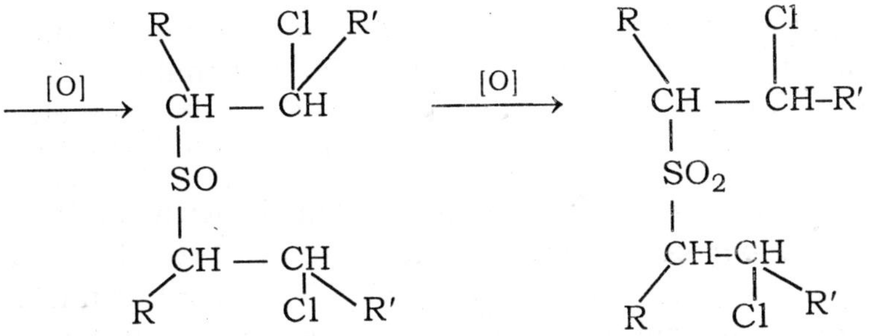

$$\xrightarrow{[O]} \begin{matrix} R & & Cl & R' \\ & CH & - & CH \\ & | & & \\ & SO & & \\ & | & & \\ & CH & - & CH \\ R & & & Cl \quad R' \end{matrix} \xrightarrow{[O]} \begin{matrix} R & & & Cl \\ & CH & - & CH\text{-}R' \\ & | & & \\ & SO_2 & & \\ & | & & \\ & CH & - & CH \\ R & & Cl & R' \end{matrix}$$

(b) (c)

Where R = $CH_3(CH_2)_7$–, –$(CH_2)_7COOR$ or –$(CH_2)_7CN$

and R′ = –$(CH_2)_7COOR$, –$(CH_2)_7CN$ or $CH_3(CH_2)_7$– respectively.

Addition of hydrogen cyanide and nitriles: Hydrogen cyanide and a variety of nitriles can be added to olefine in strong acid media to give substituted amides with good yields:

$$RCN + >C = C< \xrightarrow[2)\ H_2O]{1)\ H+} RCONHCH - CH$$

Addition of thiocyanogen: Kaufmann (5) showed that the addition of thiocyanogen to isolated double bonds is smooth and quantitative:

$$-CH = CH + (SCN)2 \longrightarrow \underset{SCN}{-CH} - \underset{SCN}{CH} -$$

The determination of thiocyanogen value is defined as the equivalent per cent iodine absorbed (centigrams of iodine absorbed per gram of sample). But this method has now become obsolete.

Addition of maleic anhydride and other dienophiles: Maleic anhydride acts as a dienophile in the Diels-Alder reaction; and in this reaction, the dienophile adds to the conjugated double bond systems of fatty acids, or of unsaturated fatty acids capable of generating them, to give compounds which on hydrolysis yield tricarboxylic acid functionality.

This reaction is utilized to determine the diene number which is calculated in terms of equivalent of iodine like the thiocyanogen number and the iodine number. The accuracy of diene number largely depends on the reaction conditions. For example, at a slightly elevated temperature (at about 200°C) maleic anhydride reacts monoenes like oleic acid in a non-Diels-Alder condensation reaction.

Diels-Alder reaction of maleic anhydride with safflower oil containing 70 per cent linoleic acid at about 110–130°C in the presence of a catalyst (sulfur) yields 51–55 per cent of mixture of adduct and maleic acid in a 60:40 ratio. The experiments were also made using other oils. Drying oils are 'maleinized' to increase their functionality and reactivity.

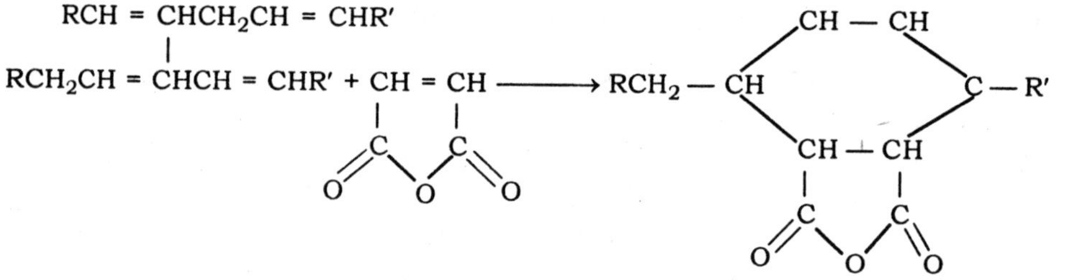

$$RCH = CHCH_2CH = CHR'$$

$$RCH_2CH = CHCH = CHR' + CH = CH \longrightarrow RCH_2 - CH \quad C - R'$$

CH — CH

CH — CH

C O C

O   O   O

Addition of phenols and cresols: Oleic acid is used for addition to the aromatic compounds through Friedel-Crafts reaction. For eg., oleic acid reacts with benzene in presence of sulfuric acid as catalyst to yield mixed isomers of phenyl-stearic acids. Like oleic acid, 10-undecenoic acid also reacts with benzene in the Friedel-Crafts reaction to give mixed phenyl undecenoic acids. Similarly, phenols and cresols also react with oleic acid smoothly.

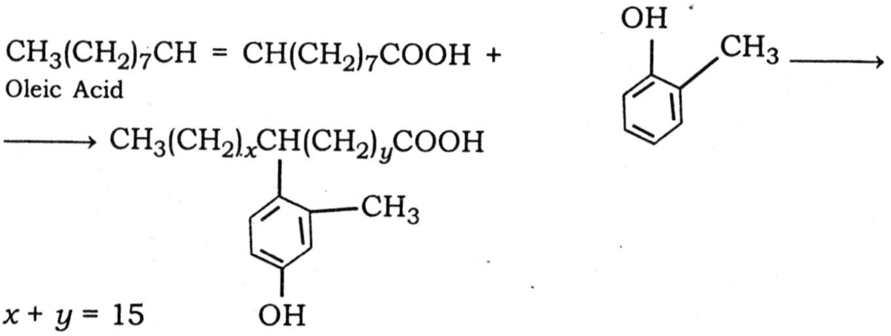

$$CH_3(CH_2)_7CH = CH(CH_2)_7COOH + \quad \text{OH, } CH_3 \longrightarrow$$

Oleic Acid

$$\longrightarrow CH_3(CH_2)_xCH(CH_2)_yCOOH$$

CH_3

$x + y = 15$ OH

(one of a series of isomers)

Reaction of long chain α-olefins with phenol occurs in the temperatures range 200–425°C without a catalyst.

Addition of hydrogen sulfide and mercaptans: Hydrogen sulfide and mercaptans add to the unsaturated bonds of fatty materials. Mercaptans are used in the elaidinization of methyl oleate to methyl elaidate. This reaction has been applied for many industrial products like lubricants, synthetic rubbers, flotation collectors etc.

Hydrogen sulfide adds to methyl oleate, methyl linoleate and soyabean oil at –70° to + 25°C with boron trifluoride. With methyl oleate at –70°C H_2 gives the product methyl 9(10)-mercaptostearate:

$$\begin{array}{c} SH \\ | \\ CH_3(CH_2)_7CH(CH_2)_8COOCH_3 \end{array}$$

$$\begin{array}{c} CH_3(CH_2)_8CH(CH_2)_7COOCH_3 \\ | \\ SH \end{array}$$

Sulfation and sulfonation: The double bonds of the unsaturated acids are attacked by sulfuric acid by the introduction of either sulfate group OSO_2OH or, at higher temperature, sulfonate group SO_2OH. If the fatty acid or glyceride contains hydroxyl group it is readily sulfated to give esters. The sulfation and sulfonation reactions are used industrially for the preparation of sulfonated oils for utilization in the textile industry.

Saturated fatty acids can also be sulfonated in α-position successfully by sulfur trioxide in dioxane. Sulfonated castor oil can be prepared by using sulfur trioxide which gives higher degree of sulfation and sulfonation than the conventional route through sulfuric acid. Hydroxyl group content of unsaturated oils, acids and other esters can be increased by partially sulfating the unsaturated bonds, followed by the hydrolysis to remove the sulfate group for replacement by the hydroxyl group. The sulfonation of an unsaturated fatty compound by sulfuric acid is shown here

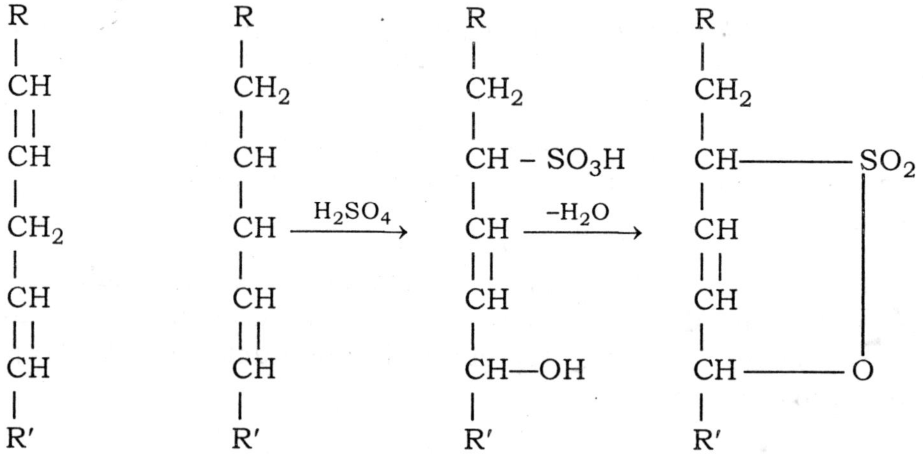

$$\begin{array}{l} R \\ | \\ CH \\ || \\ CH \\ | \\ CH_2 \\ | \\ CH \\ || \\ CH \\ | \\ R' \end{array} \qquad \begin{array}{l} R \\ | \\ CH_2 \\ | \\ CH \\ | \\ CH \\ | \\ CH \\ || \\ CH \\ | \\ R' \end{array} \xrightarrow{H_2SO_4} \begin{array}{l} R \\ | \\ CH_2 \\ | \\ CH - SO_3H \\ | \\ CH \\ || \\ CH \\ | \\ CH—OH \\ | \\ R' \end{array} \xrightarrow{-H_2O} \begin{array}{l} R \\ | \\ CH_2 \\ | \\ CH—SO_2 \\ | \\ CH \\ || \\ CH \\ | \\ CH—O \\ | \\ R' \end{array}$$

Addition of carbon monoxide: There are three modes of addition of carbon monoxide to the double bonds of fatty acids:

1. Hydroformylation:

$$- CH = CH - + CO + H_2 \xrightarrow{CO_2(CO)_8} - CH - \underset{}{CH_2}- \xrightarrow{H_2} \xrightarrow{} \begin{array}{l} -CH - CH_2- \\ | \\ CH_2OH \end{array}$$

2. Koch reaction:

$$-CH{=}CH- + CO + ROH \xrightarrow{H_2SO_4} \begin{array}{l} - CH - CH_2- \\ | \\ COOR \end{array} \quad R = H, \text{alkyl}$$

3. Reppe reaction:

$$-CH{=}CH- + CO + ROH \xrightarrow{Ni(CO)_2} \begin{array}{l} -CH -RCH_2- \\ | \\ COOR \end{array} \quad R = H, \text{alkyl}$$

The products of all three reactions are usually mixtures of many positional isomers.

Miscellaneous additions to the double bond: Formaldehyde reacts with the unsaturated fatty acids to give a variety of products:

```
-CH-CH-          -CH-CH-          -C = C
 |   |            |   |            |   |
 OH CH2           O   CH2              CH2
     |            |   |                |
     OH           CH2-O                OH

        OH
        |
       CH
     /    \
-CH          CH-              -CH - CH -
 |            |                |     |
 CH2          CH2              CH2 - CH2
    \       /                     \ /
        O                          O
```

Free radical addition of several other compounds to the double bonds of fatty acids has been achieved. Acetic acid, acetic anhyride and cyanoacetic esters add to methyl oleate and methyl undecenoate, dialkyl phosphonates add to methyl oleate and methyl undecenoate, dialkyl phospohonates add to methyl oleate and methyl undeceoate. Several terminal olefins add to the α-position of saturated fatty acids like palmitic and stearic acid by free radical addition mechanism. Nitrosyl chloride easily reacts with methyl oleate and the addition is quantitative:

```
-CH=CH- + NOCl ------> -CH-CH     +     -CH-CH
                         |   |            |   |
                         Cl  NO           NO  Cl
```

Chemical oxidation

Epoxidation: Epoxidation reaction is a specific, quantitative, reproducible and a large scale industrial chemical reaction:

```
                                       O
                                      / \
-CH=CH- + CH3COOOH ------> -CH — CH + CH3COOH
```

Peroxyacetic acid is used *in situ* as a means of oxygen transfer. The use of this reagent minimizes the hazardousness of the reaction and peroxide oxygen loss thus making the process economical.

Epoxidized oils like epoxidized soyabean oil, tall oil and oleic acid esters are used as important plasticizers and stabilizers for polyvinyl chloride resins.

Epoxidation may also be carried out by hydrogen peroxide. At 60°C, a 50 per cent yield of 10, 11-epoxyundecanoic acid is obtained in 4 hours by the *in situ* epoxidation of 10-undecenoic acid with 46 per cent hydrogen peroxide in acetic acid.

Hydroxylation: There are two important routes for hydroxylation of the double bonds of fats, fatty acids and their esters. If the conditions in the peroxy acid oxidations are less mild than in the case of optimum epoxidation, ring opening occurs and the product, after hydrolysis, is the α-glycol corresponding to the oxirane:

Cis-9, 10-Epoxystearic acid + $CH_3COOH \longrightarrow$

$$\longrightarrow \underset{\substack{|\\OH}}{R\text{-}CH}\text{-}\underset{\substack{|\\OCCH_3\\||\\O}}{CH}\text{-}R' \xrightarrow{H_2O}$$

$$\longrightarrow CH_3(CH_2)_7 - \underset{\substack{|\\OH}}{CH} - \underset{\substack{|\\OH}}{CH} - (CH_2)_7COOH$$

9,10-Dihydroxy stearic acid

$R = CH_3(CH_2)_7^-$, $R' = (CH_2)_7COOH$

Yields of glycols may be quantitative if stoichiometric amount of peroxy acid is used.

Methyl oleate and oleyl alcohol is oxidized by 70 per cent hydrogen peroxide/tungstic acid and a high yield of dihydroxylation product is obtained. Dilute aqueous alkaline potassium hydroxide may also be used for oxidation purpose.

Ozonide formation and decomposition:- Ozone adds to the double bonds of unsaturated fatty acids to form unstable ozonides which then give mixed monobasic and dibasic acids. An example is shown using oleic acid:

$$CH_3(CH_2)_7CH = CH(CH_2)_7\ COOH \xrightarrow{O_3}$$
Oleic acid

$$\longrightarrow CH_3(CH_2)_7COOH + HOOC(CH_2)_7COOH$$
Pelargonic acid   Azelaic acid

The *cis* and *trans* ozonide isomers of methyl oleate and methyl elaidate have been prepared, separated and characterized (5). The intermediate ozonides of methyl oleate and related products in the ozonization reaction have been possible to isolate

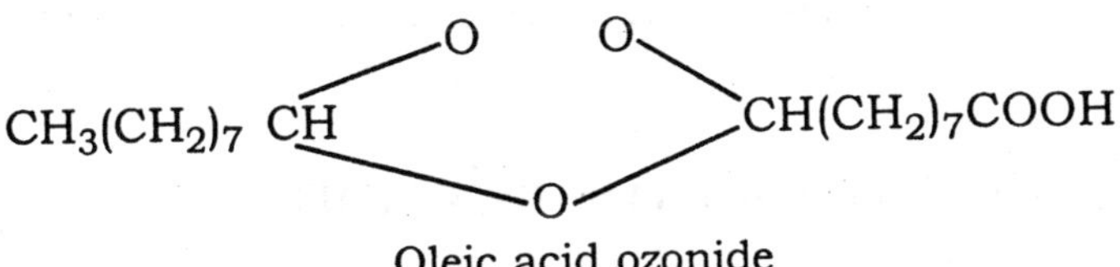

Oleic acid ozonide

It was seen that the ozonides obtained from methyl oleate gives high yield of aldehydic product after reductive ozonolysis with Pd/C at 45°C.

Other double bond oxidations: Periodic acid (H_5IO_6) is used as an important cleavage agent for vicinal (1,2)-dihydroxy compounds, but the reaction is also applicable to hydroxy keto, 1,2-diketo, and 1-hydroxy -2-amino groups:

$$RCHOHCHOHR' \longrightarrow RCHO + R'CHO + H_2O$$

$$RCHOHCR(=O) \longrightarrow RCHO + R'COOH + H_2O$$

Lead tetra-acetate also gives the same cleavage reaction. Chromic acid is a strong oxidizing agent which oxidizes fatty acids to carbon dioxide. Oleic acid can be oxidized by 85–95 per cent nitric acid at 20–25°C to give dibasic acids.

Oxidation of saturated fatty acids: Saturated fatty acids are comparatively less reactive than the unsaturated fatty acids to the action of oxidizing agents except in the presence of catalyst or at elevated temperature. They give a mixture of products.

When saturated acids undergo degradative oxidation they give ketones:

$$RCH_2CH_2COOH \xrightarrow{H_2O_2} RCOCH_2COOH \longrightarrow RCOCH_3 + CO_2$$

Oxidation of saturated acids by nitric acid gives dibasic acids, which is also degradative oxidation.

Formylation: The hydroformylation/oxidation of double bonds of fatty materials gives aldehydic intermediates which on catalytic oxidation yield dibasic acids. For example, hydroformylation of methyl oleate by carbonmonoxide, hydrogen and a rhodium catalyst gives 9(10) formylstearates which on catalytic auto-oxidation gives methyl 9(10) carboxystearates, capable of hydrolysis to dibasic acids.

ATMOSPHERIC OXIDATION, RANCIDITY

It has been difficult to preserve foodstuffs from the ancient times when man hunted and fished. Now-a-days, due to the increase in population and urbanization, foods are carried from one place to another or imported from different countries. Hence, the preservation of foods, specially fatty materials which deteriorate quickly, is very important.

Deterioration of fats and fatty materials involves oxidation either by the bacterial action or by enzyme-catalyzed hydrolysis and oxidation or direct chemical attack by the oxygen of air, i.e., auto-oxidation. Cold storage, packing, efficient transportation method and sterilization can more or less overcome the first two types of effects, but auto-oxidation cannot be prevented easily as it is a chemical reaction of very low activation energy [4-5 KCal/mole]. Auto-oxidation causes rancidity to the fats.

There are two main types of atmospheric oxidation. One is the oxidation occurring with highly unsaturated fatty acids present in drying oils and protective coatings which is accompanied by polymerization. Here oxidation is desirable. Another type of oxidation involves less highly unsaturated fatty acids which leads to rancidity, and this is the cause of most of the spoilage of fats and oils. This occurs mainly by auto-oxidation. The latter type of oxidation is discussed here.

Mechanism of rancidification: Polyunsaturated compounds of fats are more readily oxidized than the mono-unsaturated and saturated compounds. Formerly, it was assumed that during the initial stages of auto-oxidation, the products formed consist of conjugated monohydroperoxides which are transformed from *cis* to *trans* configuration by double bond shifting:

$$\underset{cis}{-CH=CH}-CH_2-\underset{cis}{CH=CH}- \xrightarrow{O_2} \underset{cis}{-CH=CH}-\underset{trans}{CH=CH}-\underset{COOH}{\overset{|}{CH}}$$

It is seen that the rates of auto-oxidation of linoleates and other methylene-interrupted polyunsaturated systems are much higher than that of mono-unsaturated systems. This is because a methylene group is activated by the surrounding double bonds. Gunstone (6) had reported that the rates of auto-oxidation of methyl oleate, methyl linoleate and methyl linolenate are 1:12:25. The auto-oxidation of unsaturated fatty compounds goes through a free radical mechanism in which hydroperoxides are formed and decomposed. Now, it is known that hydroperoxides can also be formed by photosensitized oxidation without the free radical mechanism.

Hydroperoxides formed in either route are further oxidized and decomposed into various products, such as low and medium molecular weight C_3-C_{11} saturated and unsaturated aldehydes, ketones, acids and other bifunctional oxygenated products. The compounds have strong unpleasant odours and are responsible for the rancidity of auto-oxidized fats. It has been possible to isolate and identify the different carbonyl compounds which are responsible for different types of odour of auto-oxidized fats.

GENERAL CHARACTERISTICS OF FATS AUTO-OXIDATION

By observing the auto-oxidation reaction of a fat in terms of oxygen uptake, it is seen that the course of auto-oxidation shows two distinct phases as shown in fig. 3.1 (6). At the initial stage the rate of oxidation is relatively slow and uniform but later, the second stage is reached when the rate of oxidation increases rapidly and eventually becomes many times greater than the initial rate. It was observed that the time when the fat begins to smell rancid coincides with the beginning of the second stage. The initial phase of slow oxidation is termed as induction period.

Different types of fats differ in the way at which they are oxidized and become rancid. The amounts of oxygen absorbed vary greatly according to the composition of oils, the presence of natural or added anti-oxidants, the metal content (specially copper, iron and heavy metals), the temperature and conditions under which the oxidation is carried out.

Normally, most anti-oxidants, including tocopherols increase the resistance to oxidation and also the amount of oxidation required to produce rancidity. Common anti-oxidants such as BHA, BHT and—tocopherols increase the induction period 3–8 times when as little as 0.01 per cent is added to the fat.

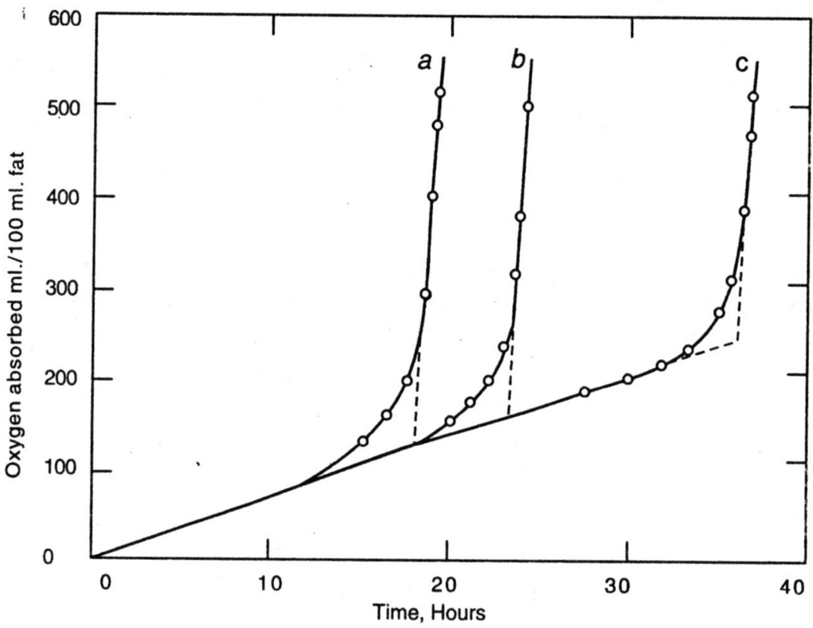

Fig. 3.1 **Oxygen uptake at 90°C of (a) corn oil, (b) corn oil plus 0.02% lauryl caffeate, (c) corn oil plus 0.10% lauryl caffeate (222).**

FACTORS DETERMINING THE RATE OF OXIDATION

The ease of oxidation of an unsaturated fat is determined by the distribution, geometry and number of double bonds in the glyceride chain. The structural features of fatty compounds are largely responsible for oxidation. The methylene group attached to the double bond is very much susceptible to oxidation. Other structural factors are: (a) conjugated double bonds are more reactive in auto-oxidation than non-conjugated ones (eleostearic acid is more reactive than linolenic acid); (b) oleic and *cis* acids oxidize more readily than elaidic acid or *trans* isomers; and (c) addition of each reactive methylene group to a structure results in a directly proportional increase in auto-oxidation rate.

The auto-oxidation is decreased by randomizing a fat with alkali and any means of affording an increase of concentration of unsaturated acyl groups at the 2-position appears to increase oxidation stability. A deep red coloured oxidation product from tocopherol, called tocored was isolated from soyabean oil and it was assumed that the product is responsible for the colour reversion of the oil.

Anti-oxidants and pro-oxidants: The ease and rapidity of oxidation of a fat depends largely on the number of double bonds and their arrangement in the compounds and these are influenced by the presence of certain compounds. The substances which favour the oxidation of fats are called pro-oxidants and which inhibit the reaction are termed as anti-oxidants or inhibitors.

All naturally occurring fats and oils contain characteristic anti-oxidants. These substances protect the oil from deterioration during the normal life of the parent seed or fruit. Vegetable oils contain a higher amount of anti-oxidants compared to the animal fats and oils and so they are more stable than the latter type of oils and fats with an equivalent degree of unsaturation.

Anti-oxidants work best in a very low concentration in the oil. For example, if tocopherol content in lard at 97°C becomes higher than 250 μg/gm, it has almost no effect on the auto-oxidative stability of lard.

There is much confusion about the mechanism of such a trace amount of anti-oxidant exerting such a powerful effect on retarding the oxidation. A possible retarding effect of hydroquinone and tocopherol may be as follows:

RH = the activated linoleic pentadiene system with H as the labile hydrogen of the methylene group, R^* = free radical,

RO_2^* = oxygenated free radical,

RO_2H = primary oxidation product (hydroperoxide) and

HO —⬡— OH = hydroquinone

$$HO-C_6H_4-OH + R^* \longrightarrow HO-C_6H_4-O^* + RH \quad ..(1)$$

$$HO-C_6H_4-OH + RO_2^* \longrightarrow HO-C_6H_4-O^* + RO_2H \quad ..(2)$$

$$HO-C_6H_4-O^* + HO-C_6H_4-O^* \longrightarrow$$

$$HO-C_6H_4-OH + O-C_6H_4-O \quad ..(3)$$

Equations (1) and (2) show that the anti-oxidant supplies hydrogen atoms to terminate both the initiating and propagating steps in the auto-oxidation chain reaction.

There are some compounds like ascorbic acid, phosphoric acid, citric acid, tartaric acid and ethylene diaminetetraacetic acid (EDTA) which can enhance the effect of anti-oxidants on vegetable oils although they have little effect on animal fats. These materials are called synergists. They actually act as metal deactivators or chelating agents and hence eliminate the effect of metal catalyzed auto-oxidation.

ACCELERATED OXIDATION TESTS

The methods of determining the stability of fats and the methods to evaluate standard or new anti-oxidants for the preservation of fat products are very important in fat chemistry.

Stability of a fat is defined as the resistance to auto-oxidation under prescribed conditions of ageing and is measured in units of time required for the product to (a) acquire a state of oxidation which can be correlated with organoleptic detection of rancid odour and flavour or (b) to reach the end of the induction period if oxygen absorption measurement or peroxide analysis is used.

For routine testing, the methods of test used is aeration test or the active oxygen method (AOM) or the Swift stability test. This test involves continuously aerating the sample at 280°F (97.8°C) and, by periodic titration, determining the time required for a specific peroxide number to develop.

In a situation where aeration for determining fat stability is inconvenient (as for many baked goods) another simple method called oven or Schall test is applied. In this method, the sample is incubated in a loosely closed container in an oven at 55°–60°C until it makes a specific peroxide number or rancid odour. No standard of this test has been developed.

The evaluation of anti-oxidants is generally described by the test used and the stability values of the control fat (no anti-oxidant) and sometimes by a factor called protection factor (PF).

$$P.F. = \frac{\text{Stability of the sample containing anti-oxidant}}{\text{Stability of the control sample}}$$

FLAVOUR REVERSION

Certain refined and deodorized oils may develop objectionable odour due to oxidation though it is not as strong as produced in rancidity. This change in flavour is termed as 'flavour reversion'. Generally, the oils containing linoleic or linolenic acid components such as linseed, soyabean, rapeseed and all marine animal oils undergo this change. The most troublesome flavour reversion occurs in soyabean oil which is one of the most important edible oil.

The exact amount of oxygen required for flavour reversion cannot be known but it is a very small amount, probably a fraction of one per cent of oxygen that is required to produce rancidity. The reverted flavour due to oxidation of oil cannot be removed by hydrogenation. The hydrogenated oil also smells the same as the unhydrogenated oil. It is evidenced that unsaponifiables are not responsible for reversion of flavour. But it is seen that nitrogenous materials are involved in the development of fishy odour in unhydrogenated vegetable oils. Some highly unsaturated fatty acids are also responsible for this type of odour.

The moisture content of the beans during milling is also an important factor for the tendency of oil to develop flavour reversion.

It was reported (7) that 2-pentylfuran is mainly responsible for the reversion of flavour for soyabean oil and it was assumed that this originates from linoleic acid. It was proposed that linolenic acid catalyzes the auto-oxidation of linoleic acid.

To prevent the flavour reversion and also oxidation, some important steps are to be taken. These include storage of crude oil rather than refined oil, good contact between oil and lye in the refining step, thorough removal of refining soap, sufficient bleaching, elimination or minimization of metal contamination by chelating agent, inert gas blanketing of products during storage and ion-exchange resins or alumina treatments of crude or refined oils etc.

Polymerization: Highly unsaturated fatty acids besides being oxidized, undergo polymerization. Co-polymerization of unsaturated fatty acids also occurs in some instances. Unsaturated fatty acids also react with alcohol to produce polymers but the type of alcohol and the reaction condition are important factors. Triglycerides also form polymers especially under the condition of oxidation. Simple fatty acid esters polymerize in themselves to produce largely dimers and sometimes trimers.

Several types of polymers are recognized. Among them the main polymers are the thermal dimers and polymers produced through conjugation followed by the Diels-Alder reaction and 'ionic' dimers or polymers produced with the aid of catalysts like boron trifluoride or other acidic catalysts etc.

Polymerization reaction is applied commercially in the 'bodying' of oils for the paint and varnish industries and in the manufacture of linoleum and other drying oil products.

Isomerization: Unsaturated fatty acids contain both geometric and position isomers. Mainly the fatty acids have *cis* configuration, such as the oleic, linoleic and linoleic acids etc.; *trans* Isomers are less soluble, react slowly and have different X-ray diffraction patterns and infra-red spectrum than the corresponding *cis* isomers.

Cis isomers are converted easily to their *trans* forms by many catalysts like sulfur, selenium, mercaptans etc. *Cis-trans* isomerization is an equilibrium reaction and the equilibrium mixture can be obtained by taking either isomer as the starting material. In the oleic-elaidic systems, the equilibrium mixture contains about two-thirds of the *trans*.

Conjugated acids — eleostearic and licanic, isomerize to the higher melting *trans* forms quite easily. Isomerization of non-conjugated fatty acids and their esters to their conjugated forms (1,4 diene to 1,3-diene systems) can be done by treating with ordinary alkalis at elevated temperatures, or with stronger base at lower temperatures. Alkali isomerization is a basis of the highly important analytical method for the determination of polyunsaturated fatty acids because the conjugated fatty acids absorb light in the *uv* region and can be estimated spectrophotometrically.

Isomerization reaction of glyceride molecules containing saturated or unsaturated acyl compounds is characteristic of them. Isomerization of the acyl radicals in a 2-acylated monoglyceride (β-monoglyceride) occurs with dilute acid or alkali and produces an equilibrium mixture containing about 90 per cent of 1-isomer.

FORMATION OF 'CYCLIZED' ACIDS

Methyl linolenate or linolenate-rich oils such as linseed oil can be cyclized with ethylene glycol and alkali at 225–295°C or with aqueous sodium hydroxide at 1200 psig and 300°C to give substantial amounts of 1,2-disubstituted cyclohexadines. If the products are heated in nitrogen atmosphere, the cyclized portions will contain C_{18} cyclohexadiene structure and if ethylene is used it enters into the reaction and the cyclized products contain C_{20} ring structure.

$$R\text{-}CH = {}^{15}CHCH_2CH = {}^{12}CH\text{-}CH_2CH = {}^{9}CH\text{-}R'$$

natural mono-conjugated triene

$$R\text{-}CH_2CH = {}^{14}CHCH = {}^{12}CHCH = {}^{10}CH\text{-}CH_2\text{-}R'$$

conjugated trience system

$CH_2 = CH_2$

$RCH_2\text{-}CH(\text{-}CH{=}CH\text{-})(\text{-}CH_2CH_2\text{-})CH\text{-}CH{=}CHCH_2R'$

$CH(\text{-}CH{=}CH\text{-})(={CH\text{-}CH}(CH_2R)\text{-})CH\text{-}CH_2R'$

Crude 'cyclized acids' are the mixtures of various polymerized products, cyclized monomer products and unchanged saturated components originally present in oil.

Reactions of hydroxyl groups: Hydroxyl groups can be easily introduced into the fats via double bonds through hydroxylation reaction.

Highly pure acetylated esters of monohydroxystearic acids are used as a good plasticizer for polyvinyl chloride at low temperature. Ricinoleic acid, the main component of castor oil contains hydroxyl group. This acid dehydrated to produce dehydrated castor oil having appreciable amount of conjugated acids.

Hydroxyl groups of fats can be readily acetylated by acetyl chloride or acetic anhydride in the presence of pyridine. By this reaction hydroxyl groups are estimated analytically. The acetyl number is the number of milligrams of potassium hydroxide required to neutralize the acetic acid produced by the hydrolysis of 1 gm of acetylated fat. The hydroxyl number is the number of milligrams of potassium hydroxide equivalent to the hydroxyl content of 1 gm of fat.

α-monoglycerides containing two adjacent hydroxyl groups in a carbon chain are cleaved between the two groups when oxidized by lead tetraacetate or periodic acid. Oxidation of the latter reagent gives the basis for analysis of pure monoglycerides and also determines the quality in commercial preparations. The hydroxyl groups of hydroxy acids can be ethoxylated or propoxylated by ethylene oxide or propylene oxide.

$$RCHOH(CH_2)nCOOH \longrightarrow R\text{-}\underset{\lfloor\_\_\_\_ O \_\_\_\_\rfloor}{\overset{H}{\overset{|}{C}}(CH_2)nC} = O + H_2O$$

(n = 2,3)

Ricinoleic acid of castor oil are cleaved at high temperature (275°C)

$$\underset{\qquad\qquad\quad |\atop \qquad\qquad\quad OH}{CH_3(CH_2)_5CHCH_2CH} = CH(CH_2)_7COOH \xrightarrow[250^oC]{NaOH}$$

$$HOOC(CH_2)_8COOH + CH_3(CH_2)_5\underset{|\atop OH}{C}HCH_3 + H_2$$

REFERENCES

1. Bailey's Industrial Oil and Fat Products, Vol. 1, Fourth edition, edited by Daniel Swern, 1979.
2. L.E. Craig, R.F. Kleinschmidt, E.S. Miller, J.M. Wilkinson, R.W. Davis, C.F. Montross and W.S. Port, *Ind. Eng. Chem.*, **47**, 1702-1706 (1955). C.S. Redfield, W.S. Port and D. Swern, *Ind. Eng. Chem.*, **47**, 1707-1710 (1955).
3. N.O.V. Sonntag, *Chem. Rev.*, **52** 237-458 (1953).
4. N.O.V. Sonntag, J.R. Trowbridge and I.J. Krems, *J.Am. Oil Chemists' Soc.*, **31**, 151-157 (1954).
5. O.S. Privett and E.C. Nickell, *J. Lipid. Res.*, **4**, 208-211 (1963).
6. R. Gilmont, H.S. Levenson and L.W. Elder, *Oil Soap*, **23**, 248-252 (1946).

CHAPTER 4

Analytical Methods

ANALYTICAL METHODS (I)

The analytical methods used in the fats and oils, fatty acids and fatty chemicals can be divided into five groups. The first group consists of unofficial procedures which are used to identify fats and oils according to their source and origin, for example, the land, animal, milk fat (animal), oil-seed or fish (animal) fat. Here no routine standard methods are available but second to fifth groups consists of methods of standardization of fats and oils.

The second or the largest group of methods consists of official methods many of which are employed by different recognized organizations like the American Oil Chemists' Society (AOCS) (2), the American Society for Testing Materials (ASTM) (3), the Association of Official Analytical Chemists (AOAC) (4) etc. for fats, oils and fatty matters. The methods of the second group are used to determine a component through a chemical reaction or to determine several physical properties, or, these are employed to evaluate the performance of a product. The methods used for fats and oils are the determination of moisture, iodine value, unsaponifiable matter, acid value etc. For fatty acids the methods are the determination of titer, iodine value, colour, unsaponifiable matter, ash etc. Other methods are employed to fat-based materials such as soap, synthetic detergents, glycerol, sulfated oils, nitrogenated fatty chemicals etc.

The third group of analytical method involves the determination of objectionable or toxic components obtained in the fat and oil products are used to control them. These objectionable components may be contaminations from raw materials or their degradation products, from extraction solvents, metal equipments or processing chemicals used to treat the product. Different procedures are used to detect these contaminations.

The fourth group of analytical methods are introduced to detect relatively small quantities of non-toxic components which have some beneficial effect both qualitatively and quantitatively. Some examples are the methods for the determination of γ-tocopherol in wheat germ oil (5) and of vitamin A in butterfat (6).

The fifth and last group of analytical methods consists of methods which are employed to determine trace quantities of components which ordinarily have little or no effect on the quality or performance of the product. These methods are complicated and are rarely used for quality control purposes. The examples are the determination of sterols in fats and oils, triterpene alcohols in fat and oil unsaponifiables etc.

The main developers of the standard methods are the industrial producers of fats and oil industries, the professional societies and associations such as the AOCS, the ASTM and the AOAC.

Sampling: Sample taken for study must be representative of the entire lot, as all the analytical evaluations become meaningless if it is not so. ASTM and AOCS have directions for proper sampling from both bulk and containers. Samples used are core samplers (used for sampling animal and vegetable fats and oils); bomb or zone samplers, triers (for solid fats) etc.

Insoluble impurities: Fats and oils must be free of any foreign matter. Most of the methods based on visual observation of settled matter in clear oil and presence of bones, paper and dirt etc. in fats. AOCS has employed a method (C3–46) which involves the determination of extraneous matter gravimetrically.

TEST METHODS BASED ON PHYSICAL PROPERTIES

Melting point: The commercial fats and oils have no sharp melting point range. Fats and oils are usually mixtures of different triglycerides and other non-oil components which melt at different temperatures. The determination of melting point depends on different conditions such as the rate of heating and other factors. Two AOCS assigned standard methods are used among many determinations, but both of them are empirical. One* such method is a capillary tube melting point method which is used for dry and filtered fats and oils in sealed capillary. The melting point is defined as the temperature at which the sample becomes perfectly clear and liquid when the batch is heated at a rate of about 0.5°C/min and agitated. Another** standard method employs a technique used for pre-chilled and solidified cylindrical disks of fat or oil. They are melted in alcohol/water mixtures in a test tube by heating at a 22°C increase per 10 min with air bubble agitation. The temperature at which the disk becomes completely spherical is taken as melting point.

There are other sophisticated methods for determination of melting point such as Bomer Test used to detect tallow in pork fat.

Softening point: Softening point of a fat is the temperature at which the fat becomes sufficiently fluid to slip or run. To determine the softening point, the dry, filtered samples are solidified in at least three 1mm inside diameter capillaries for about 16 hours at 4–10°C. These are then heated in a water bath 8–10°C below the softening point of the sample and at the rate of 1°C/min and near the softening point, 0.5°C/min. The mean temperature at which the fat rises in all tubes is taken as the softening point.

Slipping point: Determination of slipping point (softening point) is applicable to samples like lard, lard substitutes, butter, margarine, emulsions and fatty matters in their natural state but not applicable to the samples which has been melted or otherwise modified. Three brass cylinders packed with fat are heated in a brine bath at the rate of 1°C/min and near slipping points 0.5°C/

---

* AOCS Cc 1–25

** AOCS Cc 2–38

min. The average temperature at which the fat rises in the brass cups is termed as slipping point.

Cold test: This test measures the resistance of a dry, refined oil to crystallization at 0°C. Here selected sample is cooled from 25°C to 0°C in an ice bath and is rested at 0°C for 5.5 hrs. The samples which become clear after the specified time pass the test.

Titer: Titer is defined as the temperature at which solidification occurs under the condition of the test and is applied to normal animal, marine and vegetable oils and also to fatty acids and triglycerides. At the titer temperature, the cooling curve is changed as crystallization begins and heat of solidification begins.

Sample is first dried at 130°C and then it is cooled in the titer apparatus which is agitated at a specified rate. The bath temperature should be about 10°C above the titer point at the beginning of the experiment. The cooling is continued until the falling temperature rises slightly or remains constant for about 30 sec. The sample should be natural, unmodified and unaltered fat otherwise the data obtained from the test becomes useless.

Congeal point: The congeal point is the solidification temperature of the fat or oil under the condition which is in contrast to that of the titer. This test is applicable to fats and fat mixtures which solidify or congeal within the temperature range of 25–45°C. The cooling is carried out in two stages: at the first stage the sample is cooled in agitated liquid cooling bath to below cloud point and then the cooling is continued in air bath at 20°C during the congealing change. Care should be taken to perform this test to get congeal point data.

Flow test: This empirical test is related to the lowest temperature at which the sample will flow under the specified temperature. This test is not applicable to hard fat and fatty mixtures. The dried sample is cooled in a special test jar for two hours using ice bath. The jar is then transferred to an air bath at 5–8°C above the expected flow point. As the sample warms up, the temperature at which the solid fat flows when the jar is inverted is the flow point.

Cloud test: This test relates to the temperature at which cloud is formed within the dry sample, i.e. the first stage of crystallization begins at this temperature. The sample is cooled in a bottle in a water bath with stirring. The temperature at which the thermometer immersed within the sample becomes invisible, when viewed horizontally, is termed as cloud point.

Consistency-penetration method: It is an empirical method used to measure the firmness of plasticizer fat. In this test a standard weight of definite shape is dropped in the sample which penetrates within it and the distance of penetration is measured. This test uses a specially fabricated penetrometer which consists of a 45 gm aluminium cone mounted on shaft and support and penetrating device. Penetrometer is used to measure plastic fats and solid fats like butter, margarine etc. During the test, the tip or standard cone of the penetrometer is kept just touching the surface of the fat. On release the tip (the cone), penetrates into the sample for a controlled period of five secs. The depth of penetration is measured on the scale in tenths of millimeters. It is clear that the consistency-penetration test is dependent on temperature.

Liquid and solid fatty acids: This test method called modified Twitchell method involves the determination of the content of liquid (oleic, linoleic, linolenic etc.) and solid acids (stearic, palmitic etc.) of the glycerides of high C_{18} fatty acid oils. The liquid and solid fatty acid content is measured on the basis of insolubility of the lead salts of the saturated acids and the solubility of unsaturated acids in 95 per cent ethanol. This procedure is not applicable to the fats of the lauric acid group or the milk fat group or to the high molecular weight group containing rapeseed or mustard oils.

Solid fat index: This index measures the solid fat content of a sample. This method is particularly used to follow the course of ester-ester interchange reaction of mainly lard, cocoabutter and edible oils. It is calculated from the specific volumes at several temperatures using dilatometric technique.

Specific gravity: Although specific gravity is a general physical property, it does not give any exact idea about the characterization of an oil except very high density oils like castor oil. Generally, most of the oils have specific gravity in the range of 0.914–0.924 at 25/25°C, but castor oil has a range of 0.945–0.965. Different methods of determination of specific gravity are employed by AOCS, BSS and IUPAC.

Refractive index: Refractive indices of fats and oils are used for many important aspects like the identification of highly unsaturated, conjugated, hydroxy-substituted members in the sample. Also, they have application in following the progress of catalytic hydrogenation of fats and oils. The standard refractometer used for all normal oils and liquid fats are Abbe' or Butyro or any other standard equipment.

Viscosity: Almost all the fats and oils have similar viscosity except some oils like castor oil or hydrogentaed castor oil or polymerized oils. Viscosity of fats and oils are useful for designing of pumps or transfer ring systems. The viscosity of oils can be measured by instruments like Saybolt viscometer. The viscosity is determined by 'bubble-secs' at 25°C by comparison with a graded series of standards. The bubble-secs are converted into stokes. The reference standards have three series — light series : 15 tubes, 0.22 – 8.0; heavy series : 14 tubes, 10–2000; and very heavy series: 7 tubes, 250–1000.

Kinematic viscosity can be determined more accurately. Ostwald viscometers are used for this purpose, in which efflux time is to be measured at a specified temperature. The time required is measured in second and is then converted into stokes. Kinematic viscosity in stokes is equal to viscosity in poise divided by density in gm/c.c. Usually, centipoise and centistokes are used instead of poise and stokes.

Colour: Colours of natural vegetable fats and oils are due to the presence of pigments like β-carotene, chlorophyll, xanthophyll, lycopene etc. Animal fats and oils have less pigments than those of the vegetable origin. Other fats and oils which contain γ-tocopherol give dark red chroman-5,6-quinone on oxidation. Colour is developed in the fats and oils during storage and processing.

Many mcthods are employed for measurement of colour of animal and vegetable oils and fats. One of the earliest and most widely used methods for determination of colour is Wesson colourimeter technique which uses standard red and yellow Lovibond glasses. Later used equipments are the Stevenson colourimeter (7) and more recently the Lovibond tintometers to make colour comparison. The liquid fats and oils must be free from suspended matters before colour matching.

Odour: Odour is an important factor for fats and oils used in food industries, cosmetics, toilet articles, pharmaceutical etc. Odour of a fat or oil is attributed to two factors*natural or inherent odour and 'derived' odour. Bacterial action, auto-oxidation and enzyme-catalyzed hydrolysis cause more or less unpleasant odour.

Panel techniques are applicable to all fats and oils where subjective reaction to odour must be evaluated. Organoleptic evaluation is used most effectively to resolve unsatisfactory product problems. Gas-liquid chromatography developed during the last two decades has been successfully used to identify one or more constituents responsible for a particular odour.

TEST METHODS FOR CHEMICAL PROPERTIES

The test methods for chemical properties are a fundamental group dependent upon characteristic functional group reactivity.

Iodine value: This is one of the most useful analytical methods for the measurement of unsaturation of the fats and oils. The principle consists of the addition of halogen which reacts with the double bond and is added to unsaturated linkage. The halogen reagent originally proposed by Hübl in 1884 was iodine and mercuric chloride in about 100 per cent excess in 95 per cent ethanol (8). However, other methods have been developed and now-a-days iodine monobromide [(Hanus method (9)] and iodine monochloride [Wijis method] are used. The method is really an application of analytical iodimetry. An excess of reagent (ICI or IBr) is added to the dry, filtered sample allowing the mixture to react with or without a catalyst for about 30 mins. at about 25°C, treating the excess reagent with potassium iodide to convert it to equivalent iodine and titrating the latter with standard thiosulphate reagent using starch as an indicator. The reagent may be prepared in the laboratory or purchased from the suppliers. The reagent is reasonably stable in dark and dry area not above 30°C. The method provides exact directions for the sample size to ensure that the required 100–150 per cent excess of reagent is available. After using the reagents for many years their limitations have been recognized, as the reagents apart from addition, also result in substitution of the fatty acid chain. It has been appreciated now that the addition of halogenating agents to unsaturated fats depends critically on (a) the halogenating agent; (b) time, temperature, solvent and presence or absence of a catalyst; and (c) the structure of unsaturated fatty acid component of the glycerides such as proximity of carboxyl group to the double bonds, degree of conjugation and alkali branching. It has been shown that the double bond should be at least three methylene group removed from the carboxyl group for absorption of halogen to be approximately theoretical. Fortunately, the natural fats and oils are essentially free from the fatty acid components with these interfering structures. However, the use of iodine value for the new and uniquely structured fats and oils must be undertaken with care. For plant operations such as hydrogenation, Wiji's method can be carried out in as little as 5 mins., compared to official method of 30 mins., if the additional reaction is catalyzed by mercuric acetate. However, the disposal of mercury containing laboratory residues presents environmental problems. The iodine value when used and interpreted intelligently in conjunction of other data such as poly-unsaturated fatty acid content is useful in establishing the proportion of unsaturated fat. Used in conjunction with the data obtained from GLC of methyl ester it can be specific in the assay of, for example, unsaturated fat fractions. This can also be used for the same purpose with *uv* spectrophotometric data.

Determination of unsaturation by Hydrogenation: The distinct limitations of halogen addition for analysis of unsaturation of fats and oils has prompted attention to the catalytic hydrogenation method. Micro and semi-micro scale hydrogenation methods for quantitative catalytic hydrogenation have been devised using relatively simple apparatus. The methods involve hydrogen measurement manometrically or volumetrically.

Thiocyanogen number: The addition of thiocyanogen to the double bonds of the unsaturated fats was carried out in the past but his procedure is now rarely used since better methods like GLC and *uv* spectrometry have been developed.

Acid value and free fatty acids: The acid value is defined as the milligrams of potassium hydroxide required to neutralize the free carboxyl groups in one gm of the sample. Free fatty acid is expressed as equivalent of oleic acid in the sample on a percentage basis. For some oils, like coconut oil, it is measured as equivalent of lauric acid and in palm oil equivalent of palmitic acid. Specifically, free fatty acid (FFA) is expressed as percent oleic acid is about 1/2 of the acid value. Both characteristics indicate the quality of the oil with respect to acidity and have been used to follow the course of fat or oil refining. Standard methods have been prescribed by AOCS and other bodies. Most methods describe the technique as neutralization of the acids dissolved in 95 per cent ethanol with aqueous standard 0.1N potassium hydroxide and phenolphthalein as the indicator. AOCS method prescribes isopropanoltoluene as solvent.

Saponification value: Saponification value is defined as the milligrams of potassium hydroxide required to react completely with all the reactive groups in one gram of the sample. As this test is carried out in alcoholic alkali at the reflux temperature of 30–60 mins., it is apparent that alkali is consumed not only for the saponification of all the triglycerides, diglycerides and monoglycerides in the sample but also for the neutralization of all the free fatty acids in the sample as well as for other reactive ester-like components such as lactones. However, since waxes react very sluggishly under the conditions of saponification value determination, the analysis cannot be used reliably for fats and oils containing substantial quantities of waxes. The saponification value and certain related values measure the average molecular weight or more correctly the equivalent weight of the fatty materials. The saponification equivalent of a fat or an ester is the number of grams of fat or ester saponified by one mole (56.104) of potassium hydroxide. The relationship is therefore saponification value × saponification equivalent = 56104. According to one standard method (AOCS) a sample of about 4–5 gm is saponified with 50 ml of alcoholic KOH (about 40 gms/lit) for 60 mins. until saponification is complete and is judged by the clarity of the reaction medium. The blank and the sample test runs are washed down and back titrated by standard aqueous 0.5N HCl with phenolphthalein as indicator

$$\text{Sap. value} = 28.05 \times \frac{\text{(titration of the blank-titration of the sample)}}{\text{weight of the sample}}$$

Since sample sizes vary higher quantities of samples may have to be taken.

Hydroxyl value: Hydroxyl value of fatty oils such as castor oils and hydrogenated castor oils and also of the mono- and di-glycerides hydroxy acids are determined by the back titration of acetylated sample under a standard condition. Acetylation (hydroxyl and acetyl value) is normally done by reacting the sample with excess acetic anhydride in a pyridine-acetic anhydride reagent. Standard methods used by various authorities are available.

Oxirane oxygen: The official method is based on the ring opening reaction of epoxy (oxirane groups) with hydrogen bromide in acetic acid using crystal violet as indicator. There are standard methods for it used by AOCS and others. However, if there are cyclopropenoid acid such as often found in cotton seed oil, oxidized fat which may contain peroxide or hydroperoxides, unsaturated ketones and so, then there will be interferences. The test method is subjected to possible loss of HBr during the titration and the reagent is unstable on standing. It is recommended that average or duplicate determinations made in laboratories should not exceed 0.19.

Peroxide value: Detection of peroxides gives the initial evidence of the development of rancidity in unsaturated fats and oils. The peroxide value is defined as the content of the reacting oxygen in terms of millimoles of peroxide or milliequivalents of oxygen per 1000 gm of fat (1 mm = 2 meq.). There are standard methods prescribed by different authorities. Samples to be analyzed are dissolved in chloroform containing 5 per cent dimethylformamide (to assure solution of any glycerine). Aliquotes are oxidized by periodic acid. A solution of periodic acid in water extracts and oxidizes free glycerine, and a solution of periodic acid in methanol oxidizes α-monoglycerides plus glycerine. The unused periodic acid is determined iodimetrically by reaction with potassium iodide and titration with standard arsenite solution. The oxidation of periodic acid is specific to α-monoglycerides which have 1–2 vicinal glycol structure; β-monoglycerides are not oxidized.

Calculation of α-monoglyceride content is based on the amount of periodic acid consumed in the oxidation of α-monoglycerides and glycerine corrected by the amount consumed in the oxidation of free glycerine alone. Total monoglycerides as well as di- and triglycerides can also be determined by silica gel column chromatography prescribed by AOAC. It has been possible to efficiently separate 1- and 2- monoglycerides as trimethyl silyl ether derivative on packed and large bore capillarly columns of diethylene glycol succinate (DGS) polyester and apiezon L.

TEST METHODS FOR VARIOUS FAT AND OIL CONSTITUENTS

Since the constituents of fats are directly related to the quality and also the use of oil, specific test methods have been devised e.g. an idea of the nutritive value of the sample is obtained from the essential fatty acid components which can be determined by *uv* spectrophotometry or GLC. In the case of oils used in surface coatings, determination of conjugated diene may give an indication of the film-forming or the polymerization properties.

Test methods for minor fats and oils constituents: There are specific test methods for minor fat and oil constituents. These include those for moisture, unsaponifiable matter, metal content, ash, rosin acids in oils and fatty acid, free glycerol, *trans* isomers, moisture and volatile matters, soluble volatile matter as soap and oxidized fatty acids. These have been prescribed by different authorities and in case of India by the Bureau of Indian Standards.

Determination of short chain fatty acids: There are standard methods such as determination of Reichert-Meissl, Polenske and Kirschner values. These values relate to the volatility and water solubility of short chain (C_4–C^{10}) fatty acid components of low molecular weight fat and those of lauric acid group (coconut oil, palm-kernel oil etc.). The Reicher-Meissl value is a measure of the soluble volatile fatty acid (chiefly butyric-C_4 and caproic C_6 acids), the Polenske value is an index of the insoluble volatile fatty acids (mostly caprylic-C_8, capric-C_{10} and lauric-C_{12}) and the Kirschner value is a measure of butyric acid. In determining these values it is first necessary to saponify the fat, acidulate the soap and to steam distill in a special glass apparatus. A definite amount of the distillate is collected during the course of 30 mins and after filtration through filter paper, about 100 ml of the filtrate is titrated with 0.1N caustic soda solution after conducting a blank. Volume in ml. of alkali used blank V2 normality of the alkali = N Reichert-Meissel value = 1.1 N (V1-V2). Polenske value is obtained by washing the insoluble acid on the filter paper with water, discarding the washing, followed by dissolving it in three 15 ml portions in 95 per cent ethanol neutralized with phenolphthalein. The entire ethanol washings are titrated with 0.1N caustic soda. Polenske value is equal to milliliters of 0.1N caustic soda required. Polenske value represents the content of caprylic, capric and lauric acid in triglycerides.

Kirschner value: The Kirschner value is defined as the number of milliliters of 0.1N alkali solution required to neutralize the water soluble volatile fatty acids which form water soluble silver salts distilled from 5 gm of oil and fat by this method. The steam volatile water soluble fatty acids are determined by adding silver sulphate to the solution remaining after determination of steam volatile water soluble acids having carried out the titration with barium hydroxide. The water insoluble silver salts are filtered. After adding sulphuric acid to the filtrate it is distilled in the same apparatus (R.M. value) and under the same conditions. The distillate is filtered and the filtrate titrated with 0.1N barium hydroxide solution.

Determination of Polyunsaturated Fatty Acids (PUFA): Determination of PUFA is usually done by *uv* spectrophotometric method although lately GLC methods are being used. The method depends on the measurement of the conjugated components on the first sample in isooctane in the following wavelengths (nm)-dienoic 233, trienoic-262, 268, 274, tetranoic-308, 315, 322 and pentenoic-346. A second sample is isomerized (i.e. conjugated) by one of two methods depending on the nature of PUFA. Caustic potash and ethylene glycol (6.6%) is used when the sample contains only linoleic and linolenic acid and the saponification-conjugation is carried out at 180±0.5°C for 25 mins. When the sample contains linoleic, linolenic, arachidonic or pentaenoic acids, the use of 21 per cent KOH for 15 mins. at 180±0.5°C is preferred. This reaction is carried out carefully under nitrogen blanket and the blanks are run by KOH ethylene glycol solution. The reaction solution is diluted with reagent methanol and measured at wavelengths previously indicated. If the iodine value of the sample is measured, an estimate of the oleic acid content is made and saturated acids may also be calculated by the difference.

This method is applicable to determine polyunsaturated acids from dienoic to pentenoic in animal and vegetable fats containing only in natural *cis*-isomers of non-conjugated polyunsaturated acids. The method is not applicable, or is applicable with specific precautions, to modified drying oils, hydrogenated oils and other fats containing *trans*-isomers of unsaturated fatty acids, to fish oils or similar fats containing acids having more unsaturated acids than pentenoic or other oils where interferences are expected.

Diene value: The method is applicable to the conjugated oils or unsaturated oils. This value is expressed in terms of equivalents of iodine. This method employs Diels-Alder reaction in which a dienophile reacts with a conjugated system to form a bicyclic anhydride compound. There are specific methods used by different organization like AOAC, ASTM etc. In one of such methods the sample is refluxed with excess maleic anhydride-toluene reagent. After hydrolysing and extracting with water in a diethyl ether medium the extract is titrated with standard alkali solution. This value gives an approximate idea about the conjugation present in the sample.

Conjugated dienoic acid: Per cent conjugated dienoic acid determination is a spectrophotometric method applicable to dehydrated castor oil and the methyl and ethyl esters or free fatty acids. The determination is done in iso-octane solvent at 233 nm.

Fatty acid homologue distribution: Fats and oils are classified on the basis of fatty acid composition and this composition is determined by the GLC method which is a novel development and very essential today. This method gives very accurate and reliable measurements and has more or less discarded the old methods like thiocyanogen value, liquid and solid fatty acids, diene value etc. which are not so reliable.

The fatty acids are now converted into their methyl esters quantitatively without changing the structure by some standard processes like boron-trifluoride-methanol technique etc. before using it in GLC. Glass capillarly column chromatography (GCCC) is also a new technique (10) for the analysis of fatty acid distribution in fats and oils which permits the analysis of more than 1900 samples at temperatures up to 200°C with one column without appreciable deterioration.

PERFORMANCE TEST METHODS FOR FATS AND OILS

Crismer test: This test determines the miscibility of the oil mixed with the standard solvent of equal volumes of ethanol/amyl alcohol with water at 70°C to incipient turbidity with sweet almond oil. Equal volumes of standard solvent and the dried oil is cooled to show a cloudiness at which the temperature is noted and is reported as the test result. In this test free fatty acid is represented as oleic acid.

Refining loss: There are different standard methods for the determination of refining loss of different crude vegetable oils like coconut, corn, soyabean, cotton seed oils etc. All these methods have specific laboratory techniques of actual refining operations used to measure the losses. Different standard conditions are employed for different refining process. The refined oils are subjected to bleaching test, colour test etc. after refining.

THIN-LAYER CHROMATOGRAPHY(II)

Thin-layer chromatography has become an important and essential technique for the fat field. It is a very useful method because separations can be obtained within a very short time and much sharply than column or paper chromatography.

Thin-layer chromatography can be divided into three classes, adsorption, partition and ion-exchange chromatography. Reverse phase chromatography and chromatography on adsorbents modified by impregnation with reagents that form complexes with functional groups have now become important. In order to obtain good separation and detection of the components in thin-layer chromatography the thickness of the adsorbent should be as uniform as possible and silica gel, mixed with a binder, normally calcium sulphate, is widely used as an adsorbent.

The chromatoplate is developed by a solution of the mixture (0.1 – 1%) in petroleum ether or chloroform which is applied either as a spot or narrow band. The spots should be as small as possible to obtain sharp separations. The separated components can be detected by exposing the plate to iodine vapour, spraying with 2,7-dichlorofluorescein and viewing it by *uv* light or by charring at about 200°C after spraying with aqueous sulphuric acid (50%).

The glyceride composition of natural fats can be determined by thin-layer chromatography in conjunction with gas-liquid chromatography. Barrett, Dallas and Padley (12,13,14,15) fractionated triglycerides according to their degree of unsaturation on a silica gel/silver nitrate adsorbent. Glycerides having 0–16 double bonds were separated. The plates were developed by a mixture of carbon tetrachloride/chloroform/acetic acid (60:40:0.5) containing small quantity of ethanol (0-2%).

Triglycerides of castor oil, corn oil, groundnut oil, humanblood serum, lard, linseed oil etc. were separated by reverse-phase chromatography. Recently the method is used alongwith gas-liquid chromatography and enzyme hydrolysis to distinguish between cocoabutter and cocoabutter substitute and also other mixtures of fats.

DETECTION OF ANIMAL FATS IN VEGETABLE OILS AND FATS

Principle: Animal fats and vegetable fats differ from each other in their sterol composition (16). Cholesterol is always obtained in animal fats but it is generally not found in vegetable fats which contain phytosterol. So, after saponification of the oil or fat, the sterols are separated from unsaponifiable matter by thin-layer chromatography and then gas chromatography analysis is done to find whether any cholesterol (animal fat) other than phytosterols (vegetable fat) is present.

Method for isolation of the sterols by TLC: The unsaponifiable portion is isolated from the sample. Then this portion is dissolved in ten times of its amount in chloroform. 300–400 μg of the unsaponifiables are applied on the chromatoplate of thin-layer chromatography at intervals of 1 cm. A distance of 2 cm from the bottom and at least

2.5 cm from the sides of the plate are kept free. Ca 35 µg cholesterol dissolved in chloroform is used as reference and is applied at a distance of 1 cm from the sides of the plate.

A layer of chloroform, height 1 cm, is introduced in the glass developing tank, it is covered and allowed to stand for about three hours to equilibriate the liquid and vapour. Then the plate is placed in the developing tank, covered and is allowed to elute until the front of the liquid is 0.5–1 cm from the top of the plate. Then the plate is removed and the solvent is allowed to evaporate at room temperature. The plate is then sprayed with an ethanolic solution of Rhodamine 6G and the chromatogram is viewed under *uv*-light. The positions of the sterols are located and marked by means of the reference cholesterol and is scraped off the marked spots from the plate and these silica gel portions are introduced to a 25 ml conical flask. To extract the sterols from the silica gel, 5 ml chloroform is introduced into the flask and boiled in a water bath. The chloroform extract is filtered off by means of a fluted filter. The silica gel is removed from the filter and the fluted filter is reboiled with 5 ml chloroform. The extract is again filtered and this treatment is performed three times. The collected chloroform is then evaporated under a weak nitrogen current. The residue is dissolved in an adequate amount of diethyl ether and this solution is used for gas chromatographic analysis.

A suitable amount of the ethereal solution of cholesterol is injected into the gas chromatograph column and compared with the chromatogram obtained with a reference. If a peak having a retention time is identical to that of cholesterol, then it indicates the presence of cholesterol i.e. animal fat in the vegetable fat.

MASS SPECTROMETRY

Mass spectrometry is a unique technique which can offer the advantages of giving structural details of an organic compound and very often a complete identification is possible without recourse to other methods (17). Mass spectrometry is capable of giving more information about molecular structure than any other single physical technique.

Principle of the method: In a mass spectrometer the molecules of the compound under study are bombarded in the gaseous phase with electrons of sufficient energy (10–100 ev) and by this process an electron is removed from the molecule and hence it is ionized. These ions are then dissociated due to excess energy into a number of positive ions and neutral fragments. The nature and relative abundance of the positive ions depend on the structure of the molecule under observation and hence the fragmentation pattern, i.e., mass of a molecule is unique and its structure can be easily detected and can be used as a fingerprint.

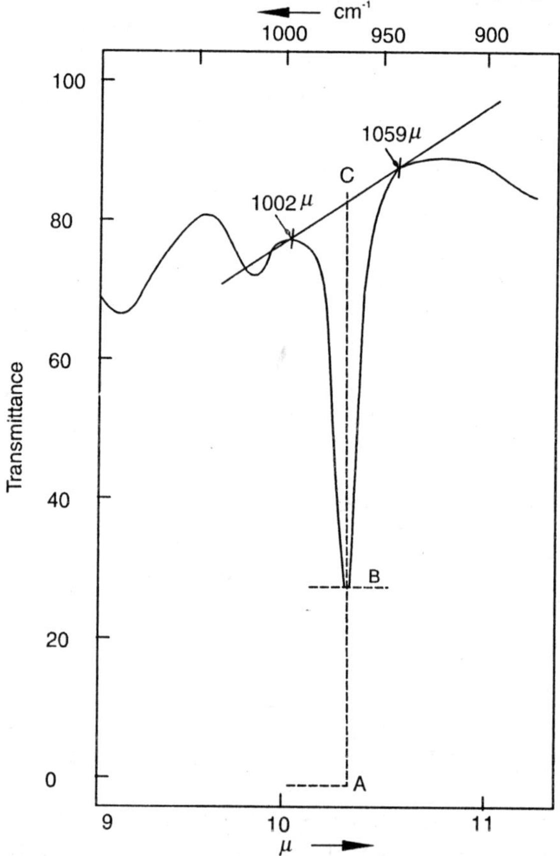

Fig. 4.1 Typical absorption curve.

The sample under study is stored in a reservoir at the pressure of the order of 0.1 mm Hg and is allowed to diffuse through a molecular leak into the ionization chamber of the source where the pressure is about 10^{-5} mm Hg. Following ionization by electron bombardment the positive ions are accelerated out of the source region towards a magnetic field at right angles to the flight path. The resultant force constrains the particle into the circular orbit which can be described by the equation

$$\frac{m}{e} = \frac{rH}{2v}$$

where 'm' is the mass, 'e' is the charge of the particle, 'r' is the radius of the orbit, 'H' field strength and 'v' the accelerating voltage.

Thus particles of increasing m/e ratio can be moved successively across a fixed exit slit either by increasing the magnetic field or by decreasing the accelerating voltage. The resultant ion current is amplified and recorded to give the mass spectrum. As most ions are singly charged the mass to charge ratio is usually equivalent to the mass of the particle.

The resolving power of a mass spectrometer must be adequate to separate adjacent masses at the molecular weight of the material under study. For fatty acids and esters this means a resolving power of at least 300. Furthermore, it must have a heated inlet system so that samples can be vaporized easily into the reservoir. A temperature of 200°C is adequate for most organic compounds. Relatively involatile or thermally unstable compound cannot be handled in this way but this limitation can be overcome by insertion directly into the source region of the mass spectrometer where a sample pressure of 10^{-5} mm Hg is adequate. In this method sample size is reduced to micrograms compared to milligrams required by the conventional method of sample detection.

Mass spectra of fatty acid esters: Mass spectrum of long chain carboxylic acid is closely related to that of the methyl esters. In 1959, Ryhage and his collaborators in Sweden had carried out a systematic survey of large number of important lipid components and had published it. Usually the mass spectra can be presented in different ways. In the conventional way, the intensity of the peak is due to most abundant ions known as the base peak which is assigned a value of 100 and all other peaks are calculated as percentage of base peak. The peak due to the molecular ions is known as the parent peak and is designated as *p*. It has been possible to recognize the methyl-esters of saturated fatty acids and unsaturated fatty acids, hydroxy, methoxy and keto acid, which show different patterns under the mass spectrometer. Detailed descriptions are available in literature (18,19,20).

Infra-red Quantitative Analysis of *Trans* Content in Fatty Acids and Derivatives

Most naturally occurring vegetable oils and fats contain only non-conjugated double bonds in the *cis*-configuration (21). These *cis* double bonds may be isomerized to the *trans* configuration during extraction and other processing procedures such as oxidation and partial hydrogenation. The non-conjugated *trans* double bonds can be detected by infra-red spectrometry. There is a specific absorption band, maximum at 10.36μ, observed due to the presence of C–H deformation about a *trans* double bond. This band is not obtained in the case of compounds having *cis* configuration or for saturated compounds. Isolated i.e. non-conjugated *trans* compound content is obtained quantitatively by comparing the intensity of the absorption band with that of a standard compound (methyl elaidate of 99+ per cent purity). The determination of *trans* content is preferably done with methyl esters.

Procedure: Solid fats are melted on a steam bath and mixed. If the sample contains any water which makes it cloudy then a small quantity of sodium sulphate is added to allow it to settle. The sample for analysis is then removed. About 0.2000 g (or 0.100 g) of standard sample (within ±0.002 g accuracy) is weighed accurately into a glass-stoppered 10 ml (or 5 ml) volumetric flask. Carbon disulphide is added to dilute it and mixed thoroughly. Concentration is to be so adjusted that a transmittance between 20 and 70 per cent at the *trans* absorption maximum will be obtained.

The cell is filled with the sample and no air bubble should be allowed within the cell. The transmittance curves are measured over the 9–11μ region. The absorbance of the sample should be measured with the same instrument with all programming controls set at identical positions as in the case of the standard sample.

The transmittance is read at the 10.36μ maximum, converted to absorbance and the absorptivity is calculated as follows:

A line is drawn from 10.02 to 10.59 μ in Fig. 4.1. The distance is measured from the zero line of the recorder chart to the absorption peak (AB). Distance to the absorption peak (AB) divided by the distance to the base line (AC) gives the fractional transmission. Absorbance is obtained by A = log 1/transmission. The so-called 'background corrected absorptivity' can be calculated by x = A/bc,

where A = absorbance = log Io/I

b = internal cell length in cm and

c = concentration of solution is g/lit.

The absorbance of the sample is obtained from the difference between the absorbance at the base-line (c) and the absorbance at the 10.36μ maximum (b), using the chart paper calibrated in absorbance.

Calculation

Calculate the % *trans* isomer as elaidate from the equation:

$$\text{per cent } trans = \frac{K \text{ sample}}{K \; elaidate} \times 100$$

where K = background corrected absorptivity.

The Application of Nuclear Magnetic Resonance Spectroscopy to Fat Analysis

Nuclear magnetic resonance spectroscopy deals with the magnetic properties of atomic nuclei to investigate the chemical environment of the nuclei and to determine the atomic structure (22). Broad-line NMR is concerned with the physical environment of the atomic nuclei and it can distinguish between liquids and solids and hence is used to measure the proportion of the liquid in a sample.

NMR has a wide application in fat and oil fields and it has use in the determination of molecular structure of the lipid components, non-destructive determination of iodine value and saponification number, determination of liquid to solid ratio in fats and investigation of the polymorphic form of glycerides.

Principle: Nuclear magnetic resonance spectra are obtained due to the possession of spin angular moments and magnetic moments of many atomic nuclei. When a proton is placed in a magnetic field its energy depends on its alignment in that field. There are two possible energy levels, the difference between them is $\frac{h\gamma}{2\pi} Ho$ where h is Planck's constant, Ho is the strength of the magnetic field and γ is a constant. This is equal to $h\gamma_0$, the energy quantum, where γ_0 is a frequency.

$$\gamma_0 = \frac{\gamma}{2\pi} \text{Ho}$$

When a sample is placed in a magnetic field *Ho*, electromagnetic radiation of a frequency γ_0 can cause transitions between the energy levels. Frequencies of 5–100 mcs and magnetic fields of 5000–25000 oersteds are involved.

So, if a number of hydrogen nuclei or protons of a sample are irradiated with electromagnetic radiation in a magnetic field of *Ho*, they absorb energy at different times. Every absorption gives a line and as the nuclei in different parts of the sample reach the resonance field value *Ho* at different times, the lines are broadened to peak.

Analysis of Fatty Acids by Gas-liquid Chromatography

Gas-liquid chromatography is based on partition theory where a volatile solute is distributed according to its partition coefficient between an involatile solvent (stationary phase) and a gas (mobile phase) (23). The stationary phase, i.e. the solvent, is coated on a solid support and is packed evenly into a column, the temperature is held constant. A small quantity of solute is injected at the top of the column and it is carried through the column by the carrier gas. The solute travels down the column at a speed which is inversely proportional to the partition coefficient of the solute in the solvent. The solute leaves the column in the form of vapour with the carrier gas and is detected by some changes in property of the flowing gas through detector and recorded in the recorder. If the solute is a mixture of different components, its components are separated as they have different partition coefficients and hence move through the column at different speeds. The chromatogram consists of a series of peaks of Gaussian form and each peak represents the corresponding component. The area under the peak is proportional to the amount of the component in the mixture.

The gas chromatogram column may be made by the user or bought complete from many manufacturers. The detectors should be sensitive and preferably have a linear response and wide linear range. An efficient column should have 'apparatus dead volume' as small as possible. 'Apparatus dead volume' is defined as the volume between the point of sample application and the exit from the detector not occupied by packing. The solid support used is mainly purified and sieved diatomaceous earth under the trade names Celite 545, Chromosorb W., Gas-Chrom etc. Stationary phase used preferably as polymer of ethylene glycol and adipic acid is referred to as polyethylene glycol adipate (PEGA) or ethylene glycol adipate (EGA).

REFERENCES

1. Bailey's Industrial Oil and Fat Products Volume 2, Fourth edition, edited by Daniel Swern, 1982.
2. Official and Tentative Methods of the American Oil Chemists' Society, 3rd ed. Including Additions and Revisions; 1979 American Oil Chemists' Society, 508 South Sixth Street, Campaign.
3. Annual Book of ASTM Standards, Part 20 (Paint, Varnish, Lacquer and Related Products etc.), April 1980, Part 22, American Society for Testing Materials, 196 Race Street, Philadelphia, PA.
4. Official Methods of Analysis, 13th edition, 1980, Association of Official Analytical chemists, 1111 North 19th Street, Suite 210, Arlington, VA.
5. W. Lange, *J. Am. Oil Chemists' Soc.*, **27**, 414*422 (1950).
6. Reference 4: Methods 43.001 and 43.008.
7. H. B. Stevenson, *Oil Soap*, **13**, 18*20 (1936).

8. A. Von Hubl, *Dinglers Polytech J.*, **253**, 281–295 (1884).
9. Reference 4, Method 28.018.
10. H.T. Slover and E. Lanza, *J. Am. Oil Chemists' Soc.*, **56**, 933–943 (1979).
11. Laboratory Handbook for Oil and Fat Analysis by L.V. Cocks, C. Van Rede, Published by Academic Press, London & New York, 1966.
12. Hilditch, T.P. and Lea, C. H., *J. Chem. Soc.*, 3106 (1927).
13. Kartha A. R.S.; *J. Am. Oil Chem. Soc.*, **30**, 280 (1953)
14. Kartha A. R.S., *J. Am. Oil Chem. Soc.*, **30**, 326 (1953)
15. Youngs, C.G.; *J. Am. Oil Chem. Soc.*, **38**, 62, (1961).
16. J.H. Recourt, Unilever Research Laboratorium, Vlaardingen, The Netherlands, in Cocks, L.V., Van Rede, C., Laboratory Handbook for Oils and Fat Analysts, Academic Press, London, 1966.
17. T. H. Shuttleworth, Unilever Research Laboratory, Port Sunlight, Cheshire, England, in Cocks, L.V., Van Rede, C., Laboratory Hand book for Oils and Fat Analysts, Academic Press, London, 1966.
18. Mc. Lafferty, F.W. In "Mass Spectrometry of Organic Ions" by F.W. Mc-Lafferty, Academic Press, New York (1963).
19. Biemann, K. "Mass Spectrometry", McGraw-Hill, New York (1962).
20. Ryhage, R. and Stenhagen, E., J. Lipid. Res. 1, 361 (1960): Stenhagen E, Z. anal. Chem. 181, 462 (1961), Ryhage, R. and Stenhagen, E. In "Mass Spectrometry of Organic Ions" by F.W. McLafferty, Academic Press, New York (1963).
21. A. A. Memelink, Unilever Research Laboratorium Vlaardinger, The Netherlands, in Cocks, L.V., Van Rede, C., Laboratory Handbook for Oils and Fat Analysts, Academic Press, London, 1966.
22. J. Clifford, Unilever Research Laboratory, Port Sunlight, Cheshire, England, in Cocks, L.V., Van Rede, C., Laboratory Handbook for Oils and Fat Analysts, Academic Press, London, 1966.
23. W. Lazarus, Unilever Research Laboratory, Port Sunlight, Cheshire, England, in Cocks, L.V., Van Rede, C., Laboratory Handbook for Oils and Fat Analysts, Academic Press, London, 1966.

Chapter 5

Dietary Fats

DIETARY FATS

The main use of fats produced in the world is as food and nutritional agents (1). Besides this, they have other important functions like lubricating agents in dressings or as heat transfer agents in frying or as carriers and protective agents for fat-soluble vitamins etc.

Fat products are consumed in the world according to the climate, food habits of people of a particular region and also on the availability of fats in that region. In some areas of Europe many oilseed crops cannot be cultivated due to adverse conditions of climate, so the inhabitants depend mainly on their domestic animals for edible fats. In India, the climate is suitable for the production of many types of oilseed crops. A hydrogenated shortening called Vanaspati is produced from vegetable origin in India. Moreover, dehydrated butter or 'ghee' is also an important edible fat product obtained from domestic animals.

There are some processing steps which are followed by people which convert crude fats of low palatability into refined products for edible purposes. The extensive treatments include alkali refining, bleaching and deodorizing.

Many fats and oils have their own characteristic flavour and they are not further purified or deodorized or processed somehow to remove their flavour. Butter and lard have their unique flavour and this is highly regarded and are not subjected to refining. Olive oil due to its delicate characteristic smell has a high price in the market.

In India, there are regional preferences for different edible oils. In the north and north-eastern region the edible oil of choice has been mustard/rapeseed group of oil which have characteristic pungent odour due to the presence of glucosinolates (allyl isothiocyanates and related compounds). In the western part of India, groundnut oil is preferred. In Kerala and nearby regions, coconut oil is the oil of choice and in other southern states like Tamilnadu, Andhra Pradesh and Karnataka, sesame and/or groundnut is used. Ghee, although used in higher quantities in the north-west and north India, is used all over India. The substitute for ghee, i.e. Vanaspati, has been used from 1934 onwards in India by people at large and particularly by Indian confectioners. In recent years, refined vegetable oils like groundnut. sunflower, safflower, soyabean have been increasingly used by the generally affluent middleclass. In rural areas, crude oils are often prepared by the 'Ghani' method which retains the flavour of the original oils intact.

NUTRITIONAL FUNCTIONS OF FATS

Fats and oils have important nutritional role for many developing countries. An insufficient intake of edible fats and oils causes a major problem. Caloric malnutrition due to lack of fats may be a more serious problem than protein malnutrition (2).

Fats have four major important roles in nutritional field. Firstly, they furnish twice (about 9 K.Cals/gm) the amount of energy which carbohydrates or proteins can give (4 K.Cals/gm each) and hence a concentrated source of energy for body. Secondly, they are carriers of oil soluble vitamins. Thirdly, fats and oils contain essential fatty acids important for many of the physiological functions of human or animal body. Fourthly fats can make foods more palatable and flavour some by adding taste, texture and good smell.

Fats and oils also have some non-caloric functions. In the case of rats besides supplying energy, fats in diet promote growth, increased work capacity, increased longevity and provide resistance against both internal and external stresses.

There are some unsaturated fatty acids which are very essential parts of diet and lack of these acids causes some diseases. For example, in the case of rats, scaly skin, loss of weight, kidney lesions and finally death (3). Human infants are also affected by loss of these essential fatty acids (EFA) in their diet and they develop eczema which can be cured by feeding them these fats (4).

Naturally occurring *cis*-9 and *cis*-12-linoleic and *cis*-9, *cis*-12 and *cis*-15 linollenic acids are identified as EFA. Body itself cannot manufacture these acids and so it is essential to take them with fats in diet (5). Soyabean oil, corn oil, cottonseed oil etc. contain a high amount of linoleic acid. In the body, linoleic acid through a series of reactions is converted to arachidonic acid. Swedish scientist Von Euler isolated a group of hormones called prostaglandins from arachidonic acid. Although the functions of these hormones are not well known they help in contraction of smooth muscles, in lipid metabolism, in the functions of central nervous system, in regulating pulse rate and blood pressure etc. So, it is seen that essential fatty acids are very important components in dietary fats.

Fat-soluble vitamins like Vitamin A, D, E and K are important and are associated with fatty diets. Vitamin E (-tocopherol) is considered as food additive to slow deteriorative processes associated with aging (6).

Fat metabolism: Fat is metabolized in the digestive tract by some digestive secretions containing lipase enzymes, i.e. fat splitting enzymes. The fatty acids and monoglycerides formed are absorbed through intestine. Some parts of them are recombined to form triglycerides and phospholipids and the very small droplets of fats are transported through lymph into the body to the point of use or storage.

Digestibility of fats: The coefficient of digestibility of a fat is obtained from a comparison of the weight of fat and fed rats (experiment is mainly done on rats) with that excreted in the faeces after allowance is made for the endogenous fat excreted on a fat-free diet.

There is almost no difference in digestibility of different fats except rapeseed oil if the melting point of fats are lower than 50°C(7). The melting point of a fat as a whole determines the digestibility of that fat. Moreover, unhydrogenated fat and partially hydrogenated fat are digested to the same extent if the difference in melting point of these two is not large.

Nutritive values of specific fats: It was thought earlier that butter fat had the highest nutritive value, but after extensive studies it was shown that there is no significant difference in nutritive value among butter fat, margarine, corn oil, cottonseed oil, olive oil, groundnut oil and soyabean oil and oils of mustard, sesame and coconut used in different parts of India. There is a wide variation of EFA content of these fats and while some fats are favoured for edible purpose others are not.

NON-NUTRITIONAL FUNCTIONS OF EDIBLE FATS

Non-nutritional functions of edible fats are obtained mainly from different physical properties. The oils and fats used for different purposes are of two types liquid oils and plastic fats. Apart from their nutritional value the fatty materials are also used for many other important functions. Liquid oil is used for dressing of green salads but for this plastic fat is not suitable. Butter is used for spreading on bread rather than an oil because oil is unsuitable to impart a mouthfeel like butter. Like butter, other plastic fats are used for icing, fillings etc. Margarine—a butter substitute, is also used as a spreading.

Bakeries use large quantities of fats in breads, cakes, pastries, biscuits etc. where fats not only contribute food value but also impart high lubricating action. Fats prevent the baked goods from becoming hard and also help the goods to keep their characteristic form and texture during the baking operation. Plastic fats impart the baked goods a crispy or spongy nature by trapping air.

Fat is an essential heat transferring medium in frying of foods. It transfers heat effectively and uniformly to the whole surface of the food and it is said that properly fried foods have almost the same digestibility as that of other foods of equivalent fat content.

Fats impart flavour and also taste to food. This is very important as foods become tastier and more acceptable by their flavour and palatability. All natural fats and oils have distinctive flavour, e.g. mustard oil has a characteristic pungent smell, butter fat has the so called ghee flavour. The flavour of a prepared dish can thus be strongly influenced by the oil used in its preparation. Food flavour promoted by a fat does not depend only on the flavour of the fat itself but to some extent on the physical properties of the fat used.

Safety of fatty foods: Safety and health aspects have long been stressed in the case of the use of different fats for different purposes although mainly for edible purposes. Safety of fat additives like anti-oxidants, silicones, emulsifiers, contamination pesticides which may be present and the effects of *trans* fatty acids in hydrogenated fats, presence of erucic acid in rapeseed oil, cyclopropene acids in cottonseed oil etc. have been studied extensively.

FATS AND CORONARY HEART DISEASE

For a long time fats have been the focus of the campaign to lower cholesterol levels. Cholesterol, a high melting point alcohol, is present in the blood serum as fatty acid ester, and to a limited extent as free alcohol. Animal fats contain relatively high levels of cholesterol. Animal fats like lard, beef tallow or butter having 40 per cent to 70 per cent of saturated acids which form high-melting point cholesterol esters are difficult to emulsify and hence transport through the arterial system. As a result, they form a solid deposit on the inner wall of the arteries. A high amount of this deposition causes high blood pressure and consequently coronary heart disease like strokes and heart attacks. Fats and oils from plant sources on the other hand, contain only trace amounts of cholesterol and they may be considered cholesterol free. Studies in early 1950s had shown that saturated fats have the tendency to raise the cholesterol levels more than that of unsaturated fats (8). Now-a-days it is seen that all saturated fatty acids are not responsible for raising cholesterol level and blood cholesterol does not always contribute towards raising the risk of coronary heart disease. There are two main forms of cholesterol in the blood—the so-called 'good' HDL cholesterol and the 'bad' LDL cholesterol. A high level of HDL cholesterol is beneficial but LDL cholesterol is responsible for heart disease. The ratio of good and bad cholesterol level is more important than the total cholestrol level. Recent knowledge suggests that when LDL cholesterol is oxidized it causes clogging in the arteries which leads to thrombosis and heart failure. Thus, it seems that cholesterol is not harmful until it undergoes oxidation.

Toxicity of oxidized fats and heated fats: Fresh fats have greater nutritional value than rancid fats. Rancidity of fats occurs due to the oxidation of unsaturated fatty material and imparts a disagreeable odour. Fats containing both oxidative polymers and thermal polymers exert a growth-retarding action on rats, and a large amount of fat in their feed, causes death of rats. Fresh fats exert a protective action when added to the diet containing oxidative polymers.

Thermal polymers do not impart any bad smell like oxidative polymers. Fats, when not heated properly, contain simple dimers, epoxides, lactones, polymers, phthalates and other compounds (9,10,11,12). Phathalates are reported to result from the plasticizer migration from polyvinyl chloride tubing or from residues in solvents.

Edible fats undergo thermal treatment during processing and these may be oxidized when heated during use, especially for frying. In these processes polymers are formed. Many methods like drop in iodine value, urea adduct formation, viscosity increase and a variety of chromatographic techniques may be employed for measurement of polymers in oils (13).

Safety of Food Additives

Emulsifiers: Food emulsifiers are important ingredients of processed foods and are used to make an emulsion between oil and water. Common emulsions are milk, butter, margarine, sauce etc.

Monoglycerides are considered as the main emulsifiers but there are a variety of other emulsifiers used for different purposes. All the emulsifiers used as food additives must be approved by some international health organisations like FAO (Food and

Agricultural Organization), WHO (World Health Organization) or the similar regulatory bodies of the country concerned. In the Indian context, the Central Committee for Food Standards (CCFS), Directorate of Vanaspati, Vegetable Oils and Fats (DVVOF), Ministry of Civil Supplies and Bureau of Indian Standards and the Ministry of Industry are the regulatory bodies.

Table 5.1 lists some 'generally recognized as safe (GRAS)' emulsifiers approved by FAO/WHO and also their acceptable daily intake (ADI):

***Table 5.1* : FAO/WHO Acceptable Daily Intake (ADI) of Selected Food Emulsifiers for Man (14).**

	Emulsifier	*ADI (mg/kg body wt. per day)*
(1)	Acetic, citric and lactic esters of mono- and diglycerides	Not limited
(2)	Diacetyl tartaric acid esters of mono- and diglycerides	0–50
(3)	Mono- and diglycerides	Not limited
(4)	Polyglycerol esters of fatty acids	0–25
(5)	Polyglycerol esters of interesterified ricinoleic acid	0–75
(6)	Polyoxyethylene and stearates	0–25
(7)	Polysorbates	0–25

Anti-oxidants: Anti-oxidants prevent the fats and oils from becoming rancid by oxidation of the unsaturated compounds present in them. It was seen earlier that the presence of a phenolic compound, tocopherol (Vitamin E), can prevent the rancidity and from them natural and synthetic phenolic compounds are used at permitted levels as anti-oxidants in fats and oils. Some widely used anti-oxidants are BHA (butylated hydro-oxyanisole), BHT (butylated hydroxitoluene), THBP (2,4,5-trihydroxybutyrophenone) etc. Recently, some harmful effects of BHA and BHT have been reported and these are being re-investigated.

A more recent anti-oxidant is TBHQ (tert-Butylhydroquinone) which has been studied extensively from the toxicological point of view. It was judged that TBHQ is safe for its intended use and safety margins were estimated to be 1,000 to 10,000 to 1.

Silicones: Silicones (Polymethylsiloxanes) are added as an anti-foaming agent to the cooking oil. The permitted range of addition is 0.1–1.0 ppm. At this low level, it does not pose any problem of toxicity. It forms a mono-layer at the oil-air interface which prevents foaming and minimises exposure of oil to air. It also serves as an anti-oxidant.

Oxystearine: In cottonseed oils or soyabean salad oils, oxystearins are used as crystallization inhibitors which are produced by blowing air into oils. If used up to a permitted level, it does not act as a toxic agent.

Pesticides: Pesticides are harmful if present in oils and fats, but they may be accidentally present. Pesticides may come from farms where these are used to protect the oilseed plants. Corn, soyabean, cottonseed and other such oilseed plants may carry trace amounts of pesticides with them upto the processing levels and occasionally the resultant crude oils contain pesticides near detection levels. Chlorinated pesticides like lindane, DDT, Chlordane etc. are highly toxic and persistent pesticides.

A more serious problem is the concentration of pesticides in the animal tissues, especially fatty tissues, if fats and oils containing pesticides are consumed by animals like broilers, cattle etc. Although there is metabolic or excretory loss, the pesticide accumulation in animal bodies is not negligible

Some chlorinated pesticides like PCB are linked to cancer (Polychlorinated biphenyls) in human being (15). DDT also has harmful effects. The pesticide levels of edible oils and fats are regulated strictly by the regulatory bodies of the country concerned so that the quantity of pesticides must be very low even below part per billion.

Safety of Hydrogenated Fats

During the hydrogenation of fat, isomerization of double bonds occurs; *cis*-double bonds are isomerized to *trans*-double bonds. Hydrogeneted fats may contain 20–70 per cent of *trans* fatty acids.

Trans fatty acids do not function as essential fatty acids. It was thought that *trans* acids are harmful to human body, but some recent investigations have revealed that *trans* acids do not have any serious effect of prolonged experiments on rats.

Safety of Specific Acids

Erucic acid: Rapeseed oil is a major edible oil used in India and it contains 45–50 per cent erucic acid. Rapeseed oil has digestibility lower than that of other vegetable oils. In experiments on rats it was seen that heart lesions developed in them if a diet containing 15 per cent of rapeseed oil by weight was fed to them (16). To reduce the heart damage risk low erucic or even zero erucic rapeseed varieties have been developed.

Cyclopropenoid acids: Cottonseed oil contains upto 0.5 per cent cyclopropenoid acids in the form of esters of malvalic acid and sterculic acid. These acids containing cyclopropene ring cause undesirable biological effects to animals when they are in the diet. It had been reported (17) that these acids cause decreased rate of egg production and reduced hatchability of fertilized eggs from laying hens and discolouration of albumen of stored eggs etc. and also have carcinogenic activity. It was investigated that these undesirable effects of cyclopropenoid acids of cottonseed oil can be removed by treatment with fatty acids at high temperature.

Polyunsaturated fatty acids: It was earlier stated that saturated fatty acids in diet present in high levels may cause cardiac disease. So, it was recommended to increase the level of polyunsaturated fatty acids in the diet which may reduce the chance of heart diseases. But it was found that if the diet rich in polyunsaturated fatty acids is fed to rats, they have a significant chance of development of tumors. This tendency is enhanced if the diet is of vegetable oil like corn oil, sunflower oil etc. rather than animal fat like margarine or lard.

A more serious problem with polyunsaturated fatty acids is that if their quantity becomes higher than the quantity of tocopherol (Vitamin E), an effective antioxidant in tissues, then the polyunsaturated fatty acids get oxidized and form a complex polymer on a protein matrix. These polymers are highly insoluble and are accumulated in most body tissues, particularly in the vascular system. So, the polyunsaturated acids which are administered to minimize heart attack are themselves responsible for lesions of the arterial wall.

GOVERNMENTAL REGULATIONS

The various regulations concerning fatty oils, their consumption and use in foods are prescribed from time to time through laws enacted by the Parliament, e.g., the Consumer Protection Act, the standards specified by the Bureau of Indian Standards (Ministry of Industries, Govt. of India), the Central Committee for Food Standards (Ministry of Health), the Vanaspati Control Order and other orders that are issued from time to time by the Ministry of Civil Supplies, Govt. of India. There are also regulations prescribed by statutory authorities, such as ministries of health at the central and state levels and orders issued by various municipal authorities etc. The oils sold by any shop or marketing agencies are sampled by the state/central governments, municipal authorities and other specified authorities, empowered by the ministries of health of the central and state Governments. Samples are regularly taken from the market and analysed in the laboratories of the Bureau of Indian Standards and the Central Food Laboratories located in Calcutta, Mumbai and elsewhere.

REFERENCES

1. Bailey's Industrial Oil and Fat Products, Volume 1, Fourth edition, edited by Daniel Swern, 1979.
2. The nutritional role of oils and fats'—PORIM, August 1991, p.4.
3. E. Aases - Jorgenson, *Physiol. Rev.*, **41**, 1 (1961).
4. A.E. Hansen, A.E. Knott, E.M. Wiese, E. Shaperman, and I. Mc. Quarries, *Am. J. Dis. Child.*, **73**, 1, (1943).
5. S. Davidson and R. Passmore, Human Nutrition and Dietetics, Williams and Wilkens, Baltimore, Md, 1966, p.109.
6. L.A. Witting, *J. Am. Oil Chemists' Soc.*, **52**, 64 (1975).
7. A.D. Holmes and H.J. Deuel, *Am. J. Physiyol*, **54**, 479 (1921).
8. 'Nutritional role of Oils and Fats' — Palm oil and human nutrition, a report by the Director General of the PORIM, PORIM, Aug. 1991, pp. 4–5.

9. R.F. Paschke, L.E. Peterson and D.H. Wheeler, *J. Am. Oil Chemists' Soc.*, **41**, 723, (1964).
10. J. Pokorny, Fette, Seifen and Anstrichm, **72**, 110 (1970).
11. B.R. Reddy, K. Yasuda, R.G. Krishnamurty and S.S. Chang, *J. Am. Oil Chemists' Soc.*, **45**, 629 (1968).
12. N.R. Artman, W.R. Michael and J. Alexander, *J. Am. Oil Chemists' Soc.*, **44**, 372 (1967).
13. A.E. Waltkins, W.E. Seery, and G.W. Bleffert, *J. Nutr.*, **52**, 96 (1975).
14. J.B. Lauridsen, *J. Am. Oil Chemists' Soc.*, **53**, 44 (1976).
15. Wall Street Journal, August 20, 1976.
16. G. Rocquelin and R. Cluzan, *Ann. Biol. Anim., Bioch. Biophs.*, **8**, 395 (1965).
17. R.A. Phelps, F.S. Shenstone, A.R. Kemmerer and R.J. Evans, *Poultry Sci.*, **44**, 358 (1965).

CHAPTER 6

Sources, Utilization and Classification of Oils and Fats

SOURCES OF COMMERCIAL FATS

Among a large number of plants and animals, a few produce oils to a commercial level (1). A small number of plants about 4,500 have been examined out of the large plant kingdom including their pulps and seeds and only ca. 100 varieties are found to have sufficient oil content in their seeds. There are about 25 vegetable oils commercially developed on a large scale now-a-days and only 15 of them are reported to constitute the maximum percentage of vegetable oil production.

Data on animal fats is quite poor. It is seen that animal fats obtained only from hogs, cattle and sheep are economically feasible.

Oil seeds are the main sources of oils today and these include mustard, cottonseed, groundnut, soyabean, sesame, rapeseed, palm fruit, palm kernel, sunflower seed, safflower seed etc. Table 6.1 shows the oil content (%) of some common vegetable oil-bearing materials.

Table 6.1 : Oil Content of a Number of Oil-bearing Vegetable Materials (2)

Ranking	*Oil-bearing material*	*Oil content (%)*
1	Copra (Coconut kernel)	65–68
2	Babassu	60–65
3	Sesame seed	50–55
4	Palm fruit	45–50
5	Palm kernel	45–50
6	Groundnut (Peanut)	45–50
7	Rapeseed/Mustard	40–45
8	Sunflower seed	35–45
9	Safflower seed	30–35
10	Olive	25–30
11	Cottonseed	18–20
12	Soyabean	18–20
13	Niger seed	35–50

Some seeds are important not only for their oils but they can also be utilized for other purposes. Soyabean seeds, besides producing oil, are used as protein-rich food. Oil is a by-product, of cotton and corn, and these crops are grown for other purposes.

The second largest source of oil production is the oil-bearing trees. Coconut, palm, olive, tung etc. are important trees supplying oil. Coconut oil is obtained from the kernel of the fruit, and palm and other oils from the pulp. Coconut and palm trees require tropical climate.

Both oil-bearing seed crops and oil-bearing trees are cultivated world-wide. Different plants require different climatic conditions and time to produce seeds and fruits. It is seen that once the cultivation has begun, the annual production of oil becomes somewhat fixed and cannot be adjusted from year to year.

Among the commercial animal fats, those obtained from hogs, cattle and sheep are important. Some other wild animals may be a source for animal fats but they are restricted to some particular areas. In the temperature climates fats mainly from hogs, cattles and sheep are produced in greatest amounts.

Milk fat is an important animal fat which may be obtained from ruminant land animals like goat, buffalo, cow etc., but world's largest production of milk fat comes from cow milk.

Marine animals and fishes contribute a considerable amount to the world's production of oils. Sea fishes are processed to obtain oil as the primary product and the residues are treated as food-stuffs or fertilizers. Whale oil is an important oil but its production is decreasing.

COMMERCIAL UTILIZATION OF OILS AND FATS (INDIAN BASIS)

The following oilseeds are grown in India on a relatively large scale and among them the most important ones like coconut, sesame, groundnut, sunflower, safflower, niger, rice bran, mowrah, cottonseed, soyabean, palm and rape-mustard are grown for edible purposes. Linseed oil and castor oil are the major industrial oils. Besides these, there are a number of non-traditional oil produced in considerable quantities. These are : Karanja, kusum, neem, tobacco seed, sal seed, kokum, dhupa, khakan, nahor, undi, palash, water melon seed, mango kernel, ambadi, silk-worm pupas, tamarind kernel, spent coffee seed, rubber seed, chilli seed etc. Some of them are by-product oils. The major by-product oils are cottonseed, rice bran, tamarind seed, tobacco seed among others. Among the animal fats, mutton tallow, beef tallow and fish oils are produced in considerable quantities. The total production of oilseeds in India has increased from 18–19 million tonnes in 1989–90 to about 22 million tonnes in 1993–94. From these oilseeds approximately 7 million tonnes of oils were produced in 1993–94. India is the world's fourth largest producer of the oilseeds. The world production in 1992–93 was about 228 million tonnes in which U.S.A. had a share of 65.3 million tonnes, China 34.3 million tonnes, Brazil about 22 million tonnes and India about 25.6 million tonnes. The basic problem facing oilseeds sector in India is that of stagnation of yield. Also very low productivity in tonnes/hectare is a constraint as against the productivity in other major oilseeds producing countries. Taking groundnut and rape-mustard which are the major oilseeds in India in terms of production, the productivity in tonnes/hectare for groundnut and rape-mustard is as follows:

	India	*USA*	*World*
Groundnut	0.62	2.05	0.82
Rape-Mustard	0.95	1.46	1.40

Apart from agricultural practices the basic reason for this is that 85 per cent of India's oilseed production is from rain-fed areas and hence there are wide fluctuations in production owing to monsoon vagaries. Oil-seeds are also grown in poor soils and compared to the progress made in wheat and rice, progress in the evolution and introduction of high yielding varieties is poor.

CLASSIFICATION OF FATS AND OILS

Fats and oils are classified here according to their uses and also their constituents. Table 6.2 lists some important vegetable oils with their principal fatty acid content.

***Table 6.2* : Main Vegetable Oil Categories (2)**

Principal fatty acids	*Oil*
Lauric	Coconut Palm kernel Babassu
Palmitic	Palm Olive Groundnut
Linoleic (medium)	Soyabean Cottonseed Sesame Corn (maize)
Linoleic (high)	Sunflower Safflower, Niger, Poppy
Erucic	Rapeseed/Mustard
Linolenic	Linseed Perilla
Conjugated acid	Tung, Oiticia
Hydroxy acid	Castor

Besides these vegetable oils, milk and other animal fats like lard, tallow and marine oils are described below:

Milk fat group: Milk fat obtained from the milk of domestic land animals like cow, goat, sheep etc. have similar composition and contains oleic, palmitic, stearic acids and also low molecular weight fatty acids like butyric, caproic, caprylic, and capric acids. Milk fat is the only fat which contains a considerable amount of butyric acid. This fat contains a peculiar combination of monoethenoid acids of 10,12,14 and 16 carbon atoms, a trace amount of highly unsaturated acids and odd and even numbered acids of 13,14,15,16 and 17 carbon atoms and also a small amount of unsaturated straight chain acids with 11,13,15,17 and 19 carbon atoms. Milk fat is fully utilized for edible purpose and is not used for other purposes due to its high cost.

Lauric acid group: Coconut oil, palm kernel oil and babassu oil have a high content of lauric acid (40–50%). They also contain a small amount of saturated acids having 8,10,14 and 18 carbon atoms and a trace amount of unsaturated acids like oleic and linoleic. These oils have lesser unsaturation than other vegetable oils and have a relatively low melting point. Hydrogenated coconut oil can be distilled to have low molecular weight fatty acids like caproic (C_6), caprylic (C_8), capric (C_{10}) etc. and these acids including high amount of lauric acid are used for the production of different fatty acid derivatives. The sodium soaps of these oils are hard, having high oxidation stability, high solubility and free lathering properties. So, they are in great demand for soap making. These oils are available and the source trees are relatively abundant in some countries of the world, particularly the tropical countries. Coconut oil is the most important oil of this group, but now the production of oil from palm tree has also flourished all over the world, particularly in Malaysia, Indonesia, Philippines and some parts of Africa.

Vegetable butter group: The group of butter is obtained from the seeds of some tropical trees. The compositions of these fats are quite similar to that of the lauric acid group and melting point is also low. This type of butters have peculiar compositions of their triglycerides, they do not contain any trisaturated glyceride, although 50 per cent of their fatty acids are saturated. These saturated fatty acids along with some unsaturated acids like oleic and linoleic are evenly distributed in their triglycerides. Cocoa butter is an example of this type of butter. These butters are utilized in the manufacture of chocolate, candies, in pharmaceutical products and are generally used as confectionary fats.

Animal fat group: Animal fats like lard, tallow etc. are the body fats of some domestic land animals and these mainly contain high amounts of C_{16} and C_{18} fatty acids along with a smaller amount of unsaturated acids like oleic and linoleic. Animal fats are characterized by the presence of some odd-numbered straight chain fatty acids in trace amount like heptadecanoic acid (2%). The presence of this acid can be detected if animal fats are mixed with any vegetable oils and fats.

Unlike vegetable butters, these fats contain fully saturated triglycerides in a considerable proportion although they are similar in fatty acid composition to the vegetable butters. They are mainly used for some non-edible purposes like soap making, manufacture of hydrogenated tallow (for lubrication purpose) and fatty acids etc.

Marine animal oil group: This group of oils contain a variety of unsaturated acids comprising C_{16}, C_{18}, C_{20}, C_{22} and longer chain acids. Higher acids contain large proportions of up to six double bonds. Fish oils, cod and shark liver oils, marine animal oils like whale oil are the important members of this group. Fatty acids with furanoid ring structure are present in cod liver oil, dogfish oil etc. (3). Marine oils, although cheaper than vegetable oils and other fats and oils, have limited use due to their poor stability, problems in storage and odour development. They can be used for edible purpose, in soap making and in coatings but these uses are not extensive. They are regarded as edible oil because they contain valuable vitamin A and D. Also, some of these oils are valued specially for their Docosahexaenoic acid (DHA) and Eicosapentaenoic acid (EPA).

Hydroxy acid group: Castor oil is the only member of this group which contains hydroxy acid, ricinoleic acid (12-hydroxyoctadec. -9-enoic acid) in large proportion (about 90%) and a small amount of 9,10-dihyroxystearic acid. This oil is exceptional among the other fats and oils due to the presence of these acids. It cannot be used for edible purpose. This oil can be catalytically dehydrated to convert it into conjugated acid oil (Dehydrated castor oil) which serves as drying oil in coating industry. Its main applications are as lubricants, in producing sulfonated oils and as a fluid in hydraulic system. Castor oil produces sebacic acid and 2-octanol on alkali fusion, undecylenic acid and heptaldehyde by pyrolysis. It is also used for the production of different oleo chemicals.

REFERENCES

1. Bailey's Industrial Oil and Fat Products, Volume 1, Fourth edition, Edited by Daniel Swern, 1979.
2. U.S. Fats and Oils Statistics, 1960–1975, Economic Research Service, U.S. Department of Agriculture, Washington, DC, September 1976.
3. O.M. Scrimgeour, *J. Am. Oil Chemists' Soc.*, **54**, 210–211 (1977).

CHAPTER 7

Composition and Characteristics of Industrial Fats and Oils

COMPOSITION AND CHARACTERISTICS OF INDIVIDUAL FATS AND OILS

The data obtained from different experiments done between 1940 and 1960 to determine the composition, properties and characteristics of fats and oils have been fairly verified in many cases by now. Modern developments in different fields of work on fats and oils have given rise to new techniques for obtaining the desired data. For example, now special methods are employed for conversion of triglycerides to methyl esters and improved gas-liquid chromatography (GLC) is used for accurate detection and analysis of trace components. Many new techniques are being developed such as nuclear magnetic resonance spectroscopy (NMR), proton-magnetic resonance spectroscopy (PMR), high performance liquid chromatography (HPLC) etc.

In the earlier days before 1939, the linolenic acid, oleic acid and linoleic acid content was measured by insoluble bromide test (for linolenic acid) and thiocyanogen value (for oleic and linoleic acid). But after that period it was detected that the values obtained from these two tests varied for different cases and, therefore, they were not reliable. Gas-liquid chromatography (GLC), which was developed after that period has solved the problem. The GLC methods have been further developed and improved and the data obtained from such tests are more accurate. Another modern technique, ultra violet absorption spectroscopy gives new opportunity to obtain the amount of unsaturation in polyunsaturated acids by converting them into conjugated acids and by detecting their absorption maxima.

Due to the recent development in agricultural and genetic fields, new crops are now cultivated whose composition has undergone changes in comparison to the previous ones, and hence the methods for determination and analysis have become more complicated. Ordinary rapeseed oil contains about 40–50 per cent erucic acid, but by employing genetic technologies the percentage of this acid is decreased even to zero (2). The linolenic acid content of soyabean oil is lowered from 15 per cent to 3.5 per cent (3). Moreover, the same seed crops grown in different areas or countries yield oils having different percentages of fatty acids in their composition due to the application of various agricultural methods and genetic development. Hence, it becomes difficult to characterize and standardize a particular oil simply due to its wide range of fatty acid composition.

Marine oils pose another problem of rapid oxidation, unstability and polymerization due to the presence of polyunsaturated fatty acids in these types of oils. The unsaturated hydrocarbon, squalene, present in the oils in appreciable quantities must be separated from the fatty acids. Other separation methods and characterization and analysis of marine oils require special techniques.

A standard methodology is to be employed for analysing fats and oils and this may be possible by comparing fats and oils with one another with respect to fatty acid content using an equivalent and also on accurate basis. Now GLC is an important technique which can efficiently separate and analyse the fatty acid composition of different fats and oils. But sometimes GLC may not be suitable for some separations like milk fats containing low carbon chain length (C_2 to C_6) fatty acids require special methyl ester preparation methods to avoid the loss of these acids by volatilization (4,5). Fish oils containing long chain polyunsaturated acids can be efficiently separated by using thin layer chromatography (TLC). However, due to its simplicity, accuracy and reliability the GLC has become a unique choice for the separation and analysis of fats and oils.

MILK FATS

Butter fat (from cow's milk): Among milk fats, butter fat generally from cow's milk is the most important. It is a complex fat having more than 500 fatty acids (6). Among these, 20 fatty acids constitute the major components and others are in trace amounts. The fatty acids believed to be present in butter fat are composed of even and odd numbered members. The carbon chain length varies from 2:0 to 28:0 and includes even and odd number monoenes from 10:1 to 26:1 except for 11:1 member. These include positional and geometrical isomers and even number acids from 14:2 and 26:2 with some conjugated geometric isomers along with polyenoic acid 18:3 to 22:6 which are of even number with some conjugated *trans* isomers. They also include mono-branch acid from 9:0 to 26:0 carbon atoms with the exception of 10:0. There are also the iso and anti-iso multi branch acids from 16:0 to 28:0 carbon atoms, both odd and even with 3 to 5 methyl branches. There are also miscellaneous Keto Hydroxy and atleast one cyclic acids. Many of these acids have not been identified positively.

The milk fat contains fat globules whose structure and composition are complex (7). A globule is approximately 2–3 μ in diameter having a 90° thick membrane surrounding a 98–99 percent triglyceride core. The composition of the membrane is quite different from that of the parent milk fat. The composition of butter fat obtained from gas liquid chromatography is presented in table 7.1.

It was later reported (9) that octadecadienoic acids in the above table are actually complex mixtures of both positional and isomeric dienoic acids. These acids are present in appreciable amounts, and traces of hexadecadienoic acid, octadecatrienoic acids, highly unsaturated C_{20} and C_{22} acids (10) dihydroxystearic acid (11) and hydroxypalmitic acid (12) have also been detected. Small amounts of octadecenoic acid, (consisting not of oleic acid but vaccenic acid– a *trans*-11,12 isomers and other monounsaturated *trans* acids) were also obtained. Butterfat contains branched chain fatty acids, some with odd numbers of carbon atoms.

Table 7.1: Composition of Butterfat (main fatty acids) (8)

Composition of fatty acids, wt%		*GLC*
Butyric	..	3
Caproic	..	1
Caprylic	..	1
Capric	..	3
Lauric	..	4
Myristic	..	12
Palmitic	..	29
Stearic	..	11
Above C_{18}	..	2
Total saturated		66
Decenoic	..	–
Dodecenoic	..	–
Tetradecenoic	..	2
Hexadecenoic	..	4
Octadecenoic (oleic etc.)		25
Octadecadienoic	..	2
C_{20} and C_{22} unsaturated	..	1
	Total unsaturated	34

To determine the triglyceride composition of butterfat an extensive study has been made. Stereo-specific analysis of both rearranged and native butterfat was done using GLC and TLC. Both native and rearranged triglycerides are separated employing silicic acid column into long and short chain triglycerides (13). These were hydrolyzed by lipase enzyme (14) and diglycerides were separated by TLC in the presence of boric acid (15). Mixed 1,2 and 2,3 diglycerides were recovered and converted into phosphatidyl phenols. These were then purified, hydrolyzed with phospholipase A (16) and the hydrolyzed products were recovered by TLC on silica gel. The fatty acid composition of triglycerides at the three positions of both native and rearranged fats were obtained by GLC of the respective methyl esters. The free fatty acids were those of position 2, the fatty acids of lysophosphatidylphenols were those of position 1, and the residual phenol fatty acids were those of positions 2 and 3. The composition of position 3 was obtained by subtracting the composition of position 2 from that of the mixed 2- and 3- positions. In the case of rearranged butterfat, butyric acid was found to be distributed randomly at 1-, 2- and 3- positions but for native butterfat, butyric acid and other short chain acids occupied the 3-position.

In the later stage it was reported (17) that the triglycerides of milk fats consist of three basic types:

I. Triglycerides of 48–54 carbon atoms, made up of long chain 1,2-diglycerides esterified with long chain fatty acids, such as oleic, linoleic and stearic acids.

II. Triglycerides of carbon numbers 36-46, made up of long chain 1,2-diglycerides esterified to short chain fatty acids in the 3-position, such as butyric, hexanoic and octanoic acids.

III. Triglycerides of carbon numbers 26-34, made up of medium chain 1,2-diglycerides combined with short and medium chain fatty acids in the 3-position.

Adulteration of butterfat by vegetable oils like hydrogenated cottonseed oil or animal fat lard has been detected by gas chromatogram which becomes possible only due to the presence of unique triglyceride structure of the butterfat.

Butterfat is different from other fats except those of the lauric acid group with low average molecular weight fatty acid content which is manifested by its high saponification number and a low refractive index. This can also be distinguished from coconut and other lauric acid fats by its high content of steam-volatile acids which is indicated by a high Reichert Meissl Value. The distinctive properties of butterfat are compared with other fats in table 7.2.

Butterfat contains small quantities of aliphatic lactones which are responsible for its flavour. δ-decalactone and δ-dodecalactone are the two major lactones which contribute the flavour (18). However, many other lactones have been identified in milk and butter which also act as flavour enhancers. Most of the lactones can be separated and detected by employing gas chromatogrpahy, thin layer chromatography or steam distillation. It was seen that the total lactone potential of butterfat is 110 ppm approximately.

***Table 7.2*: Distinctive Characteristics of Butterfat in Comparison with other Fats**

Analysis	*Butterfat*	*Coconut oil*	*Palm kernel oils*	*Other ordinary fats and oils*
Saponification number	210–250	245-260	240-250	Ca 200
Refractive index at 60°C	1.4465	1.4410	1.4430	> 1.4465
Reichert Meissl Value	22–34	6–8	5–7	< 1
Polenske Value	2–4	14–18	10–12	< 1
Kirschner Value	20–26	1–2	0.5–1	< 1.5

Some odd numbered methyl ketones containing 3–15 carbon atoms along with acetone, acetaldehyde, methyl sulfide, C_4–C_{10} free fatty acids and various lactones already mentioned are present in trace quantities in butter and are generally considered as the substance which contribute to the pleasant monoxidative flavour and odour of milk fat.

Phospholipids are present in trace amounts in butterfat out of which phosphatidylcholine and phosphalidylethanolamine are present in appreciable quantities (34.5 and 31.8 mole per cent of total '*p*' content). Sterols found in milk lipids mostly contain cholesterol esters, small quantities of lanosterols and trace amounts of dihydrolanosterol and β-sitosterol. Milk lipids also contain small amounts of carotenoids and squalene (19).

Vitamin A is an important constituent of butter from the nutritional point of view and β-carotene or other carotenoid pigments are its sources. The content of vitamin A in butter varies with the content of carotenoid substances present in the cow's diet. The yellow colour of butter is imparted by vitamin A. The vitamin A content of butter is in part due to the presence of vitamin A itself and in part due to carotene which is converted to vitamin A in the human body. One international unit of vitamin A is defined as the amount possessing the biological activity of 0.6 μg of pure β-carotene (20). The biological activity of this vitamin, either as such or its esters, is considered approximately 4,000,000 I.U./gm. On this basis the vitamin A content of butter is expressed by the following relationship:

$$I.U/g = 4.0S + 1.67C$$

where I.U = International Unit, S = vitamin A content in μg/g and C = carotene content in μg/g.

Vitamin D content of butter is much lower than the vitamin A content but it is not negligible. It varies from about 0.1 to 1.0 I.U/g.

***Table 7.3*: Composition and Characteristics of Miscellaneous Milk Fats (103-106)**

Analysis	*Goat Milk (103)*	*Sheep Milk (103)*	*Buffalo Milk (104)*	*Camel Milk (105)*	*Horse Milk (103)*
Composition of fatty acids, wt.%					
Butyric	3.0	2.8	4.1	2.1	0.4
Caproic	2.5	2.6	1.4	0.9	0.9
Caprylic	2.8	2.2	0.9	0.6	2.6
Capric	10.0	4.8	1.7	1.4	5.5
Lauric	6.0	3.9	2.8	4.6	5.6
Myristic	12.3	9.7	10.1	7.3	7.0
Palmitic	27.9	23.9	31.1	29.3	16.1
Stearic	6.0	12.6	11.2	11.1	2.9
Saturated, above C_{18}	0.6	1.1	0.9	–	0.3
Total Saturated	**71.1**	**63.6**	**64.2**	**57.3**	**41.3**
Decenoic	0.3	0.1	–	–	0.9
Dodecenoic	0.3	0.1	–	–	1.0
Tetradecenoic	0.8	0.6	–	–	1.8
Hexadecenoic	2.6	2.2	–	–	7.5
Octadecenoic (oleic)	21.1	26.3	33.2	38.9	18.7
Unsaturated above C_{18}	0.2	1.9	–	–	5.1
Octadecadienoic	3.6	5.2	2.6	3.8	7.6
Octadecatrienoic (linolenic)	–	–	–	–	16.1
Total unsaturated	**28.9**	**36.4**	**35.8**	**42.7**	**58.7**

Characteristics					
Iodine number	28.8	36.7	33.5	40.8	84.3
Saponification equivalent	233.7	244.5	240.4[a]	259.0	253.1
Reichert-Meissl value	23.2	26.2	32.2[a]	16.4	6.2
Polenske value	7.2	3.6	2.0[a]	1.6	5.9
Kirschner value	15.6	17.6	28.4[a]	14.3	2.6

Other Milk-Fats

Milk fats obtained from the milk of other grazing animals like buffaloes, sheep or goats are more or less similar in composition and characteristics to the milk fats of cows. Ghee is derived from milk of cows and its yellow colour shows the high content of carotene. Ghee obtained from the milk of other animals like buffaloes or goats is pale in colour due to a low carotene content, but is not low in vitamin A.

Lauric Acid Oils

Coconut oil: Coconut oil is obtained from dried coconut kernel called copra. Oil content of copra is usually about 65–68 per cent and it contains about 4–7 per cent moisture. Coconut oil is an important oil used in India both for edible and non-edible purposes. Its main non-edible use is in the making of soaps. It is also used in other industrial purposes.

Coconut oil (from *Cocos nucifera*) and other lauric acid oils contain high amounts of low and medium chain fatty acids. Coconut oil easily solidifies at about 18–22°C and melts at about 24–27°C. The low melting point of this oil is not due to the presence of high degree of unsaturation of fatty acids as in the case of ordinary oils but due to the presence of low average molecular weight of its glycerides. Like other solid fats, e.g. lard or hydrogenated oil, it does not exhibit a broad range of melting temperature; rather, it melts abruptly from a brittle solid to a liquid and in this respect, it resembles cocoa butter or Borneo tallow. A dilatometric curve is shown in Fig. 7.1 which illustrates the melting behaviour of coconut oil as compared to other fats.

Coconut oil contains a large variety of fatty acids of which 90 per cent are saturated. Among the saturated fatty acids lauric, myristic and palmitic acids are the main and the lauric acid content is the highest. The low range of melting temperatures of coconut oil is mainly due to the low melting points of its triglycerides composed of mainly these three acids. The melting points of lauric, myristic and palmitic acids are 44°, 54° and 63°C respectively. This fact reveals the cause of low melting points of triglycerides and hence the low melting temperature range of coconut oil.

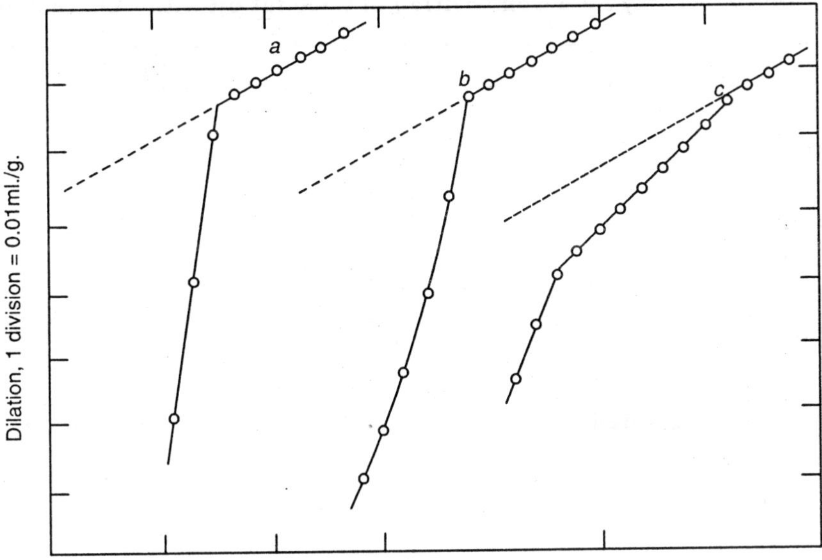

Fig.7.1 Dilatometric curves of fats, illustrating abrupt melting of (a) cocoabutter and (b) coconut oil, as compared with gradual melting of (c) lard.

By hydrogenation, coconut oil does not change its melting point or consistency to any appreciable extent. As this oil contains a major percentage of saturated fatty acids, it experiences a little change due to hydrogenation, for its small content of oleic and linoleic acid are converted into stearic acid. Completely hydrogenated coconut oil has the melting point at 45.1°C compared to the melting points range of 24–27°C of unhydrogenated coconut oil.

As coconut oil contains a very low percentage of unsaturated acids, the possibility of rancidity is very low. The stability of coconut oil in different food products was studied, and by active oxygen method (AOM) the stability of refined coconut oil has been reported (21) to be as high as 250 hours. Addition of anti-oxidants and stabilizers (BHA, Citrate etc.) increases AOM stability upto about 250 hours.

Crude coconut oil contains higher amount of free acids than ordinary vegetable oils. The fatty acid composition of coconut oil obtained by GLC is shown in table 7.4.

***Table 7.4*: Fatty Acid Composition of Coconut oil with GLC Technique**

Fatty acid	*Weight per cent range*
Caproic	0–0.8
Caprylic	5–9
Capric	6–10
Lauric	44–52
Myristic	13–19
Palmitic	8–11
Stearic	1–3
Arachidic	0–0.4
Total Saturated	**91 (Approx.)**
Oleic	5–8
Palmitoleic	0–1
Linoleic	Trace–2.5
Total unsaturated	**9 (Approx.)**

The triglyceride structure of coconut oil was investigated and it was found that coconut oil can be fractionated into 13 groups consisting about 79 individual triglycerides of carbon numbers 28-52; and out of them trilaurin (more than 10%), caprodilaurin (about 20%), caprolauromyristin (about 12%, dilauromyristin (about 11%), lauromyristopalmitin (less than 3%) and caprolauro olein (4-5%) are the major triglycerides.

Coconut oil has a high saponification number and a low refractive index due to low average molecular weight of its fatty acids. In some experiments with different samples of coconut oil the average values of saponification number; iodine number and titer have been obtained. The values are: saponification number, 268; iodine number, 10.1 and titer 23.3°C. Coconut oil contains 0.1–0.3% unsaponifiable matter consisting of sterols, tocopherol and squalene. These can be separated using TLC technique.

The flavour of the coconut oil is due to the presence of trace amount (< 150 ppm) of lactones. The average density of coconut oil is about 0.919 at 25°C and 0.893 at 60°C and the refractive index at 40°C is 1.448–1.450. If coconut oil is allowed to solidify partially and then filtered or pressed, a hard solid coconut stearin is obtained which melts at about 32°C, is used as coatings in different food products and is called confectioner's butter.

Palm kernel oil: Palm kernel oil is obtained from the kernel of oil palm which contains about 44–53 per cent oil. It is very similar to coconut oil and is used mainly in soap making and also for edible purpose.

The fatty acid composition of palm kernel oil is shown in Table 7.5.

***Table 7.5*: Fatty Acid Composition of Palm Kernel Oil**

Fatty acid	*Weight percent range*
Caproic	Trace – 1.5
Caprylic	3 – 5
Capric	3 – 7
Lauric	40 – 52
Myristic	14 – 18
Palmitic	7 – 9
Stearic	1 – 3
Arachidic etc.	Trace – 1
Oleic	11 – 9
Palmitoloic	Trace – 1
Linoleic	0.5 – 2

Palm kernel oil has iodine number 18.6 on an average. As it has a higher iodine number than the coconut oil it shows greater change in properties due to hydrogenation than the coconut oil. Other average characteristics of palm kernel oil are: saponification number 248, refractive index 1.4430 at 60°C, density 0.892 at 60°C, melting point 26°C and unsaponifiable matter 0.4 per cent. The solid stearin is also obtained from palm kernel oil by graining and pressing and it is used as a cocoabutter substitute. The liquid olein portion is used in baked goods in hydrogenation and in soap manufacture. The unsaponifiables are separated by employing thin layer chromatography.

Vegetable butters: Vegetable butters like Cocoabutter, Borneo tallow etc. have unique properties among fats and oils. They have a very short range of melting temperatures and this is caused by presence of triglycerides of unusual structure. The triglycerides are composed of mainly either single triglycerides or two or more triglycerides having very nearly identical melting point. It is seen that more than half of the triglycerides of cocoabutter are composed of oleic acid, palmitic acid and stearic acid i.e. oleo-palmitostearine with oleic acid at the 2-position (22). The consistency and melting of cocoabutter depends on the conditions at which it is solidified and stored. Some vegetable butters show a pronounced degree of polymorphism.

Cocoabutter: Cocoabutter is obtained from the bean of the plant *Theobroma cacao* grown in tropical regions, predominantly in Africa. The kernel of the cocoa bean contains about 50–55 per cent fat. The beans are cleaned, roasted and decorticated and the kernels are sent to hydraulic press to obtain cocoa butter. Its main use is as the chocolate confectioner's fat and to some extent also in cosmetics.

Cocoabutter becomes solid and of brittle nature below 27°C and it melts slightly above this temperature. It has no sharp melting point and melting temperature depends on the crystal modification of the cocoabutter. It is very suitable as a coating in confections and blending agent of chocolate as it becomes hard and non-greasy and melts at a temperature below the temperature of the human body. Unlike other fats and oils, cocoabutter is extremely resistant towards oxidation and development of rancidity. The average composition of fatty acids of cocoabutter is given in table 7.6.

***Table 7.6*: Fatty Acid Composition of Pressed Commercial Cocoabutter**

Fatty acid	*Composition (wt% mean)*
Palmitic	25.5
Stearic	34.0
Oleic	35.1
Linoleic	3.4
C_{16}/C_{18} ratio	0.75

Cocoabutter contains triglycerides out of which 2-oleopalmitin is the major triglyceride. The triglycerides of cocoabutter were separated by TLC and examined (23). It was found that there are eight distinct triglycerides present in cocoabutter of which over 90 per cent are of the following four types: 70.3 ± 0.2 per cent are SSM and SMS types (where S = saturated, M = *cis*-monoene, D = all *cis*-diene), 13.5 ± 0.6 per cent are SM_2 type and 9.7 ± 0.8 per cent are S_2D type. Polymorphism of cocoabutter is complex and by employing different techniques like X-ray diffraction method, programmed temperature. X-ray diffraction and differential scanning calorimetry (24) it was found that there are six polymorphic states which were mainly responsible for its unique melting behaviour. Some typical properties and characteristics of cocoabutter are presented in Table 7.7.

***Table 7.7*: Typical Characteristics and Physical Properties of Commercial Cocoabutter**

Characteristics	*Usual range of values*
Acid number	1 - 4
Saponification number	190 - 198
Iodine number	33 - 40
Refractive index at 40°C	1.456 - 1.458
Unsaponifiable matter %	0.2 - 0.6
Melting point °C	32 - 35
Titer °C	45 - 50

Phospholipid content of chocolate liquor was determined by extraction with a polar solvent like chloroform or methanol and it was about 0.65 per cent. The unsaponifiables can be separated using thin layer chromatography.

Due to the high price of cocoabutter it is adulterated with other fats. The foreign fats can be detected by dilatometry or other methods, as these fats alter the melting characteristics of cocoabutter. The mixing of lauric acid oils or 'stearin' obtained from them can be determined by an increase in the saponification value, Riechert-Meissl and Polenske values of the fat. When cocoabutter has been adulterated with Borneo tallow which is very similar in composition with cocoabutter, the turbidity and solidification tests (25) are recommended for its detection.

Other vegetable butters: Among the other vegetable butters, Borneo tallow obtained from the plant *Shorea stenoptera* is important. Its main producers are Indonesia and Malaysia. This fat is very similar to cocoabutter though it is several units lower in iodine number than cocoabutter and 2-3°C higher in both melting point and titer. It contains more stearic and less palmitic acid than cocoabutter and the triglycerides are mainly mono-oleodisaturated types. This fat is used for edible purposes in the countries that produced it and it is also used for making soap and candles.

Shea butter obtained from the kernels of the African plant *Butyro spermum parkii* is another vegetable butter. This butter is less desirable as a coating in confections than cocoa butter. It is more unsaturated and hence softer and less brittle than cocoabutter or Borneo tallow. It has a high content (3-10%) of unsaponifiables.

Mowrah fat obtained from the kernels of the Indian plant *Madhuca latifolia* also has almost the same composition as shea butter by having the same unsaturation and softness. Its main use is in candle and soap making with a very small percentage for edible purpose when highly refined.

Kokum butter is another vegetable butter obtained from the Indian plant species *Garcinia indica* and is mainly produced in wastern India. It shows multiple melting point behaviour which is due to the presence of large proportions of mono-unsaturated disaturated triglycerides. The triglycerides of kokum butter had been separated and it was seen that five types of triglycerides were present: StStSt (3.6), StStU (0.6), StUSt (80.9), StUU (14.3) and UUU (0.6) [St = stearic, U = unsaturated; figures within brackets are the mole percent].

Chinese vegetable tallow is obtained from the outer covering of the kernels of Chinese tree *Stillingia sebifera.* A drying oil called Stillingia oil is obtained from the kernels. Its fatty acid composition consists of mainly palmitic acid, but lauric acid and myristic acid are also present in significant amount. The glyceride composition of one sample was estimated as dipalmitostearin 13 percent, dipalmitoolein

64 percent, tripalmitin 8 percent, oleopalmitostearin 8 percent and trace amounts of other glycerides.

Nutmeg butter obtained from the tree *Myristica officinalis* or *fragrans* is rich in myristic acid (about 75 percent of its composition is myristic acid). It contains a high proportion of essential oils which are extracted by ether.

Phulwara butter is mainly a product of northern India obtained from the tree *Madhuca butyraceae*. It contains 56.6 percent palmitic, 3.6 per cent stearic and 36 percent oleic acid. The triglyceride composition of phulwara butter has been estimated to be 34.8 per cent POST, 25.2 per cent STOST and 12 per cent POP types.

Animal Fats

Animal fats are mainly lard, tallow and grease. Lard is obtained from hogs and tallow cattle as well as from sheep.

Lard: Lard is the most important animal fat used for edible purpose. Rendered pork fat is another edible fat obtained from the carcasses or edible organs of hogs.

Both lard and pork fat are obtained from clean, fresh and healthy tissues of hogs by dry or wet rendering process. Open kettle (dry) rendering process was done at 115°–120°C normally at atmospheric pressure. Recently vacuum methods are employed for rendering. Wet rendering process is done by heating the fatty tissues in presence of water usually at lower temperature than dry rendering and a milder flavoured lard results. This lard is called prime steam lard. The fatty tissues are to be ruptured to permit the lower operating temperature and continuous processes are employed for both dry and wet rendering.

The feed of the hogs and the portion of the animal used to render fat have great effect on the composition, characteristics and consistency of lard. Lard obtained from eight body sites has been examined (26) to estimate fatty acid composition and different compositions were obtained for different fats. However, the ranges of the major acids present in lard are as follows:

Palmitic acid 24.3–31.6 percent, stearic acid 11–18.2 percent and oleic acid 31.2–46.5 percent. Typical characteristics of a sample of lard is given in Table 7.8.

Table 7.8: Some Typical Characteristics of a Lard Sample

Specific gravity at 99°/15.5°C	0.858–0.864
Refractive index at 40°C	45–52
Iodine number ..	46–70
Saponification number ..	195–202
Unsaponifiable matter ..	not over 1.0 percent
Titer	36–42 percent

Fatty acid distribution of lard is diverse and it is found now that about 200 different fatty acids occur in lard and pork fat. The broad general data of fatty acid composition of lards as obtained by GLC are given in Table 7.9.

Table 7.9: Fatty Acid Distribution of Lards (8)

Fatty acid composition (wt%)	*GLC*
Myristic ..	3
Palmitic ..	24
Stearic ..	18
Total saturated	**45**
Tetradecenoic ..	–
Hexadecenoic ..	3
Oleic ..	42
Linoleic ..	9
Total unsaturated	**54**

Besides these acids, many other branched chain and odd number fatty acid homologues were detected. A fractional percentage of lauric acid and a trace of capric acid also had been detected (27). On analysis a significant amount of monoethenoid acids reported as oleic acid was also obtained which actually consisted of tetradecenoic, hexadecenoic and a small amount of dodecenoic acids. Unsaturated acids above C_{18}, mainly C_{20} and C_{22} with some C_{23}, C_{24}, may amount to 1–2 percent. Polyunsaturated acids upto five double bonds are possible.

Experiments carried out to determine triglyceride structure of lard had shown that lard contains mainly 2-palmityl glycerides (28).

By lipase hydrolysis 83 percent of the palmitic acid was found but the oleic and stearic acid at 10–12 percent and 12 percent respectively were in the second position. Later, using thin layer chromatography it became clear that the composition of lard is representated by 11 glyceride types (23) and among them three triglycerides, SMD, SM_2 and SMS (S = saturated, M = *cis*-monoene, D = all *cis*-diene), constitute about 66.5 per cent of the total fat. Table 7.10 lists 11 triglyceride types in wt per cent obtained by TLC.

***Table 7.10*: Eleven Triglyceride Types in Lard by Densitometric Determination using Argentation TLC (23)**

Triglyceride type	*Weight percent*
DMD	1.8 ± 0.3
MDD	1.3 ± 0.2
DSD	1.4 ± 0.2
SDD	1.6 ± 0.2
M_2D	7.6 ± 1.1
SMD	18.1 ± 2.4
M_3	8.0 ± 0.8
S_2D	7.0 ± 0.7
SM_2	27.9 ± 1.3
SSM	–
SMS	20.5 ± 2.8
S_2Mt	–
S_3	4.9 ± 0.5

Mt = *trans*-monoene

Presence of beef fat in lard can be detected by Bomer test (29) depending on the high melting palmitodistearins in the two fats having different isomers, though this test becomes valid only if hydrogenated lards are absent. If lard is present in vegetable oil, it may be detected by spectrophotometer method or by GLC.

Lard is suitable for many plastic fat uses due to its rheological properties and other physical properties like hardness, consistency and fat penetration. Lard has a poor oxidation stability without anti-oxidants and it becomes rancid at 100°C if it has absorbed 10–15 percent of its volume of oxygen. Addition of tocopherol improves the oxidation characteristics of lard. It was shown that the stability of oxidation of lard increases in the order: α-tocopherol, β-tocopherol, γ-tocopherol, and δ-tocopherol. But it was seen that addition of any single tocopherol had little or no effect on oxidation stability. Conventional anti-oxidants like BHT, BHA etc. and metal chelators like citric acid and malic acid, in the presence of Fe and Cu metal contaminants improve the oxidation stability to a great extent (about 3.4 to 77 hours). Anti-oxidants used are about 200 ppm. Many other anti-oxidants may be used and results have been published for them.

Greases

Grease is an inedible fat obtained from either hog or beef fat and used for fatty acid production or for soap making. Grease is characterized by its hardness. A fat with a titer below 40°C is a grease and a fat with a titer above 40°C is an inedible tallow. There are different grades of grease depending on the colour, titer, free fatty acid content and MIU content of the material. MIU is the content of the total—moisture (M), insoluble impurities (I) and unsaponifiable matter (U).

White grease is similar to lard and among the other types of greases it is the best choice. It is chiefly due to aesthetic considerations and partly due to its refined nature that it is completely free, from impurities even from inedible materials.

Yellow grease is of lower quality and less acceptable than the white grease. Its yellowish colour is due to the free fatty acid content of up to 15 per cent. Brown grease is even more low in quality, darker in colour and higher in free fatty acid content (up to 50%).

There are other types of greases which may not come from hog fat. Glue grease is one of them obtained from glue manufacture, bone grease derived from the processing of bones for gelatin, glue etc. Low grade grease are not used for industrial purposes but are used as raw material for the production of distilled fatty acids.

The greases which have physical properties like those of edible fats, as white grease, must be denatured by mixing petroleum distillates to it.

The most important characteristics of grease is the content of saponifiable matters as they are used mainly in soap making. The non-fatty materials in greases are determined by MIU. MIU together with colour and free fatty acid content determine the grade and trading of greases (M = moisture, I = insoluble impurities, U = unsaponifiable matters).

Tallow

Tallows are obtained from beef, cattle and it may be used for inedible purpose or for edible purpose after subsequent refining. Edible tallow is obtained by rendering up to a satisfactory quality, but, if its FFA content becomes more than 0.3 percent or any collagen or proteinaceous matter is present in tallow then alkali refining is necessary to increase its oxidation stability. FFA content of inedible tallow is higher than that of the edible one, and alkali refining may be done for removing FFA, nitrogeneous materials and improving the colour.

Perhaps 180-200 individual fatty acids are present in tallows, most of them in trace amounts and among them 12 are the major acids. Tallows may have different iodine numbers for different samples. Feed of the cattle, environment etc. have influence on the amount of unsaturation in beef tallow. Tallows obtained from different body sites are also different in composition.

Mutton and goat tallow are harder and less unsaturated than beef tallow. A trace amount of highly unsaturated C_{20} and C_{22} acids are found to occur in beef and mutton tallows. Mono and diethenoid acids of 20 carbon atoms and non-conjugated acids with two or three double bonds have been reported in small amounts. These non-conjugated acids are mainly normal linoleic and linolenic acids. Presence of vaccenic acid (*trans* 11-octadecenoic) and elaidic acid (*trans*-9-octadecenoic) has been reported (30) as much as 5 percent in beef fat. Branched chain acids were proved to occur in tallows by using GLC and TLC techniques.

GLC of tallow methyl ester established the fact that tallow contains a C_9 double bond. Also examining dienoic fraction in the C_{18} series showed the occurrence of the following monoene and diene isomers in trace quantities: *Trans*-11-octadecenoic, *trans*-11-*trans*-15-octadecadienoic, *cis*-11, *cis*-15-octadecadienoic and *cis*(*trans*)-11, *trans*(*cis*)-15-octadecadienoic acids.

Vander Wal had separated beef tallow by using $AgNO_3$/silicic acid TLC of the allyl esters and showed that tallow contains four molecular and isomeric types of triglycerides: S_3, S_2U, SU_2 and U_3 are the molecular types and the two isomeric types are SUS and SSU. The TLC data is shown in Table 7.11.

Table 7.11: TLC Data for Triglyceride Composition of Tallow

Molecular types (wt%)				*Isomers (wt%)*	
S_2	S_2U	SU_2	U_3	SUS	SSU
14.0	50.7	24.7	10.7	28.8	21.9

A typical set of the characteristic and physical properties of pure beef tallow is listed below:

Specific gravity at 99/15.5°C	0.860 – 0.870
Refractive index at 40°C	46 – 49
Iodine number	35 – 48
Saponification number	193 – 202
Unsaponifiable matter	Not over 0.8 percent
Titer	40 – 46°C

Tallows are very low in natural anti-oxidant level and hence have poor oxidation stability. Addition of tocopherol or other conventional anti-oxidants improve the stability towards oxidation.

Oleo stock is a high grade tallow obtained from the fresh internal fat of beef carcasses by wet rendering process. It is a light yellow coloured, good smelling fat having FFA content less than 0.2 percent. It can be pressed to separate oleo oil and oleo stearin, a hard fat. The fatty acid composition of oleo oil and oleo stearin is comparable to that of oleo stock from which they are derived.

Oleic-Linoleic Oils

Cottonseed oil: Cottonseed oil obtained from the seeds of the cotton plants *Gossypium hirsutum* or *Gossypium barbadense* is used widely as an edible oil. It is of high cost hence it cannot be used in industries directly. However, the cottonseed soap stocks are used as a raw material for the manufacture of distilled cottonseed fatty acids.

Cottonseed oil itself is used for the preparation of margarine, shortenings and as a cooking or salad oil.

The whole seed contains 15–24 per cent oil and the kernel contains 30–38 per cent oil. The oil is mainly obtained by solvent extraction but pressing to get oil is also common. The free fatty acid content and the quality of the oil depends on the weather at the time of cotton seed maturity. The free fatty acid content of the oil varies from 0.5 to 5 per cent. The crude oil has a strong characteristic odour and a dark red colour. It can be refined to reduce the colour using caustic soda treatment. As the free fatty acid content of the oil increases, a corresponding increase of the refining loss occurs.

Cottonseed oil has more saturated acids than most of the oils having similar iodine number. Table 7.12 gives the values of different characteristics, fatty acid distribution and triglyceride composition of cottonseed oil.

***Table 7.12*: Characteristics, Fatty Acid Distribution and Triglyceride Composition of Cottonseed Oil (8)**

Analysis	*Range of values*	*GLC*
Characteristic		
Iodine number	90 – 117	
Saponification number	190 – 198	
Specific gravity 25/25°C	0.916 – 0.918	
Refractive index at 25°C	1.468 – 1.472	
Unsaponifiable matter	Not over 1.5%	
Titer	30 – 37°C	
Free fatty acid (as oleic)	Not over 0.25%	
Fatty acid, wt per cent		
Myristic	0.5 – 1.5	1
Palmitic	20 – 23	29
Stearic	1 – 3	4
Arachidic	0.2 – 1.5	Trace
Tetradecenoic	–	Trace
Hexadecenoic	–	2

Table 7.12. (Contd.)

Oleic	23 – 35	24
Linoleic	42 – 54	40
Glycerides, mole percent		
Trisaturated	0 – 0.1	
Oleo-disaturated	14[a]	
Linoleo-disaturated	–	
Oleo-linoleo-saturated	–	
Saturated-dilinoleo	50 – 58[b]	
Oleo-dilinoleo	28 – 36[c]	

a = total disaturated, b = total monosaturated,
c = total triunsaturated.

It is seen that the main constituent acid of cottonseed oil is oleic and linoleic acid which covers about 65–70 per cent of the total fatty acid. The principal saturated acid is palmitic acid (25–30%). Now, small quantities of C_{20}, C_{22} and C_{24} saturated and unsaturated fatty acids have been detected. Cyclopropenoid acids have also been obtained in amounts ranging from 0.1 to 0.3 per cent. The presence of small quantities of cyclopropenoid acids has been correctly indicated by the Halphen test where a cherry red coloured complex is formed by the cyclopropenoid acids like sterculic acid or malvalic acid (31, 32).

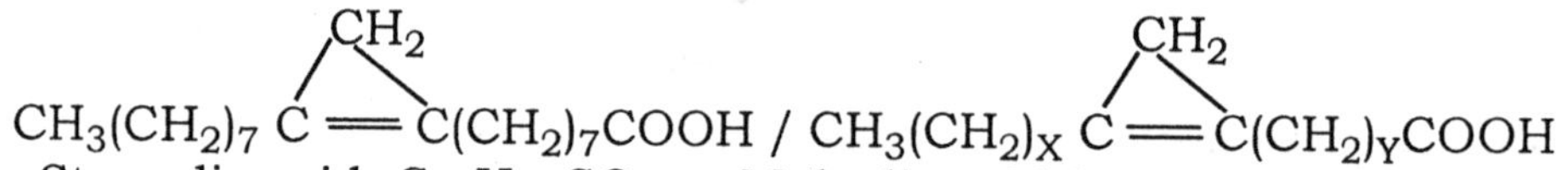

$$CH_3(CH_2)_7\overset{\overbrace{\quad CH_2 \quad}}{C=C}(CH_2)_7COOH \;/\; CH_3(CH_2)_X\overset{\overbrace{\quad CH_2 \quad}}{C=C}(CH_2)_YCOOH$$

Sterculic acid, $C_{19}H_{34}CO_2$ Malvalic acid $C_{18}H_{32}CO_2$, (x+y = 13)

The detection of cyclopropenoid acids has also been possible by means of spectrophotometric method or by GLC technique by converting the acids into its derivative by means of methanolic silver nitrate solution. Cyclopropenoid acids, specially the sterculic acid, have adverse biological effect on animals if present in the diet as low as 25 mg/day. These acids can be removed by hydrogenation, deodorization at 230–235°C in the presence of free cottonseed fatty acids and heating with cottonseed fatty acids. The fatty acids rupture the cyclopropenoid ring and form esters.

The triglyceride composition of cottonseed oil had been studied employing TLC and it was seen that the principal triglycerides of the oil were SLL 22.5, LLL 13, SLS 12.4, SOL 9.4, SLO 8.4, LOL 6.5 and OLL 6.4 (S = saturated, L = linoleic, O = oleic, figures in mole per cent). Gunstone and Qureshi (33) also used silica gel-impregnated thin layer chromatography and determined glyceride structure of the samples of cottonseed oil. Their data along with the data of other two investigators are assembled in the following Table 7.13.

Table 7.13: Observed and Calculated Proportions of Glycerides in Cottonseed oil

Reference	*Saturated acids (mole %)*	*Observed (mole %)*			*Calculated (mole %)*		
		S_2U	SU_2	U_3	S_2U	SU_2	U_3
	27	17	48	35	16	49	35
	33	21	49	29	25	50	25
	32	23	50	27	23	50	27

S = saturated U = unsaturated

Cottonseed oil contains other non-oil substances which include phospholipids, sterols, resins, carbohydrates, gossypol and pigments and these constitute about two percent of the oil. Most of these elements can be removed by alkali refining. Cottonseed oil contains an appreciable amount of tocopherols. They were separated using extraction, saponification and TLC and their proportions (mg/100 g refined oil) were found to be, α-tocopherol 32.0, β-tocopherol 31.3, and a trace of β-tocopherol. Presence of γ- and δ-lactones was observed in the oil but their amounts were quite low.

Crude cottonseed oil contains phospholipids and gossypol, both of them are undesirable with respect to edible purpose. Oxidation stability of the oil suffers due to the presence of phospholipid and gossypol which also imparts toxic effect. Both of them should be removed by refining with solvent or caustic. To enhance oxidation stability, anti-oxidant like tert-butylhydroquinone (TBHQ) is used but other conventional anti-oxidants like butylated hydroxyanisole (BHA) or propyl gallate (PG) are not effective here.

Groundnut Oil (Peanut Oil): Groundnut oil is an important edible oil all over the world obtained from the kernels (oil content 45-55%) of the plant *Arachis hypogaea*. In India, this plant is grown in large quantities. This pale yellow oil is almost free from phospholipids and other non-oil components compared to cottonseed oil. This oil, used mainly for cooking and frying purposes, is also used for the preparation of shortenings, margarines and as salad oil. A general description of the characteristics and composition of the Indian groundnut oil is given in Table 7.14.

***Table 7.14*: Characteristics and Composition of an Indian Sample of Groundnut Oil (8,34)**

Analysis	*Range of values*	*GLC*
Characteristics		
Iodine number	93.6	
Saponification number	190	
Titer °C	31.8	
Unsaponifiable matter percent	0.4	
Fatty acids, wt percent		
Myristic	0.5	Trace
Palmitic	8.0	6
Stearic	4.4	5
Arachidic	2.4	2
Behenic	3.1	3
Lignoceric	1.1	1
Total saturated	19.5	17
Hexadecanoic	1.7	Trace
Oleic	52.5	61
Linoleic	26.3	22
Total unsaturated	80.5	83
Glycerides, mole percent		
Trisaturated (GS_3)	–	
Disaturated (GS_2U)	1	
Monosaturated (GSU_2)	56	
Triunsaturated (GU_3)	43	

Besides the above fatty acid composition capric, lauric and eicosenoic acids have also been reported. Saturated acids of C_{20}, C_{22} and C_{24} are also detected by GLC. Linolenic acid is also found in a trace amount.

Like sunflower oil, rapeseed oil and safflower oil, groundnut oil is also subjected to genetic development. As groundnut oil is an important edible oil its triglyceride structure has been studied extensively. It was found (33) by Gunstone and Qureshi that a sample of groundnut oil contain 21 mole percent saturated acid, 11 mole percent S_2U, 40 mole percent SU_2 and 49 mole percent U_3 composition. Some other investigators (34) and (35) found triglyceride structure of groundnut oil using silica gel impregnated TLC and proposed that out of the total 26 triglycerides only nine constitute 80.6 mole percent of the total triglyceride present and their percentages were SLS 5.3 percent, OOO 6.2 percent, SOL 5.0 percent, SLO 15.7 percent, OOL 7.8 percent, OLO 8.7 percent, SLL 10.7 percent, OLL 17.8 percent and LLL 3.9 percent (where O = oleic, L = linoleic, S = saturated).

Normally, groundnut oil contains 0.2–1.0 percent of unsaponifiables which include sterols, tocopherols and other anti-oxidants, squalene and other hydrocarbons. The pigments of oil are decreased as the plant matures. The major pigments are carotene and lutein which amount to less than 1 μg/lit of oil from mature groundnut.

Groundnut oil contains phospholipids which are mainly phosphatidylethanolamines. The anti-oxidant tocopherols are present in the oil with the principal tocopherols being α- and β-tocopherols. Groundnut oil, along with a small amount of linolenic acid has good oxidation stability, and therefore it is a superior quality cooking oil.

Groundnut oil solidifies in a refrigerator. Its cloud and pour points are determined to be approximately 4.5°C and 1.1°C respecitvely.

Olive Oil: Olive oil is an important oil obtained from the fruits of the tree *Olea europa* by crushing and pressing. Indian production of olive oil is not significant. Olive oil is almost totally used for edible purpose, but it is also utilized in the manufacturing of soaps and in the production of textile oils and sulphonated oils.

The fruit of olive tree contains about 35–70 percent oil and the pulp contains more than 75 percent oil. The oil is obtained by pressing. The oil extracted by first pressing is of the highest grade, called 'virgin' olive oil and can be used without refining. With successive pressing the grades become lower and require subsequent processing like refining by solvent extraction, deodorization and other treatments. The oil obtained from the last pressing is called 'olive foots' or 'sulfur olive oil' as it is refined by carbon disulfide. But this solvent is now completely displaced. 'Olive foots' is inedible and it is used for soap making and other industrial purposes.

Lower grade olive oil with strong flavours may be mixed with milder flavoured oils like cottonseed and soyabean oils. Olive oil is detected by its high squalene content (36) if present in other oils. Other tests of adulteration of olive oil have been reported and the best of them is the fractional extraction of the more organic solvent soluble triglycerides (olive oil contain about 36 percent OLL and OOL triglycerides) with a methanol-acetone mixture at 0°C and then study of fatty acids of the extracted glycerides by GLC. This method can be employed successfully when olive oil is adulterated with cottonseed, sesame, corn, soyabean or linseed oils.

The general characteristics of the edible olive oil are summarized below:

Characteristics	*Values*
Specific gravity at 25°/25°C	0.909 – 0.915
Iodine number	80 – 88
Saponification number	188–196
Titer °C	17–26
Unsaponifiable matter per cent	Not over 1.8
Free fatty acids per cent	Not over 1.5

The percentages of fatty acids vary for different samples and the ranges of values of different constituent fatty acids with GLC values are given in Table 7.15.

Table 7.15: Composition of Fatty Acids of Olive Oil Samples

Fatty acid wt%	*Range of values*	*GLC*
Myristic	0.1–1.2	Trace
Palmitic	7–16	14
Stearic	1–3	2
Arachidic	0.1–0.3	Trace
Hexadecenoic	–	2
Oleic	65–85	64
Linoleic	4–15	16

The triglyceride structure of olive oil has been investigated and the principal triglycerides are found to be: POO 18.4 percent, SOO 5.1 percent, POL 5.9 percent, OOO 43.5 percent, OOL 6.8 percent.

Besides the fatty acids and glycerides, olive oil contains numerous non-oil components which were investigated extensively. C_{11} to C_{30} *n*-paraffins and trace amount of iso- and anti-iso hydrocarbons are present in the virgin oil. Benzenoid and naphthenic hydrocarbons along with carotene and an appreciable amount of squalene are also present in the oil. Esters of C_{27}–C_{32} alcohols and sterols are found in olive oil. Tocopherols with composition 88.5 percent α, 9.9 percent β- and γ- and 1.6 percent δ-tocopherol are common in the oil. Chlorophyll in the oil after decomposition yields phytol and this is the only aliphatic alcohol present freely in the oil. Hydroxytriterpenes and 4-methylsterols were also observed by investigators. Sterols are present in olive oil and the principal sterol is β-sitosterol (about 0.1–0.2 percent of the oil) and other sterols present are campesterol and stigmasterol. Phospholipids are obtained in trace amounts ranging from 40 to 135 ppm. Oleic acid is the main fatty acid of the phospholipids.

Palm oil: Palm oil is obtained from the fruits of the tree oil palm; *Elaeis guineensis*. Palm fruit yields two types of oils, one is from its pulp, called palm oil and another is from its kernel, called palm kernel oil. These two oils differ in their compositions; palm kernel oil contains significant amounts of lauric (40–52%) and myristic (14–18%) acids which is quite similar to the coconut oil whereas palm oil contains 32–47 percent palmitic and 40–52 percent oleic acid.

Oil palm trees are grown mainly in Malaysia, Indonesia and Africa. These oils are used principally as edible oils in the countries which produce them though these are also used in manufacturing shortenings, margarine and soap.

Palm oil may contain free fatty acids as high as 50 percent. Oils are classified and graded according to their geographical origin and free fatty acid content. Malaysian oils are of the highest grade containing about 3–5 percent FFA. Palm oil is deep red coloured due to its high carotene content (0.05–0.2%). This oil can be bleached to light yellow colour like other vegetable oils. The oil is very stable towards oxidation and has no drying property. The oil having a relatively small amount of FFA is called soft oil whereas high content of acidity turns it into hard oil.

Palm oil has approximately equal amounts of saturated and unsaturated fatty acids. Palmitic and oleic acids are the major component acids. The fatty acid distribution of Zaire oil, an edible red palm oil of Malaysia is given in Table 7.16.

Among the non-oil constituents, carotene is the main, as its amount varies from 0.03 to 0.15 percent by wt. within 0.2 to 0.1 wt

***Table 7.16*: Fatty Acid Distribution of Zaire Palm Oil and Malaysian Palm Oil**

Fatty acids	*Weight per cent (Zaire)*	*Weight per cent (Malaysia)*
Saturated acids		
Caprylic	0.005	–
Capric	0.005	–
Lauric	0.12	–
Myristic	1.02	2.5
Palmitic	45.5	40.8
Stearic	5.9	3.6
Arachidic	0.36	0.36
Unsaturated acids		
Palmitoleic	0.10	–
Oleic	34.6	45.2
Linoleic	11.81	7.9
Linolenic	0.29	–

percent unsaponifiables in the oil. It imparts deep red colour to the oil. Tocopherols, sterols, phosphatides are also present in the oil in appreciable amounts. These compounds can be separated by using thin layer chromatography.

Glyceride composition of palm oil has been studied and it was seen that more than 85 percent of unsaturated acids are located at the 2-positions of the glycerol molecules. Both glyceride structure and free fatty acid content affect the melting point and plasticity of palm oil.

Palm oil contains lactones but at much lower level than palm kernel oil. Palm oil is less stable to oxidation than cottonseed oil which is due to the presence of high quantity of carotene.

Sunflower Oil: Sunflower oil is the second largest produced vegetable edible oil in the world. It is obtained from the seed of the plant *Helianthus annus* which is grown on a large scale in Russia and other countries including India. The seed contains about 22-30 percent oil. The oil is obtained by pressing or by solvent extraction of the decorticated seed. It is almost entirely used as edible oil in cooking and for the preparation of shortenings and margarines.

Crude sunflower oil contains some phosphatides and mucilaginious matters. Its free fatty acid content is 0.5 percent and above, which is similar to other vegetable oils. It has a characteristic odour which is removed by steam deodorization.

General characteristics and fatty acid composition of sunflower oil are given in Table 7.17.

It is seen that sunflower oil contains about 15 percent saturated acids and 85 percent unsaturated acids with a small amount (roughly 2%) of linolenic acid.

The composition of oil in the seed is strongly dependent upon the climate, temperature, genetic factors and position of the seed in the flower head. It is peculiar that the linoleic and palmitic acid content of sunflower oil increases as the seed's location shifts from the perimeter to the centre of the flower head.

***Table 7.17*: Characteristics and Fatty Acid Compositions for Sunflower Oil (8).**

Analysis	*Range of values*	*GLC*
Characteristics:		
Specific gravity 25/25°C	0.915–0.919	
Refractive index at 25°C	1.472–1.474	
Iodine number	125–136	
Saponification number 188-194		
Unsaponifiable matter	Not over 1.5 percent	
Titer	16–20°C	
Fatty acids:		
Palmitic	3–6	11
Stearic	1–3	6
Arachidic	0.6–4	-
Behenic	trace–0.8	-
Lignoceric	trace–0.4	-
Total saturated	8.7–14.2	17
Oleic	14–43	29
Linoleic	44–75	52
Total unsaturated	85–91	83*

* includes 2 percent linolenic acid. "Composition and Constants of Natural fats and oils by Gas- liquid Chromatography" chart published by Archer-Daniels-Midland, Minneapolis, MN 1961.

It is seen that oils obtained from different areas and plants differ in composition. It was observed that seeds of north Canada give oil which was 70 percent linoleic acid whereas seeds from South Canada yield oil having less than about 20 percent linoleic acid. Sunflower oil sometimes may contain some oxygenated acids (37) if their seeds are stored for a prolonged period and this seems to occur due to deterioration in oxidation stability.

The triglycerides of sunflower oil were examined and it was found that the following types of triglycerides are present in the oil: D_3 24.6 ± 0.7, DMD and MDD 30.0 ± 0.6, DSD and SDD 18.8 ± 0.3, M_2G 8.9 ± 0.2, SMG 10.6 ± 1.0, M_3 1.3 ± 0.3, S_2D 2.0 ± 0.3, SM_2 2.7 ± 0.8 and SSM and SMS 1.1 ± 0.1 (wt percent, D = all *cis*-diene, S = saturated, M = *cis*-monoene).

Non-oil components like waxes, hydrocarbons and sterol occur in appreciable amounts in the sunflower oil but natural anti-oxidants are in less amounts. The non-oil components increase the cloud point of the oil and it is improved by solvent winterization treatment using 85 percent acetone/15 percent hexane. Oxidation stability of this oil is increased by using conventional anti-oxidants. TBHQ responds well to stabilize oil. BHA has a less effect on the stability.

Sesame oil: Sesame oil is an important vegetable oil of our country, obtained from the seed of the plant *Sesamum indicum*. Other producer countries are China, Myanmar, Turkey, Mexico, Sudan etc. This oil is of high grade for edible purpose and is used in cooking, as salad oil, for the manufacture of margarine and shortenings. This oil can also be used for soap making, in pharmaceuticals and as synergist for insecticides.

Sesame oil contains relatively high amount of unsaponifiable matters including sterols which cannot be removed by refining. These unsaponifiable matters of sesame oil contains some compounds like sesamin, sesamolin (38,39) which do not occur in other oils. These substances give distinctive colour reactions and by these reactions small quantity of sesame oil can be detected in other oils. These substances particularly sesamol imparts a high oxidation stability to the oil. The structure of these substances are shown below:

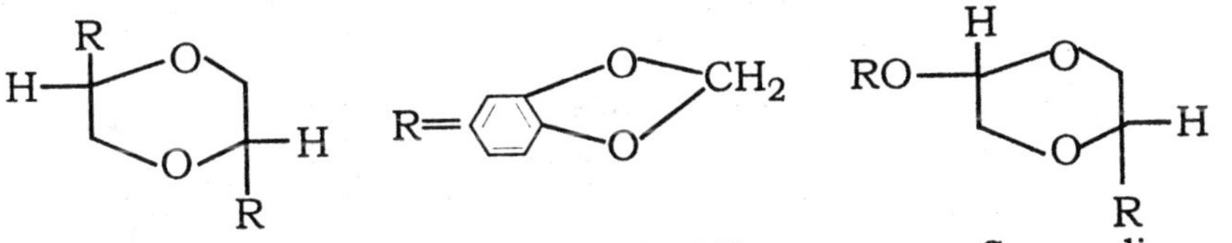

Sesamin   R= (benzodioxole ring with O–CH_2–O)   Sesamol=R–OH   Sesamolin

The amounts of these compounds present in oil are sesamin 0.4–1.1 percent, sesamolin 0.3–0.6 percent and a trace amount of sesamol. Sesamol can be generated from sesamolin through hydrogenation or in the conditions of processing or storage.

Sterols containing β-sitosterol, campestrol and stigmasterol and triterpene have been isolated and estimated using TLC and GLC. Presence of small amounts of tocopherols has also been reported. The characteristics and composition of sesame oil are listed in Table 7.18.

Table 7.18: Characteristics and Fatty Acid Composition of Sesame Oil (8)

Analysis	*Range of values*	*GLC*
Characteristics		
Iodine number	103 –116	
Saponification number	188–195	
Specific gravity at 25/25°C	0.914–0.919	
Refractive index at 20°C	1.470–1.474	
Unsaponifiable matter	Not over 1.8 percent	
Titer °C	20–25	
Fatty acid wt percent		
Myristic	trace	–
Palmitic	7–9	10
Stearic	4–5	5
Arachidic	0.4–1	–
Total saturated	–	15
Hexadecenoic	<1	–
Oleic	37–49	40
Linoleic	35–47	43
Total unsaturated	–	85

It is seen that oleic and linoleic acids are the main components of the oil and they are present almost in similar quantities. It is observed that Indian sesame oil contains higher palmitic acid (9.1%) than other oils.

The triglyceride composition of sesame oil was investigated (40) and was found as follows: 39 mole percent = 1 oleic + 2 linoleic, 37 mole percent = 1 saturated + 1 oleic + 1 linoleic, 15 mole percent = 2 oleic + 1 linoleic, 4 mole percent = 1 saturated + 2 linoleic, 3 mole percent = 2 saturated + 1 oleic, 1 mole percent = 2 saturated + 1 linoleic and 1 mole percent = 3 linoleic.

Corn oil: Corn oil is obtained from the palm plant *Zea mays* by wet or dry milling of corn. Crude oil has a dark reddish colour which, after refining, becomes somewhat light but still remains darker than other vegetable oils.

Table 7.19 shows the general characteristics and composition of corn oil.

Table 7.19: Characteristics and Composition of Corn Oil (8)

Analysis	*Range of values*	*GLC*
Characteristics		
Specific gravity 25/25°C	0.915–0.920	
Refractive index at 25°C	1.470–1.474	
Iodine number	103–128	
Saponification number	187–193	
Unsaponifiable matter	Not over 2.0 percent	
Titer :	14–20°C	
Fatty acid wt %		
Myristic	trace–1.7	–
Palmitic	8–12	13
Stearic	2–5	4
C_{20}– C_{22} saturated	trace –0.2	trace
Total saturated	12–18	17
Hexadecenoic	0.2–1.6	–
Oleic	19–49	29
Linoleic	34–62	54
Linolenic	–	–
Total unsaturated	88–82	83
Analysis	*Range of values*	*GLC*
**Triglycerides*		
Trisaturated (GS_3)	0	
Disaturated (GS_2U)	2.2	
Diunsaturated (GSU_2)	40.3	
Triunsaturated (GU_3)	57.5	

*Investigated for germs only

Composition of corn oil has shown that oleic and linoleic acid constitutes about 80 percent of the total oil in a ratio of about 1:2. Linolenic acid may also be present but in trace amount.

Corn oil contains a high proportion (1–3%) phosphatides and other non-oil components, predominantly sterols. Free fatty acids are present in large quantities than most of the vegetable oils. Tocopherol content of the oil is also high and this accounts for its high oxidation stability. A very small percentage (0.05%) of wax is present in the oils and for this the oil clouds when cooled to a low temperature.

The triglyceride structure of corn oil had been studied by Vander Waal who concluded that there were no S_3 type triglycerides present in the oil and other triglycerides were 7.3 percent S_2U, 36.3 percent SU_2, 56.4 percent U_3 in a sample of total 17 percent saturated fatty acid.

Besides the high content of tocopherol, corn oil contains a very small amount of another anti-oxidant, ferulic acid. It has excellent oxidation stability. Ferulic acid is present in the oil as an ester with dihydro β-sitosterol (41).

$$\begin{array}{c} OH_3 \\ HO-C_6H_3-CH = CHOOOH \end{array}$$

Ferulic acid

Corn oil is used as an edible vegetable oil for cooking and frying purpose.

Safflower oil: In India safflower oil is an important edible oil which is obtained from the seeds of the plant *Carthamus tincotorius*. Other producer countries are Mexico, United States, Spain, Portugal, Australia etc.

Safflower oil is used as an edible oil, in the manufacture of paints and alkyd resins. A typical analysis comprising the characteristics and composition of safflower oil is shown in Table 7.20.

Safflower oil is characterized by the highest linoleic acid content than other oils. The triglycerides of safflower oil contain high linoleic acid and most of them are triunsaturated being dilinoleoglycerides and trilinolein (42).

Various samples of safflower oil have been investigated and the limits of fatty acid content have been reported as : 45 percent maximum oleic, 45 percent minimum linoleic and 80 maximum oleic with 10 percent minimum linoleic and 10 percent maximum stearic acid. The modern 'high oleic' safflower oil contains 79 percent oleic, 13 percent linoleic, 6 per cent palmitic and 2 percent stearic acid approximately. At present 'high linoleic' safflower oil is the commercial oil and 'high oleic' oil has passed through the experimental stage (49) and is being bred and used in pharmaceuticals and certain other industrial purposes.

***Table 7.20*: Characteristics and composition of safflower oil (8)**

Analysis	*Range of values*	*GLC*
Characteristics		
Iodine number	140–150	
Saponification number	186–194	
Refractive index at 25°C	1.473–1.476	
Specific gravity at 25/25°C	0.919–0.924	
Unsaponifiable matter %	0.3–1.3	
Fatty acids, wt %		
Myristic	trace	trace
Palmitic	3–6	8
Stearic	1–4	3
Arachidic	trace–0.2	trace
Total saturated	5–10	11
Oleic	13–21	13
Linoleic	66–75	73
Linolenic	trace	1
Total unsaturated	90–95	89

Safflower oil contains 'cholestrol-like' sterols in its unsaponifiables. It also contains tocopherol. 'High linoleic' safflower oil has poor oxidation stability. Tert-Butylhydroquinone (TBHQ) responds well to stabilization whereas butylated hydroxyanisole (BHA) and propylgallate PG are not so effective. 'High oleic' safflower oil is relatively more stable than 'high linoleic' oil.

Tobaccoseed oil: Tobaccoseed oil is obtained from the seeds of two species of plants *Nicotiana tabacum* and *Nicotina rusticum*. These plants are grown in Bulgaria, Turkey, India and Greece but not to a large extent. This oil is also high in linoleic acid (about 70%). Refined tobaccoseed oil can be used for edible purpose as it contains no objectionable matter. This oil is mainly used for the preparation of non-yellowing alkyd resins. Table 7.21 lists the characteristics and composition of tobaccoseed oil (in the Indian context).

***Table 7.21*: Characteristic and composition of tobaccoseed oil (Indian)**

Analysis	*Range of values*
Characteristics	
Iodine number	140–145
Saponification number	189–190
Refractive Index 25°C	1.4742–1.4756
Unsaponifiable percent	1.9–2.5
Fatty acids, wt percent	
Total saturated	9.8–13.1
Oleic	7.8–20.1
Linoleic	68.2–73.8
Total unsaturated	87–90

The main fatty acid in tobacco seed oil is linoleic acid. Linolenic acid is actually absent but if present then only in a trace amount. Composition of glycerides of tobacco seed oil is similar to that of safflower oil. It had been reported (43) that the triglyceride composition of tobaccoseed oil is as follows: 37.4 percent trilinolein, 22.8 percent dilinolein, 4.9 percent dioleolinolein, 19.7 percent dilinoleosaturated, 9.3 percent linoleosaturated and 3.3 percent linoleo-disaturated.

Poppyseed oil: Poppyseed oil is obtained from the seed of the plant *Papaveracae*. It is mainly produced in China, India, Turkey etc. Cold-pressed oil from cleaned seed can be used for edible purpose.

The characteristics of the oil are as follows:

Iodine number	133.4
Saponification number	197.5
Density at 60°C	0.898
Refractive index at 60°C	1.4604
Unsaponifiable matter	0.4–1.2 percent
Titer ..	15–20°C
Solidification point	–15° to -20°C

The fatty acid distribution of poppyseed oil was observed by GLC as palmitic 10 percent, Stearic 2 percent, Oleic 11 percent, linoleic 72 percent and linolenic 5 percent.

Teaseed oil: Teaseed oil is obtained from the seeds of the plant *Thea sasanqua* and it has only commercial importance in China. It has almost identical composition as olive oil and can only be detected by means of Fitelson test (44) if mixed in olive oil as low as 5–10 percent concentration.

Some characteristics of teaseed oil are also similar to those of olive oil. These are: Iodine number 83–89, saponification number 193–196, refractive index at 25°C 1.467–1.469 and titer about 22°C. The fatty acid composition as detected is as follows: 74–87 percent oleic acid, 7–14 percent linoleic and 7–11 percent saturated acid. Triterpenes (0.04%) and sterols (0.6%) are also present in the oil as unsaponifiables. However, the composition of Indian teaseed oil as reported (44A) is broadly as follows:

	Palmitic	*Stearic*	*Oleic*	*Linoleic*
Assam	15.0	0.3	58.4	24.3
Darjeeling	16.1	0.4	59.0	22.3

Kapok oil: Kapok oil is obtained from the seed of the plant *Eriodendron anfractuosum*. The plant is grown in Indonesia and other tropical countries. The oil is very similar to cottonseed oil.

The characteristics of kapok oil are described as follows: Specific gravity 15/15°C, 0.920–0.933; refractive index 1.406–1.472, iodine number 86–110, saponification number 189–197, unsaponifiable matter 0.5–1.8 percent, titer 27–32°C.

Kapok oil contains a higher percentage of oleic acid than cotton seed oil and a lower percentage of linoleic acid. The fatty acid composition of kapok oil was investigated and it was found to contain about 15–20 percent saturated and 80–85 percent unsaturated acids which include 10.5–10.8 percent palmitic, 4.9–8.6 percent stearic, 1 percent C_{20}–C_{22} saturated (mostly arachidic), 46.1–56.6 percent oleic and 27.7–34.6 percent linoleic.

Rice bran oil: Rice bran oil is obtained from the rice bran of the plant *Oryza sativa* which contains about 15–20 percent. As rice is grown world wide, rice bran oil is an important oil. It is mainly used as cooking oil, salad oil and also to produce hydrogenated fat.

Rice bran oil has very good oxidation stability and this accounts not only for the presence of tocopherols but also for esters of ferulic acid which act as strong natural anti-oxidants.

$$HO-C_6H_3(OCH_3)-CH = CHCOOH$$

Ferulic acid

Rice bran oil has a high amount of free fatty acids. It is seen that oil obtained from the fresh bran contains about 4–6 percent free acid and the percentage increases at a rate of one per hour when stored at 25°C.

The characteristics of rice bran oil are summarized as follows:

Specific gravity 25/25°C	0.916–0.921
Refractive index at 25°C	1.470–1.473
Iodine number	99–108
Saponification number	181–189
Unsaponifiables	3–5 percent
Titer	24–28°C
Acid value	4–120

The composition of fatty acids as reported is as follows:

15–20 percent saturated and 80–85 percent unsaturated which includes 0.4–1.0 percent myristic, 12–18 percent, palmitic, 1–3 percent stearic, 1 percent C_{20} – C_{22} saturated, 40–50 percent oleic, 29–42 percent linoleic, 1 percent linolenic and 0.2–0.4 percent palmitoleic.

Rice bran oil contains appreciable percentage of waxes which can be removed by solvent extraction. It also contains high amount of unsaponifiables which include triterpenes, sterols etc. This oil has high oxidation stability. Tocopherol content of the oil is 38.8 mg/ 100 gm oil.

If rice bran oil is contaminated with cottonseed oil it can be detected by GLC method based on stigmasterol content as cottonseed oil is free of this sterol.

Sorghum oil: This oil is obtained from the grain of the plant *Sorghum vulgare*. Like corn oil this oil is also obtained from the germ and is very similar to corn oil.

The characteristics of sorghum oil as reported are given below:

Refractive index at 25°C		1.469–1.472
Iodine number	..	108–122
Saponification number	..	181–191
Unsaponifiable matter	..	1.7–3.2 percent

The fatty acid distribution in the sorghum oil as investigated is 10–15 percent saturated and 85–90 percent unsaturated, which include 0.1 percent myristic, 6–10 percent palmitic, 3–6 percent stearic, 0–1 percent hexadecanoic, 30–47 percent oleic, 40–55 percent linoleic and 0.1 percent linolenic acid.

This oil also contains high amount of lipids which consist of glycolipids and phospholipids.

Other Oleic/Linoleic Acid Oils

There are other oleic/linoleic acid oils, some of which are described below:

Almond oil is obtained from the kernels of the tree *Prunus amygdalus* or *communis*. This oil contains about 5 percent saturated and 95 percent unsaturated acids which consist of myristic, palmitic, oleic and linoleic acid. Oleic acid content (77%) is the highest.

Apricot kernel oil is obtained from the kernels of the tree *Prunus armeniaca* and it is mostly identical to almond oil. Its oleic acid content is about 61 percent and it is the predominant acid.

Pecan oil is also similar to the above oil, having oleic acid as the principal fatty acid. It is obtained from the kernels of the tree *Carya pecan* or *illinoensis*. Its oleic acid content is about 71–79 percent.

Cereal grains like barley, corn, oats, rye, wheat etc. are the important sources of oleic/linoleic acid oils. As the cereals are the important crops in the world, their seed oils are also important. They contain high amounts of lipids like phospholipids and glycolipids.

Other oleic/finoleic acid are grapeseed oil, tomato seed oil etc. which are not so much important except in those countries which produce and process them largely.

Erucic Acid Oils

Rapeseed oils: Rapeseed oil obtained from the seeds of the plants *Brassica napus* (rape) and *campestris* (turnip rape) is produced mainly in China, India, Japan, Pakistan etc.

Rapeseed oil can be classified into three types: (1) Rapeseed oil having high content of erucic acid (20–55%) which is mainly obtained from the species *B. campestris*, (2) Low erucic acid (0–5%) rapeseed oil called canbra or canola oil and (3) Zero erucic acid rapeseed oil obtained from the species *B. napus*. Rapeseed oil is mainly used as edible oil but it may be utilized in the manufacture of erucic acid and behenic acid which are used as lubricant additives and for other industrial purposes.

Crude high erucic rapeseed oil is dark yellow in colour but after refining it becomes pale yellow. This is the traditional rapeseed oil. But it is seen that high erucic acid content of the oil causes some physiological disturbances. Hence its use as edible oil is decreasing.

General characteristics and fatty acid distribution in the Indian rapeseed oil are compiled in Table 7.22.

Due to the high amount of erucic acid (40-55%) the rapeseed oil has a low saponification number.

The composition of rapeseed oil is amenable to change and low erucic or even zero erucic rapeseed oil is now developed through genetic changes. Rapeseed oil has an appreciable quantity of linolenic acid (15–18%) and rapeseed oil having zero erucic acid can be termed as linoleic acid oil. Experimental work was done in Canada for the genetic development of rapeseed oil whose erucic acid content is nil. High erucic acid oils are used for industrial purposes and low or zero erucic oils are regarded as edible oils. High erucic rapeseed oil also contains high proportions of glucosinolates and these cause adverse physiological effects when rapeseed meal is used as a feed to the animals. Low erucic rapeseed oil has similar composition as soyabean oil.

The composition of rapeseed oil varies with the climate. In warmer climate the percentage of linoleic acid and eicosenoic acid increases. β-sitosterol decreases. Erucic acid content of low erucic acid rapeseed oil increases when the plant is grown in warmer climate but it is opposite in the case of plants of high erucic rapeseed oil.

***Table 7.22*: Characteristics and fatty acid composition of rapeseed oil from Indian origin (Gujarat) (45)**

Analysis	*Values (Gujarat)*
Characteristics	
Iodine number	94.2
Saponification equivalent	315.2
Saponification number	178
Unsaponifiable matter percent	0.7
Fatty acids wt %	
Palmitic	1.9
Stearic	3.0
Arachidic	2.4
Behenic	1.4
Lignoceric	0.8
Total saturated	9.5
Hexadecenoic	2.6
Oleic	16.3
Eicosenoic	5.1
Erucic	45.0
Linoleic	13.4
Docosadienoic	0.9
Linolenic	7.2
Total unsaturated	90.5

The triglyceride composition of rapeseed oil has been studied and it is seen that the oil contains about 54 percent mono-oleodierucins, 28 percent dioleomonoerucins and 18 percent monosaturated mono-oleomonoerucins. It has been observed that erucic acid appears almost in all triglycerides in their one and three positions and unsaturated C_{18} fatty acids appear at the middle 2-position. Besides erucic acid one and three positions are also occupied by diunsaturated C_{20} and C_{22} acids. For zero erucic rapeseed oil, the middle position of the glyceride (2-position) is preferentially occupied by linoleic and linolenic acids.

Rapeseed oil and other cruciferae seed oils exhibit complicated phase behaviour which have been studied by X-ray diffraction methods (46). The phase transitions in cruciferae oils with erucic acid content 25 percent or higher is shown below:

(dashed arrows ——→ transformation during cooling and solid arrow ——→ transformation during heating).

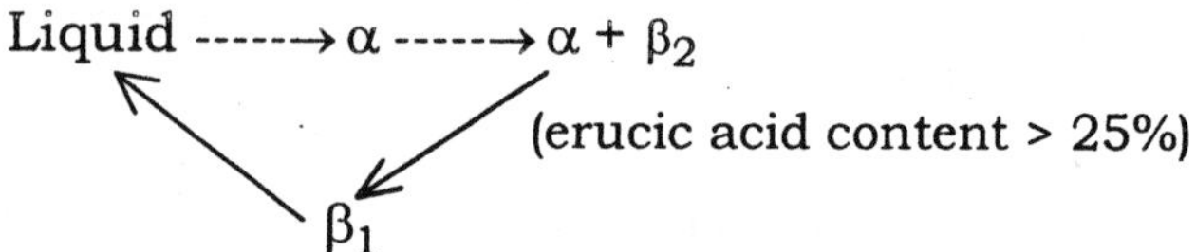

Similar phase transformation of low erucic acid oils are also shown below:

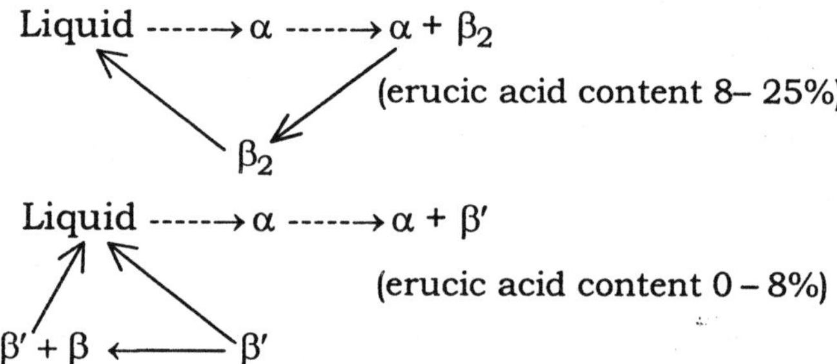

α-phases appear incidentally for oils having greater than 25 percent erucic acid but they can also be frequently detected for low erucic acid oils.

The cruciferae oils exhibit unusual polymorphism which cannot be obtained for other oils like soyabean, cottonseed, groundnut, corn or sunflower oils. The cause of it may be the presence of high amount of C_{20} and C_{22} fatty acid and preferential positioning of erucic and eicosenoic acids in one and three positions of glycerides.

High erucic rapeseed oil contains triterpene (0.03% wt) and sterols (0.5 wt%) but does not contain stigmasterol. This oil contains a low percentage of tocopherols. Though high erucic oil has low content of tocopherol it has relatively good oxidation stability. Low erucic acid oils are superior to those of high erucic oils.

Crambe oil: Crambe oil obtained from the two species *Crambe abyssinica* and *Crambe hispanica* is a high erucic acid oil (50–57%) used mainly for industrial purpose. This oil has certain adverse effect in the body as it contains high amounts (8–10%) of glycosinolates, sinapin (0.5%), but this is sought to be genetically developed.

This oil has a high C_{22} acid and low C_{20} acid content. This oil has industrial uses like splitting of erucic acid to a mixed brassylic/pelargonic acid and the brassylic acid is then utilized for the manufacture of polymers.

Other Erucic Acid Oils

Among the other erucic acid oils, mustard oil is an important oil in India. It is obtained from the seeds of the plant *Sinapsis alba* (white or yellow mustard) and has about 51 percent erucic acid content. Other countries producing this oil are China, United States, Canada etc. Mustard oil is similar in composition to the rapeseed oil. Table 7.23 gives an idea about the characteristics and composition of fatty acids of an Indian mustard seed oil.

***Table 7.23*: Characteristics and fatty acid composition of mustard seed oil (Indian) (47)**

Analysis	*Values*
Characteristics	
Iodine number	102
Saponification number	173
Unsaponifiable matter percent	3.8
Fatty acids wt percent	
Palmitic	1.5
Stearic	0.4
Arachidic	0.5
Behenic	2.0
Lignoceric	1.0
Total saturated	5.4
Oleic	22.0
Eicosenoic	7.0
Erucic	44.2
Linoleic	14.2
Linolenic	6.8
Total unsaturated	94.2

There are other erucic acid oils. Tumbling mustard obtained from the species Sisybrium altissinum yields oils containing about 25 percent erucic acid and 35 percent linolenic acid. This plant is grown in United States. Hare's ear mustard seed oil from *Coringa orientalis* and fansweed seed oil from *Thlaspi arvense* are similar to rapeseed oil but the former has about 12 per cent eicosenoic acid and the latter has about 30 percent linoleic acid. Rape oil from *Eruca sativa* is nearly like rapeseed oil except that it has a lower iodine number and a higher content of eicosenoic acid.

Linolenic Acid Oils

Linseed oil: Linseed oil is an important oil obtained from the seeds of the plant *Linium usitatissimum*. These plants are grown in cold areas of United States, India, Argentina, Russia etc. Main use of linseed oil is as drying oil in paints and coating industries.

Crude linseed oil is obtained either by pressing or solvent extraction after pressing and has a dark colour and strong odour. The odour of the linolenic acid oils is due to the presence of high amount of unsaturation and it cannot be removed fully by steam deodorization. The free fatty acid content of linseed oil is about 0.5 percent i.e. similar to other vegetable oils.

Linseed oil cannot be used as an edible oil due to the presence of highly unsaturated acids. It is used principally in the manufacture of paints, varnishes, printing inks, oil-cloths, linoleum etc. and for other industrial purposes. Due to the high rates of auto-oxidation and thermal polymerization, this oil cannot be used for cooking or other edible purposes.

A general description of the characteristics and composition of the oil is given in Table 7.24.

***Table 7.24*: Characteristics and composition of linseed oil (8)**

Analysis	*Range of values*	*GLC*
Characteristics		
Iodine number 165-204		
Refractive index at 25°C	1.477–1.482	
Titer °C	19–21	

Analysis	Range of values	GLC
Unsaponifiable matter	Not over 1.7 percent	
Fatty acids wt percent		
Palmitic	4–7	6
Stearic	2–5	4
Arachidic	0.3–1	trace
Oleic	12–34	22
Linoleic	17–24	16
Linolenic	35–60	52

Phosphatides and mucilaginous material are present in high amount in the linseed oil which are to be removed before use by alkali refining. A trace quantity of wax is also found in this oil which is separated by winterizing.

As linseed oil is used as a drying oil in coating industries, it requires a high degree of unsaturation i.e. high iodine number. As iodine number varies from oil to oil, hence to get an oil of desired iodine number, blending is necessary. Refractive index of linseed oil varies with iodine number. The ASTM specification of linseed oil (raw) is as follows:

ASTM Specification D234-48	
Specific gravity at 15/15°C	0.931–0.936
Acid number, maximum	4.0
Saponification number	189–195
Unsaponifiable matter % max.	1.50
Iodine number, Wijs, min.*	177
Loss on heating at 105–110°C, max. %	0.3
Appearance	Clear and transparent at 65°C
Colour	Not darker than a freshly prepared solution of 1 gm of $K_2Cr_2O_7$ in 100 ml of pure H_2SO_4 (sp.gr. 1.84)
Foots, max percent	
Heated oil	1.0
Chilled oil	4.0

If linseed oil of the high iodine number type is desired, the minimum iodine number should be changed to 188.0.

The Triglyceride composition as studied (46) is as follows: monounsaturated 5 percent, saturated oleolinolenins 18 percent, saturated linoleolinolenins 15 percent, saturated dilinolenins 10 percent, oleolinoleolinolenins 5 percent, oleodilinolenin 22 percent, linoleodilinolenins 24 percent and trilinolenins 1 percent. More accurate and comprehensive results are obtained by the use of GLC, paper chromatography and partition chromatography.

Soyabean oil: Soyabean oil is the vegetable oil produced in largest volume in the world. It is obtained from the seeds of the plant *Soja max* by pressing or solvent extraction method. This oil is refined, bleached and deodorized before it is used for the manufacture of margarine and shortenings by partial hydrogenation. Its other use is as drying oil.

Crude soyabean oil having dark yellow or greenish colour is reduced to light yellow colour by refining with alkali. Non-glyceride materials mainly phosphatides are present in large quantities (1.5–2.5%) in the crude oil obtained from solvent extraction. Phosphatides are removed by water washing. Free fatty acid content of soyabean oil is about 0.5 percent.

The general characteristics and fatty acid composition of soyabean oil is listed in Table 7.25 (8,49,50):

***Table 7.25*: General characteristics and fatty acid composition of soyabean oil**

Analysis	*Range of values*	*GLC*
Characteristics		
Specific gravity at 25/25°C	0.917–0.921	
Iodine number (wijs)	120–141	
Saponification number	198–195	
Unsaponifiable matter	Not over 1.5 percent	
Refractive index at 25°C	1.470–1.476	
Fatty acid wt percent		
Myristic and lower	trace–0.5	trace

Palmitic	7–11	11
Stearic	2–6	4
C_{20} and higher	0.3–3	trace
Total saturated	11–20	15
C_{16} and lower	trace–1	–
Oleic	15–33	25
Linoleic	43–46	51
Linolenic	5–11	9
Total unsaturated	83–90	85

It is evident from the table that linolenic acid content is 5–9 percent and linoleic acid is 43–56 percent in soyabean oil. For edible purpose the linolenic acid content of the oil should be lowered more. Through genetic means the percentage of linolenic is possible to be decreased to about 3.5 percent. Linolenic acid is also thought to be responsible for flavour reversion of soyabean oil. Therefore, the percentage of linolenic acid should be minimized either genetically or by selective hydrogenation or other possible methods. Lipoxygenase enzyme in the soyabean oil is responsible for the oxidation of linoleic and linolenic acid components to generate undesirable smell; so this enzyme should also be eliminated and this has been tried through genetic means. Linolenic acid also lowers the quality and usefulness of the alkyd resins and certain coatings manufactured with soyabean oil if it is present beyond its optimum level in the oil.

The glycerides of soyabean oil mainly contain two unsaturated fatty acids. Trisaturated or disaturated glycerides are almost absent in the oil. Gunstone and his coworkers (51) had investigated the glyceride structure of soyabean oil by employing $AgNO_3$-impregnated silica TLC, lipolysis and GLC and shown that 16 major triglycerides are present and about 57 per cent glycerides contain two or three linoleic acid groups.

Soyabean oil contains varieties of triterpenes, sterols and tocopherols. Tocopherol content of the oil is high in refined oil (about 600 to 1000 mg/kg oil) and the oil is moderately stable to oxidation. To increase the oxidation stability, butylated hydroxyanisole (BHA) can be used.

Perilla oil: Perilla oil is obtained from the seeds of the plant *Perilla ocymoides*, which grows in the Far East, Indonesia and India. It acts as an important drying oil and is similar to the linseed oil in odour, appearance and other general characteristics except iodine number which is much higher than linseed oil. Its phosphatide and other non-oil contents are low and free fatty acid content is similar to that of linseed oil. It dries much faster than linseed oil. The general characteristics of perilla oil are as follows:

Specific gravity 15/15°C	0.930–0.937
Refractive index, n_D^{40}	1.473–1.479
Iodine number	192–208
Saponification number	188–197
Unsaponifiable matter	Below 1.5 percent
Saturated acids	6–12 percent
Unsaturated acids	83–88 percent

The fatty acid distribution of perilla oil has not been investigated widely and the data obtained by different methods are not reliable. A perilla oil sample investigated by ultraviolet spectroscopic method (52) gave somewhat reliable data about fatty acid composition consisting of 6.6 percent saturated acids, 13.5 percent oleic acid, 14.7 percent linoleic acid and 65.2 percent linolenic acid.

Hempseed oil: Seed of the plant *Cannabis sativa* give hempseed oil which is mainly obtained in the Far East and in Russia. The oil is similar to the linseed oil in physical properties and characteristics though iodine number is lower than that of the linseed oil.

The general characteristics of hempseed oil are as follows:

Specific gravity 25/25°C	0.923–0.925
Refractive index n_D^{40}	1.470–1.473
Iodine number	140–175
Saponification number	190–193
Unsaponifiable matter	about 1 percent
Titer	15–17°C

Fatty acid composition data of this oil is not sufficient. A GLC data gives the composition as (8) palmitic 6 percent, stearic 2 percent, oleic 12 percent, linoleic 55 percent and linolenic 25 percent.

Wheat germ oil: Wheat germ oil obtained from the wheat grain is an important oil for its high content (0.3–0.5%) of tocopherols (Vitamin E) and is used as a nutritious food. The general characteristics of wheat germ oil are as follows:

Specific Gravity 25/25°C	0.925–0.933
Refractive index n_D^{40}	1.469–1.478
Iodine number	115–129
Saponification number	179–190
Unsaponifiables	2–6 percent

On investigation the fatty acid composition of wheat germ oil was found to be 11–16 percent palmitic, 1–6 percent stearic, 0–1 percent C_{20}–C_{22} saturated, 8-30 percent oleic, 44–65 percent linoleic and 4–10 percent linolenic.

Unsaponifiables of wheat germ oil can be separated using thin layer chromatography. Wheat germ oil contains a high amount of tocopherol comprising α-, β-, γ- and δ-tocopherols. Besides these, this oil contains 5,8-dimethyl tocopherol and 7-methyl and 5,7-dimethyl tocotrienol in small quantities. Ferulic acid in the form of ester with dihydro-γ-sitosterol acts as another natural anti-oxidant of wheat germ oil.

Horse Fat: Horse fat is different from the fats of other land animals being rich in linolenic acid and having a high iodine number (about 95). The approximate fatty acid composition as reported is lauric 0.4 percent, myristic 4.5 percent, palmitic 26 percent, stearic 4.7 percent, arachidic 0.2 percent, unsaturated C_{14} and C_{16} 6.8 percent, oleic 33.7 percent, linoleic 5.2 percent, linolenic 16.3 percent, unsaturated C_{20}–C_{22} 2.3 percent. Glyceride composition is obtained as (mole%) – GS_3 2.6 percent GS_2U 22.4 percent, GSU_2 55.9 percent, GU_3 19.1 percent.

Horse fat is used for the preparation of inedible greases.

Other Linolenic Acid Oils

Zero erucic rapeseed oil can be classed as linolenic acid oil. Through genetic development rapeseed oil has become free from erucic acid for the sake of edible purpose and the linolenic acid becomes its major component acid. Genetic changes in the plants *Brassica napus*' (rape) and *Brassica campestris* have resulted in zero erucic rapeseed oil.

Pumpkin and squash type vegetables yield oil from their seeds. *C. andreanna* has 39.4 percent oil in its seeds and the composition of the oil is palmitic 19 percent, trace amount of stearic, oleic 34 percent, linoleic 42 percent, linolenic 15 percent and conjugated trienoic, about 0.35 percent.

Rubber seed oil is obtained from the kernels of the seed of the plant *Hevea brasiliensis*. The oil is edible but has no commercial use. The fatty acid composition of the oil is as follows: palmitic 9–12 percent, stearic 5–12 percent, arachidic acid 1 percent, oleic 17–21 percent, linoleic 35–38 percent, linolenic 21–24 percent.

Australian candlenut or lumbang oil is obtained from the kernels of the plant *Aleurites molucanna*. The fatty acid composition of the oil is palmitic acid 5.5 percent, stearic acid 6.7 percent, oleic acid 10.5 percent, linoleic 48.5 percent, linolenic 28.5 percent. This oil has drying property similar to that of the linseed oil.

Walnut trees give nuts rich in oil. The oil is obtained from the two species *Juglans regia* and *Juglans nigra*. The oils are edible but the oil obtained from the former species is mainly used for making artists' colour. The GLC analysis of these oils as reported is 11 percent palmitic, 5 percent stearic, 28 percent oleic, 51 percent linoleic and 5 percent linolenic.

Stillingia sebifera, called Chinese tallow tree, yields two types of fats: one is from the outer covering of the seed and the other is from the kernel. The former is a hard fat with low iodine number and the latter is a drying oil called stillingia oil of high iodine number and with a large content of linolenic acid and 2,4 decadienoic acid. The fatty acid composition of stillingia oil is as follows: saturated acids 9 percent, oleic acid 20 percent, linoleic acid 25–30 percent, linolenic acid 40 percent and 2,4-decadienoic acid 3–6 percent.

The seeds of citrus fruits contain appreciable quantities of linolenic acid. Grapefruit seed oil and orange seed oil also have significant amounts of linolenic acid in their composition.

Conjugated Acid Oils

Iodine number in the case of conjugated acid oils has relatively little importance than in the case of other unsaturated acid oils. This is due to the facts that conjugated acids do not absorb iodine quantitatively except under special conditions, so iodine number of these oils has almost no significance, and moreover, for these types of oils drying property is of prime importance and the iodine number, although determined accurately, cannot give any measure of the drying property of the oil.

Ultraviolet spectrophotometer and GLC methods are used to determine the amount of conjugation and to separate the acids from their non-conjugated isomers.

Tung oil: Kernels of the tung tree, *Aleurites fordii montana* yield tung oil which is obtained in China, southern United States and also in India and Burma. The oil has pronounced drying and polymerization properties due to the presence of high amounts of the conjugated acid—eleostearic acid. The oil is mainly used in enamels and varnishes. It is also combined to other drying oil to improve water and alkali resistance.

The phosphatide or other mucilagenous materials are present in low amounts in tung oil. The oil may contain some solid or become entirely solid on exposure to light or by trace amounts of sulfur, iodine, selenium etc. This occurs due to isomerization of the liquid α-eleostearine acid to β-form. So this oil must be carefully stored. Solidifcation can be prevented if the oil is heated for 30 mins at about 200°C.

The general characteristics of tung oil are given below:

Saponification number	189–195
Iodine number (wijs)	160–175
Unsaponifiable matter	Below 1 percent

Refractive index at 25°C	1.516–1.520
Titer	36–37°C

The GLC data on fatty acid composition of tung oil as obtained are (53): palmitic 3.1 percent, stearic 2.1 percent, oleic 11.2 percent, linoleic 14.6 percent, 18:3 conjugated acid (eleostearic acid. *cis*-9, *trans*-11, *trans*-13) 69.0 percent

Oiticica oil: Brazilian oiticica tree *Licania rigida* gives oiticica oil from its nuts and Mexican oiticica tree *Licania abrorea* also gives oil having about the same composition as the former. The Mexican oiticica oil is usually called cacahuanache oil.

Oiticica oils are similar to tung oil. They dry little more slowly than tung oil but are used in the same fields as tung oil. Oiticica oils solidify more readily through isomerization than tung oil and the solidification can be prevented by prior heating at or above 225°C.

The main acid component of oiticica oil is a C_{18} conjugated trienoic acid having a carbonyl group on C_4 position, called Licanic acid. The fatty acid composition of the oil obtained from a GLC analysis (8) is as follows: palmitic 7 percent, stearic 5 percent, oleic 6 percent, licanic 78 percent and hydroxy acids 4 percent.

The triglycerides of oiticica oil were separated by column chromatography and the triglyceride composition was obtained as 27.3 percent monoketo acid triglycerides, 32.1 percent diketoacid triglycerides, 12.7 percent higher polar triglycerides and 15.9 percent other triglycerides.

Oiticica oil contains about 0.5 percent unsaponifiable matters and about 0.02 percent triterpenes.

Other Conjugated Acid Oils

Cucurbita species plants (pumpkin, squash etc.) contain oils in their seeds to a moderate degree and among them several oils contain conjugated trienoic acids up to 16–26 percent. Some species such as *C. digitata* having 18.3 percent conjugated acids and *C. cordata* having 9.0 percent conjugated acids are known. These oils act as mild semi-drying oils.

Marine Oils: Marine oils are important from commercial point of view and are obtained from whale and cod, herring, anchovy etc. have significant use. Some liver oils of fishes like cod, halibut, shark etc. are also very important from nutritional point of view as they contain valuable Vitamins A and D.

Marine oils are mainly consumed as edible oils. But these have inedible uses also. Marine oils contain a wide variety of polyunsaturated fatty acids with differences in chain length and unsaturation. The degree of unsaturation cannot be determined easily. There are some methods which are employed to determine the average degree of unsaturation. One of these method expresses the degree of unsaturation in terms of the number of hydrogen atoms to make the acid completely saturated. Marine oils can be easily detected in other oils by alkali isomerization and *UV* spectral examination as they contain fatty acids with more than three double bonds. Now-a-days GLC analysis is a method of choice.

Whale oil: Whale oil is the most important marine oil obtained from sea mammals but its production is decreasing. The main uses of whale oil are in the manufacture of soap, margarines and shortenings. There are various grades of whale oil having different free fatty acid and volatile matter content and they may also differ in colour. The general characteristics of whale oil are indicated as follows:

Specific gravity at 25/25°C	0.910–0.920
Refractive index at 25°C	1.470–1.477
Iodine number (Wijs)	110–135
Saponification number	185–202
Unsaponifiable matter	Not over 2.0 percent

There are different species of whales from which oils are obtained. Among them sperm whale oil is important. Its characteristics and composition are stated as: sp.gr. at 15°C 0.844–0.88 percent, iodine number 70.4–96.4, saponification number 120–150.3, unsaponifiable matter 17.5–44.0, unsaturated fatty acid 81–90 percent and saturated fatty acid 10–19 percent.

Hilditch analysed the triglyceride composition of Antarctic whale oil and reported that the oil contained 66 percent glycerides with one unsaturated C^{18} and one saturated and one unsaturated C_{14}, C_{16}, C_{20} or C_{22} acid group, 12 percent of glycerides with one unsaturated C_{18} and two unsaturated C_{14}, C_{16}, C_{20} or C_{22} acid groups, 8 percent of glycerides with one unsaturated C_{18}, one myristic and one palmitic acid, 6 percent of glycerides with two unsaturated C_{18} acid groups and one saturated acid group and 4 percent of glycerides with three unsaturated C_{14}, C_{16}, C_{20} or C_{22} acid group (54).

South Atlantic (California) Sardine or Pilchard oil: This oil is derived from the fish *Sardinops caerulea* and is known as sardine oil in United States and pilchard oil in Canada. The sardine oil is rendered from the whole fish and so it contains some protein decomposition products and non-oil materials. It is inferior in quality than whale oil and is higher in free fatty acid content. Sardine oil after hydrogenation is used for the manufacture of soaps, lubricating greases, sulfonated oils and fatty acids. The oil itself is used in paints, varnishes printing inks and other drying oil applications. The characteristics of sardine oil are stated below:

Iodine number	170–188
Saponification number	188–199
Refractive index at 25°C	1.4785–1.4802
Unsaponifiable matter	0.1–1.25 percent
Free fatty acid	0.1–13.0 percent
Specific gravity at 25/25°C	0.914–0.921

The fatty acid composition of sardine oil is complex and it was reported that the oil contains about 22.7 percent saturated fatty acids and the rest are unsaturated acids comprising C_{14}, C_{16}, C_{18}, C_{20}, C_{22} and C_{24} fatty acids.

Japanese sardine oil: Japanese sardine oil is obtained from the sardine *Clupanodon melanostica*. It is almost similar to the American sardine oil except in the content of C_{24} acids. Japanese sardine oil contains lesser amount of C_{24} acids. By modern techniques it was seen that the previously supposed C_{24} acids are virtually polyunsaturated C_{22} acids. So, now the earlier data of the percentage C_{24} acid present in oil cannot be taken as true. Japanese sardine oil also contains about 22 percent saturated acids and the rest are unsaturated acids of C_{16}, C_{18}, C_{20}, C_{22} chain length and up to about 6 double bonds.

Menhaden oil: Menhaden oil is obtained from the fish menhaden (*Brevoortia tyrannus*). Its iodine number varies from about 150 to 185. Its other characteristics are – saponification value 192–199, slip point 22–28°C, refractive index n_D^{60} 1.4590–1.4623. This oil also has a series of saturated and unsaturated acids of C_{14} and C_{22} chain length in its composition. Menhaden oil is used for inedible purposes like in paints and varnishes, fatty acids and in printing inks. Edible menhaden oil is also obtained which is used for the manufacture of margarine or shortenings.

Herring oil: Herring fishes (*Clupea harengus*) found in the Atlantic Ocean are used to prepare oil when the fishes are not used as food. The general characteristics of herring oil are given below:

Iodine number	115–160
Saponification number	180–192
Refractive index at 25°C	1.473–1.478
Unsaponifiable matter	0.5–1.7 percent
Free fatty acids	0.2–5.0 percent
Titer	25°C

This oil contains a high amount of unsaturated C_{22} acids and low amount of unsaturated C_{18} acids along with low degree of unsaturation in C_{22} acids and high degree of unsaturation in C_{16} and C_{18} acids. The glyceride composition of herring oil was reported to be GS_2U 4 percent, GSU_2 62 percent, GU_3 35 percent.

Herring oil contains a high amount of hydrocarbons which can be separated by using GLC. Aliphatic hydrocarbons are predominant in the oil.

Anchovy oil: Anchovy oil is obtained from the fish anchovy (*Clupeidae engraulidae*) by first cooking anchovy at 98–100°C for 15 minutes and then by pressing and centrifugation. This oil is mainly produced in Chile, Peru and South Africa. The average characteristics and properties of the oil are as follows: Free fatty acid 2.2 percent iodine number 184, unsaponifiable matter 1.0 percent, impurities 0.04 percent. This oil contains saturated acids (C_{12} to C_{18}) and unsaturated acid (C_{18} to C_{22}) in about equal amounts.

Fish liver oils: Among the fish liver oils, cod liver oil is the most important. This oil is obtained from the cod fish (*Gadus morrhua*). High grade cod liver oils are used for medicinal purpose. Low grade cod oil is used in industries mainly in the manufacture of leather. In Japan, vitamins are extracted from cod liver oils by autolysis or alkali digestion of the livers. After extraction of vitamins the residual oils are used for making margarines and shortenings. The general characteristics of cod liver oil are:

Iodine number	118–190
Saponification number	182–191
Refractive index at 25°C	1.477–1.481
Unsaponifiable matter	0.9–1.4 percent
Vitamin A	550–10,000 U.S.P. Units/gm.
Vitamin D	20–300 I.U./gm.

The fatty acid composition of a sample of cod liver oil having iodine number 172 was reported and it was seen that this oil contains about 19 percent of saturated fatty acids, and among the unsaturated acids it contains monoene 45 percent, diene 4.5 percent, triene 1.8 percent, tetraene 2.6 percent, pentaene 11.8 percent and hexaene 15.0 percent.

Besides cod liver oil there are other liver oils; among them dogfish liver oil and shark liver oil are well known. Dogfish liver oil contains a high amount (7.62%) unsaponifiables. Shark liver oil also contains Vitamin A and D.

Hydroxy Acid Oils

Castor oil: Castor oil is the only oil which has a significant amount of hydroxy acid and is of commercial importance. This oil is obtained from the seeds of the castor tree *Ricinus communis* which is grown mainly in Brazil, India and Russia. It is not used as an edible oil except for medicinal purpose. The general characteristics of castor oil are given below:

Specific gravity at 25/25°C	0.945–0.965
Refractive index at 25°C	1.473–1.477

Iodine number, Wijs	81–91
Saponification number	176–187
Unsaponifiable matter	Not over 1.0 percent
Acetyl value	144–150

The major component of castor oil is hydroxy acid, ricinoleic acid. It amounts upto 85–95 percent of the fatty acids in the oil. Other fatty acids present are linoleic 4.5–5.0 percent, oleic in trace amount, saturated acid 1 percent and also an optically active isomer of 9,10-dihydroxystearic acid in trace amount. GLC data of the methyl esters of fatty acids of castor oil obtained from various investigations (8,55,56) gave similar data with some variations.

Castor oil is more viscous and denser than other vegetable oils. Its hydroxy or acetyl value is quite high and unlike other oils it is soluble in alcohol but insoluble in petroleum solvents. Hydrogenated castor oil has a high melting point (86–88°C).

Unsaponifiable matter of castor oil contains triterpenes and different sterols like β-sitosterol, stigmasterol and campesterol. It also contains hydrocarbons, aliphatic alcohols and tocopherols.

Castor oil is mainly used in the production of dehydrated castor oil (DCO), which is utilized in the manufacture of paints and varnishes. Castor oil is also sulfonated to produce sulphonated castor oil. Sebacic acid is obtained by alkali fusion of castor oil which is used in the production of plasticizers and polymers. Pyrolysis of castor oil yields undecylenic acid and heptaldehyde.

Miscellaneous

Oils obtained from plants of *Thalictrum genus* contain acids with double bonds in *trans*-5-position. These oils are mainly used in Bulgaria especially as semi-drying oil.

Other oil which do not make a broad class are the oils from the species *Vernonia anthelmintica* (Indian iron weed) containing high amount of *cis*-12,13-epoxyoleic acid (Vernolic acid). Two other oils include the oil from the species *Onguekoa Gore* called isano or boleko oil being rich in C_{18} acid with two acetylenic and one ethylenic bonds and the sterculia oil obtained from the species *Sterculia foetida* or Java olive having major component sterculic acid, a C_{19} cyclopropenyl fatty acid etc.

REFERENCES

1. Bailey's Industrial Oil and Fat Products, Volume 1, Fourth edition, Edited by Daniel Swern, Published by John Wiley and Sons, pg 183, 1985.
2. R.K. Downey, *J. Am. Oil Chemists' Soc.*, **41**, 475–478 (1964).
3. R.M. Howell, C.A. Brim, and R.W. Rinne, *J. Am. Oil Chemists' Soc.*, **49**, 30–32 (1972).
4. S.W. Christopherson and R.L. Glass, *J. Dairy Sci.*, **52**, 1289 (1969).
5. G.F. Spencer, S.F. Herb and P.J. Gorminsky, *J. Am. Oil Chemists' Soc.*, **53**, 94–96 (1976).
6. R.G. Jensen, *J. Am. Oil Chemists' Soc.*, **50**, 186–192 (1973).
7. S. Patton, *J. Am. Oil Chemists' Soc.*, **50**, 178–185 (1973).
8. "Composition and Constants of Natural Fats and Oils by Gas-liquid Chromatography", Chart published by Archer-Daniels-Midland, Minneapolis, MN, 1961.
9. A. Rauramma, M. Kreula, E. Hukari and H. Personen, *Meijeritieteclinen Aikauskirja*, **34**, 143–151 (1976).
10. J.L. Henderson and E.L. Jack, *Oil Soap*, **21**, 90–92 (1944), E.L. Jack and J.L. Henderson, *J. Dairy Sci.*, **28**, 65–78 (1945), E.L. Jack, J.L. Henderson and E.B. Hinshaw, *J. Biol. Chem.*, **162**, 119–128 (1946).
11. C.A. Brown and T.S. Sutton, *J. Dairy Sci.*, **14**, 125–135 (1931).
12. A.W. Bosworth and G.E. Helz, *J. Biol. Chem.*, **112**, 489–492 (1936).
13. M.L. Blank and O.S. Prirnett. *J. Dairy Sci.*, **47**, 481 (1964).
14. W.C. Brekenridge and A. Kuksis, *J. Lipid Res.*, **3**, 388 (1968).
15. A.E. Thomas III, J.E. Scharoun, and H. Ralston, *J. Am. Oil Chemists' Soc.*, **42**, 789–792 (1965).
16. H. Brockerhoff, *J. Lipid Res.*, **6**, 10 (1965).
17. A. Kuksis, L. Marai and J.J. Myher, *J. Am. Oil Chemists' Soc.*, **50**, 193–201 (1973).
18. M. Keeney and F.J. Doan, *J. Dairy Sci.*, **34**, 728 (1951).
19. R. Ristow and H. Warner, *Fette Seifen Anstrichem*, **70**, 273 (1968).
20. Although international units and U.S.P. units are ordinarily taken as equivalent this is not strictly justified.
21. C.W. Fritsch, V.E. Weiss and R.H. Anderson, *J. Am. Oil Chemists' Soc.*, **48**, 392–393 (1971).
22. D. Chapman, A. Crossley and C.A. Daries, *J. Chem. Soc.*, 1957, 1502–1509, E.S. Lutton, *J. Am. Oil Chemists' Soc.*, **34**, 521–522

(1957).

23. Chobanov, R. Tarandjiska and R. Chobanova, *J. Am. Oil Chemists' Soc.*, **53**, 48–51 (1976).
24. C. Litchfield, R.D. Harlow and R. Reiser, *J. Am. Oil Chemists' Soc.* **42**, 849–857 (1965).
25. H.R. Jensen, The Chemistry, Flavoring and Manufacture of Chocolate Confectionery and Cocoa, Blakiston, Philadelphia, 1932.
26. G.K. Chacko and E.G. Perkins, *J. Am. Oil Chemists' Soc.*, 42, 1121–1124 (1965).
27. P.B.D. de la Meara and F.B. Shortland, *Analyst*, **69**, 337–339 (1944).
28. F.E. Luddy, R.A. Banford, S.F. Herb and P. Magidman, *J. Am. Oil Chemists' Soc.*, **45**, 549–552 (1968).
29. A. Bomer, R. Limprich, R. Kronig and J. Kuhlmann, Z. Untersuch Nahr-Genussm, **26**, 559–618 (1913), V.C. Mehlenbacher, editor Official and Tentative Methods of the American Oil Chemists' Society, Chicago, 1957, Official Method Cb 5-40.
30. D. Swern, H.B. Knight and C.R. Eddy, *J. Am. Oil Chemists' Soc.*, **29**, 44-46 (1952).
31. J.R. Nunn, *J. Chem. Soc.*, 1952, 313.
32. E.L. Schneider, S.P. Loke and D.T. Hopkins, *J. Am. Oil Chemists' Soc.*, **45**, 585–590 (1968).
33. F.D. Gunstone and M. Ilyas Qureshi, *J. Am. Oil Chemists' Soc.*, **42**, 961–965 (1965).
34. T.P. Hilditch and J.P. Riley, *J. Soc. Chem. Ind.*, 64, 204–207 (1945), T.P. Hilditch, M.B. Ichaporia and H. Jasperson, *J. Soc. Chem. Ind.*, **57**, 363–368T (1938) B.G. Gunde.
35. G. Jurriens and A.C.J. Kroesen, *J. Am. Oil Chemists' Soc.*, **42**, 9–14 (1965).
36. J. Fitelson, J. Assoc. Offio. *Agric. Chem.*, **26**, 499–506 (1943).
37. K.L. Mikolajczak, R.M. Freidinger, C.R. Smith, Jr. and I.A. Wolff, Lipids, **3**, 489–494 (1968).
38. For a review with numerous literature references, see P. Budowski and K.S. Markley, *Chem. Rev.*, **48**, 125–151 (1951).
39. P. Budowski, *J. Am. Oil Chemists' Soc.*, **41**, 280–285 (1964).
40. M.M. Chakrabarty and T.P. Hilditch, *J. Soc. Food Agric.*, **2**, 255 (1951).
41. T. Tamura, N. Sakaedani and T. Matsumoto, Nippon Kagaku Zasshi, **79**, 10111 (1958).
42. T.H. Applewhite, *J. Am. Oil Chemists' Soc.*, **43**, 406–408 (1966).

43. A. Sengupta and U.K. Mazumdar, *J. Am. Oil Chemists' Soc.*, **53**, 680–683 (1976).

44. J. Fitelson, J. Assoc. Offic. *Agric. Chem.*, **19**, 493–497 (1936).

44A. S.C. Charkrabarty and M.M. Chakrabarty–*Ind. Soap* **20**, 16, (1954), SC & CULT (India) **20**, 186, (1954).

45. M.N. Baliga and T.P. Hilditch, *J. Soc. Chem. Ind.*, **67**, 258–262 (1948).

46. U. Runer, *J. Am. Oil Chemists' Soc.*, **47**, 129–133 (1970).

47. S.L. Kapur and B.F. Daubert, *J. Am. Oil Chemists' Soc.*, **26**, 472–475 (1949).

48. H.J. Dutton and J.A. Cannon, *J. Am. Oil Chemists' Soc.*, **33**, 46–49 (1956).

49. See *Oil Soap*, **17**, 151 (1940).

50. F.I. Collins and R.W. Howell, *J. Am. Oil Chemists' Soc.*, **34**, 491–493 (1957), F.I. Collins and V.E. Sedgwick, *J. Am. Oil Chemists' Soc.*, **36**, 641–644 (1959).

51. F.D. Gunstone and F.B. Padley, *J. Am. Oil Chemists' Soc.*, **42**, 957–961 (1965).

52. B.A. Brice, M.L. Swain, B.B. Schaeffer and W.C. Ault. *Oil Soap*, **22**, 219–224 (1945).

53. F.D. Gunstone, R.J. Hamilton, F.B. Padley and M. Ilyas Qureshi, *J. Am. Oil Chemists' Soc.*, **42**, 965–970 (1965).

54. T.P. Hilditch and L. Maddison, *J. Soc. Chem. Ind.*, **61**, 169–173 (1942).

55. R. Sreenivasan, *J. Am. Oil Chemists' Soc.*, **46**, 119–120 (1969).

56. R.G. Binder, T.H. Applewhite, G.O. Kohler and L.A. Goldblatt, *J. Am. Oil Chemists' Soc.*, **39**, 513–519 (1962).

CHAPTER 8

Handling and Storage of Oils and Fat Products

INTRODUCTION

Crude oils and fats may contain substantial quantities of non-glyceride materials as a result of their separation from the parent source (1). These materials include phosphatides, gums, carbonaceous materials, moisture, seed fragments, meal etc. These components must be removed prior to bulk storage to avoid hydrolysis of triglyceride by lipases in the presence of moisture. The handling and storage of oil seeds and finished product affects in a large measure the quality of final product and this has to be a very important concern for the processor.

OIL SEEDS

Oil bearing materials may be received in bulk through trucks or in bags at the processing site. The free flowing oil seeds like soyabean, sunflower and linseed could be handled easily, more or less in the same manner as foodgrains like wheat. However, bulkier materials like cottonseed, copra (dry coconut kernel) etc. should be handled in a different manner. The truck loads may be dumped at a platform and fed to conveyers for storage and processing. Depending upon the scale of operations seed materials are also received through railway wagons or trains or river borne shipments. For the various types of oil seeds the usual conveyor system such as *screw conveyor, pump conveyor* or *chain conveyor* and *bucket elevators* are reasonably satisfactory. The bags may be stored in airy godowns before use. Loose seeds may be stored in concrete silos or steel storage tanks with provision for control of moisture, temperature and aeration facilities to avoid auto-ignition through hot spots. In some cases an initial drying is required which may be done by sun drying or usually available grain dryers. Later, cleaning of these seeds is conducted by the use of vibrating screens or shaker type sieves depending on their nature combined with aspiration to remove dust etc. to prevent *air pollution.* Air from the aspiration system should be discharged into collectors and cyclones. As the availability of oil seeds is seasonal it is usual to provide sufficient amount of storage capacity at the farms besides substantial amount of storage at the unloading point near the factory site. Many oil seeds such as cottonseed and soyabean or materials such as rice bran are prone to deterioration in storage (2) due to enzyme activity accompanied by digestive breakdown of oil carbohydrates and protein in the kernel. Free fatty acid level of the seed is a good indicator, sometimes as in cottonseed, gossypol and similar pigments present gives rise to reactions which serve to set the colour of the oil produced. If at storage point seeds have a high FFA then neutral oil obtained from these seeds will be accordingly reduced. Again, moisture present, as in the case of soyabeans, as well as temperature prevailing in the storage unit affect the quality of the oil. Therefore, dryers are employed to bring down the moisture to the minimum acceptable level. Similarly, seeds are cooled by using fans so that the air is circulated. Usually arrangements to measure temperatures such as thermometer or thermocouple are kept at various levels at the storage bins to monitor the rise of temperature. Again, oil seeds from dryers have to be cooled before storage for preventing heat damage.

The safe moisture constant of different oil seeds for storage varies according to their composition, such as presence of saturated or unsaturated acids and also fat content. In the case of kernels like palm kernels, copra etc. this is about 6–8 percent, preferably 6 percent. In the case of most oil seeds the limit is between 9 percent and 12 percent, preferably 9 percent. Storage is ordinarily done in bins of silo type which are built in reinforced concrete brick tile or steel and suitable arrangements are made to have conical bottoms so that they can be emptied mechanically through conveyors or pneumatically which is cheaper and more convenient. In general, seeds should always be stored under low moisture and temperature conditions in order to minimize fermentation and prevent the formation of water and carbondioxide which develop as a result of the presence of sugar and starch in the seed. Crude storage and handling result in bad oil quality. Depending upon the material being processed through expellers or through solvent extraction some preparatory pre-treatment is called for. This may involve one or more operations, (Fig. 8.1) such as *cracking, decorticating, dehulling, delinting, flaking* etc. before actual extraction of the oil from the source material. Other preliminary treatments such as *caking, conditioning, crisping, pelleting* etc. are carried on. These are necessary to prepare optimum extractibility, filterability, proper size distribution and minimum fines content.

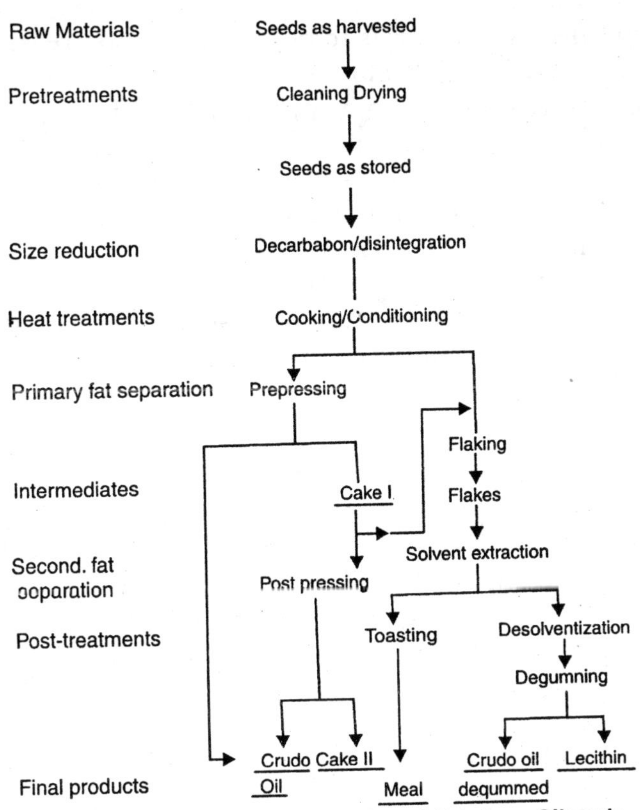

Fig. 8.1 Processes for the Separation of Fats from Oilseeds

OILS AND FATS

Fats and oils are generally received in tankers, lorries or drums. Oils and fats may be in liquid or semi-liquid state. The compounds usually are heated through the medium of gas, steam or coal (in the case of semi-solid or solid fats) and the material obtained is sampled. As the oily materials normally contain suspended impurities, trace materials, moisture etc. an attempt should be made to reduce the onsetting of rancidity which is hastened by contact with atmospheric oxygen and water to the formation of free fatty acid. Unsaturated oils are readily susceptible as they react more rapidly with water or oxygen. It is necessary to ensure that fat and oil is as free as possible from water. The sampling procedure is prescribed by the Bureau of Indian Standards or similar bodies. A representative sample should be taken as this may give a clue to future recovery of pure oil minus the moisture and *foots*.

Again, the processor has to be careful to ensure that metals like copper and bronze which may act as catalysts to promote the rancidity are avoided in fillings or pump parts which come in contact with oil or fat. The usual practice is to store oils and fats in huge covered steel tanks located in the open. Each storage tank should be provided with suitable drainage system capable of containing any possible spillage. It is necessary to weigh the oil coming to the storage facility and that which is withdrawn from storage to find out the loss. It is necessary also to store only such quantity of oils and fats which can lead to unhampered production. A long storage in tank is to be avoided as free fatty acid of the oil increases with the length of storage tank.

Storage of the finished oil should be normally done in the range of 60–85°C in order to reduce *auto-oxidation*. Preferably, the oil should be cooled and if possible it should be blanketed with nitrogen to avoid oxidation. The tanks should be provided with heating and cooling coils as also agitation arrangements. Sunlight being a promoter of *rancidity*, exposure of the refined oils to sunlight as also contact with air should be avoided. In the case of oil or fat which has to be heated for unloading, the use of high pressure steam or the heating coil should be avoided as this may result in colour fixation. It is also necessary to restrict the storage of oils at a high temperature. Contamination or unwanted mixing with other oils should be guarded against. It is preferable to have a few big tanks instead of numerous small tanks.

The pumping and unloading systems usually employ centrifugal or positive displacement pumps, the latter being preferred. The oil transfer lines should be proper with good insulation. Although carbon steel has been used for the construction of pumps, a relief valve must be installed to safeguard against blockage or operating against closed valves in order to prevent damage to the pump and to the operating personnel like blow off. Arrangements should also be made to interpose a strainer ahead of the pump in order to prevent damage to the pump from metal pieces and in the discharge line a filter should also be installed. Usually a drain valve is provided at the lowest point in the transfer line to take care against accumulation of fat and blockage of pipes. Processing of crude oils and fats comprises of various steps which include storage, degumming, refining, bleaching and hydrogenation *fractionation* or *winterizing*. Each step constitutes an independent operation and removes specific components from the oils. Care has to be taken to ensure efficient removal of naturally occurring contaminations and to minimize the formation of undesirable new compounds and precipitance.

Deodorization is usually the last stage through which oils and fats are processed for human consumption. For example, refined and hydrogenated oils, those for shortening and margarine and confectionery oils are always deodorized. The properties of the processed fats undergo considerable change during deodorization. If all the operations stated earlier have been conducted properly, deodorization should yield a light coloured product free from bitter taste, hydroperoxides and contaminants. The primary purpose of deodorization operation is to distill off residual free fatty acids and other objectionable materials contributing to the odour and flavour of the finished product. Usually quality criteria for good oils calls for absence of water, foreign odour, products due to deterioration, hydrolysis, thermal decomposition, and packaging material. As a rule, finished oil should be used or packed as soon as possible after deodorization. However, it is an important target for finished oils, prior to packaging, to avoid oxidation which occurs under the following condition: (1) Increase of temperature and oxygen pressures; (2) increased concentration of peroxide, aldehydes, *lipo oxidants* and *hematin compounds*; (3) heavy metal which catalyzes oxidation; and (4) increase in the intensity of any ionizing radiation such as ultraviolet and visible light. For these, handling temperature should be as low as possible. As pumping range depend upon the viscosity of the fat, close control of temperatures with the handling system is necessary as viscosity depends on temperature. The polished and filtered oil is cooled to 50–65°C for storage in order to reduce auto-oxidation. To avoid contact with oxygen and minimize atmospheric contamination the entire handling operation may be conducted under vacuum or under the blanket of a protective gas, normally nitrogen, under a pressure of 1–2 kg per sq.cm. The oil leaving the deodorizer is delivered to the cooler or tank under complete nitrogen blanket. As a rule, atmospheric adulteration is avoided by employment of completely closed storage tanks with provision for nitrogen blanketing. Sparging and blanketing with nitrogen provides a practical method for protecting finished edible oils from oxidation.

DAMAGE DURING TRANSPORT

Copper is an oxidation catalyst and generally increases the rate of oxidation. Copper metallic subjects are strong pro-oxidants. Use of copper therefore should be completely avoided as the material of contraction in oil processing. Iron is also not preferred. The best material to use would be stainless steel. However, it is expensive.

Anti-oxidants, *synergists* and metal reactivator are used to improve the quality of deodorized oil. Citric acid is the most common reactivator employed for the purpose in quantities varying from 0.005 to 0.01 percent. Citric acid is generally added to the deodorizer in the form of alcoholic or aqueous solution in the cooling stage. Citric acid improves filter stability and functions as a *metal's scavenger* or *chelator* by inactivating traces of metal contact of refined oil. Sunlight should be avoided as far as possible (3).

Freshly refined and deodorized oil contains no living organisms. Externally uncleaned transport devices may lead to introduction of such organisms which may produce lipase enzymes causing hydrolysis. Uncleaned tanks are generally the source of dirt and foreign particles. As fats and oils easily absorb taste and odour resulting from solvents, care should be taken when airing newly painted tanks.

FATTY ACIDS

Handling Requirements

Fatty acids in general have some tendency to combine with oxygen to form impurities having colour and odour. This tendency is lower for long chain fatty acids, excessive at the higher temperature, but increases in the more unsaturated short-chain acids. Traces of metals like iron and copper which the seeds may take up from storage tanks or containers catalyze this reaction. The possibility of contamination of fatty acids below temperatures of 100°C acts as a limiting factor for the choice of materials.

The methods of handling have to be adapted to a particular seed as fatty acids vary widely in their characteristics. Titer, acid value, iodine value, colour and odour, are quality factors which greatly influence the handling requirements (4). Although most higher fatty acids are not corrosive to carbon steel, they are subject to oxidation degradation in the presence of iron and copper that may be present if carbon steel is used. Being mild acids, the fatty acids react chemically with certain metals like iron and form *metallic soaps*. These acids again act as catalytic agents giving rise to colour deterioration and accelerating oxidation, particularly in the case of unsaturated fatty acids. For most applications, therefore, type 304 or 316 stainless steel equipment and pipe are used in producing and handling fatty acids. Transfer lines can be of 304 stainless steel, but most process lines and equipments are made of 316 stainless steel. Fatty acids can be stored in aluminium, stainless steel, resin-lined steel, reinforced plastic or glass lined tanks. Type 304 or 316 is very versatile, but if moisture and air are not circulated, aluminium is also quite satisfactory. To prevent freezing of the fatty acid, storage tank may be heated but care should be taken not to over-heat and the tank temperature should be limited to 50°C for unsaturated fatty acids and 65°C for saturated fatty acids.

Packing and Transportation

The most important criterion for selection of proper packing for transport of fatty acids is their physical form. Oleic acid (an unsaturated long-chained acid) and short-chained saturated acid such as caprylic acid comprise medium and titer groups that can be packed and transported in metal drums and tankers depending on their bulk and quantity. Fatty acids with titer below 20°C are liquid and those which are having titer between 20 and 40°C are semi-solid.

Maximum fatty acids are normally liquid and semi-liquid which are generally packed in mild steel drums with polythene liners. They are usually of 160 kg capacity to 180 kg capacity. The drums should be stored vertically. Distilled fatty acids should be preferably taken in corrosion resistant drums as they are likely to be broken with iron contamination on melting. Some of the solid fatty acids are available with free flowing flakes which can be easily handled. The containers are either constructed with aluminium, stainless steel 304 or they are resin or glass-lined. The tank tracks may be insulated to minimize the heat loss. Sometimes hard fatty acids such as stearic acid and other high titer acids which are solid at 50°C and above, are produced and sold in flake bead, powder or cake form. Flaked material can be supplied in multi-wall papers bags or polyvinyl or jute bags, whereas fibre drums, paper bags, polyvinyl or polyethylene liners have proved satisfactory for a number of materials. Commercial stearic acid (55/45 palmitic to stearic acid) has a crystalline structure in solid form. These cakes or slabs are most commonly taken in paper board cartons, wooden containers or boxes.

Low pressure steam (10–15 psig) should be used for melting the fatty acids at the point of unloading to minimize breaking of the material. This may require many hours of steaming. The acid is heated to about 20°C above the melting point. When it is completely melted it should be allowed for pumping to prevent solidification and blocking of existing line. It is usual to feed all pumping lines with copper steam tubing. The pumping lines are later cleared of the laden liquid by clearing out with steam or inert gases like nitrogen or carbondioxide. Air blowing is avoided to eliminate remelting of the solid or semi-solid fatty acids with steam at the point of destination. These may be transported in tankers provided with stainless steel or aluminium jacketted insulated seals. This helps to avoid any metallic contamination and preservation of colour. The tank and lorries may also be supplied with an ejector or pump to empty the bulk contents at the point of destination. Sampling from tank lorries has been properly codified in the Bureau of Indian Standards. Similar quantities of solid or semi-solid fatty acids are generally transported in lined or unlined drums. Where preservation of colour is important, phenolic or epoxy resin-lined drums are used. A better medium would be stainless steel or aluminium drums, but they are costly.

Sampling and Storage

The sampling procedure ensure that the sample is representative of the material under reference. The procedure of sampling should be conducted as mutually agreed upon between the buyer and the supplier. Samples when taken should be placed in dry solid containers to avoid contact with air, and may be properly levelled. In the case of liquid and semi-solid materials, it should be liquefied by heating and thoroughly mixed before drawing samples. In the case of solid, flakes or powdered fatty acid not less than 10 percent of the packages be sampled at random and it should be thoroughly mixed to get the average sample which may be packed in cloth bags or card board containers.

Storage

Tanks for bulk storage of fatty acids may be designed horizontal or vertical, round or rectangular in keeping with the floor space available. Cylindrical tanks are preferable, because they permit the use of thinner shell sections. Stainless steel or aluminium are the preferred construction materials. Synthetic resin-lined tanks are satisfactory at normal temperatures, but alkalies cannot be used in cleaning such tanks. Glass-lined tanks although resistant to attack by fatty acids present maintenance problems due to brittleness of the glass lining besides being costly. Copper and ordinary steel tanks as also wooden tanks should be avoided as far as possible. For the bulk transfer and storage of fatty acids, aluminium tankers and tanks are satisfactory, provided, no moisture or mineral acid is present. However, they cannot be cleaned with alkalies although they are more economical as compared to stainless steel. Resins linings have also been used with success.

Heating coils for storage tanks are used successfully where very low melting fatty acids are to be stored. These coils should be constructed in stainless steel 316 in order to obtain best results against deterioration of the quality. It is preferably to maintain the fatty acids at the lowest temperature at which they are completely liquid which also permits for their safe handling. The stored material, therefore should be maintained in the liquid phase rather than remelting periodically for use. Saturated fatty acids can be handled at the maximum temperature of 65°C and unsaturated fatty acids at a maximum temperature of 50°C. Fatty acids are poor conductors of heat and tend to discolour rapidly when subjected to local overheating. In the case of low titer acids there is a tendency to 'seed out' in cold weather. They are steamed or heated by means of closed steam heating coils. The correct design of heating coils is important and in the fatty acid tank, internal heating coil should be avoided. It is desirable to insulate heated tanks. Stearic acid bags should always be stored in a ventilated room away from hot pipes or surfaces. Only low pressure steam 8–10 psig should be used for melting of fatty acid in order to minimize darkening of the material. Bags should not be stored at a temperature which exceeds the melting point of the acid or near hot pipes or radiators etc. The bags should always be stored in cool dry place and care should be taken to prevent tearing of bags. Drums containing fatty acids should preferably be stored at a reasonably constant temperature, preferably at not much above the room temperature. Most fatty acids being mixtures of two or more fatty acids, there is a tendency for selective phase separation of higher melting from lower melting fatty acids. To guard against this, the contents of partially used drums should be completely liquefied and thoroughly mixed prior to usage to ensure complete uniformity.

GLYCERINE

Handling of glycerine does not pose much of a problem, except in very viscous state. The temperature range 38–50°C is optimum as regards its viscosity for the purpose of its pumping and handling. Handling above 50°C should be avoided as the colour of glycerine is deteriorated. Glycerine holding tanks are usually provided with steam heating coils and only low pressure steam (about 10 psig) should be employed for this purpose. Hot water may also be used. Piping is built from galvanized iron, aluminium or stainless steel. Material for construction of pumps and furnaces should be preferably cast iron, bronze or stainless steel. Spent lye and sweet water are normally stored in mild steel tanks and concentrated glycerine is stored in tanks made of 304 or 316 stainless steel. Resin linings have also been employed. These storage tanks are sealed from atmosphere to prevent absorption of moisture and the resultant corrosive effect. For this purpose the storage tanks are usually fitted with an air breather trap or pipe provided with a desiccant system.

SOAP

In many operations which use high temperature and pressure, care should be taken to avoid introduction of contaminants. Raw materials for soap are mainly oils and fats, caustic soda, salt and water. The handling and storage requirements of oils and fats have already been discussed. However, care has to be taken to prevent oily material congealing on the walls of pipes, tanks, pumps and other equipments where there are points of restrictions as valves, bends or dead legs. Adequate drain and other provisions are necessary to prevent temporary overloading. Attempts should be made to ensure that pipe lines and plant are cleaned and comparatively dried since fats and oils are susceptible to oxidation and rancidity. It is necessary to take care to store fat in covered tanks and minimize the period of storage which would reduce this tendency. For the production of common quality toilet soap, high FFA fats are preferably good. A preliminary alkali refining treatment is followed by bleaching through absorbents in order to minimize the deleterious effect of high free fatty acids on the quality of toilet soap. The main ingredients used are alkali, caustic soda and soda ash. While handling alkalies, the introduction of oils and grease from over-head stirrer driving gears and feeding arrangements, which allow picking up dirt and atmospheric dust, should be avoided. Caustic soda is generally received in solid form in drums or in the concentrated liquid state in tankers. While unloading liquid caustic soda through pumps safety precautions as protective devices should be used. The handling equipments and instruments are washed immediately after use. All caustic lines are identifiable, having been usually painted in grey colour. Liquid caustic is diluted to the desired strength whenever necessary in an intermediate tank before actual use.

Common salt: The salt is used for graining of soap in the kettle boiling operation and may be received in bags or the tank from where the salt is carried manually or by screw conveyor when it is dissolved in water to form the concentrated brine solution for use.

Water: Water is used in many stages of soap-making operation and may contain the impurities in low concentration which may not be troublesome. However, in high pressure hydrolysis these impurities may act as catalytic poisons. Therefore, water which would be added should be boiled feed water.

Other additives to the soap include resins, colouring materials, perfumes, bleaching agents and preservatives etc. Appropriate quality control should be resorted to. Numerous raw materials are used in production of synthetic detergents. Handling and storage of important ones are as follows:

Oleum

Acid *oleum* used in the *sulphonation process* is received in tank, lorries etc. The contents of these tanks are unloaded into mild steel storage tanks placed below ground level but located near the sulphonation plant. The oleum is transported through the centrifugal pump to the first stage of sulphonation process by the suction of a proportioning pump. The temperature of feed is maintained at 27°C through a heat exchanger for setting a constant rate and also adjusting the flow rate. Instrumentation is provided to cover the direct pressure control valve, the flow totalizer and a flow indicator. Safety precautions are very important while handling oleum. All personnel handling oleum must wear protective clothing, safety goggles, gloves etc.

There are prescribed procedures for handling of materials used for different detergent formulations. In general, it may be said that the cost benefits to a large extent depend on efficiency, handling and storage of raw materials and finished products in oil and fat industry as in any other industry.

PACKAGING OF EDIBLE OILS (5)

Current packaging systems: In India a large part of edible oils is distributed and sold in loose form amounting to about 64 percent (2). Of the remaining 36 percent, about 17 percent is packed in 15 kg tin plate containers whereas 19 percent is sold in consumer packages of different capacities, 5 kg, 2 kg, 500 gms, etc. Consumer packages used are tin plate containers (11–12 percent), plastic containers (3.5

percent), paper packed bag in boxes (about 3 percent) and flexible pouches (about 0.7 percent). In recent years the preparation of consumer packs in flexible pouches have a tendency to increase. In the case of imported oil which is distributed through public distribution system, about 2/3 is packed and distributed to the districts in 15 kg tin containers from where it is sold loose to the consumers. The remaining oil is sold in consumer packs of 2 kg tin plate containers (about 0.9%), 2 kg PVC containers (about 0.1% and 1 kg flexible pouches (about 1.2%).

The stability of edible oil may be dependent on the packaging condition. The various factors are light, atmospheric gases like oxygen, temperature and moisture. Of these, oxygen is the most critical factor affecting the stability. Atmospheric oxygen may be present in the oil, in the head space of the package, or it may enter the oil through the body or seals of the package. The presence of oxygen causes formation of hydroperoxide which gives rise to rancidity. These reactions increase in rate and intensity in the presence of factors like light and heat. Moisture is another degrading agent, which may enter into the pack by permeation, resulting in hydrolysis of the oils. Light and heat is also to be avoided as they act as initiator to the oxidation process.

Consumer packs and alternatives: Although many types of materials are available, a processor has to be prepared to face a large number of situations. These are: the packaging material should be non-toxic, should be environmentally protected, be leak-proof and transport-worthy, have printability and be shiftable. Design of consumer packs are such that consumer interest is protected which are hygiene, protection against adulteration, assurance of actual weight, retailers' ability to quick supply and avoidance of consumers' own containers. Packaging system may be rigid, semi-rigid and flexible. They may be made through tin plate of different sizes. Semi-rigid containers such as high density polyethylene, polyvinyl chloride (PVC, HDPE) and PET (Polyethylene terephthalate). These ensure moderately long shelf-life and are economical as against tin plate containers. There are specifications decided by the Bureau of Indian Standards about various types of packaging systems for edible oils. In recent years paper bags of 5 layer structure have been used. Also flexible pouches made from laminates or multi-layer films of different composition.

There are manufacturers who manufacture laminates or multi-layer filaments. The Indian Institute of Packaging with its headquarters at Mumbai has laid down various specifications and also offers guidance to the processors in the case of packaging materials.

Cost of packages vs product cost work out to be as follows:

Packaging material	*Cost of package vs product cost percentage*
Tin plate container	20
Plastic container: HDPE	15
PET	13
PVC	10
Flexible pouches	3

Since these packaging material cause problems of waste management, the processor has to think of these problems before distributing the consumer material to a particular area. Although tin plates are preferably as packaging material its consumption is around 1,25,000 tonnes of which about 20,000 tonnes of tin plate valued at Rupees 50 lakhs have to be imported for edible oil. By switching over to flexible pouches, it is possible to save considerable foreign exchange.

REFERENCES

1. Treatise on Fats, Fatty Acids and Oleochemicals, O.P. Narula, Editor-in-Chief, 2nd Volume, Industrial, 1995, Chapter U-3 'Handling and Storage of Oils and Fat Products' by O.P. Narula.
2. Stokes Robert and Reid, L.N., *J. Am. Oil Chem. Soc.*, **33** (1956).
3. Paul, G., Radtke R. & Becker K., *Fette Seifen Anstrichm.*, **74**: 359 (1972).
4. Sutker, L.L., *J. Am. Oil Chem. Soc.*, **31**, 11 (1954).
5. Treatise on Fats, Fatty Acids and Oleochemicals, O.P. Narula, Editor-in-Chief, 2nd Volume, Consultant (India), 1995, Chapter U-1, 'Packaging of edible oils' by M.C. Dordi (Ms).

CHAPTER 9

Extraction of Fats and Oils

INTRODUCTION

The vast multitude of oil-bearing materials of plant origins are featured by a well knit combined matrix of oil, protein, carbohydrates including cellulosic materials, ash and moisture. In comparison, the tissues from animals (land and marine sources) are less well-knit. They consist principally of fat or oil, water and non-fatty solid residues. The oil-bearing materials of plant source vary in the relative proportions of their oil and non-oil constituents, which in turn significantly influence their physical state widely. The choice of the technology of extraction for recovery of oil and non-oil constituents is made accordingly. The extraction technology has developed over the decades and newer technologies are still emerging. The extraction aim to obtain the fatty material in as much quantity as possible, in as much good quality as possible, and to provide residual oil cake or animal fat residue in good condition and quality.

The oil content of some common oil-seeds[1] is shown in Table 9(a).

The proportions of hulls and kernels and oil content[1] in kernels or whole seeds are included in Table 9(b).

Table 9(a): Average Yield* of Oil from Commercial Processing of Common Oil-seeds (per cent) Oil from Seed of Normal Moisture Content)

Babassu (kernel)	63	Perilla seed	37
Castor beans	45	Poppyseed	40
Coconut (copra)	63	Rapeseed	35
Corn (germs)	45	Rice bran	14
Cottonseed	18	Safflower seed	28
Flaxseed	34	Sesame seed	47
Hempseed	24	Soyabeans	18
Kapok seed	20	Sunflower seed	25
Oiticica (kernels)	60	Tea seed	48
Palm, African (kernels)	45	Tung	35
Peanuts	35		

*Yields are by mechanical expression for all except soyabeans, cottonseed, and rice bran which are by solvent extraction, and refer to whole of undercorticated seeds, unless otherwise stated.

Table 9(b): Approximate Proportions of Hulls and Kernels in Different Oil-seeds and Oil Contents of Whole Seed, Kernels and Hulls

Oil seeds	Per cent Kernel	Per cent Hull	Percent Oil in Whole Seed	Percent Oil in Kernel	Percent Oil in Hull
Usually decorticated					
Oil Palm	25	75	–	48	–
Babassu	9	91	–	67	–
Cohune	10	90	–	67	–
Tucum	30	70	–	47	–
Murumuru	40	60	–	–	–
Tung	60	40	30	50	–
Oiticia	65	35	38	58	–
Cocoa beans	88	12	50	–	–
Castor beans	70-80	20-30	40-50	–	–
Cottonseed (delinted)	62	38	19	30	1-2
Peanuts	75	25	38	50	0.5-1
Sunflower seed	45-60	40-55	22-36	36-55	1-2
Kapok	60	40	20-25	40	-

Safflower	50	50	28-33	55-65	1.5-2
Soyabeans*	93	7	18	19	0.6
Usually not decorticated					
Flaxseed	57	43	–	58	22
Perilla seed	68	32	34	–	–
Hemp seed	62	38	31	–	–
Rapeseed	82	18	42	–	–
Mustard seed	80	20	–	–	–

*Soyabeans are now being dehulled at many mills so as to produce a 49 percent protein soyabean meal especially suitable where low fiber content is important in a feed.

Nearly all animal fats are recovered by rendering, while vegetable fats and oils are obtained by crushing or by solvent extraction, or by both.

EXTRACTION OF ANIMAL FATS (1,1a)

Animal tissue, like oil-seeds, contain varying fat content. Fatty animal tissues free from muscle or bone are usually 70-90 per cent fat; the remainder consists of water plus a small amount of connective tissue. The latter is made up largely of proteins; hence the residue from rendering ("stick", etc.), like the residue from the processing of oil seeds, is essentially a protein concentrate that is used principally as animal feed.

The product of highest fat content (92-95%) obtained in meat packing establishments is leaf fat from hogs. The internal fat from cattle used for the manufacture of oleo stock contains 60-80 per cent fat. Back fat and other so called cutting fat from pork has 80-85 per cent fat. A considerable amount of lard and tallow is obtained, however, from bone stock and other low fat material, which may not contain more than 10-15 per cent fat.

Most of the fish oil produced comes from the rendering of whole small oily fishes, such as sardines and herring, which contain 10-20 per cent oil. However, whales, which give an average oil yield of near about 15 tonnes per animal, are trimmed of their fatty tissues or blubber, which contains about 70 per cent fat and is rendered separately from bones or flesh.

Broadly, there are two general methods of rendering tissues; wet rendering and dry rendering.

Dry Rendering (1a)

Dry rendering is normally carried out in horizontal steam-jacketed tanks with a large charging opening in the centre of the top and equipped with an agitator. The tanks hold from 2 to 4 tonnes per charge. The agitator has paddles attached by arms to a horizontal shaft. The steam pressure in the jacket is from 50 to 75 lbs. p.s.i. The operation is carried out by charging the tank and by continuing the agitation of the contents until the product is dried down to the right consistency. The entire contents are then discharged into a steel box equipped with a perforated liner, and all possible free liquid is drained off. The residue is pressed, and the pressed liquid fat is combined with the drained fat. After settling, centrifuging, or filtering, it is ready for marketing. The residue is ground as a protein supplement for animal and poultry feed.

The cooking or drying operation may be conducted either under internal pressure, atmospheric pressure, or under vacuum. If no grinder is available for the bones, internal pressure is necessary. The best yields are produced under vacuum, but most plants operate at atmospheric pressure.

For years the agitators have been designed with an r.p.m. of 15 to 20, and this was considered normal. Within the last few years this has been increased to from 34 to 40 r.p.m., with a decrease in cooking time of about 25%. Recently r.p.m.'s as high as 50 to 60 have been tried with even better results.

Wet Rendering (1a)

Wet rendering in general is used for edible products where colour, flavour, and keeping qualities are of prime importance and the relative percentage of residue is small. In wet rendering the fat cell walls are hydrolyzed by steam under pressure until they are partially liquefied and the released fat floats on the surface of the water.

The wet rendering process is carried out in vertical tanks holding anywhere from 2 to 10 tonnes. Means are provided for drawing off the fat just above the interface. When the interface is not in line with the draw off outlet, it is raised or lowered by admitting or drawing off water at the bottom of the tank. In practice, after the tank is filled, the head is fastened on and the steam is admitted to the bottom of the tank. Thus it serves for both heating and agitation. The steam outlet at the top of the tank is left open until all air has been displaced. It is then closed, leaving a small by-pass open so that steam will continually feed in at the bottom and provide agitation during the cooking period. The pressure used may be from 40 to 75 lbs. of internal pressure.

The latest practice is towards the higher pressures to cut down on the time. This reduces hydrolysis of the fat and maintains a lower free fatty acid. Cooking usually takes from three to five hours, depending on the material rendered, the size of the tank, the size of the load, and the steam pressure. After the material has settled, a small amount of cold water is introduced at the bottom of the tank to cool the water solution in the lower part of the tank under the liquid fat. This is done to prevent flashing while the fat is being drawn off and the pressure on the liquid is reduced.

After the fat is removed, the residue, consisting of a liquid containing hydrloyzed protein, and the solids are dropped from the tank and the solids are separated from the liquid. The liquid portion is usually evaporated to a concentration of 50 to 60% of solids in a double-effect evaporator, recombined with the solid portion, and dried in a steam-jacketed drier.

Several modifications of the above process have been introduced. In one the liquid and solid material, after dropping from the tank, are agitated and conveyed or pumped to a hammer mill, which disintegrates the material and discharges it into a jacketed tank equipped with an agitator. After heating to about 85-95°C it is pumped through a horizontal solid bowl centrifuge for the continuous separation of solids from liquids. The liquids are screened, reheated, and again centrifuged through a clarifying centrifuge that discharges solids to recover the grease. The liquid and solids are then handled as previously outlined. This process can be used on edible or inedible products.

Another modification is used for edible product only but is somewhat related to this. The fats are ground or hashed before loading in the tank, and cooking takes place under pressure for about an hour. The liquid and solids are drawn off at the bottom of the tank until the liquid fat is reached. The fat is then drawn off, cooled and settled or filtered. The liquid and solids are filtered or centrifuged as described above. This process has the advantage of cutting down the time, increasing the capacity of the tank and producing a milder lard of low free fatty acid.

Most of the animal fat produced is generally rendered by the steam process. The lard produced by this method of rendering is known as "prime steam lard". In addition to lard, tallow and whale oil are also usually steam rendered.

In the steam rendering of high fat stock, 99.5 per cent or more of the fat in the raw material is ordinarily recovered. The fat that is not recovered consists of a small residue in the tankage plus a very small amount that remains in the "stick water". The usually "killing" and "cutting" fats (from service body fat or tallow from cattle body fats) yield about 80 and 70 percent lard, respectively, plus 2-3 per-cent each of dry tankage and dry "stick" or solid residue from the evaporation of "stick water". The dry tankage and stick will ordinarily contain about 10-12 percent and 1.5-2.0 percent fat respectively. Tankage from good-stock may contain as high as 70-72 percent protein, and stick may contain as high as 90 percent or more.

The advantages of steam rendering are that an efficient recovery of fat is obtained in relatively simple equipment and that it is adapt-able to a wide variety of materials. There is little tendency for proteins and other substances to dissolve or disperse in the fat in the pres-ence of water; hence the fatty stock may contain a large proportion of non-fatty tissue. Bony stock can be handled by this process since it is effectively disintegrated by prolonged treatment with steam under elevated pressure.

Steam rendering is less rapid and less efficient than dry render-ing from the standpoint of heat consumption. However, a large amount of water must be evaporated to recover the non-fatty residue in a concentrated form. Some hydrolysis of fat occurs during steam ren-dering; the free fatty acids content for prime steam lard is seldom less than about 0.35 per cent. At about 25 kg pressure, development of free fatty acids is at the rate of about 0.06 per cent per hour. The acidity in any case depends on the rendering time and temperature and the storage temperature and duration of storage of the fatty stock before it is processed.[2,3] By careful scheduling of operations, killing fats may be rendered reasonably soon after the animals are slaughtered, but carcasses are chilled to 18-20°C before cutting fats are available. The stability of lard toward oxidation bears no relation to the acidity and appears to depend principally on processing and handling subsequent to rendering[3].

One of the major developments in rendering has been the discovery that, anti-oxidants added before rendering greatly enhance the stability of the fat produced[4,5]. Sims and Hilfman[6] studied the stabilization of lard and edible beef fats during pressure steam rendering. Anti-oxidants tested included butylated hydroxyanisole (BHA), butylated hydroxytoluene (BHT), propyl gallate and citric acid combinations and a mixture of BHA and BHT. Best results for a given stabilizer level were obtained with the individual phenolics BHA and BHT. Poor results were obtained with the mixtures in propylene glycol.

There are several modifications of continuous wet-rendering operations in which attempts are made to obtain a better product and a better protein residue than the usual pressure tank produces.

The Kingan Continuous Rendering System (7). This is an interesting development in that it has been one of the first successful continuous wet-rendering systems. It is a system that was originally devised to handle beef fats for eleo stock without chilling but has been expanded to handle pork fats for lard. It has the advantage of simplicity and can handle either hot or chilled fats.

The Titan Process (8). This method of rendering originated in Denmark and is used largely for the production of lard. In South America it has found favour for the production of oleo stock from beef fats.

The equipment is designed to fit the job. The fat is fed through a hopper into a cutter and discharged into a preheater. Hot water is added, and the temperature is raised to ca 85°C. From the preheater the fat is pumped into a vertical boiler, where with live steam the temperature is raised momentarily to about 115°C. The product is then discharged through a reducing valve to atmospheric pressure, which causes the final rupture of the fat cells. It is then screened, and the screenings are pressed in a continuous press. The fat, water, and fine scrap are put through a centrifuge which separates the fat from the water and solids and discharges the solids periodically. The fat is finally polished through a high-speed centrifuge. Passage through the unit takes about 10 min. and it is continuous. The capacity is rated from 2,200 lbs. to 2,800 lbs. of raw fat per hour.

Delaval Centriflow Process (,10): Cell rupture is accomplished by mechanical disintegration (first minimizing the temperature as required) in a specially designed disintegrator. Then cracklings are removed from the fat mass by a "desludger" centrifuge after which the liquid phase is heated, deodorized and centrifuged to produce purified oil and glue water.

The Sharples Rendering System (11): this system is also a continuous wet-rendering process, where the fats are separated from the protein material before high temperatures are applied. It operates at a capacity of about 4 tonnes per hour, and the put through time is also about five minutes. Its aim is to render the fat and free it from the protein residue without high temperatures and at the same time to avoid the formation of an emulsion layer that limits the capacity of systems using clarifying centrifuges.

The fat is first ground through a grinder with about a 1/2 in plate. It is then heated to 85°C in a jacketed tank provided with an agitator or a heat exchanger. From the heater it passes through a horizontal solid bowl centrifuge built for the continuous operation of solids and liquids. The solids, which are deposited against the wall by centrifugal force, are moved by a screw conveyor to solids discharge ports located at one end of the unit. The solids consist of approximately 61% water, 20% protein, and 13% fat. The liquid fat, with some unrendered material, is then run through a comminutor to complete the rupture of the fat cells and insure satisfactory operation of the centrigue valves on the subsequent clarification. The comminuted fat is heated to 90°-95°C and pumped to an autojector clarifier centrifuge. This is a centrifuge which is equipped with a series of automatically operated valves located around the periphery of thc bowl to dischargc solids automatically. Thc accumulation of solids in the bowl causes the valves to open. When the solids have discharged through the open ports, the valves close. The water and protein are removed from the fat, and the fat is discharged to storage.

The solids accumulated from both centrifuges may be collected and dry-rendered for final recovery of fat and protein.

Impulse Rendering (12,13,14): This process is used mainly for preparing and defatting bones for glue. Fat stock, especially crushed bones, is continuously disintegrated in a high sped hammer mill under an excess of flowing cold water. The intense impact sets the fat free. The discharge from the mill is allowed to settle in a cold water tank from which the fat is continuously skimmed off. The ground bone is continuously removed from the bottom and transferred to another tank containing hot water (20-30°C) where more fat is separated. It is claimed that better quality fat and higher protein residue are obtained by this process.

The production of marine oils is rather similar to animal fat rendering (15). It varies with the type of fish processed and whether vitamin A and D oils or high quality fish meal are the prime objective. With the synthetic production of vitamins D_2 and D_3 as well as vitamin A, the current trend is directed to fish meal and growth factor production. Details of commercial processing methods are given in the literature cited (15).

Digestive Rendering Processes: Considerable attention has been given to wet rendering of animal fats with the assistance of added chemicals or enzymes that promote the separation of fat by hydrolysing and dissolving the connective tissue.

Deatherage (16) has described in detail the laboratory and pilot plant experiments on the alkali rendering of lard and beef fats. The best results are obtained when the fat is digested at 85-95°C for 45 min. to 1 hour with a 1.75 per cent sodium hydroxide solution. After digestion is complete the fat is separated from the aqueous liquid, which contains a small amount of undigested solids, by centrifuging and washing, first with 2-5 percent salt solution and then with water. Fat recovery is equivalent to or better than that obtained by steam rendering without significant hydrolysis or darkening of the amount of hydrolysis has occurred are difficult to process because of the excessive formation of soap in the aqueous phase. Soap is derived only from free fatty acids in the fat. Under the mild conditions of digestion there appears to be no appreciable saponification of neutral fat. Of course, the fat is alkali refined as it is rendered and, therefore, it is produced substantially free of acidity. A typical lard has a free fatty acid content of 0.01 per cent and Lovibond colour of 2 yellow and 0.3 red. Other reports indicate that 1.2 and 2.0 percent sodium hydroxide (30% of fatty materials) give improved yields over conventional wet rendering (9, 17, 18, 19).

The rendering of slaughterhouse waste by ammonia plus ammonium diacid phosphate under pressure for peptonization and separation into aqueous and fat phases has been patented (20).

Alkali rendering has been found better than steam, water, or acid digestion for recovering vitamin A from fish livers with an oil content of 30 per cent and upward. Partial removal of anti-oxidants does not impair the stability of the vitamin (21).

The use of proteolytic enzymes in rendering is described in a number of patents. It does not, however, appear to have been used commercially, except perhaps in the recovery of fish liver oils (21). The patent of Perfentjev (22) describes the digestion of fish livers with pepsin at a low pH and a low temperature. The process of Keil (23) for the recovery of lard or other animal fats involves digestion of the fatty stock with a proteolytic enzyme of vegetable origin, for example, 0.005-0.20 per cent papain at a pH of 6.0-7.5, followed by heating to 60°-85°C to separate the fat. Helmbacher (24) has patented the use of an enzyme, such as papain or ficin, with a cysteine activator to decrease digestion time while increasing the yield.

EXTRACTION OF FATS

Introduction

The vast multitude of oil-bearing materials of plant origin are featured by a well knit combined matrix of oil, protein, carbohydrates including cellulosic materials, ash and moisture. The oil-bearing materials vary in the relative proportions of their oil and non-oil constituents which in turn significantly influence their physical state widely. The choice of the technology of extraction for recovery of oil and non-oil constituents is made accordingly. The extraction technology has developed over the decades and newer technologies are still emerging.

Extraction of oil from oil-bearing materials is usually done by pressing, solvent, and by prepressing and solvent combined depending on their oil content.

The oil-bearing materials with oil content ca. 20 per cent are known as the low oil content materials and those with oil content above 20 per cent are the high oil-bearing materials. Generally, oilseeds with a low oil content, such as soyabean, rice bran, sal, mango kernel etc. are subjected to solvent extraction. High oil-bearing materials, such as rape/mustard, groundnut, cottonseed, sunflower seed, and copra are extracted by pressing and in two stages by prepress and solvent extraction. The extraction of high oil-content materials by exclusively solvent process is also an interesting commercial approach. Among the newer technologies, mention may be made of super critical extraction technology and biotechnology.

PRE-TREATMENTS REQUIRED FOR PREPARING OIL-BEARING MATERIALS

Irrespective of the process adopted for oil recovery, the oil-bearing materials invariably require pretreatments like cleaning, drying, dehulling and decortication, size reduction and heat treatment (cooking). The pretreatments are briefly as:

Cleaning: The oil-bearing materials are cleaned by removing stems, leaves, and other trash materials like sand and dirt by using vibrating screens. Oil-bearing materials are often contaminated with iron particles (tramp) and these are removed by setting up permanent or temporary electromagnets over a conveyor belt.

One of the benefits of cleaning is the removal of certain kind of foreign matter that causes deterioration of oil-bearing materials by their decomposition. The removal of foreign matter is often mandatory for production of high protein meals.

Drying: Moisture content is a critical factor in the processing of oil-bearing materials. Drying is often necessary for processing reasons such as to facilitate cracking and dehulling. It is also required to adjust moisture and conditions for prepressing and/or solvent extraction.

Drying is done in a specialized equipment. The two main kinds of equipment are based on either direct (gas fired) or indirect (steam heat exchange) air heating.

Dehulling and Decortication: The oil-bearing materials are preferably decorticated before actual extraction to increase the total yield of oil and increase the capacity of the extraction equipment and produce low fibre content cake or meal. Dehulling is very important for high hull containing seeds (25–40%) such as sunflower and cottonseed and even for rape/mustard. The seeds like soyabean and shelled groundnut are processed as hull content is relatively low (9–11%) unless the cake/meal has a specific requirement as for example, soyabean meal for human consumption.

The dehulled/decorticated materials are subjected to sieving and air classification for separation of hulls and cotyledons/meats. Usually 8–12 per cent of hull are left in the feed to facilitate expelling and recovering more oil. Dehulling increases the cost of oilseed processing. Two types of hullers, namely bar hullers and disc hullers are used in the case of medium to big sized oil-bearing materials (seeds).

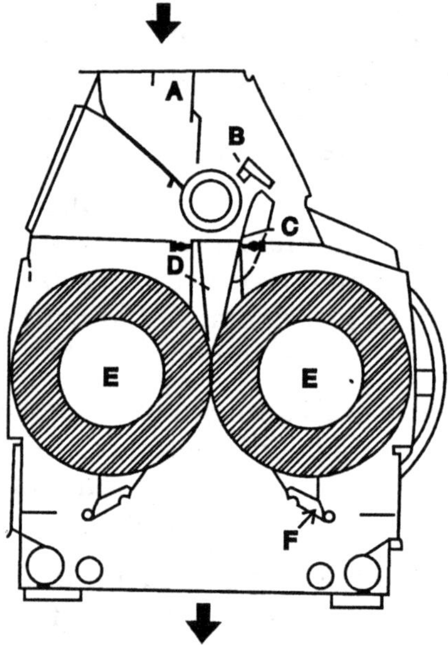

Fig. 9.1 Cross Section of a Flaker

Size Reduction: The oil-bearing materials are subjected generally to size reduction (small particles) of specific thickness and height to facilitate the recovery of their oil. This size reduction is achieved by converting the crushed seeds to flaking, pelletising or extrusion.

a) *Flaking*: Flaking of oil-bearing materials to the desired flake thickness (0.25–0.35mm) is accomplished in mills by bringing crushed/

cooked materials between two smooth rolls running against each other on a horizontal plane.

The choice of the flaker is guided by the type, size, temperature and moisture of the oil-bearing materials, and on the final flake thickness desired. Fig. 9.1 shows the cross section of a typical flaker.

b) *Pelletizing*: Certain oil-bearing materials such as rice bran, dry-milled corn germ, etc. are usually pelletized to reduce the production of fines and thereby facilitate efficient solvent extraction.

Pelletizing of the properly steam cooked, cooled and lightly dried bran is achieved through the action of rolls upon a die face. The pelletized bran is a crispy material and suitable for solvent extraction. The pellets are screened when cooled to eliminate the smaller particles, which in turn are sent back to the pelletizer.

However, the pelletizing machines require great energy consumption and heavy die requirement.

c) *Extrusion* (25): Extrusion refers to the compression of oil-bearing particles in an extruder at temperatures well above the boiling point of water. The oil-bearing particles expand after being extruded through the holes of die due to the release of compressed water vapour at atmospheric pressure.

The expansion process has not been widely used in processing high oil content seeds. It is mostly used in the extraction of soyabeans, cottonseed and rice bran.

The main advantages for extraction of oil-bearing materials using extrusion include:

(i) Increased extractor capacity because of the increased capacity of conveyors etc. due to higher density of extruded materials versus flakes.

(ii) Better extraction percolation because of less fragility of the expand material and the incorporation of fines therein.

(iii) Improved drainage of solvent in the last stages of extraction with less solvent hold up in the marc and hence less carry over to the DT (Desolventisation-Toaster) resulting in energy savings.

(iv) Less solvent used

(v) Lower energy consumption (solvent recovery)

(vi) Lower residual oil in meal even with increased capacities;

(vii) Reduced operator attention.

Optimum operating conditions vary between the oil-bearing materials but are generally within the ranges of 10–15 percent moisture and 105–121°C die exit temperature. The crushed seeds or prepressed cake material is fed into the extruder (Fig. 9.2) at one end of the barrel from where it is conveyed through the barrel by a shaft

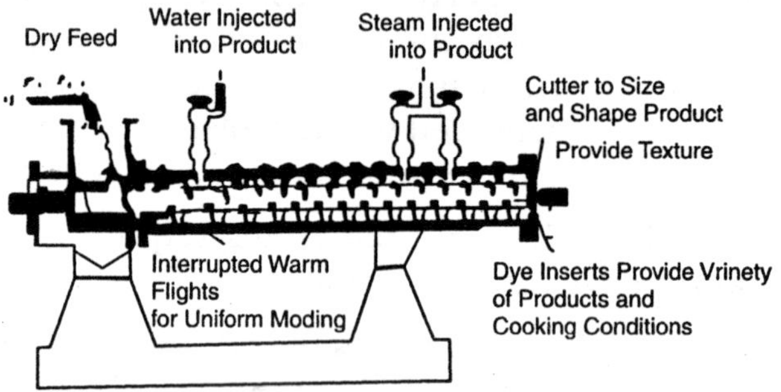

Fig. 9.2 Anderson Extruder

fitted with flights. This flighting is interrupted by breaker screws along the length of the barrel. These breaker screws interrupt the material as it is moved along the barrel causing a mixing action. The higher shear produced by this process mixes the steam which is injected into the barrel with the cake. The pressure created by the injected steam and the heat from the friction of the rotating shaft raises the temperature significantly further forcing the steam into the mixed mass. As this mass is moved to the discharge end of the barrel, it is forced against the small die openings in the end plate. This produces the high pressure which is released as the material exits the die. When the pressure is released, the moisture in the hot material vapourises instantly rupturing the mass and producing thousands of openings. This mass which is generally discharged in a rope-like fashion now has exposed oil spherozomes. Some free oil is released immediately but this oil will reabsorb on the surface as the material cools. However, the pores or openings will generally remain, and it is this expanded material that is cooled and dried before being sent to the extractor.

Relationships between flake thickness and residual oil content of soyabean meal, for example, when directly extracted and used in making expander collets, are shown in Fig. 9.3.

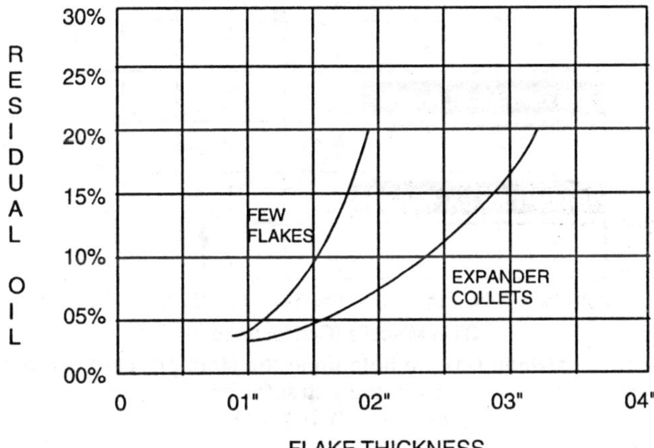

Fig. 9.3: Effects of Thickness of Direct Extracted Flakes and Flakes used in Making Expanded Collets Residual oil Content of Extracted Soyabean Meals

For example, a residual oil content target of 0.7 percent in the meal can be attained either by flaking to 0.3 mm (0.012") for direct extraction or to 0.5 mm (0.020") for extrusion. Some savings through reduced electrical costs, as well as increase of the throughput by approximately 50 per cent by flaking to 0.5 mm (0.020") for extrusion can be made.

Soyabean flakes 0.3 mm (0.012") weigh approximately 401 kg/m^3 and extruded collets approximately 609 kg/m^3. Approximately 52 per cent more material in collet form can be processed in the same extractor, provided other factors do not become limiting.

Relationships between hexane hold up in the marc, per cent oil content in the miscella and gallons of solvent used in the extractor per ton of soyabeans processed are shown in Fig. 9.4. Soyabean flakes typically have a hexane hold-up of 33 per cent in the marc, as compared to 20 per cent hold-up for extruded collets. Typical soyabean mills without extruders require about 296 gallons (1120L) of hexane for each ton of flakes extracted and produce miscellas with oil contents of approximately 30 per cent, while soyabean mills using expanders require about 194 gallons (734.1) of hexane per ton of collets extracted and produce miscellas containing approximately 35 per cent oil.

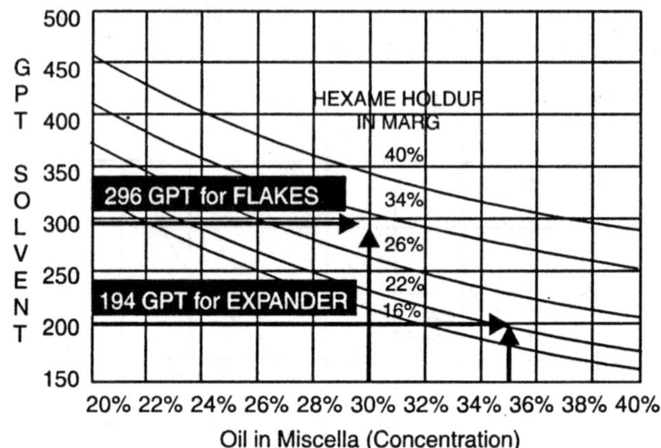

Fig. 9.4 Relationship between Hexane hold up in the Marc per cent Oil Content in the Miscella and Gallons of Solvent used in the Extractor per Ton of Soyabeans Processed [Ton = 0.9072 Tonnes]

Fig. 9.5 shows that more energy is required to concentrate the 30 per cent oil containing miscella from extracting flakes to 65 per cent oil content in the first effect evaporator than the 35 per cent oil miscella extracted from collets (92 and 69 btu/ton, respectively). Generally, however, this heat is recaptured from DT vapors and is "free".

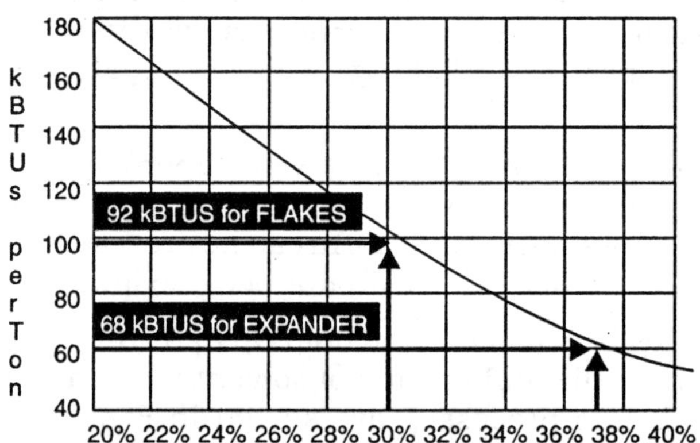

Fig. 9.5 Effects of Oil Concentration of Miscella from Extractors on kBTUs Required to Concentrate Miscella to 65 oil in to the Effect Evaporated per Ton of Soyabean processed

With the DT operating with vapors at 77°C (170°F, Fig. 9.6), the greater hold-up of solvent in extracted flakes require approximately

230 KBTUs for desolventising each ton of soyabeans processed as compared to 117 KBTUs for extracted collets.

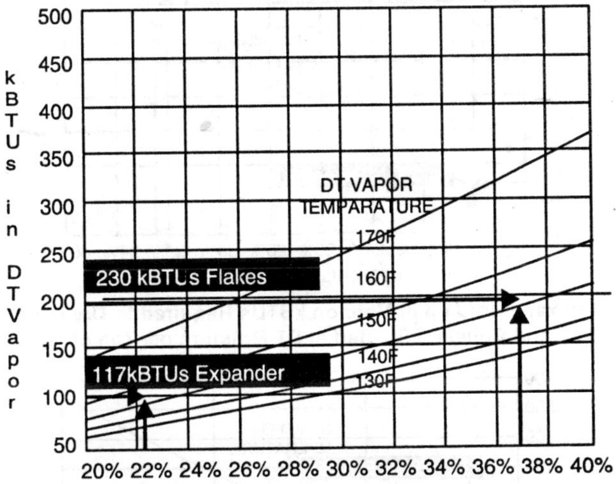

Fig. 9.6: Relationship between DT Vapour Temperatures, Hexane Hold-up in the Marc and kBTUs Require to Vaporize Solvent from Soybean Flakes and Extruded Collets per Ton of Soybeans Processed.

The energy for desolventising flakes and collets is obtained from direct steam injection into the DT. However, the steam condensates in the meal must be removed later in a meal dryer, with the desolventised flakes and collets requiring 97 and 41 kBTUs, respectively, per ton of soyabeans processed if entering the extractor at 10 per cent moisture level (Fig. 9.7).

Table 9.1 shows that savings in electricity in cracking, flaking and breaking water balls when converting from a flake extraction process are approximately offset by increased use of electricity for running the expander and collet cooler fan. However, the expander process uses approximately 149 lbs less steam than the flake extraction process. On an overall basis (Fig. 9.8), approximately 327 kBTUs in the form of steam are required for solvent recovery, desolventising and meal drying per ton of soyabeans processed in flake forms, as compared to 183 kBTUs for collets. Depending on the local cost for producing steam (Fig. 9.9), the savings in energy from processing expander produced collets instead of direct extraction of flakes can vary significantly per ton of seed. More crude oil is recovered per ton of cottonseed processed by the expander collet method than by the flake Direct Extraction method.

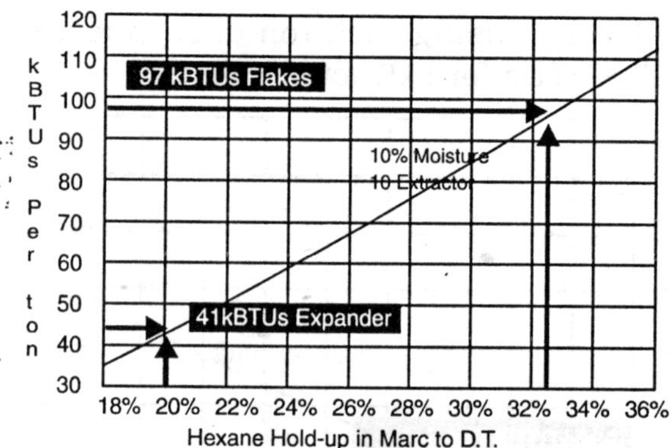

Fig. 9.7: Effects of Hexane Hold up in Marc on kBTUs Required to Dry Desolventized Meals from Flakes and Collets Extracted at DT Moisture per Ton of Soyabeans

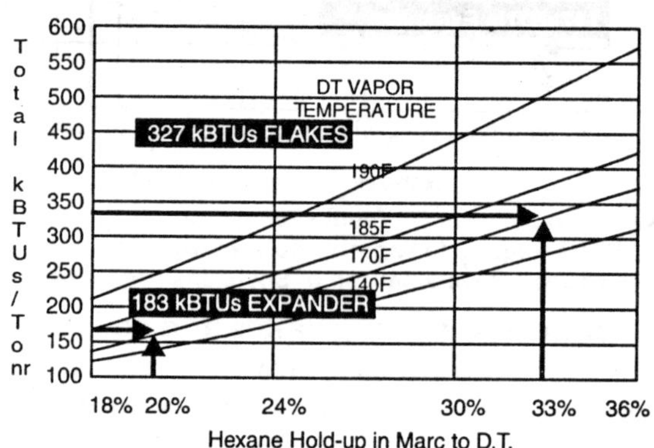

Fig. 9.8: Relationships between DT Vapor Temperature Hexane Hold up in Marc and Total kBTUs for Solvent Recovery Desolventizing Toasting and Meal Drying of Flakes and Collets per Ton of Soyabeans Processed

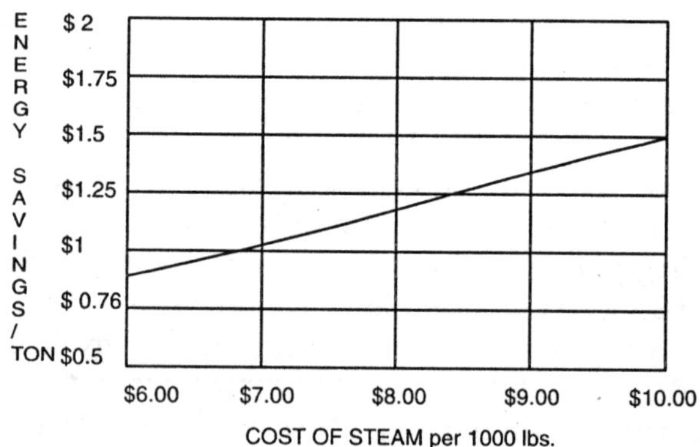

Fig. 9.9: Relationship between Steam Costs and Savings in Energy per Ton of Soybeans Extracted by Direct-flake and Extruded Collet Processes.

Table 9.1: Comparative Energy Requirements Per Ton of Soyabeans Processed by Direct Extraction of Flakes and Expander Produced Collets.

Process	Flakes		Collets	
	Steam (lbs)	Electricity (kwh)	Steam (lbs)	Electricity (kwh)
Cracking	–	3.0	–	2.0
Conditioning	60	–	30	–
Flaking	–	8.0	–	4.0
Expander	–	–	–	5.0
Collet Cooler	–	–	–	1.0
Desolventizer	230	–	117	–
Water Ball Breaker	–	1.0	–	–
Meal Dryer	97	–	11	–
	387	12.0	238	12.0

Expander saves 149 lbs of steam per ton of soyabeans processed with no difference in electrical usage

The ALCON Process subjects flaked soyabeans to a prolonged holding period at high moisture and temperature. It has been reported to achieve 99 per cent reduction of urease activity and phosphorous contents of 7–16 ppm in water degummed crude soya-oil. In comparison, extrusion of soyabean collets, including direct injection of steam at 40–60 psi during the 10–15 second processing time in the expander, is an effective high temperature-short time treatment for inactivating phospholipase and can produce water-degummed crude oils with 6–15 ppm phosphorous, provided the cracked/flaked soyabeans had not previously been held for any appreciable time above 65°C (150°F). However, the short term expander treatment had little effect on reduction of urease or trypsin inhibitor activities, and additional heating in the DT (Desolventuser Toaster) would be required to deactivate these enzymes (Table 9.2) as is practiced currently in processing flaked soyabeans.

High oil-content seeds (like peanuts, rapeseed and dehulled sunflower seeds) typically do not make cohesive flakes and are processed by a 2-step prepress solvent extraction process. The Anderson International Corporation has incorporated a horizontal bar drainage cage in its Solvex Series Expanders (TM) for partial removal

***Table 9.2*: Effects of Heating Cracked Soyabeans Followed by Immediate Flaking and 200°F Extrusion on Enzyme Activity. Nitrogen Solubility Index and Non-Hydratable Phosphatides Content of Water Degummed Crude Oil.**

Conditioning Treatment	*T.I.U.*	*Urease pH Units*	*N.S.I.*	*Non-Hydratable Phosphatides as ppm*		
				Phosphorous	*Calcium*	*Magnesium*
135°F (57°C)						
12% moisture	38,821	2.02	36.8	6.1	0.2	0.1
150°F (65°C)						
12% moisture	34,387	1.91	37.2	13.6	2.2	1.4
180°F (82°C)						
8% moisture	71,545	2.10	54.9	151.9	5.0	7.0
12% moisture	54,548	2.08	40.5	16.5	0.8	0.5

of oil from seeds, and has named this new line its Hivex-Series Expanders (TM). The manufacturer's literature indicates that this machine (Fig. 9.10) is intended to reduce oil content of high oil-bearing seeds to less than 30 percent before extrusion of collets. Anderson Expander with oil removal cage has been successfully tried with essentially all major oilseeds.

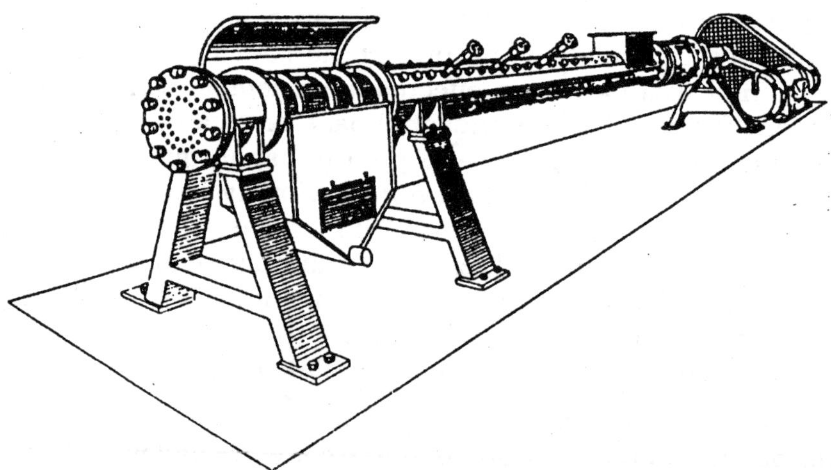

Fig. 9.10: Anderson International HIVEX SERIES EXPANDER TM with Horizontal Bar Drainage Cages for Partial Removal of Oil from High Oil Bearing Seeds Contest of Anderson International Corp.

The expander oil drainage cage combination is likely to replace many prepress operations for processing high oil content seeds. This configuration has a capacity 4–5 times that of a prepress machine but only requires the horsepower of one prepress and has maintenance costs estimated at 1/8 to 1/10 that of prepress requirements. Residual oil levels of collets made by expanders equipped with oil drainage cages are less than those for prepress cakes because the collets are porous and solvent extraction is more rapid and complete. The cooking requirements before extrusion are less than those for prepress and less steam is used. However, since recovery of much of the oil is transferred to the solvent extractor, the load on the solvent recovery systems is increased

Cooking or Heat Treatment of oil-bearing materials: Oil-bearing materials are generally subjected to heat treatment with intact moisture or with additional moisture in order to recover their oil and non-oil portions more readily and in better quality. The process is known in commerce as cooking. The principles involved in cooking are as follows:

(i) The small oil droplets distributed throughout seed/oil-bearing materials when subjected to heat treatment in cooking operation coalesce into large droplets which tend to flow more readily out of the cells of the seed/oil-bearing materials.

(ii) The heat denaturation of proteins and similar substances occurs in the cooking process resulting in their coagulation and causing the emulsion of oil droplets to break and therefore facilitate thereby the separation and flow of the oil from the solid portion in the seed/oil-bearing material. Before the proteins become coagulated through denaturation, the oil droplets are virtually in the form of an emulsion.

The cooking process achieves a number of objectives like (a) coagulation of proteins of the seed materials, effecting agglomeration of oil droplets, (b) reduction of the efficiency of the oil for the seed surfaces thereby facilitating the oil recovery readily by pressing, (c) plasticity of seed mass for efficient pressing, (d) hydration and insolubilisation of phosphatides, (e) destruction of molds and bacteria and other toxic materials.

Important secondary effects of cooking are drying of the seeds to give the seed mass the proper plasticity for efficient pressing, insolubilization of phosphatides and possibly other undesirable impurities, destruction of molds and bacteria, increase of the fluidity of the oil through increase in temperature, and, in the case of some oilseeds containing toxic compounds, deactivation of such compounds.

All oilseeds have an enzyme system which can influence oil quality. The enzymes are inactivated by heat treatment in the cooker. In rape/mustard, the enzyme myrosinase can influence quality by catalyzing the hydrolysis of glucosinolates to produce compounds that act as catalytic poisons during hydrogenation of oil. This problem remains though not as great for the new varieties of rapeseeds that have lower glucosinolate levels. Therefore, it is very important to keep the glucosinolates intact by effectively inactivating myrosinase through proper cooking (conditioning).

To do this, moisture in the cooker (conditioner) must be kept low enough, approximately 6 to 10 per cent and cooking (conditioning) should be at 80° to 90°C. Too extensive a heat treatment can have other negative effects on the oil and meal quality. These enzymes can influence the final oil quality significantly and therefore it is important to evaluate the time, moisture and temperature of the flaked seed with regard to formation of non-hydratable phosphatides. Considerable work has been done in this area with regard to soyabean phosphatides using processing methods (Alcon process, etc.) to minimise the creation of non-hydratable phosphatides. The moisture content of the oil-bearing material controls the cooking operation. The materials which are very dry cannot be efficiently freed of their oil. The oilseeds with a moisture content as low as 1–2 per cent yield their oil with greater difficultly than similar seeds having a much higher moisture content, e.g. 10 per cent. The cooking process produces a film of water on the seed surfaces. This film of water favours the diffusion of the oil out of the seed.

Figure 9.11 shows some significant data during the extraction of oil from soyabean flakes under different moisture conditions, but at the same temperature. The curves clearly indicate that there is an optimum moisture content (9%). Each seed has an optimum moisture content which should be found out if efficient extraction is desired. In cottonseed, for example, 5–6 per cent moisture is best for hydraulic pressing, whereas about 3 per cent is optimum for expellers or screw presses; at moistures of 4 per cent and higher, excessive amounts of oil are left in the cake. Soyabean are ordinarily dried to 2.5–3 per cent moisture before pressing in expellers; copra and sesame seed require moisture levels of about 2 per cent.

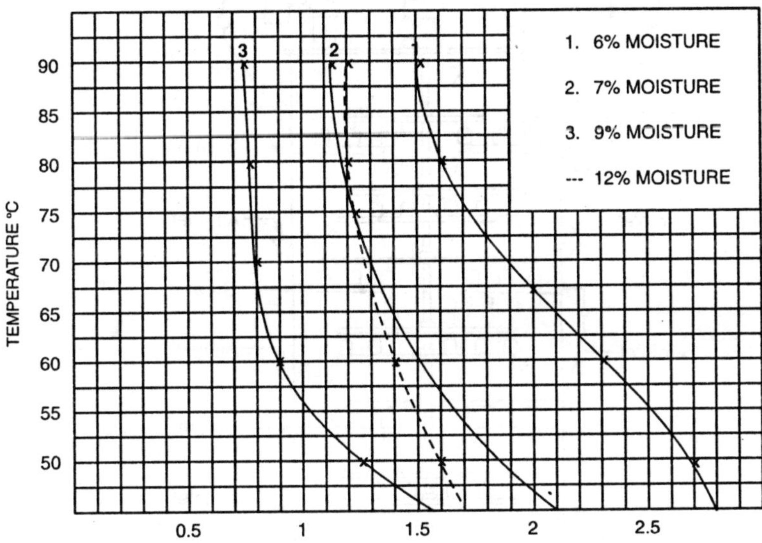

Fig. 9.11: Relationship between Moisture and Temperature and Residual Oil after Extraction from Soyabean Flakes

Cooking is usually done in either a multistage vertical stacked cooker or a horizontal rotary unit. The majority of conditioning in the industry takes place in cookers employing stacked steam-jacketed trays (Fig. 9.12). The stacked trays located one on top of the other are used either to add or to remove moisture from the seed and to control the temperature of the seed to meet the needs of the prepess. A slowly turning vertical shaft with sweeps attached mixes the rolled or flaked seed as it feeds over the heated trays. These sweeps scrape the tray floors to prevent overheating of the product and to ensure proper feeding from the top kettle down through the other kettles to the bottom of the cooker. One shortcoming of the stack cooker may well be the lack of movement of material through the kettles on a first in first out basis (plug flow). This may mean that some material is cooked for longer or shorter periods of time than necessary for best results.

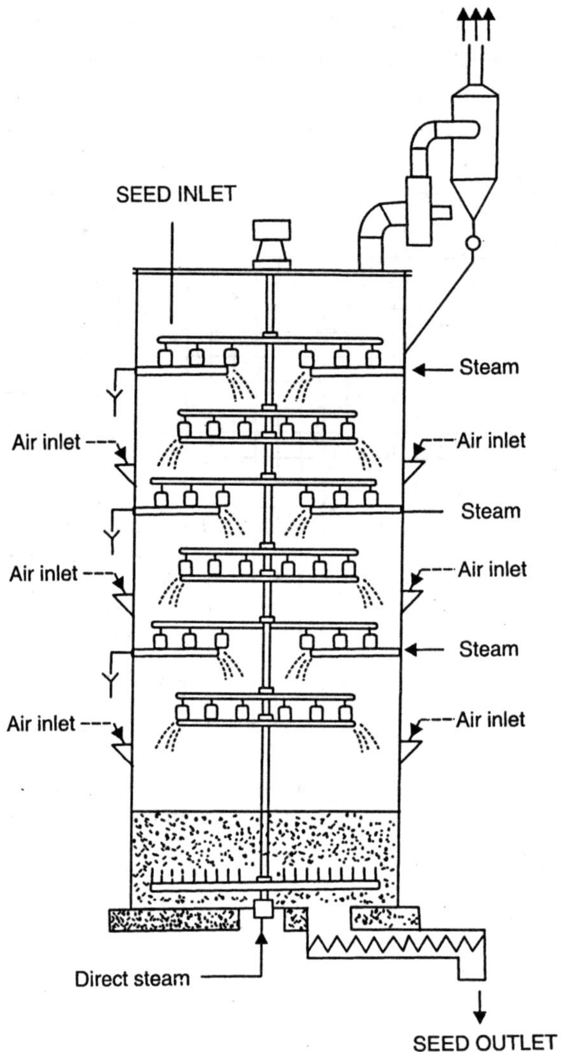

Fig. 9.12: Superimposed Pan Cooker

ENZYME-TREATMENT OIL SEEDS

Sosulski *et al* (26) reported a reduction in extraction time and increased oil yield by treatment of canola with carbohydrases compared to untreated canola. The optimum treatment before hexane extraction included flaking, autoclaving, adjustment to 30 per cent moisture content, and incubation with 0.12 per cent enzyme for 12 hr at 50°C followed by drying to 4 per cent moisture. The relative order of enzyme efficiency in enhancing oil extraction was: mixed-activity enzyme > beta-glucanase > pectinase > hemicellulase > cellulase.

Sosulski and Sosulski (27) reported that pretreatment of canola with cellulase, protease, beta-glucanase and two mixed-activity enzymes increased throughput, oil flow rate and efficiency of oil extraction in fullpressing. Enzyme treatment improved the quality of meal in respect of digestibility and available lysine content.

Holding a large amount of oilseeds in an enzyme treatment process requires considerable space and investment, and it is too early to predict whether an economically competitive carbohydrase treatment process to enhance oil extraction can be developed for large scale operations like current oilseeds extraction mills.

EXTRACTION OF OIL BY MECHANICAL PRESSING

Mechanical pressing process technology is adopted generally for recovery of oil from high oil content seeds (oil content > 30 per cent). Among the pressing techniques, hydraulic pressing of oilseeds was developed. The process remains in vogue in many countries where the volume production of oils is low. In many countries the process has been found to be uneconomical primarily because of increased mechanization and labour costs.

The hydraulic pressing consists of pressing the oil seeds in its simplest form, in bags, cloths, cages or other suitable devices. The factors such as pressure, pressing time, and temperature influence the recovery of oil. Batch presses may be divided into two main classes: the 'open' type and the 'closed' type. The open type press is one where the mateiral is pressed in press cloths that are cofined in either plate presses or box presses. The 'closed' type presses operate the materials in specific closed cages and include in practice pot presses or cage presses.

CONTINUOUS PRESSING (FULL PRESSING)

Continuous expellers or screw presses are the most efficient mechanical press and now replace hydraulic presses as the principal means of extraction of high oil content seeds (e.g. rape and mustard, peanuts, sunflower, seasame etc. and of copra, palm kernels, flaxseed and many other oil seeds). A continuous screw press comprises the essential points like feeder, cage (usually called the pressing barrel) and adjustable cone for cake discharge. The cage can be either vertical or horizontal or inclined or their combination.

A screw press is a continuous device for gradually increasing the pressure on material fed to it as the latter progresses inside a closed barrel, with provisions for the oil to drain out as it is squeezed from the feedstock. A column of compressed meal is formed at the discharge end of the barrel acting like a hydraulic press head with new cake being formed at the end as cake is expelled past a choke device. Fresh feed is forced in by feed worms against the frictional resistance of the plug at the choke, thus creating a hydraulic pressing equivalent to that of a hydraulic press ram. The operation of the expeller press is as follows:

(i) Seeds/the prepared seeds fall from a hopper into a screw feeder.

(ii) The screw feeder pushes the material into the fore part of the pressing cage.

(iii) The worm flights convey the seeds to the discharge point, the seeds are subjected to increasing pressure, and the oil is squeezed out.

(iv) The cage is made up of a number of special steel bars, which are spaced apart by shims to provide adequate drainage of the oil. The oil passes through the bars and flows out of the cage.

(v) Following their path along the pressing cage, the de-oiled seeds reach the discharge end of the rotating screw shaft where a choke mechanism is provided.

The first continuous screw press, called an expeller, was manufactured by the Anderson International Company. The Anderson expellers utilise a vertical cage to express the most easily removable oil, followed by a horizontal cage (Fig. 9.13) for attainment of the high pressure necessary for removal of most of the remaining oil. The French Screw presses are only horizontal cages where pressure is gradually built up to a maximum. Another point of difference in machines of the two manufacturers is in the method of cooling. Expellers are required to be cooled and it is done by pumping the pressed oil at an appropriate rate after removal of 'foots' in a screening tank and cooling in heat exchangers to reduce the temperature to approximately 120°F (ca 50°C). Screw presses, on the other hand, are equipped with water-cooled shafts and water-cooled ribs in bar cages. In both types of machines the pressure necessary to force the oil out of the cooked flakes is obtained by means of continuously rotating worm shafts and worms with a choke mechanism by means of which cake thickness is controlled. The main worm shaft and worms are designed to exert a pressure of 5–8 tons/in^2 on the seed being processed and at the same time to convey the seed through and out of the pressure chamber. Several different worm shafts may be employed depending on the material to be processed and whether expression is to be complete or merely prepressing preliminary to later solvent extraction.

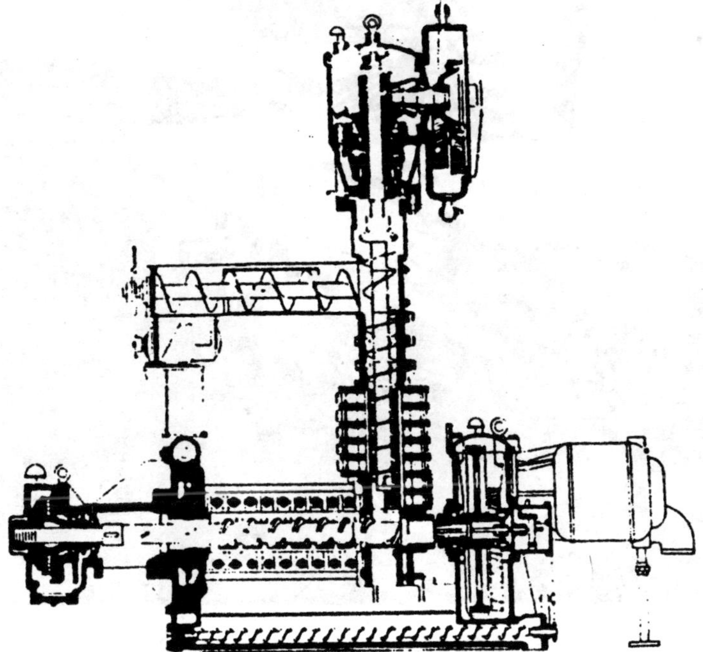

Fig. 9.13: Anderson Super Duo Expeller

The drainage barrel is made up of rectangular bars which fit into a heavy barrel bar frame (Fig. 9.14). The individual bars in the drainage barrel are separated by bar spacing clips; the specific spacings depending on the type and preparation of the material being processed. For example, in an expeller processing cottonseed, the spacing of bars in the main barrel may be 0.001 in., in the feed section,

0.0075 in., in the centre section and 0.010 in., in the discharge section. The same sections processing copra may have bar spacings of 0.030, 0.020, and 0.010 in. The spacing of the bars not only permits the drainage of oil from the material being processed but also acts as a course filter medium for the solids.

Continuous-screw presses are available with a 24 hour prepressing capacity in the range of several hundred tonnes depending on oil and hull contents. The residual oil left in the prepressed cake is around 16–20 per cent and about 8 per cent in the full press cake.

Fig. 9.14: Assembly of Drainage Barrel

The Krup Co., Hamburg, has developed an expeller press for cold pressing of oilseeds and therefore eliminates sizing and cooking. This has been possible due to the special design of the wormshaft which ruptures nearly all the oil-bearing cells of the untreated seeds. The power requirement of the plant is much less and the capacity of the plant is also somewhat lower (reported to be about 180 tons/day of sunflower seed) than the normal expeller pressing.

THEORY OF SOLVENT EXTRACTION

The process of extracting oil-bearing materials by solvent is complex and the evaluation of the rate at which the oil is removed by the solvent is very much involved. The rate of penetration of the solvent into the solids (solid-liquid extraction or leaching) and the diffusion of the oil through the cell membranes are the two factors on which the mechanism of extraction depends. Diffusion of oil through cell membranes is comparable with that through porous slabs. Nearly ideal conditions are set up for testing the theory of diffusion by solvent extracting slabs of porous clay impregnated with soyabean oil. This diffusion may be defined by Frick's Law. The following conditions were selected for the experiments for diffusion to take place:

(i) Diffusion taking place through both faces of a slab and along one axis only, there being no diffusion sideways.

(ii) Thickness of solid being uniform.

(iii) There being a uniform concentration of solute in the solid at the beginning of the experiment.

(iv) The solvent being maintained at a consistent concentration throughout the experiment.

(v) The diffusion coefficient being constant.

(vi) The temperature being constant.

(vii) The transfer of material at the surface of the solid being sufficiently rapid not to interfere with the internal diffusion process.

(viii) The porous solid being rigid and inert, and also uniform, i.e., having a structure which is the same in reference to any axis.

In the extraction process the transfer of oil from the seed/oil-bearing material on to the solvent takes place mainly by diffusion and the equation which best represents the rate of diffusion is that of Fan and co-workers, which is as follows:

$$\log_{10} \frac{q}{q_0} = -0.0911 - 4.286 \frac{D}{(2L)^2} \times t$$

in which q = the quantity of oil in unit volume or weight (quantity of oil per unit of weight after extraction)

q_o = initial quantity of oil in the solid – that is, at $t = t_o$ (quantity of oil per unit of weight prior to extraction).

D = Coefficient of diffusion (diffusion constant) at a given temperature

L = thickness of slab

t = extraction time

This is an equation for a straight line when $\log_{10} q/q_o$ is plotted against t. The slope of the line is used to evaluate the diffusion constant.

The diffusion constant, D, is not however constant, as it varies with extraction time and the type of oil-bearing material and its physical state.

The dissolving power of a solvent is related to its viscosity and the diffusion coefficient is a function of the product of the viscosities of solvent and oil and can be represented by the formula

$$D = 12.96 \times 10^{-6} (\mu_o \mu_s) - 0.46$$

where μ_o and μ_s refer to the viscosities, in centripoises, of oil and solvent, respectively.

The moisture contained in the slices influences the rate of extraction. It is observed also that from some slices diffusion is more rapid than predicted by theory. This appears to be due to ruptured cells on the surface of the slices, though low moisture content may have the same effect.

In a diffusion process, one can expect a linear relationship between the logarithm of the residual oil content and the extraction time after a short period has elapsed, when a major quantities of the oil is extracted very rapidly vide Fig. 9.15.

The proportion of easily extractable oil is found to decrease rapidly with increase in the slice thickness. From the Curve B (Fig. 9.15), it would appear that the extraction curve for peanut slices 0.026 in. thick becomes linear after about 76 per cent of the oil is extracted. With flakes of similar moisture content there is linearity with 0.032 in flakes after about 51 per cent of the oil is extracted, and linearity with 0.040 in flakes after about 30 per cent is extracted. With flake of constant thickness, there is a progressive decrease in the content of rapidly extractable oil with increase in the moisture content.

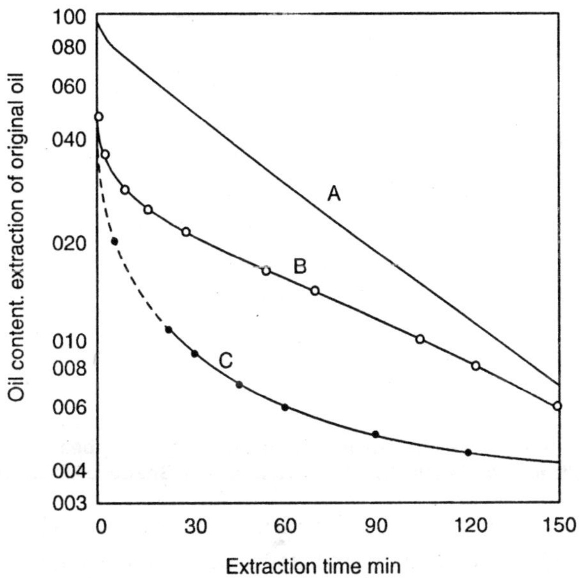

Fig. 9.15: Solvent Extraction Curves (A) Theoretical Curve for Homogeneous Oil-impregnated Platlets of Uniform Thickness J.C. Brier and J.O. Osburn, Trans. Am. Inst. Chem. Eng. 38, 967-993, 1942 and M.R. Wingard and W.C. Shand, JAOCS, 26, 422-26, 1949

According to Fan *et al* (29) the diffusion coefficient is related with the moisture content. It decreases considerably with increase in the moisture content. At 10-25% moisture, the decrease is reportedly 0.4 cm2/sec per one percent moisture. The diffusion coefficients are around 7 × 10–9 cm^2/sec in the case of peanut slices having 12% moisture content and extracted with hexane.

In the extraction of oil in practice, extraction in the range of about 0.5–5.0 per cent residual oil (on the dry, solvent-free meal basis) is so slow that it actually controls the overall extraction rate and extractor design. On a semi log plot the concavity of the extraction curve is so great in this region that actually a nearer approximation to a straight line is obtained with a log-log plot as in Fig. 9.16. The time required to reduce oil seeds to one per cent residual oil content varies with a power of the temperature that ranges from -1.9 to -2.4 with cottonseed, soyabean, and flaxseed flakes and hexane as solvent. A log time/log temperature plot yields a straight line.

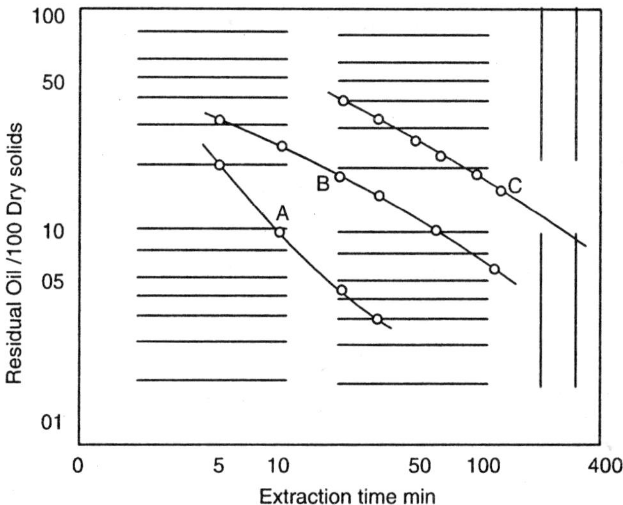

Fig. 9.16 Representative Curves for Solvent Extraction of (*A*) Soyabean Flakes (*B*) Flaxseed Flakes and (*C*) Cottonseed Flakes M.R. Wingard and W.C. Shand, JAOCS, 26, 422-26, 1949.

An empirical equation is developed to predict the relationship between extraction temperature and extraction rate when extraction is carried out to a specified residual oil level:

$y_b / y_a = (T_a / T_b)^2$ where at 0.5 per cent residual oil specified,

Ta = temperature at 120°F, Tb = Temperature at 145°F. (63°C) (49°C)

Ya = minutes extraction time at 120°F and

Yb = minutes extraction time at 145°F.

A much longer time and hence a larger extractor is required to obtain the same specified residual oil content with "cold" vs. "hot" extraction temperature. For this particular case, the extraction time at 145°F is only 68 per cent of that for 120°F.

According to Coats and Wingard (30) the time required to extract flakes or grits, of soyabeans, cotton seed, flaxseed and peanuts to a one per cent residual oil (dry basis) can be expressed as $T = KD^n$ where T equals the time, K and n are constants, and D related to the flake or particle size. This means the smaller the particle, or thinner the flake, the faster the extraction.

K in the equation $T = KD^n$ is a measure of the ease of extraction (of flakes 0.010 in thick), whereas n is a measure of the influence of flake thickness on the extraction rate. Thus soyabean flakes extract more readily than cottonseed flakes of equivalent thickness and cottonseed flakes, in turn, extract more readily than flaxseed flakes (Fig. 9.17).

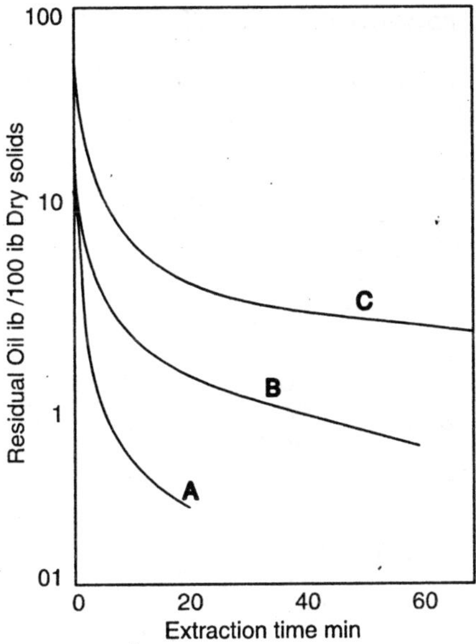

Fig. 9.17: Extraction Rates (Residual oil vs extraction time in min.) of Three Oil Seed Flakes, (a) Soyabeans 0.0075 in thick (b) Cottonseed 0.0095 in thick and (c) Flaxseed 0.0075 in thick *

*(G. Karnofsky, Proceedings of a Six Day Short Course in Vegetable Oils, American Oil Chemists' Society, University of Illinois, 1948, pp. 61-69)

Besides the relation of Coats and Wingard, we can add that of Othmer and Agarwal (31):

$$-\frac{dc}{dt} = 0.173F^{3.17}C^{3.5}$$

As this equation shows, the rate of extraction (-dc/dt) is proportional nearly to fourth power of the flake thickness (F) and to the 3.5 power of the residual oil (C). Decreasing the flake thickness to 1/3 of its former value will increase the extraction rate by 80 times.

Extraction time to a specified residual oil level varies inversely as flake thickness is raised to the 2.2 power as shown

$$\frac{Ya}{Yb} = \left(\frac{Xb}{Xa}\right)^{2.2}$$

where 0.5 per cent residual oil specified

Xa = 10 MIL thick flakes

Xb = 20 MIL thick flakes

Ya = Minute extraction time for *Xa*

Yb = Minute extraction time for *Xb*

Table 9.3: Effect of Pellet Size on Extraction Time and Residual Oil Content in Rice Bran (32)

Pellet size	*Weight of pellets in gm.*	*Extraction Time in minutes*	*Surface area in sq. cm.*	*Percentage residual oil content*	*Surface area sq. cm./gm*
		330		4.14	
16 mm Dia × 9 mm	1.600	440	8.540	2.20	5.330
length		600		1.45	
		90		5.00	
16 mm Dia × 4 mm	0.668	120	6.031	2.80	9.028
length		240		0.45	
		180		3.25	
13 mm Dia × 9 mm	0.999	270	6.230	2.10	6.330
length		600		0.76	
		40		13.00	
13 mm Dia × 4 mm	0.423	100	4.288	1.53	10.137
length		180		0.364	
		180		2.30	
10 mm Dia × 9 mm	0.612	270	4.400	1.29	7.333
length		420		0.70	
		40		7.00	
10 mm Dia × 4 mm	0.252	60	2.828	2.70	11.320
length		150		0.29	
		30		7.737	
Rice Bran		40		3.00	
		100		0.19	

On an approximate basis, it will take four times to extract 20 MIL thick flakes than 10 MIL flakes to the same specified residual oil level.

SOLVENT FOR OIL EXTRACTION

The type of solvent used in extraction plays a decisive part in the design, economy, and efficiency of extraction plant. Not every solvent of fats and oils is suitable for solvent extraction. The ideal solvent must possess the following characteristics to be of any use of solvent extraction:

(i) It must be non-inflammable and its vapour non-explosive.

(ii) It must be stable, it should have a uniform composition and a constant boiling point which should have a uniform composition and a constant boiling point which should not be too high, it should have a low specific heat, small latent heat and a low freezing point (considerably below 32°F).

(iii) It should have a selective solvent power for oil, that is, it should not extract pigments and other matter together with the oil. It should penetrate easily into the flakes, but it should also be easily removable from the residual meal and the oil.

(iv) It must have no corrosion action on the usual materials of construction.

(v) It must not be toxic and must have no deleterious effect on either the residual meal or the oil. Its vapours, too, must not create any toxic effects when inhaled.

(vi) It should not be miscible with water, nor form any constant boiling point mixtures with it.

(vii) It should have a low specific gravity.

(viii) It should be easily available, it should not deteriorate in storage and should be easily transportable.

(ix) It must be cheap.

There is no solvent that would comply with all these demands, but petroleum ether comes nearest to it. Its greatest disadvantage is its inflammability, but the risk involved is not great. Among the petroleum solvents, normal hexane has received maximum preference and use throughout the world in solvent extraction of oils.

ALTERNATIVE SOLVENTS

Research in alternative solvents has concentrated primarily on ethyl and isopropyl alcohols and mixed solvent systems, methylene chloride, supercritical carbon dioxide and water (aqueous extraction).

Abraham *et al* (33) developed a computer model for solvent extraction of oilseeds and reported the following effectives in extracting lipids from cottonseed flakes: hexane > ethanol > isopropanol. In their study, hexane required 7 extraction stages to reach minimum residual lipids in the flakes and ethanol and isopropanol required eight stages.

High petroleum prices renew the interest in ethanol and isopropanol (IPA) as extraction solvents. It may be noted that (i) mixtures of alcohols and oils are not completely miscible at all rations as with hexane, (ii) solubilities of oils increase considerably with alcohol concentrations and extraction temperatures, and (iii) latent heats of recovering alcohols by distillation are so large that it is more energy efficient to reclaim alcohols by chilling and removing the separated oil by gravity settling or coalescers.

SOLVENT EXTRACTION PROCESS

Oil-bearing materials are solvent extracted by different process, depending on how the solvent and the oil-bearing materials are brought in contact with each other. The contacts can be by percolating the solvent through the bed of oil-bearing material, conveying the material through the solvent (Fig. 9.18), or by combination of the two contact systems.

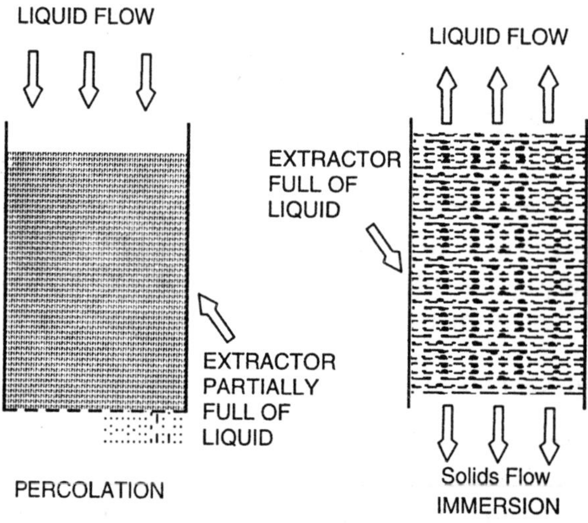

Fig. 9.18: Liquid Flow of Percolation Extractor Versus Immersion Extractor

Accordingly the solvent extraction processes have been classified as:

(1) Percolation process

(2) Immersion process

(3) Combined percolation and immersion processes.

The extraction plants are also classified accordingly as Percolation Extractors, Immersion Extractors, and Combined Percolation and Immersion Extractors.

The solvent extraction plants can be batch and continuous ones. The batch solvent extraction plants are used for the recovery of oil from expeller cakes or from low oil oleagineous materials such as rice bran, sal seed etc.

A standard batch solvent extraction plant (34) consists of four or more stationary extractors of cylindrical design with dished bottoms provided with a charging door at the top and an unloading door at the bottom (Fig. 9.19).

The material to be extracted is charged into the extractor by gravity until it is filled to desired height. The extractor is equipped with an internal perforated metal plate covered with a filter cloth. After the material loading, the extractor is filled with solvent and set in cycle with other extractors which are already extracting materials. The miscella enters in succession the extractors containing material of progressively increasing oil content, and the spent material is finally extracted with fresh solvent before being discharged. Upon completion of washing, which generally varies from 8 to 10 hrs, live steam is admitted to the extractor to distill the solvent retained by the spent material.

After desolventising, a pressure of 0.5–1.0 kg/cm^2 is built up in the extractor, the unloading door is immediately opened, and the extractor contents are automatically discharged by the effect of pressure. After discharging the material the door is closed and a new batch of material is charged into the extractor, which is again set into the operating cycle.

The complete plant is equipped with four extraction tanks as shown in Fig. 9.19 plus the ancillary equipment including miscella

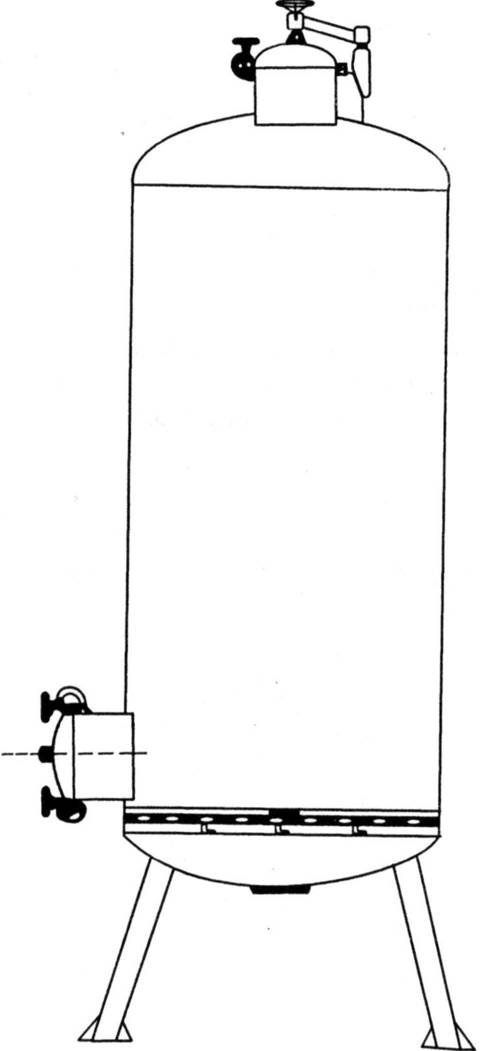

Fig. 9.19: Batch Extraction Unit

filters, which are continuous distillation facilities to remove solvent from the miscella, and a system to recover the solvent. Utility and material requirements per ton of material processed in this type of plant are usually the following: steam, 600–700 kg; power, 6 KWH; water, 15 m^3 and solvent, 5 kg.

PERCOLATION EXTRACTORS

The extractor used most commonly even today is the percolation extractor. In this type, solvent or miscella is distributed over a bed of oilseed flakes or cake, where it percolates down through the bed and exits the bed at the bottom through some type of supported filtering device such as a perforated plate, a mesh screen, or a wedge wire screen bar system.

Percolation extractor can be classified into the following categories according to extractor design and path that the material to be extracted follows during the washing process:

(i) Carousal or rotary type extractor

(ii) Perforated belt type extractor

(iii) Vertical and horizontal basket extractor

All these extractors basically utilise similar principles and are counter-current extractor. However, the Carousal type is the most widely used percolation type extractor. In India, continuous extraction plants are mostly equipped with belt type extractors, although many new ones have opted for Carousal type. Belt type extractor had found large application before the advent of the Carousal type. Now-a-days, the latter is preferred because it has a lower space requirement and is more easy to use.

Vertical basket extractor is no longer in use, as it does not allow adjustment of miscella percolation and the baskets and subject to channelling. That is why it has been replaced by the horizontal type, which is similar in operation to Carousal extractor, as it allows control of flow of the miscella in each stage by means of a number of pumps or by a combination of pumps and special arrangement of the basket travel. This extractor is widely used in Germany. Another variation is the drag chain type which is also in use. All these extractors follow continuous process of extraction, have low solvent loss and high production capacities.

(i) *Carousal Extractor/Rotocel Extractor*: Fig. 9.20 illustrates a conventional Rotocel percolation extractor. This extractor comprises a series of cells or baskets which rotate about a vertical axis. Each cell has an automatically operated outlet door fitted with a drainage screen. The raw material enters the extractor through a specially

1. Dry raw material.
2. Feed conveyor.
3. Material plug seal prevents leakage of solvent vapours and minimses of entering extractor.
4. The extractor is simply driven by a fractional horse-power motor with flood speed control.
5. Automatic vent seal provides additional safety device to protect against excess pressure.
6. Convenient sight glasses facilitate observation of the process.
7. Special clean solvent control can be lifted to ensure flooding even with low solvent to meal ratio.
8. Adjustable door dropping gear controls impact of door on stop to assist in cleaning door drainage screen.
9. Simple agitator discharge ensures vapour seal.
10. Variable speed discharge conveyor.
11. Large mesh screen allows free drainage and recycling of fines.
12. Stage pumps recycle miscella (solvent and oil) to food meal at each of four or five stages.
13. Fines automatically recycled and screen continuously washed clean.
14. Self cleaning screen ensures high clarity of final such miscella.
15. No rubbing between meal and extractor wall.
16. Meal in cell acts as natural filter.
17. Controlled flooding for uniform extraction without channelling.
18. Miscella is distributed over whole cell instead of spraying thus permitting fine solids to be put back into meal which acts as filter.

Fig. 9.20: Sectional View of Extractor

designed sealing feed screw and travels in the cell round the extractor. It is discharged through another automatically operated sealing device. During its passage round the extractor the material is continuously soaked in a countercurrent segregated system of diminishing miscella strengths and finally washed with clean solvent. Soaking periods are adjusted to suit the material being processed. The rotational speed of the extractor can be varied and due to the skilful arrangement of the solvent flooding devices the position of each miscella wash can be adjusted very simply by the operator to suit different materials or different circumstances. These design features and the simplicity of adjustment make the extractor the most reliable and flexible now available.

Because of the flooding facility created by the cells it is not necessary to use fine sprays to wet the material. Even distribution is achieved by an adjustable device. This solvent distributor can easily pass a large percentage of fine solids and, as illustrated, it is possible to use the bed of material in the extractor as a filter for the miscella. Any fine particles which drain out of the cell bottom with the miscella are returned via the stage pump to the flooding device at the top of the extractor. The filtering effect is continuous throughout the process. This unique arrangement made possible by separated miscellas not only permits the recirculation of fines but also prevents hidden channelling. The Rosedowns extractor makes maximum use of the space in the extractor for flooded soaking extraction and produces a very clear miscella.

As an added precaution a fine mesh screen is sited above the final rich miscella outlet. This double-inclined large-surface screen clad with stainless steel mesh is washed clean by the miscella which is returned via the pump to the top of the extractor. Only screened miscella is discharged to the evaporation system. The advantages of this system will be very clear to experienced solvent plant users. It can be readily appreciated that this clean miscella greatly assists evaporation and stripping lecithin recovery and subsequent processing.

Ample interstage miscella storage is provided below the rotor of the extractor and the simple flow pattern facilitates efficient processing of all percolating beds.

The depth of material in each cell is easily adjusted in relation to the throughput of this pre treatment plant and at the turn of a dial the speed of the rotor can be varied to give the desired suitable time of contact between the material and the solvent.

The simple flooding facility greatly simplified process observation and it promotes efficient extraction control even with variable raw materials.eg.

1. Dry raw material
2. Feed conveyor

3. Material plug seal prevents leakage of solvent vapours and minimises oil entering extractor.
4. The extractor is simply driven by a fractional horsepower motor with variable speed control.
5. Automatic vent seal provides additional safety device to protect against excess pressure.
6. Convenient sight glasses facilitate observation of the process.
7. Special clean solvent control can be fitted to ensure flooding even with low solvent to meal ratio.
8. Ajustable door dropping gear controls impact of door or stop to assist in cleaning door drainage screen.
9. Simple agitator discharge ensures vapour seal.
10. Variable speed discharge conveyor.
11. Large mesh screen allows free drainage and recycling of fines.
12. Stage pumps recycle miscella (solvent and oil) to flood meal at each of four or five stages.
13. Fines automatically recycled and screen continuously washed clean.
14. Self cleaning screen ensures high clarity of final rich miscella.
15. No rubbling between meal and extractor wall.
16. Meal in cell acts as natural filter.
17. Controlled flooding for uniform extraction without channelling.
18. Miscella is distributed over whole cell instead of spraying thus permitting fine solids to be put back into meal which acts as a filter.

(ii) *Perforated Belt Type Extractor/Endless Band Type Extractor*: In this extractor the material is carried horizontally by means of an endless band and sprayed on top with miscella. A most successful example of this kind of extractor is the De Smet Extractor (Fig. 9.21). It consists essentially of a horizontal endless band made of articulated perforated draining plates, on which the material to be extracted travels as one compact bed. Each draining plate is covered with monel metal gauze and controlling shutters permit the regulation of the thickness of the material between about 24 inches and 28 inches. Under the travelling bands are compartments for collecting miscella which drains from the band. To each compartment belongs one pump which circulates the miscella back to the travelling band. There are altogether six pumps, feeding six atomizers with miscella. One other pump sprays fresh solvent on to that portion of the material which has arrived near the discharge end of the extractor. Between the miscella compartments are baffle plates, arranged in such a way as to allow miscella to overflow in the direction of the material inlet end. With the help of this device and the way the pumps are operated, the fresh solvent enriches itself continuously with oil as it progresses towards the meal inlet end of the extractor. The full miscella from this end flows to the concentration section via the filters. Very high extraction rates are thus obtained and concentration of miscella is high.

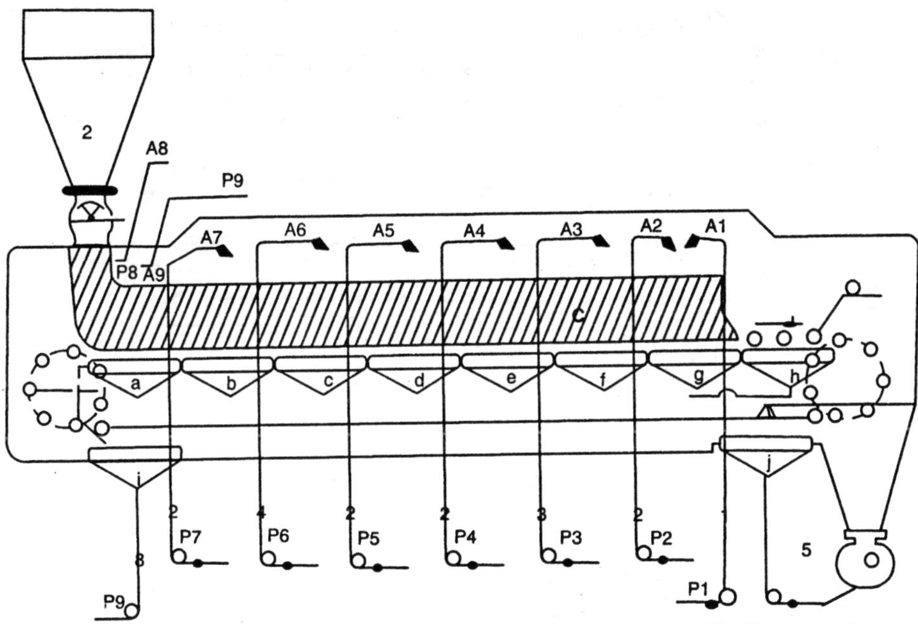

Fig. 9.21: De Smet Extractor. The Material Enters at 2 and Meal is Discharged through Chamber F. Solvent is Circulated through the Material C and Emerges as Full Miscella at Pump p.8.

The advantage of the endless band type extractor is that it is horizontal, an advantage shared by some of the screw conveyor types. Where floor space is cheap and maximum saving of building and structure costs is essential, it is a somewhat cheaper proposition that vertical designs. Where existing crushing mills have installed rows of presses in long one-storey buildings, it is often possible to make use of the existing building when it is required to exchange the older system for continuous solvent extraction.

(iii) *Vertical and Horizontal Basket Extractor*: This kind of extractor is characterized by small buckets or baskets with perforated bottoms and fixed to endless chains which carry the solid particles through a vapour-tight housing, while solvent is sprayed into the baskets. The early patents covered both horizontal and vertical forms of the inventions. It was the horizontal type that appeared more promising (Fig. 9.22).

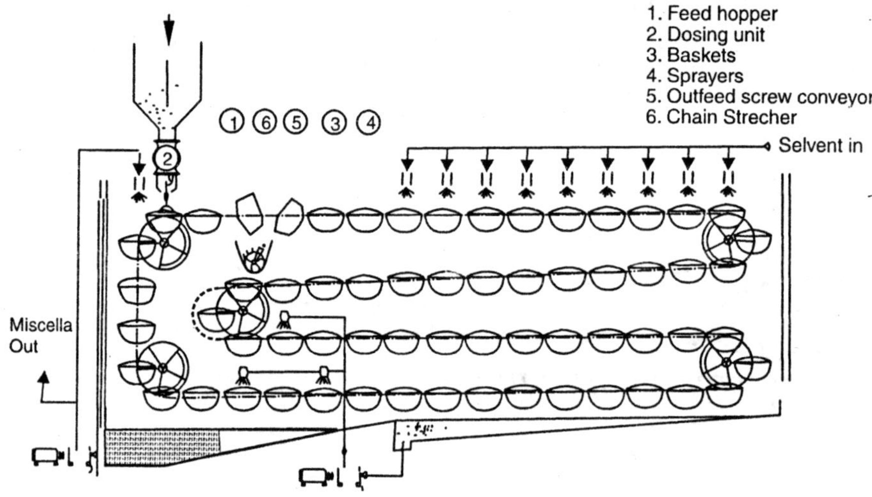

Fig. 9.22: C.M.B. Extractor (Simplex-40)

The material to be extracted enters through a hopper (I), and its constant level is ensured by an adjustable volumetric device. The material is dumped into the basket near the top of the descending side of the chain. As the baskets are filled with material and start down the descending side the solvent is sprayed onto them and an extraction process by percolation of solvent through the flake mass begins. This is a multiple extraction process carried out under programmed countercurrent conditions so that the miscella of low oil concentration is brought into contact with material with the least oil in it and vice versa. Under these conditions, ca. 40 washings are given by only two pumps. The special shape of the baskets and the volumetric dosing system ensure a uniform and shallow bed of material, which is never higher than 50 cm. Under these conditions, there is no possibility of channeling solvent or miscella flow through the flakes and consequent incomplete extraction of some areas. In the extractor the solvent or miscella percolates by gravity through four superimposed beds of material each 50 cm thick. Thus a 2 m thick bed of material is extracted without risk of uneven compacting of the material and consequent channeling and incomplete extraction.

IMMERSION EXTRACTOR (35)

Immersion extractors are very useful and are in service under the following conditions:

(i) For extraction of small volumes or low capacities of conventional seeds.

(ii) For direct extraction of high oil content seeds such as copra, safflower, peanut, cottonseed, and other seeds which tend to disintegrate and form fines upon extraction.

(iii) For extraction in continuation with percolation extractors in two-stage extraction systems.

The advantages of the immersion extractor are:

(i) *Its simplicity of operation*: It is not as sensitive to establishment of precise preparation conditions as is the percolation extractor. The immersion extractor, when properly designed, is able to extract efficiently very fine particles which could choke the bed of a percolation extractor.

(ii) *Ease of operation*: The liquid to solids ratios are not very critical, and the liquid flow rates are set to prevent backflow or carry over of fines. Since the immersion extractor easily accommodates a wide range of liquid and solids rates, it is not much affected by natural variations in these rates. It is also easily able to accommodate finely ground materials.

A typical Immersion Extractor with a continuous system where the diffusion process occurs most intensively is shown in Fig. 9.23 (36).

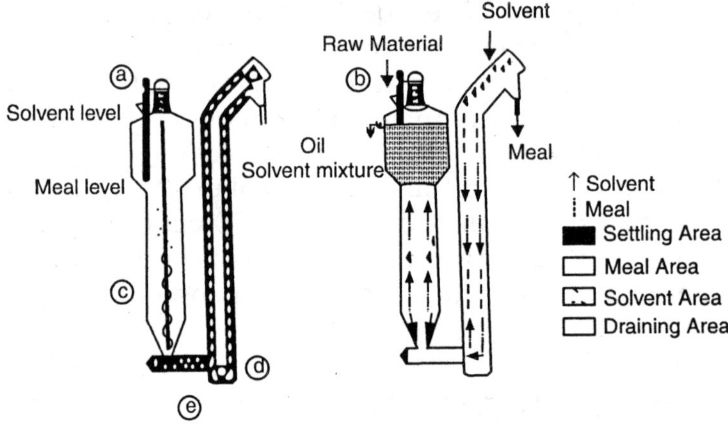

Fig. 9.23: Cross Section of the Immersion Extractor

The material to be extracted enters the immersion extractor through the dosing unit (A). The extractor housing is divided into two areas. The top acts as a setting area for the miscella and the lower part as an extraction area. In the latter area the material is generally agitated by the screw stirrer (C) which provides an intimate and homogeneous contact between material and solvent. The action of the screw stirrer also prevents packing of the material as this tends to ascend to the top of the cylindrical housing and then to fall by gravity along the housing walls. The extracted material coming out of the extractor bottom is removed by the metering screw conveyor (B) and conveyed to the elevator (D). The level of material in the extractor is maintained constant. The meal is taken by the elevator (D) and lifted up in a path counter current to the fresh solvent, which is admitted at the top of the elevator. While progressing in its ascending path, the meal leaves the solvent bath, is drained off as it is held in a series of baskets with perforated bottoms and is finally dumped into the desolventizer. The solvent flows in a counter direction, as indicated by the arrows, and thus counter-current extraction takes place.

This type of extractor is specially designed to handle materials which are difficult to prepare and to extract, principally the materials which are in a comminuted form. These materials cannot be handled in standard percolation extractors because of poor drainage through the fines unless provision is made for special pelletizing equipment. Even this often does not work and is very expensive.

COMBINED PERCOLATION AND IMMERSION EXTRACTORS

This process was first developed and applied on an industrial scale by CMB Company and has been patented under the name "Direx". By this system it is possible to remove completely the oil contained in the oilseeds, even if they have a high oil content, with no need for prepressing. The sequence of the processing steps is the following:

(i) Preparation of the seed, which is generally very simple in as much as it consists of cleaning, decortication, where required, cooking and reduction of the material.

(ii) Partial extraction of the oil in the percolation extractor until lowering the oil content of the seed to values ranging from 10 to 15 per cent, even if the incoming oil content is as high as 60 per cent.

(iii) Wet flaking of the partially defatted seed as it comes out of the percolation extractor. This operation is carried out in a special solvent-tight flaker operating in a solvent vapour-saturated atmosphere.

(iv) Final extraction of the flaked material in a second extractor of the immersion type, as described previously.

EXTRACTION UNIT

The operation of the "Direx" unit is described in Fig 9.24 (37).

The fresh solvent enters at point *C* descends through the elevator, and ascends to the top of the immersion extractor (3) containing the partially defatted material. By overflow the miscella passes into the first spouts of the percolation extractor (1) where it is brought in succession in contact with seed of increasingly higher oil content. In this way a highly concentrated miscella is obtained. The final miscella is then pumped to the subsequent filtration, preconcentration and distillation stages. The seed coming from the preparation room enters the percolation extractor through the metering unit (A), is partially extracted down to a residual oil content of 10–15 per cent for some kinds of seed even 3–4%), is then conveyed in to the flaker (2) and finally into the immersion extractor (3) where it is completely extracted. The totally defatted meal is dumped into the desolventizer-toaster (4) for solvent removal.

The fundamental step for an efficient performance of the "Direx" process is carried out in a flaker operating in a solvent-tight housing saturated with solvent vapours, which affords maximum reliability of service, as confirmed by the safe operation of such plants. This is further proved by the fact that the centrifugal pumps normally used in solvent extraction plants do not present any danger, even though they revolve at a speed of 1,400 rpm, as they operate in a solvent saturated atmosphere. The flaker runs at a low speed and the rolls always revolve at the same number of revolutions, thus operating only by flattering and never by grinding. The flattering pressure is extremely low as the material to be flaked is very soft, being impregnated with solvent. Cell breakage is therefore very easy and complete, thus facilitating the subsequent solvent extraction step.

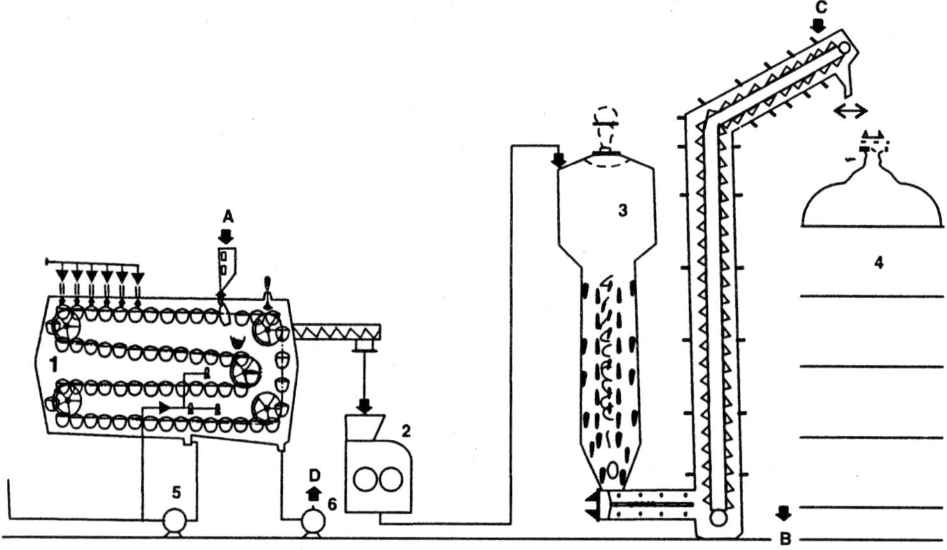

Fig. 9.24: Flow Sheet of the "Direx" System

Table 9.4: Comparison of Solvent Extraction Systems

Requirements per ton	*Prepressing and batch solvent extraction system*	*Prepressing and continuous solvent extraction system*	*"Direx" system*
Steam kg	100	280	290
Power KWH	45	55	29
Water m^3	14	12	15
Solvent kg	5	4	4
Labor hr	0.8	0.5	0.2

Some of the main advantages of the direct extraction system are lower capital investment, reduced maintenance costs, lower utility and material requirements, minimum labour requirements, and better quality of the finished products.

QUALITY OF THE FINISHED PRODUCTS

In the "Direx" plant the seed is extracted at a low temperature (maximum 50°C), and the operating temperature never exceeds 105°C in the subsequent meal desolventising stage. The latter factor is extremely important, especially when the meal is to be further processed into protein concentrates and isolates.

Such low temperatures cannot be maintained in the prepressing process in as much as the material is to be subjected to increasingly higher pressures in the expeller cage and the operating temperature ranges from 150 to 180°C, which causes some deterioration of the meal.

Table 9.4 lists the utility and material requirements per ton of seed processed which are recorded in the direct extraction system, conventional prepressing and batch solvent extraction procedure, and prepressing and continuous solvent extraction system.

A comparison between the traditional prepressing solvent process and the cold pressing-solvent process, as shown below, points out that the cold pressing-cum-solvent extraction has some advantages over the prepressing and solvent extraction in respect of power consumption and raw material cost:

Table 9.4A: Comparative Power Consumption (kwh/T)

Operation	*Prepressing & Solvent Process*	*Cold Pressing & Solvent Process*
Mechanical preparation	10	–
Cooking	65	–
Prepressing	15	35
Solvent Extraction	1	1
Desolventising	50	50
Transport and other costs	50	60
Total	191	146

Table 9.4B: Costs (U.S. Dollar) of Raw Material/Ton

Kind of costs	*Prepress & Solvent Process*	*Cold Pressing & Solvent Process*
Depreciation	18.80	17.49
Instalments	26.85	24.99
Hexane	1.24	1.23
Power	11.46	8.76
Oil in Meal	3.57	2.74
Total	61.92	55.21

The great advantage of solvent extraction over mechanical oil production methods is the fact that only one per cent (usually lesser) oil is left in the meal. In the case of Soyabeans oil recovery is over 97 per cent when solvent extraction is used against 80 per cent when pressing. In the case of seeds richer in oil than soyabeans, oil recovery is over 99 per cent with solvent extraction. The comparative yields between pressing and solvent extracting of soyabean seeds is shown here:

Comparative Yields between Pressing and Solvent Extracting

	Yields, per cent of total	
	Continuous Screw Press	*Continuous Solvent Extraction*
Oil	14.0	17.4
Meal	83.1	79.0
Moisture, dust, scale error and other losses in processing	2.6	3.6

Operation cost comparison per unit of production between screw pressing and solvent extraction indicates, as shown below, that the advantages are on the side of the solvent extraction rather than pressing:

	Per cent of total cost	
	Solvent Extraction	*Screw Press*
Labour	35.0	22.7
Power	17.5	30.0
Steam	32.0	21.7
Maintenance, repair	13.0	23.6
Supplies, expense	2.5	2.0
	100.00	100.00

SUPERCRITICAL EXTRACTION WITH SUPERCRITICAL CARBON DIOXIDE (38)

Beyond its critical temperature (T_c) and critical pressure (P_c), a gas enters a "supercritical fluid" state. For example, carbon dioxide above 31°C and 72.9 atmospheres exists in supercritical (SC-CO_2) state. The supercritical carbon dioxide is capable of solubilizing oils from oil-bearing materials and it has resulted in the development of extraction technology for recovery of vegetable oils.

Yields, colour, taste and odour of oils extracted with SC-CO_2 are dependent upon pressure and temperature during extraction as well as size and shape of the seed particles. Freidrich *et al*, (39) extracted full fat soya flakes with SC-CO_2 at pressure of 204–680 ATM (3,000–10,000 psi) and 50°C, and reported that the oil was lighter in colour and contained less iron and about one tenth the phosphorous of hexane extracted oil. Refined oils from the hexane and SC-CO_2 extraction initially had equivalent odour and flavour scores and after 4 days of storage at 60°C. In general the SC-CO_2 extracted oils had lighter colour, required less caustic refining and showed less tendency to undergo colour fixation in storage.

The various characteristics of SC-CO_2 process are:

(1) Higher pressures (8,000 psi vs 5,000 psi) extract more oil per unit of CO_2 consumed (Fig. 9.25).

(2) Higher temperatures increase SC-CO_2 extraction of oil at any pressure (Fig. 9.26).

(3) As in hexane extraction, phosphorous contents increase in later oil fractions extracted by SC-CO_2 (Fig. 9.27).

(4) Nonsaponifiable contents increase in later oil fractions extracted by SC-CO_2 (Fig. 9.28).

(5) Free fatty acid contents in oils first decrease and then increase (Fig. 9.29).

(6) Moisture levels within the range of 3–12 per cent have little effect on extractability of oil, but smaller particles improve extractability of oil by SC-CO_2 (Fig. 9.30) (40).

(7) Palmitate, linolenate and linoleate contents of later extracted oils decrease while oleate and stearate contents increase (Fig. 9.31).

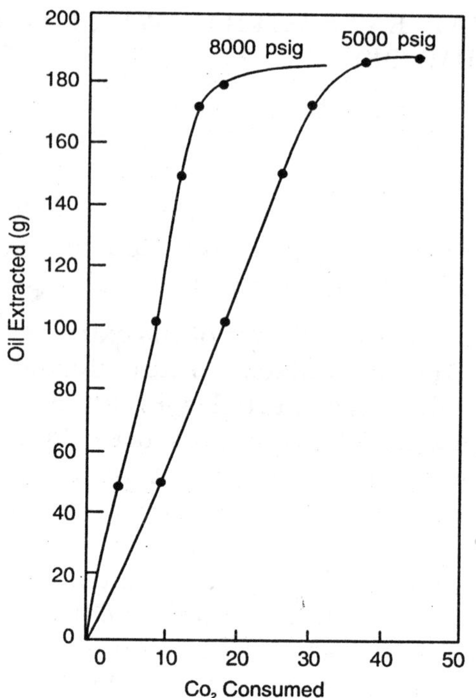

Fig. 9.25: Effects of Pressure on SC-CO_2 Extraction Efficiency of Soyabean Flakes at 50°C (39)

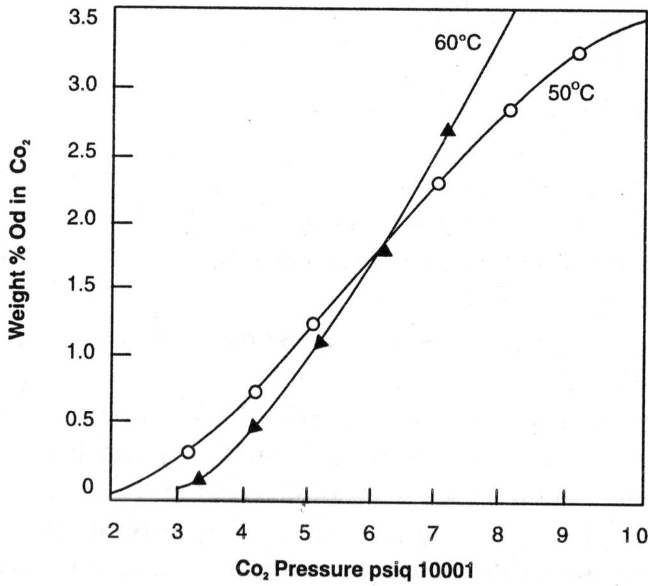

Fig. 9.26: Effects of Temperature and Pressures on Solubility of Soyabean Oil in SC-CO_2 (39)

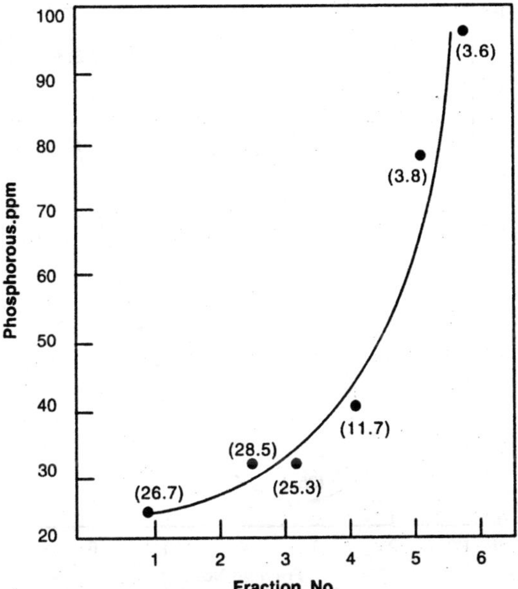

Fig. 9.27: Phosphorous Content of Soyabean Oil Fractions Extracted with SC-CO_2 at 8.000 psi and 50°C Numbers in Parenthesis are the Percentages of the Total Oil Represented to each Fraction (39)

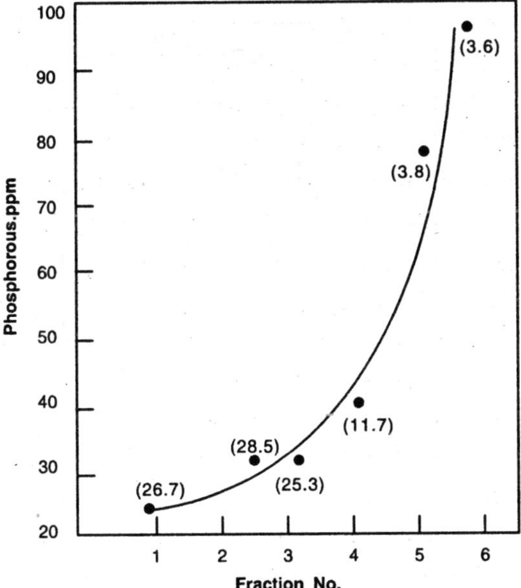

Fig. 9.28: Unsaponifiable Contents of Soyabean Oil Fractions Extracted with SC-CO_2 at 8.000 psi and 50°C Numbers in Parentheses are the Percentages of the Total Oil Represented by each Fraction

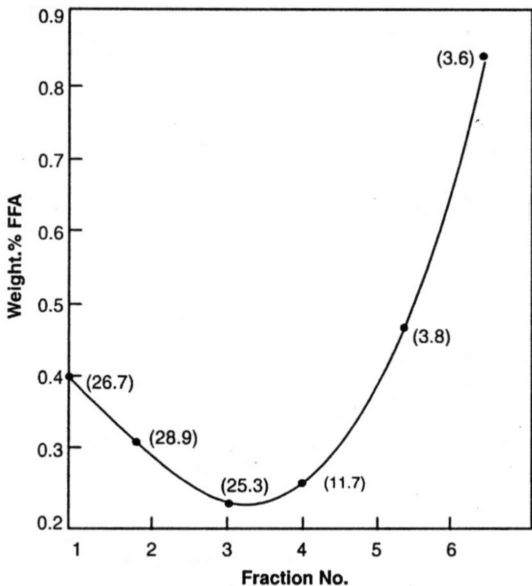

Fig. 9.29: Free Fatty Acids Content of Soyabean Oil Extracted with SC-CO_2 at 8.000 psi and 50°C. Numbers in Parentheses are the Percentages of the Total Oil Represented by each Fraction

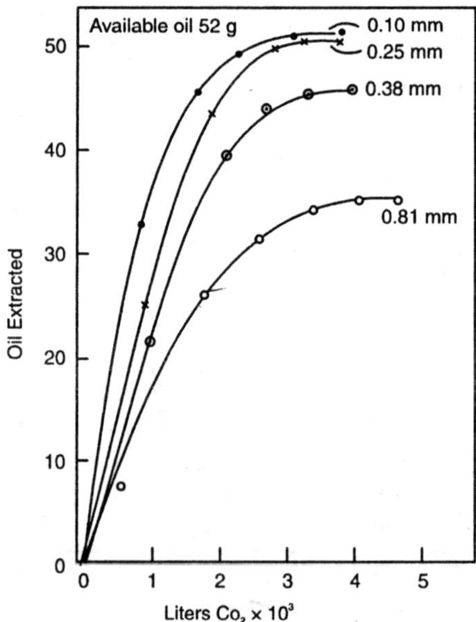

Fig. 9.30: Effect of Soya Flake thickness on Oil Extraction by SC-CO_2 Conditions 8.000 psi 50°C Available oil 52 g

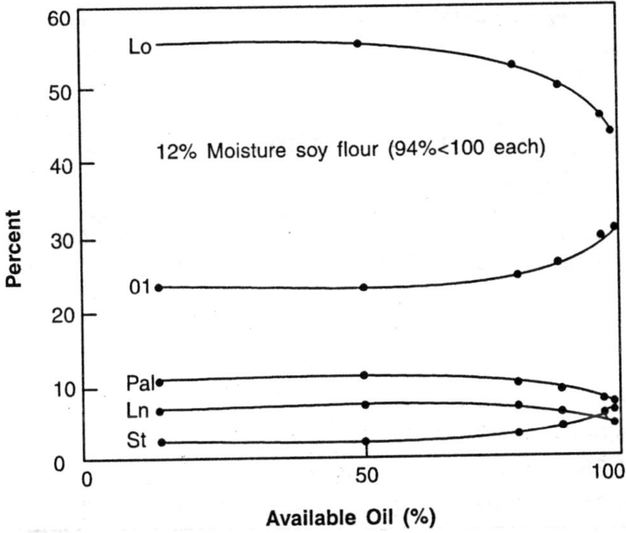

Fig. 9.31: Changes in Fatty Acid Composition During SC-CO_2 Extraction of Soyabean. Conditions: 8,000 psi, 50°C, Soy Flour with 12% Moisture. Ln = Linolenate; Lo = Linoleate; 01 = Oleate; St = Stearate; Pal = Palmitate.

(8) Hexane extracted degummed soyabean oils exhibit better oxidative stability because phosphatides, which are natural antioxidants are essentially absent in SC-CO_2 extracted oils although levels of tocopherols are unchanged (41).

(9) SC-CO_2 extraction does not dectivate amylase, chymotrypsin, trypsin or trypsin inhibitor, and additional heat treatment is needed by soya flakes to obtain optimal feed efficiency.

(10) Quantity of deodorized SC-CO_2 extracted oils is equivalent to commercial salad oils, as determined by a newly designed, laboratory-scale deodorizer.

Supercritical extraction technology has been succcssfully applied to fractional extraction of rice bran oil, extraction of jojoba and other seeds which are cold pressed for cosmetics and other uses commanding premium prices.

Although technical feasibility of extracting oilseeds with SC-CO_2 has been well demonstrated, applications in the oilseeds industry are delayed by the challenges of designing (1) extractors that will handle significant amounts of oilseeds (1000–4000 TPD range) at pressures of 6,000–10,000 psi and (2) continuous systems for loading and unloading oilseed flakes against the difference in pressures.

Some typical extraction data with supercritical carbon dioxides are included in the table below:

Oil-bearing material	*Pressure (psig)*	*Temp. (°C)*	*Time (hr)*	*Moisture (%)*	*Residual oil(%)*	*Oil recovered (%)*	*Oil description*
Cottonseed	8,000	50	8	8.6	0.5	30.8	Light
Groundnut	10,000	70-75	8	9.0	0.9	48.0	Light
Rice bran	9,500	70	6	6.5	0.9	19.2	Waxy
Soyabean	8,000	50	8	11.4	0.7	19.4	Light
Rapeseed	4,350	50	–	–	–	32.8	Light
Sunflower	4,350	40	–	–	–	46.2	Light

A comparison of Supercritical carbon dioxide and Hexane Extracted Crude Soyabean oil is shown below:

Analysis	*Hexane*	SC–CO_2 50°C, 8000 psig
Yield (%)	19.0	18.3
Residual oil (%)	0.7	2.1
Chromatographic Refining Loss (%)	1.9	0.6
Free Fatty Acid (%)	0.6	0.3
Peroxide value	<0.1	0.1
Unsaponifiables (%)	0.6	0.7
Iron (ppm)	1.45	0.3
Phosphorous (ppm)	505	45

AQUEOUS EXTRACTION (38)

In an aqueous extraction process, crushed seeds are dispersed in appropriate volume of water at a suitable temperature (~60°C) for a definite period of time at selective pH and centrifuged in a three-phase centrifuge when oil, whey and proteins are recovered.

Aqueous processes have been developed for producing food grade proteins from major oilseed like soyabeans, low gossypol, cotton, peanuts, sunflowers, coconut etc. Products analogous to defatted flour, protein concentrate and protein isolate have been made by aqueous extraction processing. Industrial type ultrafiltration and reverse osmosis membranes have been applied to assist separation of protein isolate and concentrates (42).

The aqueous process has a major limitation in that the protein concentrate or isolate contains residual oil in much greater proportion than the corresponding solvent process. The phosphatide level in an oilseed is directly related to residual oil content. In fact, it is difficult to produce dry peanut protein isolate with less than about two per cent residual oil, cottonseed and sunflower seed isolates with less than six per cent residual oil, and soyabean isolate with less than nine per cent residual oil.

DESOLVENTISATION OF MISCELLA AND MEAL (MARC) FOR RECOVERY OF OIL AND MEAL

In the extraction of oil-bearing materials by solvent process, the following stages are carried out after the miscella (oil in solvent) and marc (meal with solvent) are discharged from the extraction plant to respective outlets for the recovery of solvent-free oil and solvent-free meal.

The stages are as follows:

1. Miscella filtration
2. Miscella evaporation
3. Meal desolventisation and toasting
4. Meal drying and cooling (conditioning)

MISCELLA FILTRATION

The solvent extraction invariably leaches out some fine particles (fines) of oil-bearing materials, which have got to be rid of since these cause refining loss and colour and odour problems. These fines can be removed in industry by filtration only.

Filtration of miscella needs special precaution in view of the inflammability of the solvent. Another problem associated is the handling of high volume of miscella (several thousand liters per hour depending on the quantity of materials extracted by solvent). This step is therefore carried out in closed type filters.

THE RECOVERY OF SOLVENT FROM MISCELLA (MISCELLA DISTILLATION)

The miscella after filtration is separated into oil and solvent that is recycled in the extraction plant. The miscella from continuous extraction contains from 15–30 per cent of oil. Modern and efficient equipment yields a miscella with higher concentration. This is important with regard to the amount of solvent in circulation and to heat economy. The more diluted the miscella the more costly it is to evaporate all the solvent.

Equipment for miscella concentration varies somewhat but the principle consists of providing adequate steam-heated surface to evaporate the solvent. Vertical evaporator with central down-falls, climbing film evaporators, horizontal stills with withdrawable calendria, flash stills, and other designs are in use.

Miscella distillation systems have developed over the years into three basic designs. The first is simple distillation without steam savings. This process is generally utilised in the small capacity plants processing 100–200 tonnes seed per day.

The second is steam saving (economising) evaporation, using desolventiser-toaster vapours to vapourise part of the solvent in the miscella. This system uses single stage evaporation for obtaining fully concentrated miscella (90 per cent or more by weight) in those cases where the cooling water temperature available is sufficiently low to allow operation at high vacuum. This is followed by single stage oil stripping for final removal of solvent.

The third miscella distillation system also utilizes desolventiser-toaster vapours but with two-stage evaporation, followed by two-stage oil stripping and oil drying in those cases where the cooling water temperature is high or air-cooled condensers are used. The use of air cooled condensers is getting more widely adopted because of the need for water conservation and also because of the unavailability of cold cooling water in many parts of the world, both of which result in increasing the reliance on cooling towers or air-cooled condensers.

In the earlier practice, the recovery of solvent is achieved in 3-stages (19). The miscella is concentrated by 50% by distillation in two steam-heated pot stills. Later the miscella is further concentrated in a falling film evaporator when the oil contains 5-10% solvent. This last portion of the solvent is removed by steam stripping the oil in one or two passes.

In many plants, the pot stills have been replaced by miscella preheaters and rising film evaporators with entrainment separators. In some cases the falling film evaporator has been retained or replaced with a small rising evaporator operated with sparging steam; in others, miscella from the first large rising film evaporator goes directly to the stripping column. Usually the stripping column is maintained under reduced pressure with a steam ejector.

One of the most widely used special continuous equipment unit is shown in Fig. 9.32 that runs as follows:

(i) The highly concentrated miscella containing traces of solvent enters at point A, descends as a falling film over a series of trays and finally collects at the bottom of the column.

(ii) In the bottom area live steam is injected which flows upwards countercurrent to miscella because of the action of a water condenser and strips off the last traces of solvent form the oil.

(iii) Vacuum is maintained on the whole system during operation.

MEAL DESOLVENTISING, TOASTING, DRYING AND COOLING

Meal (marc) contains a considerable amount of solvent 25–40 per cent (W/W). The solvent must be completely removed and recovered for return to the extraction plant. Most extraction systems provide a mechanism for reducing the solvent content of the material before it enters the heated desolventising section. The material may be drained by lifting it in an elevator or it may be pressed to release that portion of the solvent which separates rather more easily.

Some oil seed meals after desolventisation require cooking or toasting to destroy antidigestive factors and inactivate undesirable enzyme activity in order to become suitable for use as cattle feeds or for human consumption. The cooking, which is most effective with moist heat, is most conveniently done in the same equipment used for desolventising.

The desolventised meal has to be dried to a final moisture content of 12–14 per cent preferably under 12 per cent and cooled to below 40°C for storage before shipping.

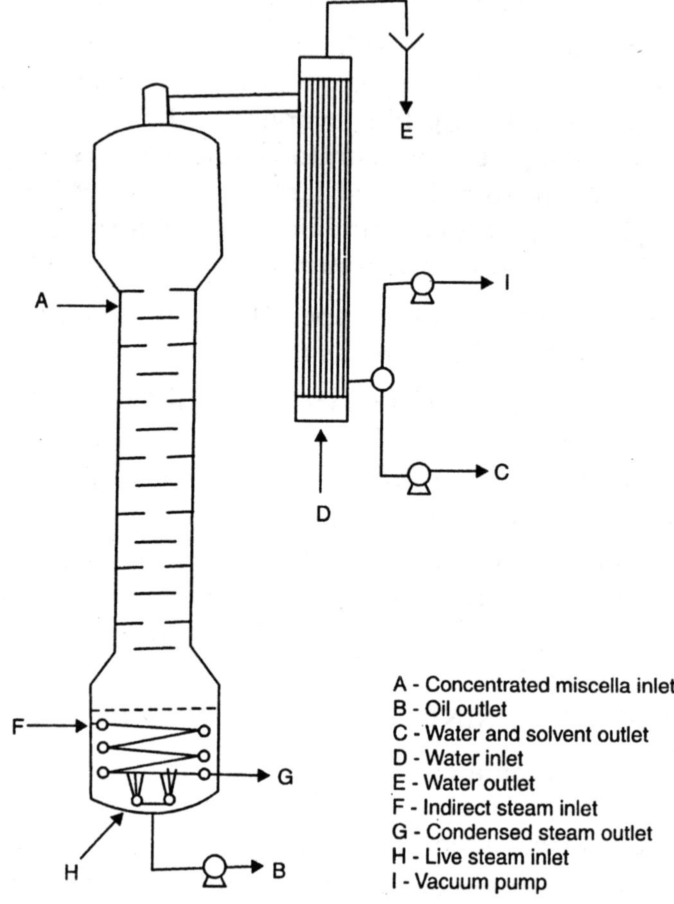

Fig. 9.32: Stripping Column

In practice the desolventisation is done by evaporation by steam in the equipment designed for indirect heating, or by direct steaming or a combination of the two.

The earliest desolventisers (indirect heated) consisted of one or more banks of steam-jacketed horizontal ladle conveyors arranged in vertical banks. The meals are sometimes discharged to a deodorizer vessel swept with live steam to remove final traces of solvent.

A special desolventiser (C.M.D. design), shown in Fig. 9.33, operates on a shallow meal layer and almost without live steam addition thus enabling the total protein-soluble protein ration to be kept almost unchanged. The operation of this desolventiser is as follows:

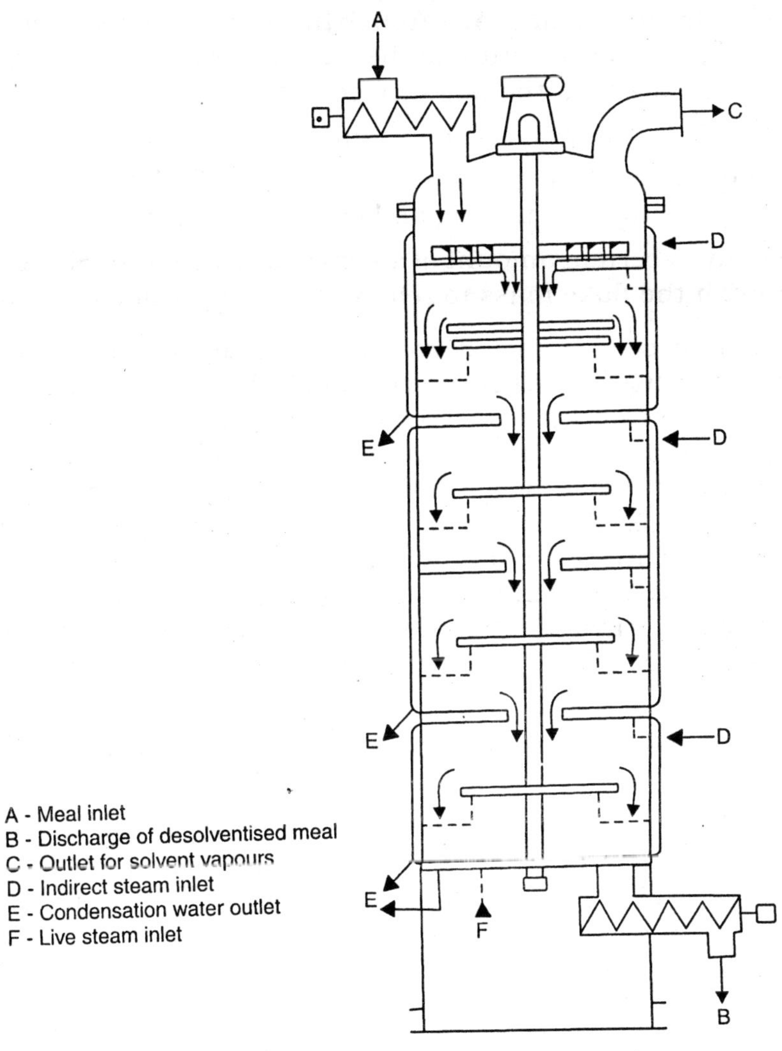

Fig. 9.33: Superimposed Tray Desolventiser (C.M.B.)

(i) The solvent-impregnated defatted flakes enter point A through a screw conveyor-feeder and fall over the first tray of the desolventiser.

(ii) A system of rotating paddles moves the flakes over the surface of the tray causing them to fall on the tray below, from where they drop to the next tray until they reach the bottom tray.

(iii) From there they are taken out by a metering screw conveyor.

(iv) During the downward flow through the trays of the desolventising column, the temperature of the flakes is increased because of their contact with the hot surface of the steam-heated trays.

(v) In addition, the walls of the column are also heated by steam.

(vi) Because of heating, the solvent evaporates and its vapours are conveyed into a scrubbing and condensing unit.

(vii) Usually, a certain amount of steam is injected in the lower tray through the flake mass to remove the last portions of solvent.

The type of seed extracted plays an important role in the designing of desolventiser. Certain oil seed meals deteriorate during contact with live steam that is injected through the meal bed during desolventisation. The number of trays and their dimensions may vary with the nature of oilseed meal. Thus for a plant processing 240 tonnes per day oilseeds and using live steam injection in the last tray during desolventisation but without any action on meal, the desolventiser should be 2.20 meter in diameter and it would require 8 steam heated trays. For peanut meal, the diameter remains the same but at least 14 trays would be required for desolventisation.

MEAL DESOLVENTISING-TOASTING (LIVE STEAM OPERATED) (43)

The meal desolventiser is largely replaced by 'desolventiser toasters', or in short D–T units. The toasting stage is very much required particularly in the case of soyabean meal to deactivate its urease enzyme content especially when the meal is to be used in ruminant feeds containing urea.

Urease hydrolyses urea into ammonium carbonate with decrease in the nutritional value of the meal. Therefore urease has to be deactivated in soyabean meal but at the same time keeping the total protein soluble protein ratio nearly unchanged. This deactivation of urease is done by toasting the meals. The toasting process is most efficiently accomplished by the addition of large quantities of sparge steam. The urease activity of a soya meal is expressed in terms of (i) milliequivalents of NH_3 (ammonia) produced by one gram of soyabean meal; and (ii) pH increase of a meal-urea solution under given conditions of temperature, usually 30°C.

A good-grade soyabean meal should have a urease activity of 0.4–0.6 expressed in milliequivalents of NH_3 and a pH increase ranging from 0.05 to 0.10 and the nitrogen solubility index (NSI) of 15–20.

The design of an early desolventiser-toaster is shown in Fig. 9.34. The machine consists of a series of circular steam-heated trays arranged vertically inside a vapour-tight shell. A central vertical shaft has been attached to it sweep arms which rotate with a close clearance to the trays. The top tray sweep arms have perforated pipes through which live steam is admitted to the solvent-wet bed of flakes. The level of material on each tray is maintained by chutes or gates on each tray, and the overall level in the machine is maintained by variation in the speed of the discharge screw. In operation, the solvent-wet flakes enter at *A* and drop into the bed of flakes on tray 1. Live steam from the rotating pipe C vaporizes most of the solvent from the flakes, and the condensing steam increases the flake moisture content to over 20 per cent. The process is explosive, which ruptures the internal flake structure and renders it more susceptible to uniform cooking for optimum nutrition. Vapour leaves the top and is scrubbed of fines before being condensed and recovered. The flakes then pass from tray to tray down through the machine where they are cooked and dried. The vapour passes upward through grids in the steam-heated trays.

This type of design is still used today with variations. The French-Oil Mill D-T is equipped with individual level control gates on each tray and can also be furnished with chutes. The Dravo D-T is similar to the Central Soya design, and the EMI D-T is equipped with internal gates and chutes for meal level control by variable speed discharge. This type of D-T design is also furnished by De Smet, H.L.S. Ltd., Bernardini, Extrakionstechnik, Crown Iron Works and Simon-Rosedowns.

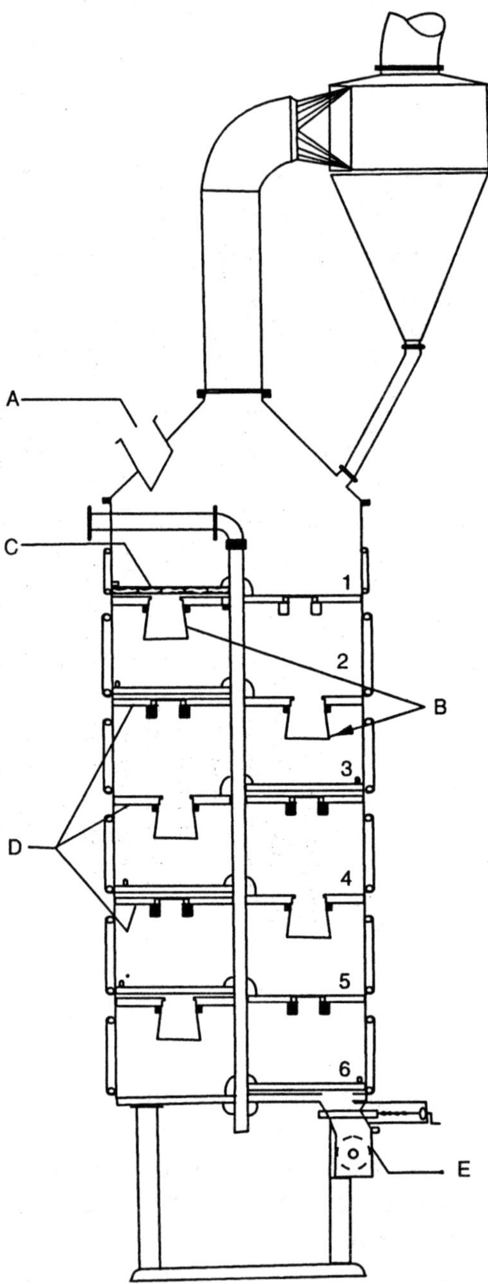

Fig. 9.34: Early Desolventizer-toaster Design

A number of Desolventiser-Toaster equipments are commercially available with variations in the mechanical and the process features of the D-T systems.

Basically, the processes involved in any of the desolventising toasting systems are similar. The D-T unit is a vertical shell vessel containing steam heated trays, sparking steam sections, sweep arms and a drive system, and a means for discharging the vapour at the top and the meal at the bottom. Hexane wet flakes enter the top section and are sparged with steam on one of the upper trays. Steam condenses in the flakes vaporizing the hexane so that the moisture content is raised to about 20 per cent. In the lower tray sections, flakes are heated to about 225°F (107°C), and the moisture content of the flakes is reduced several per cent. Desolventiser-toasters are commonly used to produce toasted meal for ruminant and poultry feeds. A typical equipment of high capacity 1000 TPD is shown in Fig. 9.35. Typical PDI products range from 10 to 30 per cent.

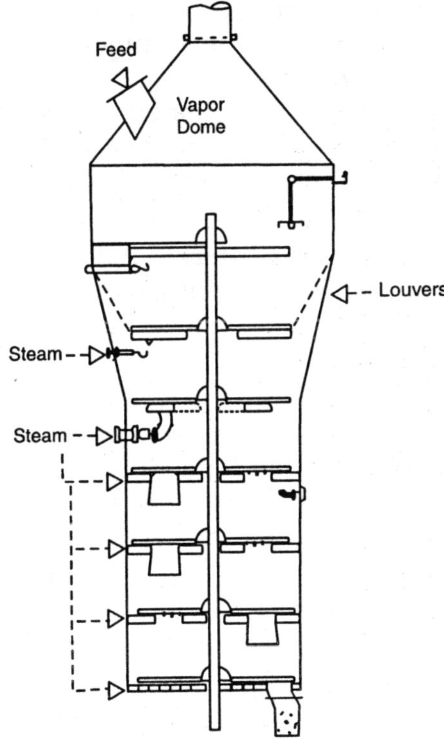

Fig. 9.35: Improved Model Desolventiser-toaster

An expanded vapour dome is added to reduce vapour velocity and minimize entrainment of flakes and fines. The top tray serves as a hot plate for partial desolventising of the flakes. The third tray is perforated for addition of live steam, which is admitted into the flake bed in the tray above by means of louvered panels, thus permitting counter current passage of live steam and flakes with good contact between them. The three upper trays form the desolventising zone, the lower four trays form the cooking zone. Normally, with this type of D-T optimum cooking is achieved within 10–12 min retention time and the D-T is designed to discharge meal at over 20 per cent moisture and temperature of 100–110°C requiring subsequent drying and cooling.

The extremely large capacity processing plants being built today employ the same basic design with modifications which include additional sparging trays, live steam addition through sweep arms and pots in the annular space of the tray, increased diameters of trays and vapour domes, and drive horsepower up to 250 HP.

Another development in the conventional D-T is the DTDC (desolventiser-toaster-dryer-cooler) machine which performs desolventising, toasting, drying and cooling continuously. The desolventising-toasting is accomplished on a perforated live steam tray. The drying is done by contacting hot air with the toasted meal on another tray. Finally, the cooling is accomplished by blowing cold air through the meal on the bottom tray.

This process ruptures the internal flake structure and renders it more susceptible to uniform cooking for optimum nutritive value. Meal toasting is continued on the lower trays where meal moisture is reduced by indirect steam heating.

The solvent vapours leaving the top of the unit are scrubbed of fines, then fed to the steam economizer in order to preconcentrate the miscella, and are finally condensed. Therefore, three zones may be distinguished in this D-T, namely: (i) Upperpreheating zone; (ii) central desolventising and humidifying zone; and (iii) lower cooking zone.

VAPOUR DESOLVENTISER-DEODORIZER

Figure 9.36 illustrates the Dravo vapour desolventiser deodorizer system, which is designed to provide a broad range of PDI products. This system is designed to allow time, temperature, and moisture to be adjusted at will during operation. All but one per cent of the solvent is removed from the meal by the superheated vapour in the desolventiser. The meal then passes through a lock to the deodorizer. The steam pressure within the deodorizer can be regulated between 1/2 atmosphere and 2 atmospheres. By adjusting this steam pressure and by minor adjustments to the height of the discharge dam, the PDI of the meal can be adjusted over a range of 10 to 90 with the upper limit approaching one or two per cent below the PDI of the extracted flakes.

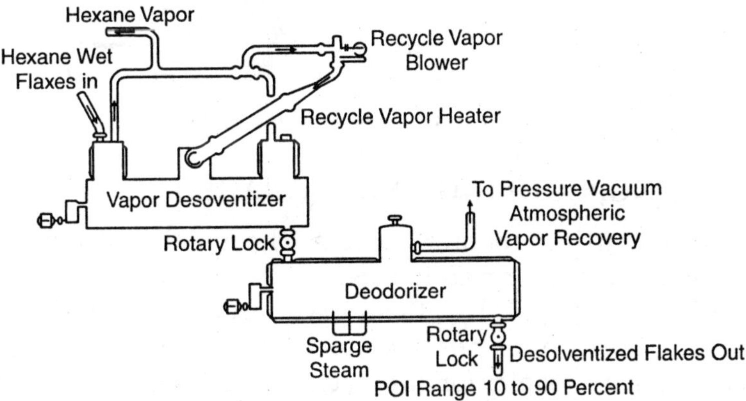

Fig. 9.36: Vapour Desolventizer Deodorizer System U.S. Patent No. 3,392,455

In some cases desolventising is accomplished by superheated solvent vapours. This and "Flash" desolventising are of interest mainly where it is desired to avoid toasting and to minimize denaturation of the protein, as for example, when undenatured or partially denatured flakes are required because of protein specifications.

Basically, a flash desolventiser (Fig. 9.37) consists of a pneumatic-conveying desolventising tube, a cyclone flake separator, a hexane blower, and a hexane vapour heater arranged in a closed-loop system in which superheated vapour is constantly circulated. Solvent-wet flakes are fed continuously into the system through a variable-speed feed conveyor to the circulating vapour stream where the solvent is removed down to about one per cent. The desolventised flakes are then discharged into a cooking system wherein the variables of time, temperature and moisture can be adjusted to allow flakes to be produced in a closely controlled PDI range as required; low range 10–25, medium range 40–60, and high range 70–85 flakes can be produced at will by adjusting the variables of time, temperature and moisture during cooking.

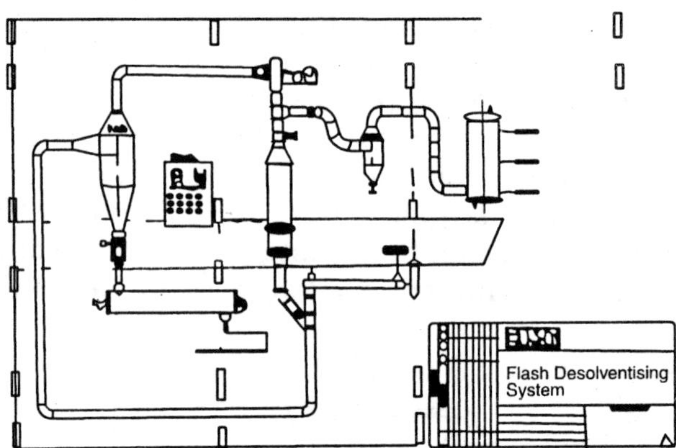

Fig. 9.37: Flash Desolventising System

MEAL DRYING, COOLING AND STORAGE

From a desolventiser-toaster system, the meal is dried first in a rotary steam tube dryer and is then cooled by air before being sent to storage. Meal from DTDC, flash dryer and vapour desolventiser-deodorizer systems does not need to be dried. Buhler-Miag offers a fluidized bed dryer-cooler for capacities of 2,000 metric tons per day of beans which is a self-contained unit with heating and air handling equipment.

Typically, the desolventised meal will be stored at 12 per cent or less moisture and about 40°C maximum temperature.

MEAL PELLETIZING

The defatted cool meal leaving the extraction plant can be sacked and stored as it is. However, to enhance handling characteristics through improved flow and less dust, it is normally compacted in pellets with greater bulk density. This also allows for a greater utilization of storage and shifting space.

RECOVERY OF SOLVENT VAPOURS (SOLVENT VAPOUR RECOVERY SYSTEMS)

Solvent losses must be kept to a minimum to make solvent extraction a paying proposition. In a good continuous extractor, losses are about 0.3 weight per cent of solvent (calculated on the material (solvent) as it enters the reactor.

Solvent losses may occur in three main ways: in the extracted meal, in the oil and from the plant. Since modern equipment ensures practically solvent-free oil and meal, the only source of loss is the "breathing" of the plant itself.

All parts of an installation in contact with the solvent, including the storage tanks, are therefore vented to a solvent vapour recovery system. There are two main types of solvent vapour recovery systems: one, which is in general use today, uses liquid absorption, and the other uses absorption on solids. Vegetable oils, or chemicals such as tetraline, may be used in the former. More often, however, a refrigerated brine is employed which gives very good results. Hexane has a relatively high vapour pressure, and consequently about a quarter of the solvent is lost at an air temperature of 60°F. It is therefore advisable to work at about 40°F solvent-air temperature. Solvent recovery by absorption of solids calls for relatively expensive plant and absorbent, usually activated carbon. Some designs are using extracted meal as absorbent in addition to the solvent upon absorption system. When the meal is soaked with solvent, it is merely returned to the desolventising.

Great care must be taken particularly with regard to solvent vapour from the meal desolventising section. This vapour usually contains a certain amount of fines which it is desirable to eliminate from the vapour before it enters the vapour recuperation plant.

When inflammable solvents are used—and this is the case today in most solvent extraction plants—the solvent vapour recuperation plant is intended not only to reduce solvent losses, but to eliminate bad effects on health and fire hazards.

CALCULATION OF NUMBER OF STAGES OF SOLVENT EXTRACTION (TAKING FLAKED SOYABEAN AS AN EXAMPLE)

Flaked soyabeans are to be leached with hexane to remove the soyabean oil. A 0.3 m thick layer of the flakes (0.25 mm flake thickness) will be fed onto a slowly moving perforated endless belt which passes under a series of continuously operating sprays. As the solid passes under each spray, it is showered with liquid which percolates through the bed, collects in a trough below the belt, and is recycled by a pump to the spray. The spacing of the sprays is such that the solid is permitted to drain 6 min before it reaches the next spray. The solvent also passes from trough to trough in a direction countercurrent to that of the moving belt, so that a truly continuous countercurrent stage-wise operation is maintained, each spraying and draining constituting one stage. Experiments show that the flakes retain solution after 6 min drain time to an extent depending upon the oil content of the solution, as follows:

wt % oil in soln.	0	20	30
Kg soln retained/kg insoluble solid	0.58	0.66	0.70

It will be assumed that the retained solution contains the only oil in the drained flakes.

The soyabean flakes enter containing 20 per cent oil and are to be leached to 0.5 per cent oil (on a solvent-free basis). The net forward flow of solvent is to be 1.0 kg hexane introduced as fresh solvent per kilogram flakes, and the fresh solvent is free of oil. The solvent draining from the flakes is generally free of solid except in the first stage: the rich miscella contain 10 per cent of the insoluble solid in the feed as a suspended solid, which falls through the perforations of the belt during loading.

How many stages are required:

Solution: The tie lines are vertical, $x = y^*$, Rearrange the drainage data as follows:

Per cent oilinsoln = 100y*	$\frac{\text{kg soln retained}}{\text{kg insoluble solid}} = \frac{1}{N}$	N	$\frac{\text{kg oil retained}}{\text{kg insoluble solid}} = \frac{y^*}{N}$
0	0.58	1,725	0
20	0.66	1,515	0.132
30	0.70	1,429	0.210

Basis: 1 kg flakes introduced.
Soyabean feed B = 0.8 kg insoluble; F = 0.2 kg oil;
N_F = 0.8/0.2 = 4.0 kg insoluble solid/kg oil;
Y_F = 1.0 mass fraction oil, solid-free basis.
Solvent R_{Np} + 1 = 1.0 kg hexane; X_{Np} + ! = 0 mass fraction oil.

Leached solids kg oil/kg insoluble solid = 0.005/0.995 = 0.00503

By interpolation in the equilibrium data,

N_{NP} = 1.718 kg solid/kg soln.

Insoluble solid lost to miscella = 0.8(0.1) = 0.08 kg

Insoluble solid in leached solids = 0.8(0.9) = 0.72 kg.

$$E_{NP} = \frac{0.72}{1.718} = 0.420 \text{ kg soln. retained} = 0.420 \text{ kg soln. retained}$$

Oil retained = 0.00503(0.72) = 0.00362 kg

Hexane retained = 0.420 – 0.00362 = 0.416 kg

$$ENp = \frac{0.00362}{0.420} = 0.0086 \text{ mass fraction oil in retained liquid Miscella}$$

Hexane = 1 – 0.416 = 0.584 kg;

oil = 0.2 – 0.00362 = 0.196 kg

R_1 = 0.584 + 0.196 = 0.780 kg clear miscella;

X_1 = 0.196/0.780 = 0.252 mass fraction oil in liquid.

N_{R_1} = 0.08/0.0780 = 0.1027 kg insoluble solid/kg soln.

The operating diagram is shown in Fig. 9.38. Point R_1 represents the cloudy miscella and is therefore displaced from the axis of the graph at N_{R_1} Point ΔR is located as usual and the stages determined with the N = 0 axis for all stages but the first. Between four and five stages are necessary. Adjustment of the amount of solvent or the amount of unextracted oil, by trial, will provide an integral number.

THE PRESENT STATUS OF OIL MILLING INDUSTRY IN INDIA

The crushing of oil seeds has been practised in India since the time immemorial with the use of bullock-driven ghani. It was after World War I that large-scale crushing of oil seeds with modern appliances was started. The Second World War gave further impetus to the oil milling industry. With this the export of oil seeds from India declined. Presently almost the entire crop is crushed within the country. Except the hand picked ground nut (HPS), sesame and niger seeds, the current export of oil seeds is marginal and the two principal feedstocks for industrial oils, i.e. castor seed and linseed are now entirely crushed in India. Castor oils and its derivatives are now exported. Generally castor seeds are not exported.

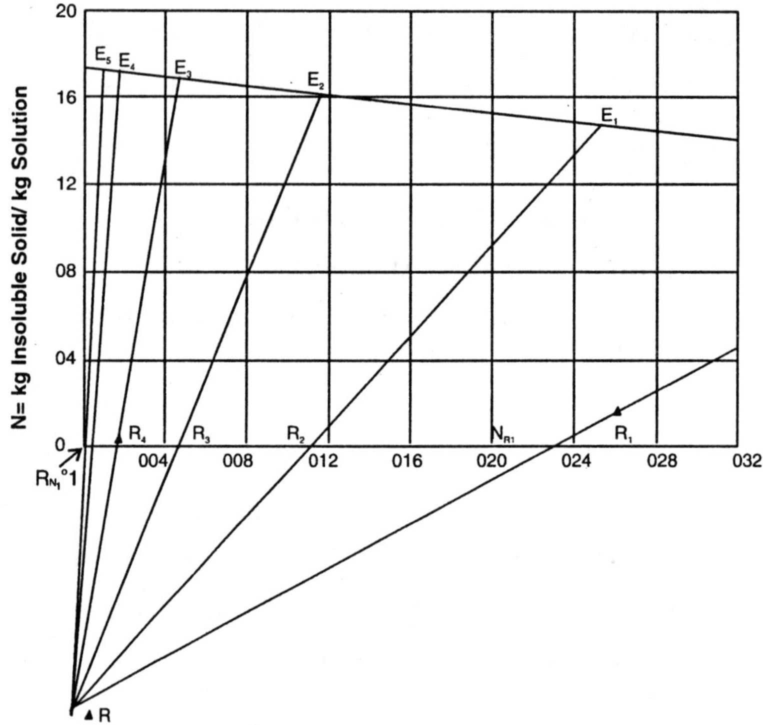

Fig. 9.38: Solution to Illustration

PRODUCTION AND PROCESSING CAPACITIES

	No. of units	*Installed capacities in Million Tonnes*	*Utilised percentage*
Ghanis	2,50,000	20.00	10 per cent
Expellers	50,000	30.00	30 per cent

The oil milling industries may be grouped under the three following heads depending on the method of milling:

(i) Ghani—manual/animal driven/or small power driven;

(ii) Larger oil mills registered under the Factories Act using presses and expellers;

(iii) Plants using solvent extraction technique.

The authentic data on the number of Ghanies operating in India at present and their total capacity are not available readily. However, it has been assessed by the trade and industry source that there are 2,50,000 ghanies all over India including 50,000 of improved power operated ghanies. It may be noted that the total number of ghanies used to be about 4.5 lakh in 1951 which declined to about one lakh by 1977, but the numbers increased to 2,50,000 by 1993. These figures have been attributed to the following factors: (a) Rural development programmes for generating rural employment; and (b) technology upgradation of ghani such as rotary ghanies which led to their utilization by large oil mills for processing groundnut, mustard and rapeseed to maintain the typical aroma having consumer preference.

Ghanies or rotary ghanies (kolhus) are mostly used for processing edible oil seeds. The castor seed is processed 100 per cent for industrial oil and it is almost entirely processed in expellers. Linseed oil, although an industrial oil, has some edible use amounting to about 20 per cent which is produced by ghanies. The following are the proportions of processing by village ghanies and power mills regarding the extraction of edible oil seeds.

	Processed by village ghanies %	*Processed by power mills %*
Ground nut	20	80
Rapeseed/Mustard	37	63
Sesame	83	17
Coconut	20	80

The ghanies are mostly concentrated in Central India and four neighbouring States of U.P., Bihar, Madhya Pradesh and Rajasthan which together contain nearly 70 per cent of the ghanies. The only other State having a fairly large number of ghanies is Tamil Nadu. Thus these five States share 80 per cent of the ghanies in India.

THE CONSTRUCTION OF GHANIES

Bullock-driven: The ghani is entirely constructed of wood with the exception of a small iron band fixed around the top edge of the mortar which measures 35 cm in size at the top tapering to about 30 cm at the bottom. It is about 75–100 cm in height of which 45 cm is normally fixed in the ground to keep the machine set when in operation. A pestle made from hard wood about 150 cm long and 60–80 cm in circumference is placed inside the mortar. It is held firmly in position by a cone like construction enabling the pestle to revolve with force inside the mortar by means of a lever about 5 meter long which on its extremity is loaded with stones. The pestle is turned around at a speed of 5 to 7 revolutions per minute by a pair of oxen fastened with the free end of the lever. At the same time it makes an up and down movement crushing the material in the mortar. A small hole at the bottom of the mortar allows expressed oil to fall in the small earthenware pot from which the oil is transferred to another from time to time. A ghani driven by two bullocks takes about three hours to crush the charge of 10 kg mustard. The cake contains 11–16 per cent of oil depending on the oil seed. The resultant pressure exerted in oil seed in ghani is about 150 lb/sq. in.

Power-driven ghani: It has all the features of a bullock-driven ghani except that the power is supplied to the pestle by an electric motor. The Khadi and village Industries Commission (KVIC), Mumbai, has technologically upgraded the traditional ghani. The improvements are: (1) the time taken per charge is reduced from 15 per cent to 30 per cent and (2) the percentage of yield of oil is improved by 1-2 per cent.

The mortar in this ghani is also made of wood lined with mild steel. The lower end of the pestle which is inserted in steel pipe crushes the seed in the mortar. The other end is fitted in the housing with ball bearing to revolving arm connecting level gears which transmit power from a 3 HP motor. The motor and the level gear arrangement are fitted on top of the ghani on a framework which can be made of wood or steel. The wooden frame occupies the space of 11 sq. meter while the pipe structure requires much reduced space. The pestle revolves along the periphery of a stationary mortar at 10 revolutions per minute and rotates about its end axis at two revolutions per minute due to friction with seeds.

It takes two hours to crush one charge of 10 kg of Mustard and the oil drains out of the 24 mm hole in the bottom of the mortar. Ten kg of Mustard yield about three kg of oil and 100 gms oil mud is recovered. The pestle needs replacement after every two months and mortar every four months. The life of this ghani is about 10 years.

Kolhu: The rotary mill is also designed on the basic idea of bullock-driven ghani. Both the pestle and the mortar are made of wrought iron. A bucket made of 1-3 mm steel sheet is fitted on the mortar serving as seed container. A ring 25 mm diameter is seated on a saucer commonly known as taper ring to serve as an oil bowl. The periphery of the mortar is fitted with vertical pegs 15 cm long forming replaceable scrapers. The pestle is obliquely placed on the taper ring in a way that there is a clearance of 0.125 to 0.5 mm between the round ring depending on the seed. The pestle rotates due to the friction with the rotation of the taper ring. The mortar is made to rotate from a shaft by means of a pinion working in the beveled wheel fixed to its lower portion. A rotary rotates at a speed of 14–15 revolutions per minute. The bucket is 45 cm diameter at the top tapering to about 30 cm at the bottom. The length of the pestle is about 75 cm and the total height of the rotary mill is around one meter. It requires 5 HP motor whereas a 7.5 HP motor can serve about two such mills. The charge consists of 12.5 kg of seeds leaving the cake with an oil content of 10–12 per cent after two runs of crushing.

The functioning of these 3 types of crushing mills is briefly compared as follows:

Capacity in the bullock-driven ghani in kg is about 6–10, overhead type power ghani and of KVIC is about 12–15 kg crushing capacity depending on the oil seeds is about 60–80 kg. In bullock-driven ghani 80–100 kg, overhead power ghani and KVIC ghani 90–115 kg. Crushing time decreases from bullock-driven ghani 90 minutes, 60 minutes in overhead ghani and 50 minutes KVIC ghani. Recovery of oil depends to some extent on seeds and percentage of recovery of oil varies from 35–38 percent in the case of oil seeds and 58–62 per cent in coconut. Floor space in square meter is reduced from 36 to 7.2–3.24 in the other types of ghanies.

In recent years, the bullock-driven ghanies are becoming extinct. However, the mechanical power ghani operated as designed by KVIC is still being used in India particularly as a means of generating rural employment and may continue to survive to crush many minor oil seeds of tree origin as local suppliers are too small to set up viable oil mills. In recent years, because of rise in the price of oil seeds and the loss of oil in the cake, modern methods are gradually replacing these comparatively outdated methods. With the change in the consumers preference as they get used to filtered, refined, deodorized oil, the survival of these types of oil mill technology is doubtful in the long run.

PRESSURE EXTRACTION BY EXPELLERS

Since the conventional ghani leaves about 12–18 per cent oil in the oil cake, the use of improved mechanical extraction processes which allow better recovery of oil leaving only 6–7 per cent oil in the cakes have been gradually gaining ground in India. Solvent extraction is used for extracting oil from the oil-bearing materials like cotton seed, rice bran and soyabean containing upto 20–22 per cent of oil. Oil seeds containing more than 22 per cent of oil, for example groundnut, sunflower, and safflower, are first extracted by expeller after which the oil is extracted by solvent.

EXPRESSION

Expression is the separation of the liquid from a solid-liquid system by compression in conditions that allow the liquid to escape while retaining the solid between compressing surfaces. The equilibrium condition of expression can be described as the condition after constant pressure has been maintained until no further flow occurs.

MACHINERY FOR OIL EXPRESSION

There are several methods of expressing liquid from solid-liquid matrix. The available designs may be placed in two groups: (1) batch process and (2) continuous process. Under the batch process there are box presses, platen presses, pot presses, cage presses etc. The continuous type of presses are screw presses or expellers. Now-a-days mostly continuous screw presses or expellers are being used. These presses were first invented in 1903 by Anderson in USA. At the present time, oil expellers used in India are old Anderson type which are manufactured with some modifications introduced by local manufacturers. The main feature of an expeller is a rotating pressure worm fitted closely inside a slotted cage or barrel. Each expeller contains three main elements: (1) screw or worm, (2) cage or barrel in which the worm rotates, and (3) choke and dischargeable pressure orifice at the end of the barrel constricting discharge of the cake. Besides these, expeller is the necessary device for turning the worm or screw such as motor transmission, thrust bearing, and hopper or the cooking kettle or steam treatment. It has a receiving pan for the oil. Improved expellers are equipped with the special cooling system comprising hollow worm shafts or hollow barrel frame through which cooling water or oil is pumped.

EXPELLER OPERATION

The seeds enter after cooking stage to the fore part of the pressure worm and fall at the helical threads of the worm which propels the seeds through a specially shaped cage or barrel. During the forward movement the clearance between the cage and the worm is gradually reduced subjecting the seed mass to increasing pressure. The cage is made up of a number of steel bars which are spaced apart by shims to provide adequate drainage of oil. The oil passes through the bars and flows out of the cage. Following their path along the pressing cage, the de-oiled seeds reach the far end of the worm where a steel cone is mounted. The cone can be moved along the shaft of the expeller and the space between the worm and the cone can be regulated. This permits easy control of the pressed seed flakes and of the degree of pressure to which the seed mass is subjected.

The expeller is a self-contained unit and has its own heating kettle where cooking is performed by direct steam. It is necessary to have control of the cooking operation. If the meal is too wet or too dry, imperfect extraction of the oil will result and consequently the yield will be poor. The moisture content allowed to remain in the collected oil seed may vary from seed to seed which is predetermined.

TYPES OF EXPELLERS IN USE

Expellers are designed to operate under two pressures—low and high. Low pressure expellers remove only a portion of oil present in the oil meal. They were used in combination with hydraulic press previously and now with solvent extractors to extract the maximum oil from high oil containing seeds, from which complete extraction at a stretch is not possible. In high pressure expellers the seed is subjected to extremely high pressure. As a result considerable heat is generated which may exceed 160°C inside the expeller cage. This may affect the quality of the oil. In efficient expellers the amount of oil left in the cake is about 5–7 per cent requiring power of not less than 45 KWH per tonne of seed. In the advanced countries this technology is now obsolete. It is still continuing in India because of the small scale operations which would make the use of solvent extraction plant uneconomical.

EXPELLER OIL MILLS

There are a large number of expeller mills estimated to be about 15,000 in number. These mills have expellers of various sizes. There are mills in small scale sector having two expellers, medium-scale 3-6 expellers and large-scale seven or more expellers. The installed capacity has now been estimated to be 30 million tonnes of which 30–50 per cent is being utilized. Besides Anderson type, which has been described, there are other types of screw presses depending on the manufacturers. In the developed countries the expellers are used for pressing of high oil content seeds before solvent extraction. The expellers of German, American and other origins have new designs.

EXPANDER—EXTRUDER PROCESS

In recent years the expellers have been upgraded to provide for better extraction of seed and have led to the development of extender extruder process. The extender is a design which is a two-stage expulsion process. The Anderson Co. in USA have been the first to develop the Extender extruder Process.

(a) The foregoing indicates the technological upgradation of Expeller process that has taken place in the developed countries as also the efforts to improve the efficiency of Indian expellers. The emerging technology in oilseed pressing is the Expander-Extruder process which has the potential to eliminate screw presses.

Expanders have been in use for the preparation of Rice Bran and other oilseed materials for efficient solvent extraction. Oil seeds of low oil content are flaked, then expanded for solvent extraction. Prepress cakes are also expanded for the formulation of porous collets to facilitate their subsequent extraction in a solvent extraction plant. The expander technology is now being applied to pretreat oilseeds for solvent extraction. It has overcome the problem of expanding oilseeds by way of the release of excessive free oil which accumulates within the barrel and interferes with acceptance of feed by the expander. Such releases of oil also interfere with the formulation and flow of collets through the dies.

The second generation Anderson Expanders have overcome the problem by providing a drainage cage. The drainage section is placed at a position where the pockets of free oil, once they are liberated, can escape from the barrel before they get a chance to interfere with the formation of collets. The oil that is drained is of good quality and reasonably low in solids and moisture. Colour and free fatty acid are as good as or better than in the oil from the screw presses. The collets are strong and porous and extract well, even in a deep bed extraction. The ultimate capacity in the expander goes up. In this context the most important advantage is that there is potential to eliminate screw presses from the processing plant and to replace them with the new second generation vegoil expanders.

The Expander produces a collet which is quite porous and yet high in bulk density. Expanders work by injecting steam to the flakes or cake while under pressure. Then the collets are foamed through plates and released to atmosphere. The collect expands as it is released. As the bulk density is increased, the capacity in the solvent plant is enhanced. Extractor capacity can be improved upto 25–30 per cent. Better drainage is achieved in extraction, reducing solvent carry-over to desolventiser to approximately 28 per cent, reducing steam requirement in the desolventiser. The better drainage is specially beneficial to deep bed extractors where flakes compact tightly. The press and expander have been consolidated into one unit so that a portion of oil is removed and porous pellets are formed.

The outstanding advantage of the integrated expander system is that it is now possible to extract high oil-bearing materials such as cotton seed, sunflower seed, mustard seed etc., directly without the use of expellers. This results in remarkable reduction in electrical and steam energy consumption for complete processing of a tonne of oil seed when compared to conventional prepressing-cum-extraction process. The expansion process provides sufficient cooking to denature proteins. Thereby proteins, which are enzymes, lose their enzymatic activity, e.g. the enzyme "Urease" in Soyabean and "Lipase" in rice bran are inactivated. Expanders are not high capacity expellers. They are essentially a preparation technology for preparing oil seeds and oil-bearing materials for a solvent extraction plant.

REFERENCES

1. Bailey's Industrial Oil & Fat Products, Vol. 2, pp. 176–177, Wiley-Interscience Publication, John Wiley & Sons, New York.

1a. H.C. Dormitzer, *JAOCS,* **33,** 471–473, 1956

2. Progress in the Chemistry of Fats and Other Lipids, Vol. 5, p. 9, Pergamon, New York 1958.
3. Morris, C.E., *Oil & Soap*, **13**, 60–62, 1936.
4. L.R. Dugan et al, *JAOCS,* **31**, 46–49, (1954).
5. W.M. Gearhart and B.N. Stuckey, *JAOCS,* **32**, 287–290 (1955).
6. R.J. Sims and L. Hilfman, **33**, 381–383 (1956).
7. W.R. Dayen Wol, et al, *Food Technol,* **7**, 421–423 (1953).
8. J.E. Thompson, *Food Eng.*, 71–73, 102–15 (1954).
9. Delavel Technical Bulletin No. 5504E.
10. P.E. Sullivan, *JAOCS,* **36**, 70 (1959).
11. Sharples Corpn. Philadelphia, USA.
12. I.H. Chayen, U.S. Pat. 2635, 104 (1953).
13. *Chem. Eng: News* **30**, 5266–68 (1952); ibid, 37–42, 43 (1959).
14. H. Chayen, U.S. Pat. 2635, 104 (1953).
15. H.N. Brocklesby, J.R. Patrick, Progress in the Chemistry of Fats and Other Lipids, Vol. 5, pp. 140–148, Pergamon, New York (1958).
16. F.E. Deatherage, *JAOCS,* **23**, 327–31 (1946); U.S. Patent 2,456,684, 1948.
17. K.E. Youssef et al. *Food Sci. Technol. Abstr.*, **8**, 6N 218 (1976).
18. K.F. Youssef, et al., Acta Agronomica Academiae Scientarum Hungaricae, **23**, (1/2), 178 (1974).
19. E. Nosova, Myasnaza Ind. S.S.S.R., **30**, (1), **14** (1959), *Chem. Abstr.* **54**, 6992i.
20. H.Haneschka and R. Helingr, Austrian Patent, 159, 739 (1940).
21. L. Hartman, *JAOCS,* **27**, 409–411 (1950).
22. I.A. Parfentjev, U.S. Patent, **2**, 355, 790 (1946).
23. H.L. Keil, U.S. Patent **2**, 423, 102 (1947).
24. P. Halmbacker, U.S. Patent 2, 527, 305 (1950).
25. E.W. Lusas, L.R. Watkins and K.C. Rhee, in World Conference Proceedings, Edible Fats and Oils Processing, American Oil Chemists' Society, Edited by D.R. Erickson, pp 58-60, 1990.
26. E.W. Lusas and L.R. Watkins, in Reprints of Papers of the 38th Holiest Processing clinic, USDA-ARS Southern Regional Research Centre, New Orleans, LA March 13-14, 1989.

27. K. Sosulski, F.W. Sosulski and E. Coxworth; *J. Am. Oil Chem. Soc.*, **65**, 357 (1988).

28. K. Sosulski, and F.W. Sosulski, Quality of oil and meal from Enzyme Treated Canola Seeds, 5th International Congress of Engineering and Food, Federal Republic of Germany, May 25 - June 3, 1989.

29. H.P. Fan, J.C. Morris and H. Ivakeham, *Ind. Eng. Chem.*, **40**, 195-199 (1948).

30. Coats, H.B. and M.R. Wingard, *J. Am. Oil Chem. Soc.*, **27**, 93 (1950).

31. Othmar, D.F. and J.C. Agarwal, *Chem. Eng. Prog.*, **51**, 372 (1955).

32. A. Shah and V.K. Shrivastava, Hand Book on Rice Bran Processing & Utilisation of Products, compiled by B.M. Seth & B.V. Mehta, The Solvent Extractors' Association of India, p. 285, June 1987.

33. G. Abraham, R.J. Mron, Sr. and S.P. Kolton, *J. Am. Oil Chem. Soc.*, **65**, 129 (1988).

34. E. Bernardini, *J. Am. Oil Chem. Soc.*, **53**, 278 (1976).

35. E.D. Milligen, *J. Am. Oil Chem. Soc.*, **53**, 287 (1976).

36. E. Bernardini, *J. Am. Oil Chem. Soc.*, **53**, 278 (1976).

37. E. Bernardini, *J. Am. Oil Chem. Soc.*, **53**, 278 (1976).

38. E.W. Lusas, L.R. Watkins and K.C. Rhee, World Conference Proceedings, Edible Fats and Oils Processing, American Oil Chemists' Society, Edited by D.R. Erickson, pp 72-75 (1990).

39. J.P. Friedrich and E.H. Pryde, *J. Am. Oil Chem. Soc.*, **61**, 223 (1984).

40. J.M. Snyder, J.P. Friedrich and D.D. Christianson, ibid, **61**, 1851 (1984).

41. G.R. List and J.P. Friedrich, ibid, **66**, 98 (1989).

42. Koseoglu, S.S. and E.W. Lusas, 'Processings of the World Congress on Vegetable, Protein Utilization in Human Foods and Animal Feedstaffs, T.H. Applewhite, (ed.), American Oil Chemists' Society, Chempaign, IL, p. 528, 1989.

43. E.D. Milligen, *J. Am. Oil Chemists Soc.*, **53**, 288-290, June 1976.

44. Technology Evaluation and Norms Study in Industrial Oils and Fatty Acids—A Study Prepared Under Technology Absorption and Adaption Scheme (TAAS), Government of India, Department of Scientific and Industrial Research, Ministry of Science and Technology, Technology Bhawan, New Delhi – 110 016. February, 1993.
    (a) Rendering of animal fats (pp. 289-290),
    (b) Extraction of fats (pp. 363-364) and
    (c) Present status of oil Milling Industry in India (p. 377) of chapter IX.

CHAPTER 10

Refining of Oils and Fats

GENERAL CONSIDERATIONS

Crude oils and fats are derived from oilseeds and other oil bearing materials by expelling, rendering, or solvent extraction. Crude oils, thus obtained, contain many non-glyceride components broadly termed as 'impurities' or 'contaminants'. These may be classified as follows :

(1) Gums, mucilaginous matters, phosphatides and other phosphorous bearing materials.

(2) Proteinous materials, sugars and glycolipids.

(3) Trace metal contaminants.

(4) Free fatty acids.

(5) Pigments and colouring bodies.

(6) Oxidized materials.

(7) Stearines and high melting waxes.

(8) Tocopherols.

(9) Flavours and odoriferous compounds.

(10) Chlorinated pesticide residues.

In production of edible oils, a number of purification steps are carried out to remove these impurities. Because of their wide applications in agricultural practices, chlorinated pesticide residues are found in vegetable oils and fats. These residues are effectively removed during purification of oils and fats. While the presence of most of other impurities have deleterious effects on nutrition, keeping quality, product appeal and ease of processing, the presence of some of these is, in fact, beneficial to the oil quality. For example, exhaustive removal of tocopherols, which are powerful natural anti-oxidants, may adversely-affect the keeping quality of the oils. However, during removal of other undesirable impurities such friendly compounds are also partially removed. Therefore, the underlying objective of the various purification steps are to retain the desirable components to the maximum extent with minimum possible loss of oil.

Combination of various purification steps to achieve such objectives is termed as 'refining' process. It includes several operations, namely degumming, neutralization, bleaching, winterization/dewaxing and deodorization. Oils and fats are usually processed in the described order of sequence for optimum product quality and yield. In practice, neutralization alone is also sometimes referred to as refining. Table 10.1 gives the different operations of refining processes and the impurities removed during each operation.

Refining of oils thus carried out is also termed as chemical refining as it utilizes alkali to neutralize the free fatty acids. In a recently developed refining process, free fatty acids are removed from the oils by distillative deacidification using steam under vacuum. Deodorization is also accomplished along with deacidification. This is known as physical refining as the use of alkali is dispensed with. Major requirements for the physical refining is more or less complete removal of phosphatides, lipoproteins, sugars, traces of heavy metals and organic pigments. Effective degumming and bleaching operations, should, therefore, precede physical refining process.

The task of refining process is governed by crude oil quality. Sound practices of handling and storage of crude oils protect this quality and, therefore, are just as important as any other operation in a refining process. It is imperative that handling and storage of crude oils should be treated as one of the component operations in any refining process. Fig. 10.1 presents the schemes for classical, chemical and physical refining processes.

Table 10.1: Refining – Unit Processes

Stage	*Principal impurities removed or reduced*
Degumming	Phospholipids, trace metals, carbohydrates, lipoproteins.
Pre-treatment for Physical Refining	Non-volatile impurities like phospholipids, trace metals, carbohydrates, lipoproteins, pigments, colouring bodies and oxidation products.
Neutralization	Fatty acids, phospholipids, colouring bodies, pigments, oil insoluble as well as oil soluble impurities, trace metals.
Bleaching	Pigments, oxidation products, traces of soap, sulphur, traces of phospholipids.
Winterization/Dewaxing	Waxes and stearines.
Deodorization	Fatty acids, partial glycerides, oxidation products, pigments decomposition products and odoriferous compounds.
Physical refining	Volatile impurities like free fatty acids and odoriferous compounds.
Polishing	Oil insolubles like carbonaceous deposits, traces of bleaching earths and insoluble citric acid.

STORAGE AND HANDLING OF CRUDE OILS

After a crude oil of stated quality has been received at the refinery, proper practices of storage and handling should be adopted to prevent any oil loss and to maintain the quality of the oil till any further processing commences.

Loss of oil occurs from the defective valves, pumps and other fittings connected to the storage tanks. On this count, proper maintenance can help reduce the losses due to leakage. Oils may also be lost by occlusion with foots which settle to the bottom of the tank. Foots are a mixture of dirt, gums and other impurities which are too viscous to be pumped and therefore, must be removed manually. Side mounted agitators should be provided to prevent such settlings.

Segregation of oils by class is important for multi-oil refineries to avoid any inter-oil contamination and consequent loss of value to the oil. It is a standard practice to separately handle lauric acid oils like coconut etc. from high iodine value oils. Soft fish oils should, as far as possible, be handled at a separate installation, as a very little contamination of it may produce the characteristic fishy odour in unhydrogenated oils.

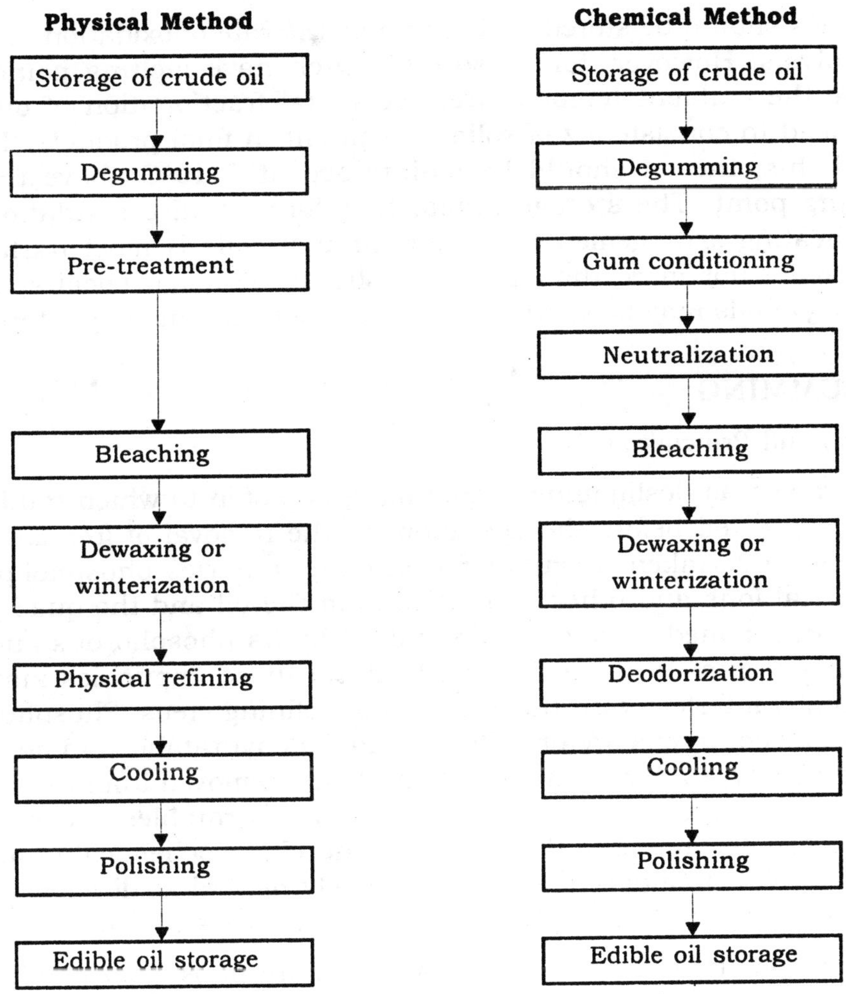

Fig. 10.1: Various Stages of Edible Oil Refining.

Moisture and temperature have very pronounced effects on the quality of oils and fats. High level of moisture content increases the free fatty acid content of oils by hydrolysis. In mild steel tanks water vapours may condense on cool surfaces and form rust, which will result in the increased iron content of oil. Iron acts as a pro-oxidant and the bleachability of the oil may be impaired. Inner side of the storage tanks, therefore, should be coated with epoxy paints.

Oils should be stored cool to reduce the rate of oxidation and hydrolysis. However, for those fats which have melting points above the ambient temperature, accidental fractionation of oils may lead to consistency of solid fat content in final products. To avoid this, the oil should be maintained at 5–10°C above the melting point. The storage tanks, therefore, require insulation and heating arrangements. Hot air through coils is most useful but steam not more than 2–3 bars should be used. Local spot heating of oils may be prevented by using side mounted agitators.

DEGUMMING

Theory and Principle

Also known as deslimming, degumming is a step to which crude oil is subjected before the operation for the removal of free fatty acids is undertaken. Degumming primarily removes phospholipids, metal ions and other impurities from the oil and the quality of the degummed oil is generally judged by its phosphorous and metal contents. If not removed effectively these impurities may eventually interfere with the subsequent refining steps. Phospholipids present in oils are broadly classified as hydratable and non-hydratable types. While hydratable types are removed from the oil by a simple water degumming step, non-hydratables are not. Knowledge pertaining to the structure and physicochemical properties of phospholipids has been extremely helpful in developing effective industrial processes for refining of oils and fats.

Phospholipids present in oils and fats are phosphoglycerides which contain two fatty acid molecules esterified to the first and the second hydroxyl groups of glycerol. The third hydroxyl group of glycerol forms an ester linkage with phosphoric acid. In addition, phosphoglycerides also contain a second alcohol, which is also esterified to phosphoric acid. This alcohol is, therefore, located on the polar head of the phosphoglyceride molecule. Different types of phosphoglycerides are named according to the alcohol in their polar heads. The parent compound of phosphoglycerides is phosphatidic acid, which has no head alcohol. Phosphoglycerides present in oils are mainly phosphatidyl ethanolamine, phosphatidyl choline, phosphatidyl ionositol and to a lesser extent phosphatidyl serine and they all contain alcohols ethanolamine, choline, ionositol and serine respectively, on their polar heads. Phosphoglycerides from oils and fats are often called phosphatides. Structure of phosphatidic acid and other phosphatides are given in Fig. 10.2. Phosphatides thus have two very different kinds of groups. A polar, hydrophilic head and a non-polar hydrophobic tail.

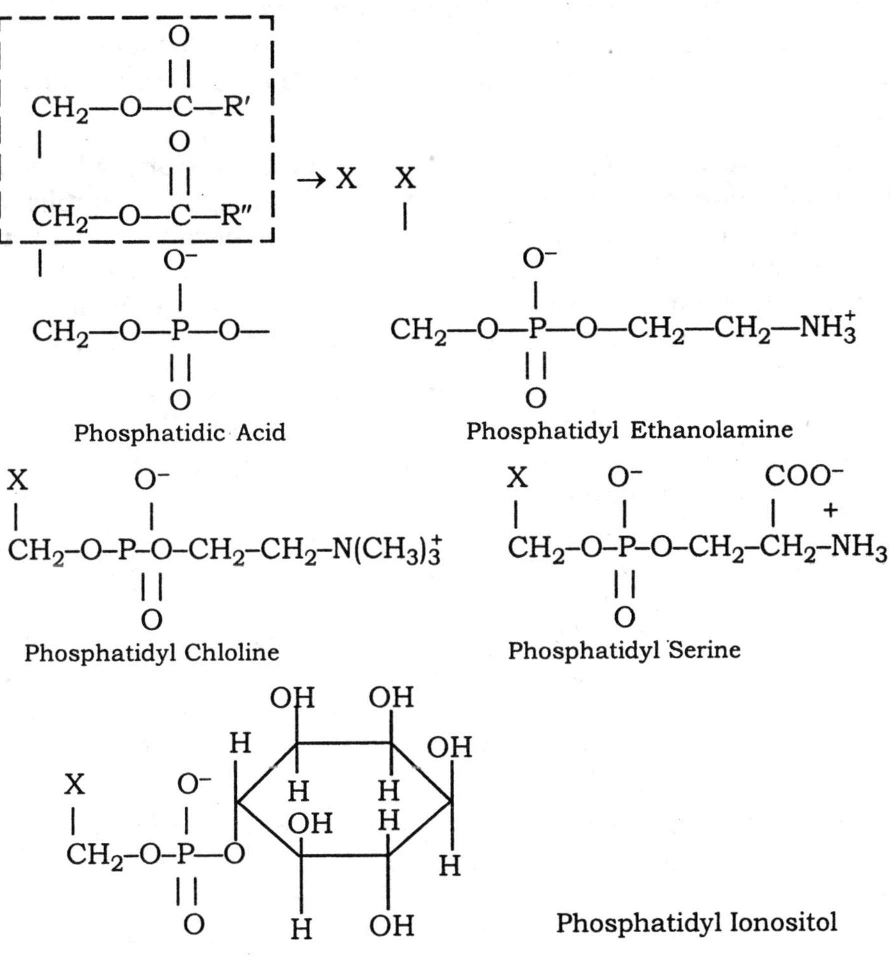

Fig. 10.2: Phospholipid Structures

As can be seen from Fig. 10.2, all phosphatides have a negative charge on the phosphoric group at pH 7. In addition, the head alcohol group may also contribute one or more electric charges at pH near 7. This indicates that phosphatidic acid and other phosphatides have cation (metal ion) binding capacity. When phosphatides are complexed with monovalent ions like sodium or potassium, they are hydratable. But binding with bivalent metal ions like calcium or magnesium may lead to non-hydratability (1). Due to enzymatic reaction during the extraction of oilseeds, a part of the phosphatides is complexed with calcium and magnesium ions. These salts are strong emulsifying agents and, if not removed are responsible for higher losses during alkali refining. Also due to their presence, water degumming step alone is not sufficient for preparing the oil for physical refining.

Phosphatidyl choline exists in oil as Zwiterion, which is hydratable. Phosphatidyl ionositol may exist in oil as a complex of monovalent or divalent cation but is hydratable because of its very highly polar ionositol group. Similarly, phosphatidic acid and phosphatidyl ethanolamine are hydratable when complexed with monovalent cations and non-hydratable when complexed with divalent ones.

Binding of calcium and magnesium ions to phosphatidic acid and phosphatidyl ethanolamine is very strong at pH 7. For example it has been found that the binding of bivalent ions is 1000 times stronger than the binding to sodium and potassium ions at pH 6 and pH 7 (2). For this reason, under the conditions of water degumming, such compounds are not hydratable. Addition of an acid, sufficiently strong, liberates free phosphatidic acid and phosphatidyl ethanolamine, and thus renders them hydratable. Bivalent metal ions bind with the added acid and can be removed with aqueous phase. Phosphoric and citric acids are commonly used in practice. However, as they do not dissolve in oil, these acids must be thoroughly mixed to achieve the desired result.

Acids like acetic acid and anhydride are also able to dissociate phosphatidic acid and Phosphatidyl ethanolamine from bivalent metal ions. It is extremely important that metal complexes formed with added acid must not be oil soluble. Effective removal of pro-oxidant metals like iron and copper is more critical than that of phosphorous for an acceptable quality of degummed oil. Acetic acid and acetic anhydride (3), although sufficiently strong, are not used for degumming of oil as the metal acetates are oil soluble. An excellent review on principles of degumming has recently been published (4).

Degumming by Hydration or Water Degumming

When water is added to a crude oil, most of the phosphatides present in the crude oil are hydrated and made insoluble in the oil. This forms the principle for the water degumming process. The crude oils are usually water degummed at the extraction plant itself.

Fig. 10.3 shows a flow diagram of a water degumming plant. The incoming oil usually has a temperature close to the degumming temperature of 50–70°C. Temperature of degumming has an important bearing on the quality and efficiency of separation. At higher temperatures, darkening of the lecithin products takes place and degumming is less complete due to the increased solubility of phosphatides in oil. At lower temperature the increased viscosity makes the separation of the phosphatides more difficult.

Water or in some cases live steam is added to the oil at the level of about the same quality as the gum content of the oil. Intimate mixing is carried out for 20-30 min in order to allow hydration of the phosphatides and agglomeration of the particles to take place. Due to higher specific gravity of the hydrated gums, they can be separated from the oil by centrifugation. Since the gum lecithins are collected over several hours, dilute hydrogen peroxide is added to stabilize the wet gums from bacterial spoilage.

The degummed oil can either go directly to subsequent processing steps or it can be vacuum dried and cooled for storage. Major oils which require water degumming are soyabean, rapeseed, rice bran, corn and sunflower.

Gum Conditioning

This is the step in chemical refining which generally precedes the short-mix process of alkali refining (to be discussed later). Its

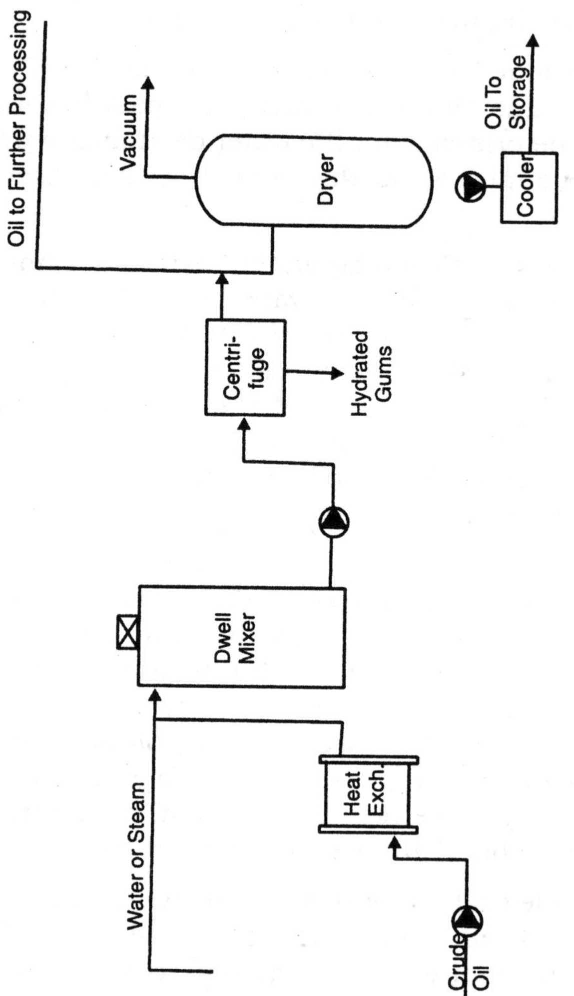

Fig. 10.3: Water Degumming Plant

purpose is primarily to remove the non-hydratable phosphatides. The most common chemical for this is phosphoric acid in concentrated (75–85 percent) form. Phosphoric acid decomposes the non-hydratable phosphatides to form a dark brown sludge which can be removed along with the soapstock. It also decomposes the iron salts and has a noticeable effect on the reduction of chlorophyll in the oil. Because of the decomposition of calcium and magnesium salts, the soapstock is more easily separated.

Addition of phosphoric acid can be employed in batch, semi-continuous or continuous methods of refining. Phosphoric acid is used in proportion to the phosphorous content of the crude or water degummed oil, i.e. from 0.05–0.2% of the oil weight. In case of rapeseed oil it can be as high as 0.5% because of the high chlorophyll content of the oil. The acid is intimately mixed with oil. As mentioned earlier, the intimate mixing of the acid with oil is extremely important and, therefore, mixers to give high shear effect (as the one given in Fig. 10.4) are recommended. The mixture is allowed to stand for 5-10 min with no or little agitation. This permits the agglomeration of denatured phosphatide particles so that they could be easily separated along with the soapstock.

Pre-treatment for Physical Refining/Dry Degumming

The process of physical refining has been in use for many years for crude oils containing high levels of free fatty acids, e.g. coconut, palm oil and tallow. As a result of considerable volume of research and development in recent years it is apparent that this process may not be applicable to all oils. The quality of crude oils and quantity of impurities present therein are of paramount importance. The process of physical is restricted to only those cases where crude or water degummed oils have very low level of phosphorous and iron content. A suitable pre-treatment for almost complete removal of these impurities is the basic pre-requisite for physical refining.

The principal requirement of a pre-treatment process is that the phosphorous content should be reduced to a maximum of 10 ppm and iron below 2 ppm. Oxidation products and non-heat degradable products should be reduced to a level such that the product after distillative-deacidification is acceptable.

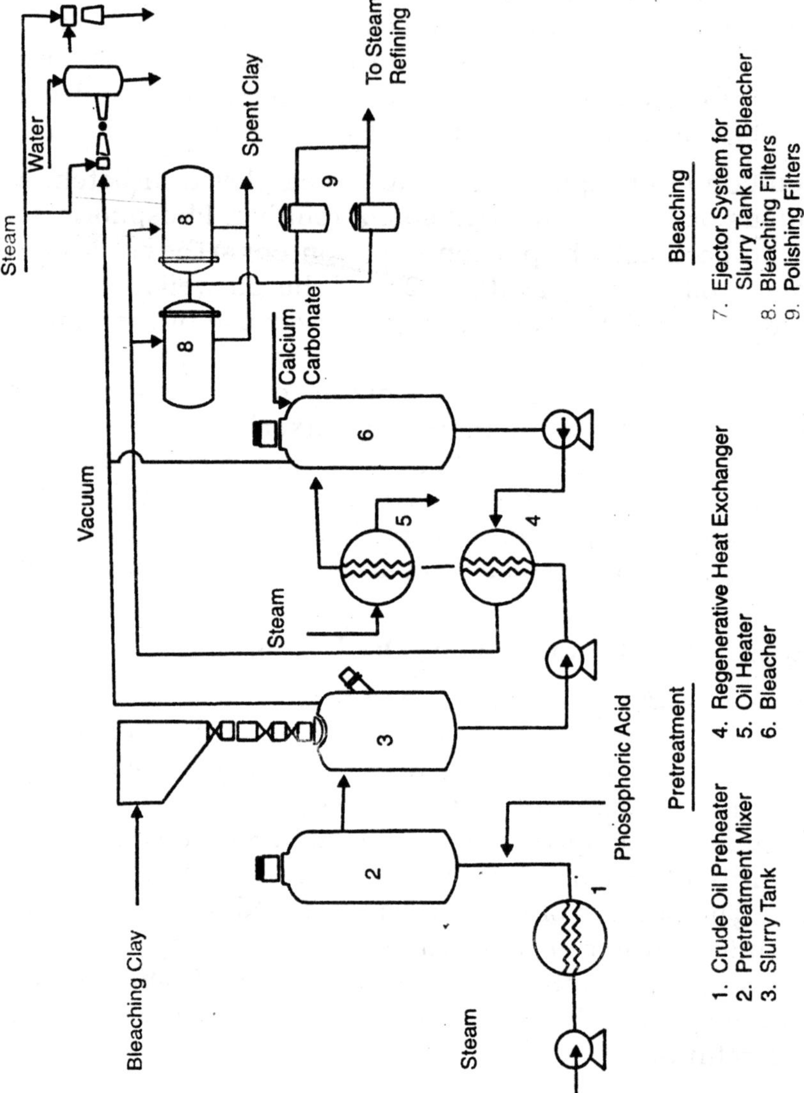

Fig. 10.4: Pretreatment and Bleaching for Physical Refining

Fig. 10.4 gives the flow diagram of such a pre-treatment which prepares the oil for physical refining. Crude or water degummed oil is treated with phosphoric acid at about 90–100°C for 10–15 min and then bleached with 1–2% of activated earth suitable for phosphatides and trace metal removal. The process is carried out under a vacuum of 28" of mercury. Temperature of 90–100°C is most suitable to give minimum oxidation problem. The earth usually adsorb the excess phosphoric acid but it is not always the case as any phosphoric acid escaping into the oil after filtration can result in colour deterioration at high temperature of physical refining. Therefore, a stoichiometric quantity of calcium carbonate is added based on the amount of phosphoric acid employed. After providing a bleaching time of 15 min the oil is cooled and filtered. The final oil contains less than 4 ppm phosphorous with almost no trace of iron and is most suitable for physical refining. Palm oil is usually pre-treated in this manner.

Acid Degumming

This process is very similar to water degumming except that non-hydratable phosphatides are precipitated by the addition of acid and water instead of only water. The acid most commonly used is citric acid although phosphoric acid can also be used. The aqueous phase is then separated in a centrifugal separator. Acid degummed oil can be taken for physical refining after a bleaching step. As acid degumming reduces the phosphorous and iron level in the degummed oil, correspondingly the requirement of bleaching earth is also reduced. Gums obtained by this process are not suitable for production of lecithin as they contain the degumming acid.

Acid Refining

In this process a small amount of alkali, like caustic soda or sodium silicate, is added to the oil after addition of an acid. Addition of the alkali ensures the hydratability of phosphatides. The amount of alkali used is such that it only partially reduces the free fatty acid content and also the soap formation is prevented. The phosphorous content can further be reduced by adding a washing step. The degummed oil can also be taken for physical refining after bleaching with earth.

Alcon Process (5)

The process, developed by Lurgi GmbH, primarily works on seeds to produce an oil in which the non-hydratable phospholipids are kept at the minimum. Due to enzymatic reaction during the extraction process, a part of the phosphatide is transformed to a non-hydratable form, which means that these cannot be removed by water degumming. As mentioned earlier, these non-hydratable phosphatides are calcium and magnesium salts of lysophosphatidic acid and phosphatidic acid. Major parameters controlled in the Alcon process are moisture, temperature and time for the preparation of seed prior to extraction. As the enzymatic activity is maximum at 40–70°C, temperature is quickly raised to 100°C maintaining the moisture content to 15–16%. Conditions are maintained for 15-20 min so that the enzyme activity is effectively inhibited, as illustrated in Fig. 10.5. Alcon treated oil can be physically refined after water degumming and bleaching. Sometimes, dry degumming using acid can also be done prior to physical refining.

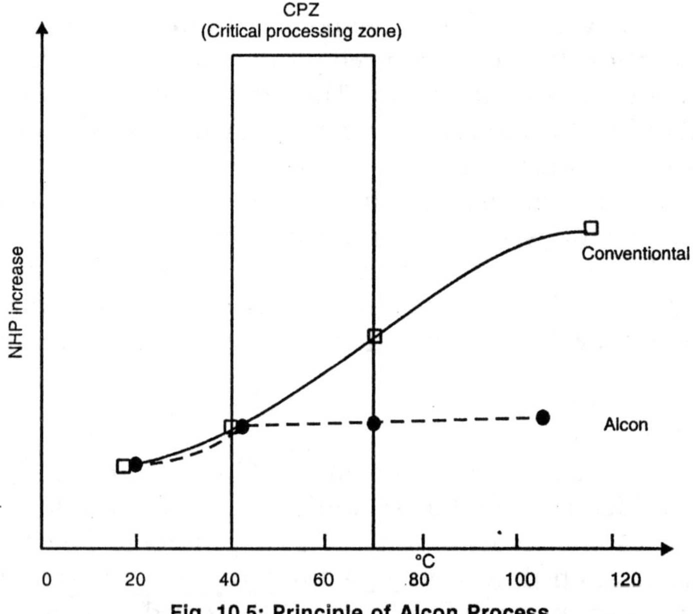

Fig. 10.5: Principle of Alcon Process

Refining

The word refining is used in the sense of caustic refining, chemical refining or alkali refining and is aimed at removing the free fatty acids present in the oil. In the refining step, the water degummed or phosphoric acid conditioned oil is mixed with enough alkali to neutralize the free fatty acids present and also to provide some excess. This converts essentially all the free fatty acids to oil-insoluble soap, which can then be separated from the neutral oil. The refined oil is washed with water to remove the residual soap.

Apart from removing the free fatty acids in soap-stock formation, refining step also reduces the colour of the oil considerably and enhances the bleachability. Pigments and other colouring bodies combine with alkali and are removed by absorption on the soap formed in the operation. Gossypol, a pigment in cottonseed oil is effectively removed during alkali refining. The alkali treatment also saponifies some neutral oil and this is taken as process loss. It is the optimum oil loss. The important process variables for the refining process include:

(1) The type of alkali used, e.g. sodium hydroxide (caustic soda), sodium carbonate (soda ash), ammonium hydroxide or sodium silicate.
(2) The strength of the alkali solution.
(3) The excess used over the stoichiometrically required amount.
(4) The temperature of agitation.
(5) The degree of agitation.
(6) The oil flow rate through the plant (in continuous operation).

The most commonly used alkali in refining of oils and fats is sodium hydroxide. Other alkalies mentioned above have also been used but none of them has been found to be as effective as sodium hydroxide.

The minimum amount of alkali required for neutralization can be calculated from the free acids (FFA) content of the oil to be refined, using the formula.

$$\% \text{ NaOH} = \% \text{ FFA} \times 0.142$$

Where the FFA is expressed in terms of oleic acid (molecular weight = 282). For any desired excess of NaOH, the calculation is

$$\% \text{ NaOH} = \% \text{ FFA} \times 0.142 + \% \text{ excess NaOH}$$

Alkali concentration is usually expressed in terms of Baume Scale (Table 10.2) rather than per cent. The excess caustic soda is usually expressed in terms of dry sodium hydroxide on a percentage basis, calculated on the weight of the oil. The alkali strength varies widely with the oil processed, its free fatty acid content and contact time.

***Table 10.2:* Alkali Concentratins in per cent NaOH and in Degrees Baume**

Degrees Baume at 15°C	*Per cent NaOH*
10	6.57
12	8.00
14	9.50
16	11.06
18	12.68
20	14.36
22	16.09
24	17.87
26	19.70
28	21.58
30	23.50

A strong caustic soda, for instance 20 Be′, is effective for the removal of impurities, especially gums and pigments, but saponification losses are higher due to oil occulsion in the soapstock. On the other hand, emulsion losses are higher when using dilute solutions. Similarly a large excess has a high purifying action but with a greater tendency to saponification. Emulsions are more likely to be encountered with low excess lye usage. A long contact time, high mixing intensity and high temperature are responsible for high saponification losses; but a long contact time helps to remove pigments and high temperature improve phase separation.

A double neutralization technique is often used for the treatment of poor quality oils. In the first stage, e.g., a 20 Be′ caustic soda is used and after soap separation the oil is given a second lye treatment using a strength of 10–20 Be′ depending upon the quality of the oil.

The generally practised methods of alkali refining are batch, continuous, semi-continuous and miscella refining. There are other less common refining processes and these are discussed in the miscellaneous refining methods.

Batch Refining

It is the oldest method of refining of oils and fats (6) and is still in use because of some advantages:

(i) low investment cost,

(ii) equipment can often be locally manufactured,

(iii) suitability for low capacity operations and

(iv) for certain poor crude oils, it produces a refined oil of good quality.

However, it also suffers from certain disadvantages :

(i) often high refining losses,

(ii) high operation cost due to high consumption of steam and water and

(iii) oils with high FFA cannot be refined.

There are usually two variations of batch refining–dry and wet methods. In the dry method of batch refining, water degummed or phosphoric acid conditioned oil is taken in a large, open cylindrical vessel with conical bottom. A two speed agitator is provided for mixing. The oil is allowed to stand for a sufficient period of time so that any occulded air bubble should rise to the top and escape through the surface of oil. The calculated amount of caustic soda with requisite excess and in appropriate dilution is sprayed on the surface of oil and intimately mixed at high agitation. This is usually done at ambient temperature. At the end of the mixing period, temperature is raised to 80–90°C to get the desired 'break'. The stirring is continued but now at a lower speed sufficient to agglomerate the small soap particles. After some time stirring is turned off and the bottom soap layer (soap-stock) is sucked off and dropped into another tank. The neutral oil is washed with hot water to make it free from soap. The refined oil is dried and is ready for bleaching.

In the wet method, the heated oil is mixed with the alkali and the soap is washed with sprays of hot water directed at the surface of the oil. This method is practised most where crude oil is relatively high in free fatty acid content and produces soft soap-stock. Considerable skill is required on the part of operator to get a good quality oil with optimum losses.

Continuous Refining

This method differs from batch refining in that mixing of oil and alkali is continuous and the separation of refined oil from soap-stock is carried out in centrifuges. Thus the contact time between oil and alkali is reduced, thereby reducing the possibility of saponification of neutral oil. This method of refining is at its best when large quantities of oil are to be refined. The modern fully continuous refining line allows for many variations of throughput, mixing speed, contact time and temperature so that the process variables can be made use of to optimize quality and yield for a wide variety of feed stocks.

Centrifugal separators (centrifuges) are the most important components of modern refining plants with continuous operation. Centrifuges are used for desludging, degumming, separation of soap-stock from oil and in washing of oil to remove soaps. By fat the separation of soap-stock from oil (heavy phase from light phase in liquid-liquid separation) is the most important area in the application of centrifuges in refining. Centrifuges commonly used in oil refining industries are either tubular bowl or disc bowl type. However, selection of a particular type depends primarily on throughput and the amount of solids present in the feed. General description of different centrifuges is given below.

Tubular Type

The tubular design has a relatively high length to radius ratio (approximately 4.35 to 7.25) and it is made to rotate at a relatively high speed. Fig. 10.6 shows the diagram of a tubular bowl supercentrifuge. It consists of a long hollow cylindrical bowl of small diameter suspended from a flexible spindle at the top and guided at the bottom by a loose fitting bushing. The motor driven units can generate a centrifugal force of about 15,000 xg-value.

With turbine driven design it can be run at 50,000 r.p.m. and the g-value can be as high as 62,500, although such situations are generally not encountered in edible oil refining industry. Inside the bowl an acceleration device is provided whose function is to ensure rotation of the liquid at bowl speed. Any solid present in the oil-soap-stock mixture is accumulated on the inner wall of the rotor tube and is removed manually by disrupting the operation. These designs are normally available in comparatively smaller capacities and extreme care should be taken to ensure thorough filteration of the oil prior to the addition of alkali to facilitate a continuous operation over a prolonged period.

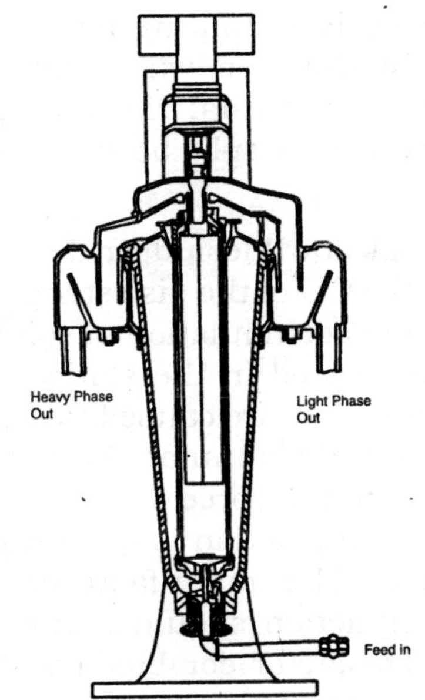

Fig. 10.6: Tubular Bowl Centrifuge (separator)

Disc Bowl/Type

This type is designed to maximise the surface area for separation and is quite cost effective for large scale operations. In this configuration a stack of discs are hermetically enclosed in a bowl shell. Large bowl size provides a longer residence time and, therefore, the separation can be effected at a comparatively lower r.p.m. These centrifuges are operated at about 6000 r.p.m. providing a centrifugal force at 7000 xg. The discs are actually truncated cones, flanged at their OD and ID to provide strength and rigidity. The inclined discs aid the separation by a mechanism similar to the plate settlers. The feed may enter either from the top or from the bottom on the axis. The light liquid phase moves to the top of the discs and travels toward the centre of the bowl. The heavy liquid phase and any solid move to the inner wall of the bowl. Both liquid phases exit at the top in separate conduits. Since capacity of the bowl is a limiting factor, even a small amount of solids present in the feed will eventually fill the bowl necessitating stoppage and manual cleaning. Self-cleaning devices are, therefore, provided as split bowl desludger and nozzle discharge centrifuges.

In split bowl desludger design (Fig. 10.7) a space is provided between the OD of the disc stack and the slopping walls of the bowl shell for the accumulation of solids. When a suitable load of solids has accumulated in the sludge space, the upper and lower sections of the bowl are caused to separate under hydraulic pressure. This exposes an annular slot through which the solids discharge under centrifugal force assisted by the head of the oil in the bowl. The desludging action is quite rapid, requiring not more that 1 or 2 seconds. The manufacturers usually recommend that the desludging action should occur no more frequently than once a minute. Thisfactor combined with sludge holding volume should be considered in sizing such a centrifuge for a given task.

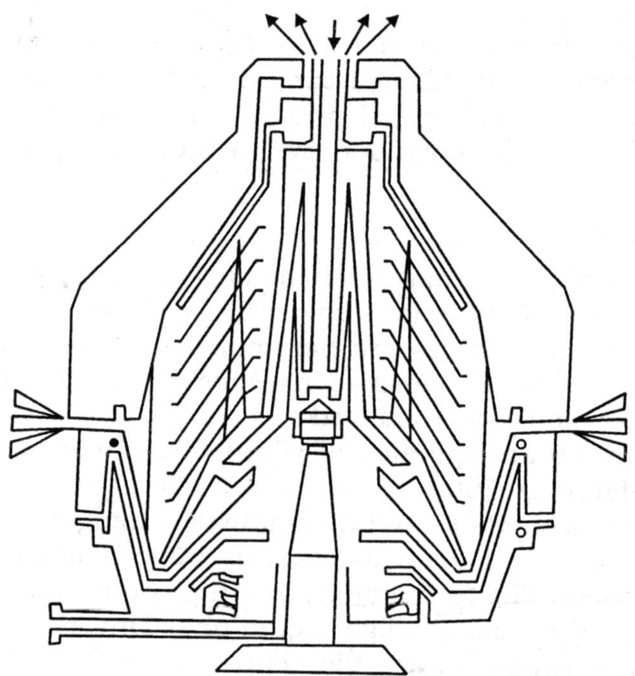

Fig. 10.7: Split-bowl Centrifuge (separator)

The mechanical complexities of the split bowl desludging centrifuge can be reduced by discharging the sedimented solids continuously from the annular band-through parts or nozzles in the outer perimeter of the bowl shell (Fig. 10.8). For good operation, the minimum allowable nozzle size is twice the diameter of the largest particle to be discharged through it. As the discharge nozzles encounter very high erosion due to high velocity of abrasive particles through them, the cost of material for construction is generally very high. For oil refineries such equipments are seldom used.

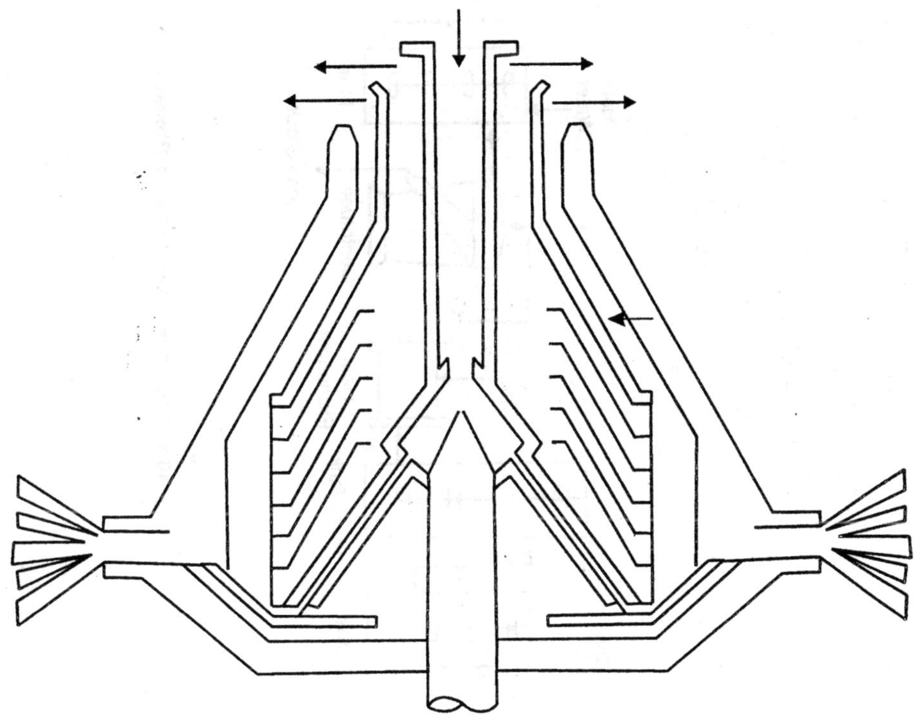

Fig. 10.8: Peripheral-nozzle-discharge Centrifuge

Process of Continuous Refining

Two basic systems are used for continuous refining. The essential differences between them are the time and temperature of mixing with caustic.

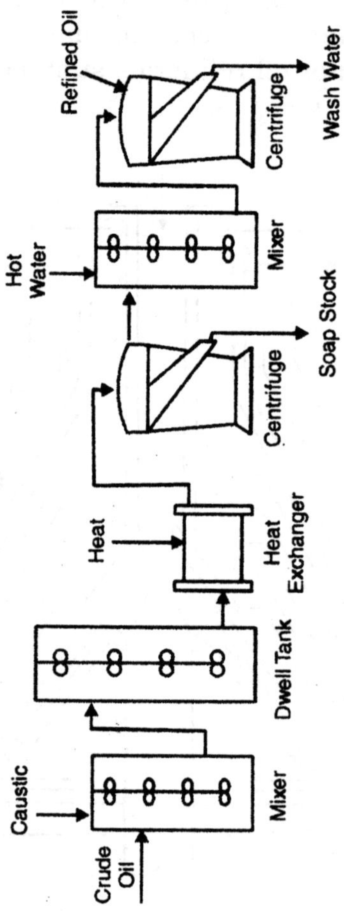

Fig. 10.9: Long-mix Alkali Refining Process

The system most commonly used in USA and Canada is shown in Fig. 10.9 and is known as 'long-mix' process. Depending on the type and quality of oil the contact time with caustic is varied (typically 5–15 min.). A high speed inline mixer is generally used and for the oils like soyabean and cottonseed, a series (upto 4 numbers) of high contact retention mixers (dwell mixers) are used. Long contact time at ambient temperature allows addition of large excess of alkali to facilitate hydrolysis of phosphatides to hydratable form. At the end of this period, the mixture is heated to 70–80°C to coagulate the soap-stock and is then immediately separated by centrifugation. For soap-stock separation solid disc bowl centrifuge is used, and to aid the removal of soap-stock from the bowl it is equipped for flushing the bowl periphery with water. This type of line may be used for the treatment of clean, low phosphatides, high quality oils which can be refined easily, e.g. good quality soyabean and palm oils.

The alternative system used in Europe and other parts of the world is short-mix process and is shown in Fig. 10.10. In this process the oil is heated to 90°C, treated with phosphoric acid for gum conditioning and then mixed with lye for 1-15 sec before centrifugation. Short-mix process is more suitable for high acidity oils as short contact time between oil and alkali will give less saponification loss. Dirt in the crude oil and denatured phosphatide sludge from gum conditioning may hamper efficient line operation by accumulating on the surface of the bowls. It is, therefore, a frequent practice for soap-stock separation to use a split bowl desludging centrifuge with programmed periodic discharge of dirt.

Short-mix process lines can also be provided with an optional re-refining step. While in the first refining step, free fatty acids are removed, a strong alkali addition in the re-refining step will reduce the colour to an acceptable level. This type of process has been especially suitable for rice bran and cottonseed oils with high acidity (upto 5%).

For oils having very high acidity (10% and above), short-mix process can further be improved by introducing caustic directly to the centrifuge through a mixing device located near the feed point. Contact time is reduced to as low as a fraction of a second and

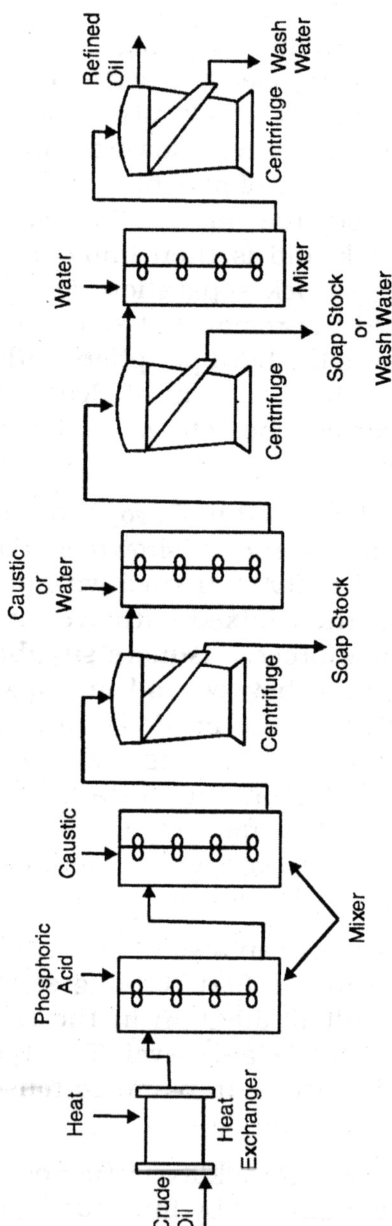

Fig. 10.10: Short-mix Alkali Refining Process

excellent results can be obtained with high acidity palm and other oils. In continuous operations, metering of crude oil and alkali are of utmost importance to yield a good quality oil with optimum refining loss. Accuracy of oil and caustic flow meters is crucial for the control of refining loss. For high volume refineries, a very small increase in refining loss means a substantial monetary loss over a period. Caustic flow meters should be highly reliable over the required range of caustic flow (6).

Whether the refining process is batch or continuous, the neutral oil should be made free from soap and subsequently dried before bleaching operation is carried out. After the separation of the soap-stock from oil, the oil contains about 2000 ppm of soap from a batch process using simple gravity separation and generally between 500 to 1000 ppm following centrifugal separation. The soap is removed primarily by water washing. The water for this purpose and for lye make up should be soft. Steam condensate is suitable and should be hotter than oil, preferably over 90°C. The quantity of water used is about 8% of oil weight per wash in batch refining and 10–20% of oil flow in the continuous process. To break emulsion formation in batch treatment, salt can be added either to the oil prior to the first wash or dissolved in the first wash water to make a brine solution of about 5% strength.

In continuous refining line, the water washing of the neutral oil to make it free from soap is carried out in another centrifuge where two washing stages are used. In continuous refining, refiners prefer a counter current washing system in which the spent wash water from the final centrifuge is used again to wash the oil in the first washing step. A counter current washing system is shown in Fig. 10.11. The wash water is mixed with oil in high shear knife mixers. The intimate mixing in such type of mixers give savings in the wash water needed.

The importance of thorough washing lies in the fact that soap is adsorbed on bleaching earth, thus reducing the earth's surface available for the removal of other impurities. Soap also has a catalytic action in the hydrolysis of triglycerides. The efficiency of water washing is to a large extent dependent on the removal of phosphatides with the soap-stock. However, with a well refined oil, the soap content in the batch process will be about 100 ppm after

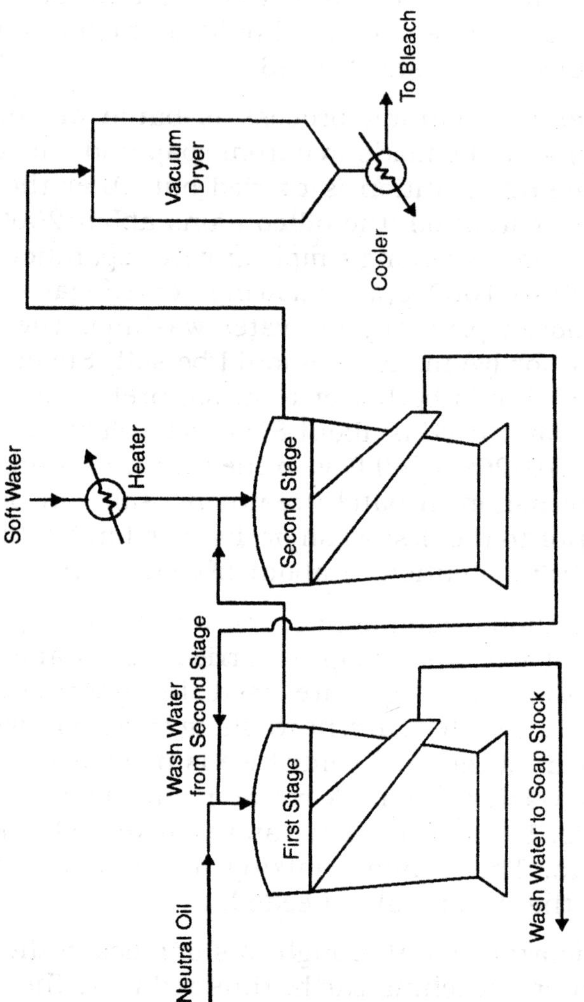

Fig. 10.11: Two Stage Counter-current Washing System

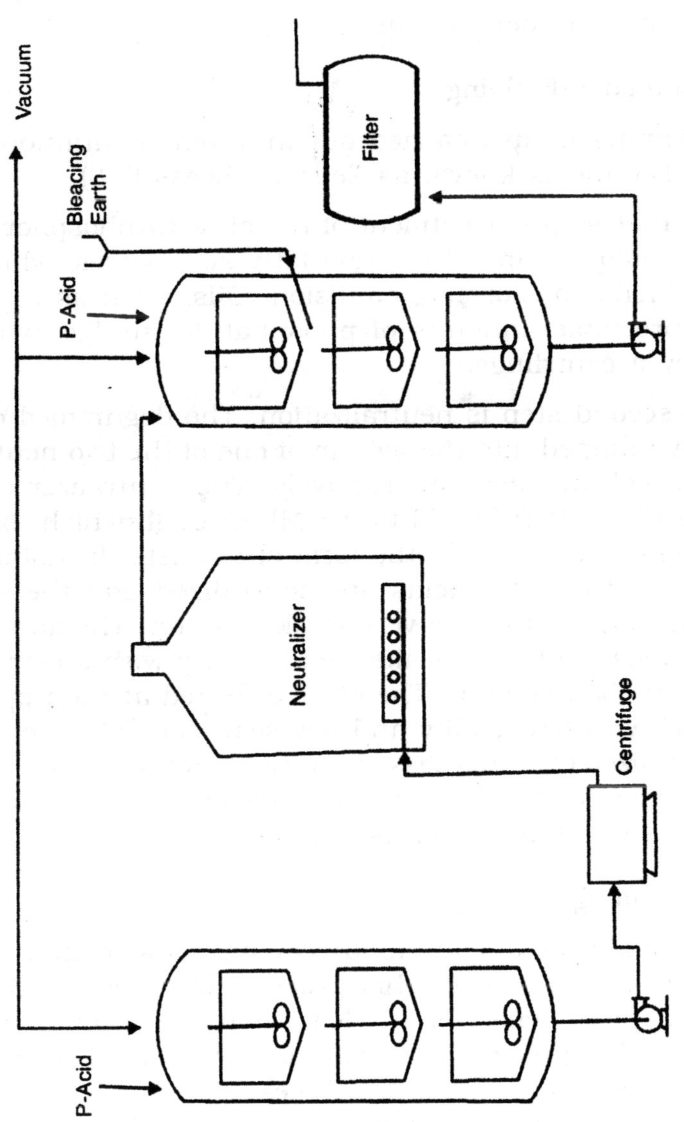

Fig. 10.12: Semi-Continuous Alkali Refining (Zenith Process)

three washes and below 50 ppm after two washes in a continuous plant. In order to eliminate soap from the oil, the refined oil may be mixed with citric acid solution to decompose the remaining traces of soap. The oil is then dried in a drier either by cascading or by spraying under vacuum.

Semi-continuous Refining

Alkali refining is also carried out in a semi-continuous manner (Fig. 10.12) and is known as Zenith process (7.8).

The first step is treatment of the oil with phosphoric acid, in order to remove some of the non-fatty substances, which have a great influence on forming emulsion. Oils, such as rapeseed oil, form appreciable amounts of phosphatidic sludge, which is removed by a centrifuge.

The second step is neutralization. The degummed oil is continuously pumped into the bottom of one of the two neutralization vessels which are used alternatively. The neutralizer contains a weak alkali solution (0.3 M to 0.8 M), through which the oil rises by difference in gravity in the form of droplets. In rising through the lye, the free fatty acids are neutralized and the soap thus formed is dissolved in the very weak solution. The aqueous solution becomes gradually weaker in alkalinity with a corresponding increase in soap content. The oil is collected at the upper part of the vessel with low acidity and low soap content. The oil leaving the neutralizer contains less than 100 ppm soap and, therefore, does not need to be washed. The traces of soap are decomposed by citric acid before the oil is bleached.

Miscella Refining

It is a special case of continuous caustic soda refining. The most favourable conditions for this process are those of the solvent extraction plant where it is also desired to refine the oil and perhaps further process it by winterization. For this reason, at one time cottonseed oil was mostly processed in this manner (9). The miscella from the extraction plant is evaporated to the desired oil/hexane proportion (approximately 65/35 in the case of cottonseed). The corrected miscella is then pre-treated with phosphoric acid and neutralized by the addition of requisite alkali. Difficulties are encountered in mixing hexane and aqueous alkali. Extreme care, therefore, should be taken to ensure a thorough dispersion of the two immiscible phases into each other. Soap-stock separation is very efficient and the miscella does not require washing. Solvent is evaporated from the refined oil and the oil is bleached and deodorized. The process is capable of giving higher yields of refined oil with improved colour of bleached oil in comparison with non-miscella refined oils. However, it would be uneconomical to add solvent at a refinery just for the advantages gained by extraction in solvent.

In another variant of miscella refining a binary solvent system is used. Rice bran oils having high acidity have been refined by this process. The process employs a double solvent system like hexane and ethyl alcohol such that after addition of alkali, soap and other impurities are separated in alcohol phase whereas the triglycerides are retained in hexane. Crude rice bran oil is refined in a sequence of degumming, hexane dewaxing and binary solvent deacidification. Refined oil after evaporation of solvents can be bleached and deodorized.

Miscellaneous Refining Methods

Non-saponifying alkalies like soda ash and ammonia have also been tried, as they neutralize only free fatty acids without saponifying any oil. In a second refining step, a strong caustic alkali refining can be carried out to improve the colour. Overall losses are less than the conventional full caustic refining. However, sodium carbonate produces carbon dioxide (and causes foaming) which has to be removed by a vacuum assisted degassing step. This causes the soap to loose moisture and as a result a re-hydration step usually follows. At one time combination of soda ash and caustic soda was extensively used for cottonseed oils but now-a-days for large sized plants such processes are not feasible. Usage of ammonia will give a soap-stock which on heating will release ammonia (reusable) and no acidulation step is required for soap-stock disposal. However, a completely closed system is required and capital investments may be prohibitive. Mainly for these reasons, caustic soda continues to be the alkali of choice. Likewise, liquid-liquid extraction using furfural, propane and organic acids have not gained acceptance for a variety of reasons.

Refining Losss and its Measurement

In order to promote effective removal of pigmented substances, the amount of alkali added is in excess of that required for neutralization of free fatty acids. This excess alkali reacts with some of the neutral oils to produce additional soap. Some more neutral oil is lost by occlusion with soap-stock. Thus, the difference between the weight of crude and refined oils is the simplest measurement of refining loss. However, some indirect methods are more commonly used. The quantity of neutral oil obtained after refining is compared with the neutral oil that was present in the crude oil as determined by adsorbing the impurities and free fatty acids on alumina. This is termed as refining efficiency. The theoretical loss, which is a combination of impurities and free fatty acids, is also termed as Wesson loss or absolute loss. This is a value below which refining loss cannot be reduced.

BLEACHING

Bleaching or decolourization is a step in refining process, which consists of contacting the oil with a suitable adsorbent material. The colouring bodies and pigments like chlorophylls, carotenoids, gossypol and related compounds are adsorbed on the adsorbent and are removed from the oils through filtration. The adsorbents used may be neutral clays, activated earths, synthetic silicates and activated carbon. Of these, activated earth is most widely used for its economy and efficiency. The only other adsorbent used in this process is activated carbon which has higher oil retention and because of its particle size is more difficult to filter. Activated carbon combined with bleaching earth in the ratio of 10-20 parts of earth to one of carbon can be more effective than bleaching earth alone. The process of bleaching by adsorption also removes the traces of soap, trace metals, residual phosphatides and compounds containing sulphur.

Theory of Bleaching

A fundamental empirical equation developed by Freundlich (10) correlates the capacity of the adsorbent with the residual concentration of the material adsorbed. It is where x = amount of solute (pigment) adsorbed

m = amount of adsorbent (bleaching earth)

C = amount of residual solute

and K and n are constants.

K is a measure of decolourizing power of the adsorbent for the pigment involved and n indicates the range of decolourization where the adsorbent is most effective. Using the Freundlich equation, bleaching earths can be compared to one another under various bleaching conditions.

Pigments of rapeseed and palm oils like chlorophyll and carotene respectively are in the state of stable colloid. Primary function of a bleaching clay is to break the stability of the colloid. This can be followed by a rate constant formula (11).

$$\ln \frac{C}{C} = -k\sqrt{t}$$

where,

t is the time from addition of the clay

C is the concentration of pigment at time t

C_0 is the concentration of pigment at $t = 0$

and k is the rate constant.

Adsorption bleaching thus appears to be a physical phenomenon only. However, bleaching also involves chemisorption and subsequent chemical reactions proceed on the surface of activated clays (12). Fig. 10.13 illustrates one such concept of decomposition and dehydration, or pseudo-neutralization of peroxides (13, 14). The primary function, therefore, of the bleaching process is to reduce peroxides and remove the secondary oxidation products. Owing to the bleaching action which occurs during modern high temperature deodorization, it is the latter properties of activated clays which have become more important.

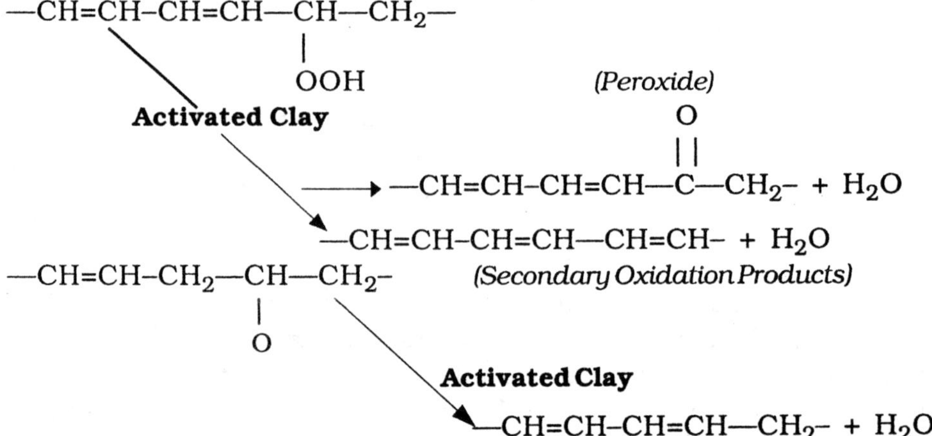

Fig. 10.13: Decomposition and Dehydration of Peroxides and Secondary Oxidation Products

For fatty oils, colours are determined by the Lovibond method, which involves matching the colour of a 51/4 inch column of oil or melted fat against standardized yellow and red Lovibond glasses. The colours are usually reported in terms of red units. Because of the subjective nature of colour determination by visualization, a photometric method has been developed by American Oil Chemists Society (15). It is based on measuring the optical density of oil at 460, 550, 620 and 640 mμ in a 21.8 mm cell.

Photometric Colour

$$= 1.29\ A_{460} + 69.7\ A_{550} + 41.2\ A_{620} - 56\ A_{640}$$

where A is the optical density or absorbance.

Apart from these, there are other methods of colour measurement like Gardner colour standards for drying oils and FAC colour standards for Tallow and Greases.

In application of colour standards, it is extremely important that the appropriate empirical relationship have been established. An under-bleached oil, e.g. to a 4 red colour on Lovibond, easily heat bleaches in a deodorizor to a less than 1 red colour, but produces an oil of substantial lower quality. Good oils have good colour, but not all good colours represent good oils. It is therefore, a necessary condition for an efficient bleaching operation to obtain

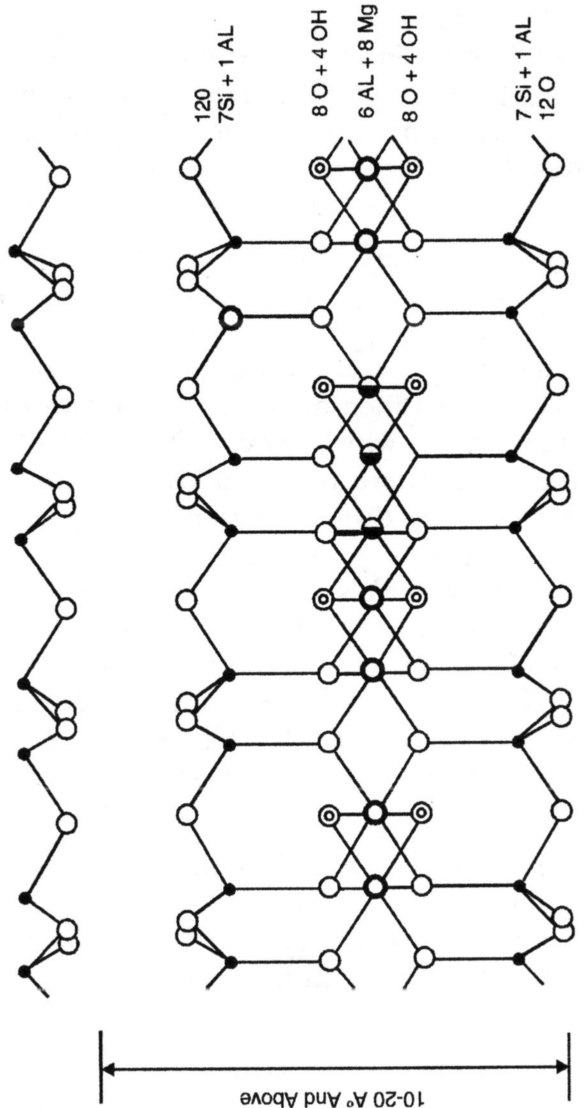

Fig. 10.14: Strcture of Montmorillonite

a zero peroxide value oil or more appropriately a low totox value along with satisfactory colour.

Bleaching Earths

Bleaching clay is produced from a bentonite, which is defined as a clay containing not less than 85% of the mineral montomorillonite on a dry basis. Montomorillonite belongs to the hydrated aluminium silicates. An idealized atomic structure of montomorillonite as projected on a plane has been shown in Fig. 10.14.

In this structure there is a central layer containing aluminium atom in octahedral coordination. Each aluminium atom is associated with four oxygen atoms and two hydroxyl groups. Above and below the central aluminium layer there are layers containing silicon in tetrahedral coordination with oxygen. There is an inter-planer space, which may contain water and then the structure is repeated. There is also a trioctahedral series when magnesium, iron or other divalent cations replace aluminium and the structure is equivalent to a further silica hexagonal network condensed on a single layer structure. A bivalent atom replacing a trivalent one leaves the crystal lattice negatively charged by an amount equivalent to one valency unit. Positive ions such as calcium, magnesium, potassium, sodium etc. become associated with the structure to neutralize this lattice charge. There are two basic types of commercial bleaching clays natural and activated.

Natural clay is produced from a bentonite that possesses natural bleaching power, in which some of the aluminium of the crystal lattice has been replaced by H ions in nature by the leaching action of the hydrolysed pyrite or organic acids' water while the clay was in deposit. In manufacture it is merely ground and dried.

Activated clay is made from a bentonite that has little natural bleaching power, i.e. the aluminium ion in the montomorillonite are relatively intact. In manufacture it is first treated with acid (hydrochloric or sulphuric) which replaces aluminium and hydrogen to a higher degree than that of natural clay. The excess acid is removed by washing with water and then the clay is dried and milled (16, 17). In treatment of the montomorillonite clay with acid, the concentration, temperature and time are all important for final activities.

The important characteristics of activated bleaching earths are total acidity, pH, moisture, apparent bulk density, effective surface area and oil retention capacity.

Total acidity or 'boil-out' acidity is determined on a water extract of the clay by titration and pH is determined on a slurry of clay in water (18). Bleaching power appears to be more a function of a clay's bound acidity. Therefore, those clays with high total acidity and a reasonable level of free acidity are preferred. If an activated earth is washed completely free of residual acid, its bleaching power is drastically reduced. Similarly, high level of free acidity will tend to increase the free fatty acid content of the oil appreciably.

Moisture plays an integral role in the physical structure of the clay. Bleaching clays normally contain 10–18% moisture. If the clay is completely dried prior to use, its internal structure collapses resulting in reduced bleaching power because of decreased surface area.

Apparent bulk density expressed as weight per unit volume, effective surface area and oil retention capacity of an activated earth are inter-related. Density depends on the amount of void space in the clay's structure. The more is the void space the lower is the density and higher the oil retention. Acid activation proportionately lowers the natural clay's density by increasing the surface area and increases its oil retention (19, 20). However, because significantly less amounts of the more activated clays are usually required, the net oil loss is usually less for the more effective bleaching earths.

Bleaching Conditions

Important bleaching criteria or considerations include dosage, temperature, time, agitation and operation under atmosphere/vacuum.

The dosage or amount of activated clay to be used should be the minimum amount needed to effect removal of impurities as measured by peroxide reduction with zero peroxide as a goal. The quantity of earth required is also dependent on the quality of the oil, the activity of the earth and the process conditions. Normally this requires 0.3–1.0% clay on a weight basis. Apart from the cost of the earth and oil retention on the spent earth, it is advisable to use as little earth as possible when bleaching caustic soda refined oil. The remnants of mineral acid in the earth tend to increase the free fatty acid of oil.

Adding clay to the hot oil reduces its adsorptive capacity because the moisture in the clay is driven off too rapidly. This causes collapse of the clay's lattice structure and subsequent reduction in its effective surface area.

Temperature of bleaching under atmospheric conditions are some what higher (around 100°C) but lower temperature (around 85°C) can be used if bleaching is carried out under vacuum.

Time is not as critical as top temperature. A 10-15 min contact time at top temperature is common. This is considered sufficient to drive off moisture and complete the bleaching reaction. Plant practices indicate that the presses should be made available after this minimal contact time.

Agitation should be sufficiently vigorous to achieve intimate contact as quickly as possible between clay and colour pigments. Violent agitation is avoided as it may incorporate air.

Although bleaching proceeds well under atmospheric conditions, vacuum bleaching improves the oxidative and flavour stability. During atmospheric bleaching, with air present, some of the pigments tend to become 'fixed' by oxidation. The bleaching process, is therefore preferably carried out in the absence of air. There are indications that activated clay may act as a catalyst for oxidation if oxygen is present during the process.

Commercially, bleaching of edible oils and fats is carried out as batch atmospheric, batch vacuum or continuous vacuum operation.

Batch Atmospheric Bleaching

In atmospheric bleaching, open cylindrical vessels with conical bottoms are employed. These are provided with agitators and closed coils for heating and cooling the oils. Oil is charged to the vessel and the temperature is raised to about 75–80°C. At this point bleaching earth is added either as powder or in the form of a slurry prepared in the same oil. Temperature is raised to about 100–104°C and held constant for a period of 10–15 min. After this the oil is filtered and recycled until the filtrate is clear. It is then cooled and sent to storage or for further processing. Most common type of filter is that of plate and frame type although others like pressure leaf filters are also used.

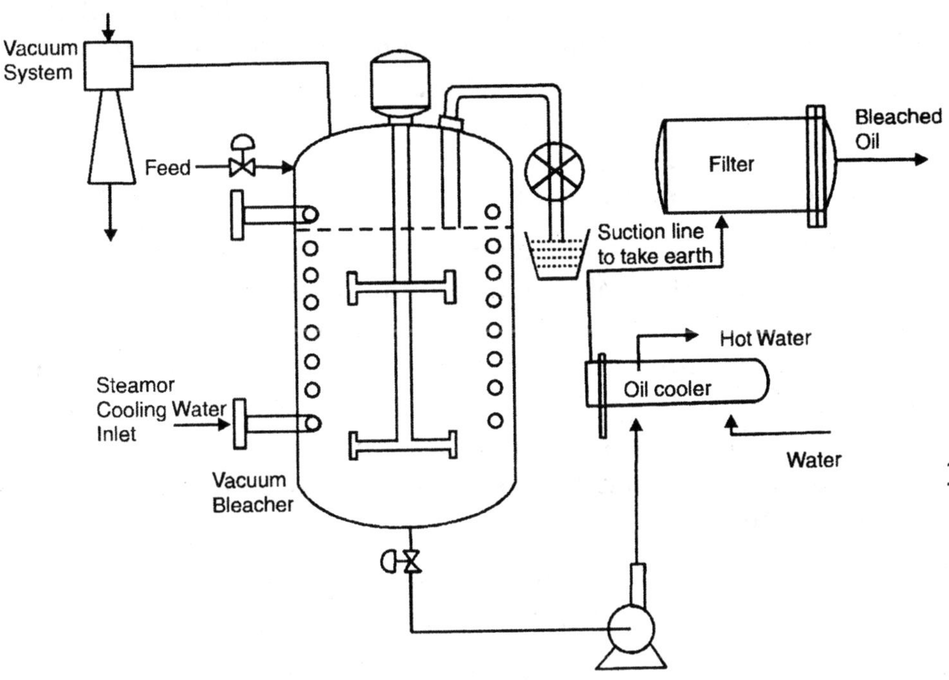

Fig. 10.15: Batch Vacuum Bleaching

During filtration in the filter press, the build up of filtration medium by earth provides a very large concentration of adsorbent in relation to oil. As a result, the plant bleached oils are usually superior to laboratory bleached ones employing the same level of adsorbent. This is known as "press bleaching" effect.

Batch Vacuum Bleaching

In vacuum bleaching, closed cylindrical vessels with dished ends are employed. A vacuum of 26–28" is attained either by a mechanical pump or by steam ejector system. The oil is deaerated by spraying into the bleacher. Addition of adsorbent under vacuum is important as it is added to the heated oil near the bleaching temperature. Adsorbent may be taken inside the bleacher with the help of a suction line. After holding the oil at bleaching temperature of 90–95°C for 10–15 min the oil is slightly cooled and filtered at 70–75°C and finally cooled to 40–45°C before leaving for storage or further processing. Flow diagram of a batch vacuum bleacher is given in Fig. 10.15.

Continuous Vacuum Bleaching

In continuous vacuum bleaching, the efficiency of bleaching is even better than that of batch vacuum. It requires less amount of bleaching earth and, therefore, for acid activated earths the rise in free fatty acid is minimum. Since the contact time between oil and earth is greatly reduced, the oxidative stability improves. A concurrent continuous vacuum bleaching which is representative of a modern installation, is supplied by EMI corporation and is given in Fig. 10.16.

Crude oil is fed in a metered quantity to the slurry tank. A mixture of activated clay and filter aid is fed to the slurry tank from the clay bin. The flow rate of the adsorbent mix is determined by the characteristics of the oil to be bleached. The slurry of oil, adsorbent and filter aid is sprayed into the vacuum bleacher to obtain complete deaeration. The slurry is then maintained for the

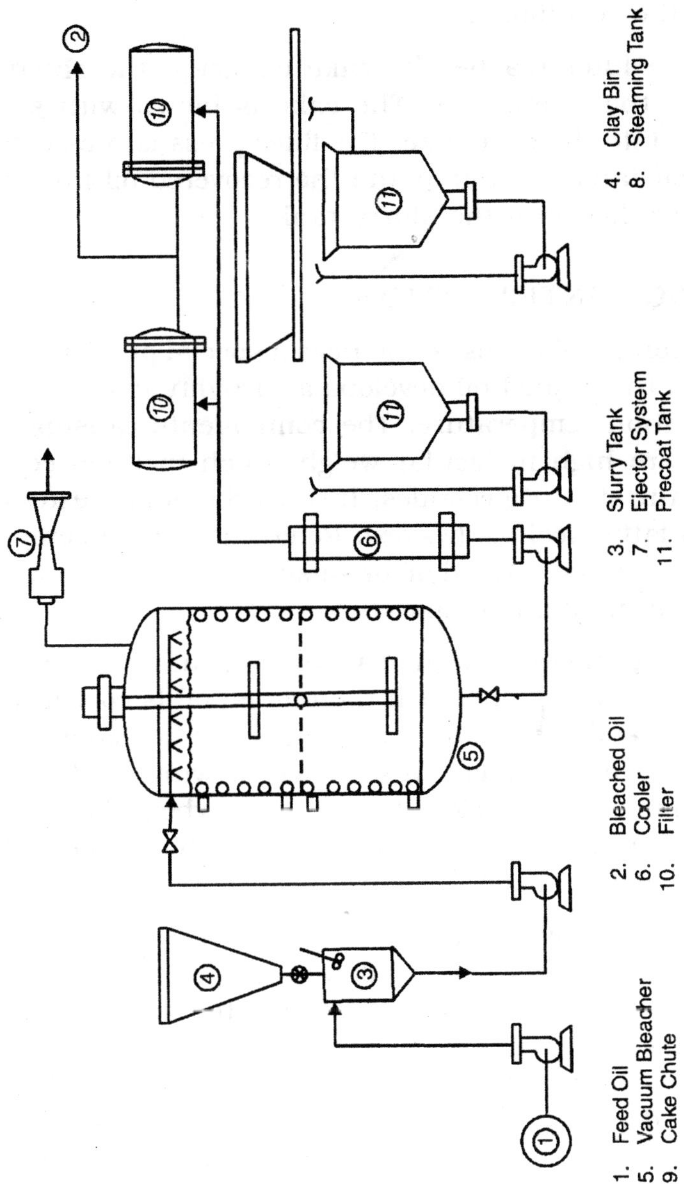

Fig. 10.16: Continuous Vacuum Bleaching System

desired length of time. Intimate contact and adequate mixing time is provided by the two stages of the vacuum bleacher. The slurry from the vacuum bleacher is passed through a cooler and filtered in one of the two filters.

When a filter reaches its cake capacity, the slurry flow is diverted to the other filter. The cake is blown with steam and discharged into the cake cute. The filter press is pre-coated using the pre-coat tank and pump and the recovered oil from steaming tank is reworked into the slurry tank.

DEWAXING/WINTERIZATION

In production of salad oils, sometimes a special problem is encountered when the refined oil develops a turbidity (haze) on standing at refrigeration temperature. The components causing turbidity to the oil are high molecular weight (high melting) compounds such as saturated triglycerides, hydrocarbons and esters of fatty acids with fatty alcohols (waxes). Removal of these compounds is necessary to obtain a salad oil or salad dressing which will remain clear at refrigeration temperature.

Dewaxing and winterization are two similar thermo-mechanical separation processes, in which the refined oil is cooled to a lower temperature and then filtered to remove the components responsible for haze development. A dewaxed or winterized salad oil by definition must remain clear for 5.5 hrs., when placed in a bath of melting ice at 0°C in a standard cold test specified by the American Oil Chemists' Society.

As the name implies, dewaxing is process to remove waxes from unsaturated oils like sunflower, grape, corn and rice bran oils. These waxes are extracted from the hulls or oil-bearing bodies themselves. Winterization is the process to remove the high melting stearines present in certain oils like cottonseed and partially hydrogenated soyabean oils. Winterization differs from fractionation in the sense that the latter process separates a fairly large quantity of stearines from liquid oils, which is widely practised in palm oil industry.

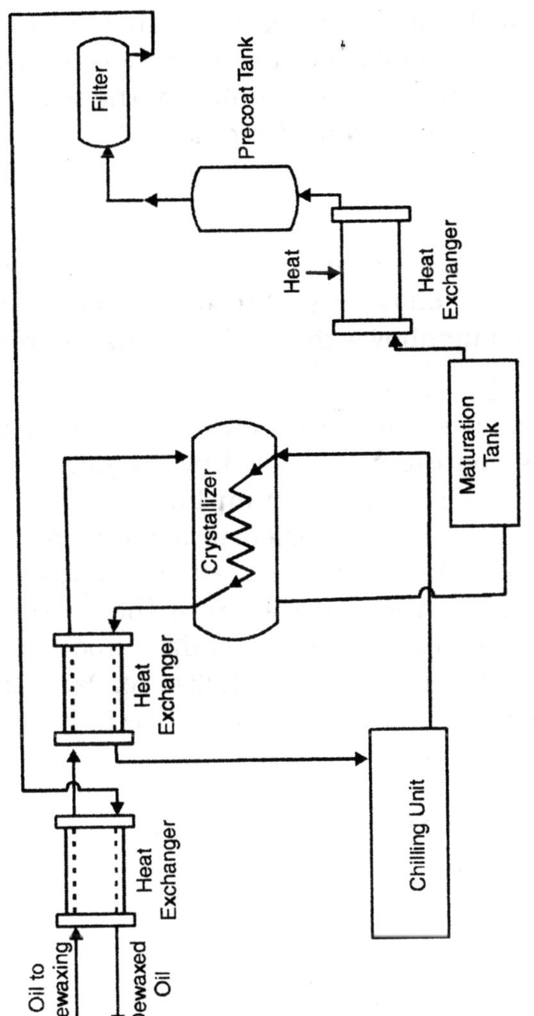

Fig. 10.17: Flow Diagram of Dewaxing Plant

The processes of dewaxing and winterization are primarily crystallization operations and should be applied on oils which are well refined and as far as possible free from any impurities. Oils should be especially free from phospholipids, as they are effective crystal inhibitors and may hinder the crystal growth. The quantity of wax present in the crude oils varies from a few hundred ppm to over 2000 ppm. Efficiency of dewaxing process is highly critical, as to get an oil with sufficient cold stability, the wax content has to be reduced to a level of about 10 ppm.

Conventional Dewaxing

In its classical form, dewaxing is carried out on either bleached or deodorized oil. The oil is carefully cooled so as to avoid any super-cooling. Slow cooling allows the wax crystals to grow to give easy filterability.

Fig. 10.17 shows a modern dewaxing plant. The bleached or deodorized oil is continuously cooled by the dewaxed oil and then further cooled by a cooling medium like chilled brine in a heat exchanger. The oil is slowly cooled to a temperature of 6–8°C in a crystallizer and held at that temperature for 4 hrs. Another 6 hrs. time is given for the maturation of the crystals. Some filter aid is added to the crystallizer, depending upon the wax content of the oil. Filter-aid facilitates crystallization and filtration. After the maturation time is over, the oil is carefully heated to about 18°C and filtered in either pressure leaf or plate and frame filter presses. A body feed approximately equal to the wax content of the oil is used to prevent blinding of the filter leaves. Without a body feed, the waxes slime over and blind the screens almost immediately.

This dewaxing process is well suited for oils with low wax content, e.g. upto 500 ppm. With new type of high oil content sunflower seeds the wax content is crude oils may go up more than 1500 ppm resulting in a lower filtration rate and higher filter aid requirement. Similar problems are encountered with rice-bran oil which contains upto 1–1.5% of high melting waxes. A pre-dewaxing operation is therefore required for such oils.

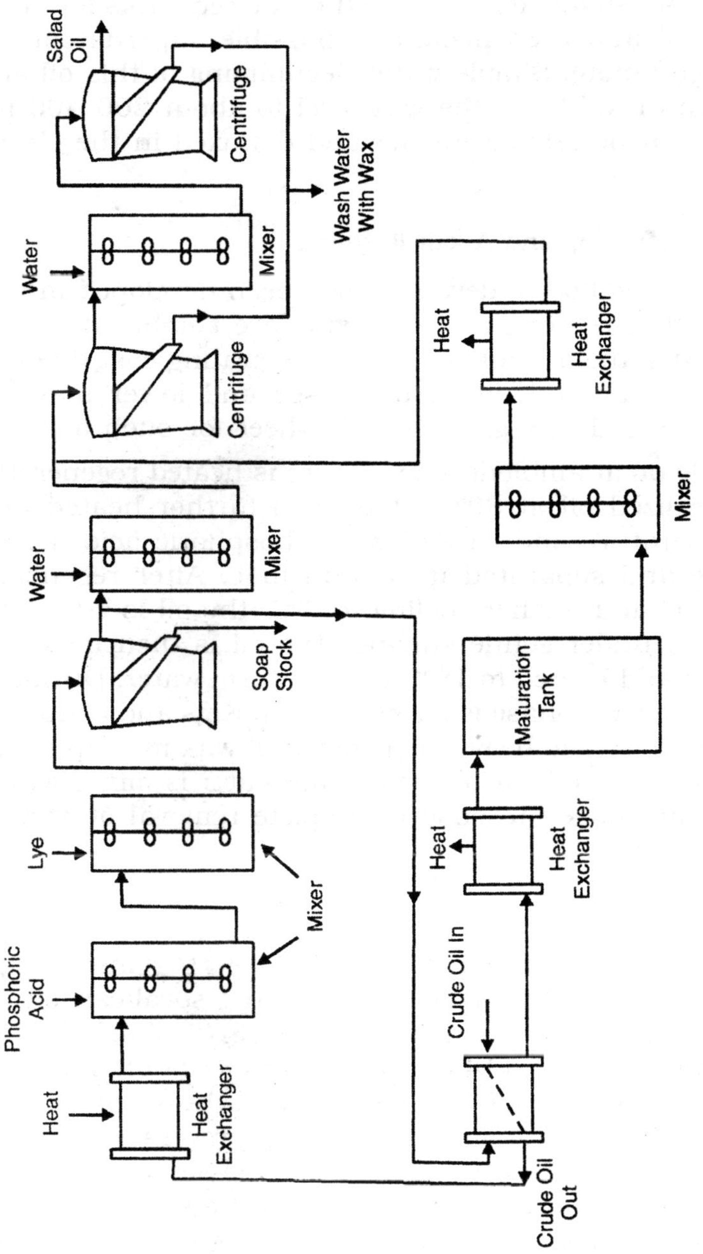

Fig. 10.18: Simultaneous Alkali Refining and Dewaxing

Pre-dewaxing and Degumming

The crude oil is usually allowed to settle in large tanks for at least 24 hrs at 25°C and then maximum possible supernatent oil is taken for degumming. Simple water degumming of this oil at the same temperature reduces the wax level to about 200–400 ppm. The oil then can be refined further and dewaxed in the classical manner.

Simultaneous Dewaxing and Alkali Refining

Recently a new method of dewaxing has been developed in which continuous alkali refining and dewaxing are combined together. This method has the advantage of lower operating cost due to the elimination of filter aid, lower oil losses and lower manpower requirements. Fig. 10.18 gives the flowsheet for such a process.

The crude oil at ambient temperature is heated regeneratively by hot neutralized oil at 80°C. It is then further heated to the refining temperature and treated with phosphoric acid, neutralized with lye and separated in a centrifuge. After regenerative cooling to 30°C and further cooling to 8°C, the oil is retained for about 4–5 hrs. under gentle stirring. The oil is then mixed with 4–6% of water and heated to 18°C by lukewarm water. During this mixing a soapy water phase is formed, which wets and extracts the small wax crystals. The heavy suspension of wax in soap water is centrifuged from the oil. Once again some water is mixed with the oil which is centrifuged to ensure complete removal of wax from the oil.

Winterization

Oils which are subjected to winterization process to remove stearine are cottonseed oil and partially hydrogenated soyabean oil. Cottonseed oil winterization normally yields 75-80% of salad oil with an iodine-value of 110–114. Partial hydrogenation lowers the linolenic acid content of soyabean oil, whose presence in unhydrogenated oil is responsible for flavour reversion characteristics. Soyabean oil hydrogenated to an iodine value of about 115 on winterization yields a stable salad oil in the range of 75–85%. A flowsheet of a modern winterization plant is shown in Fig. 10.19. In the graining tank suitable means are provided to heat, chill and agitate the oil. The oil is cooled at about 0.5°C/hr. to about 7°C,

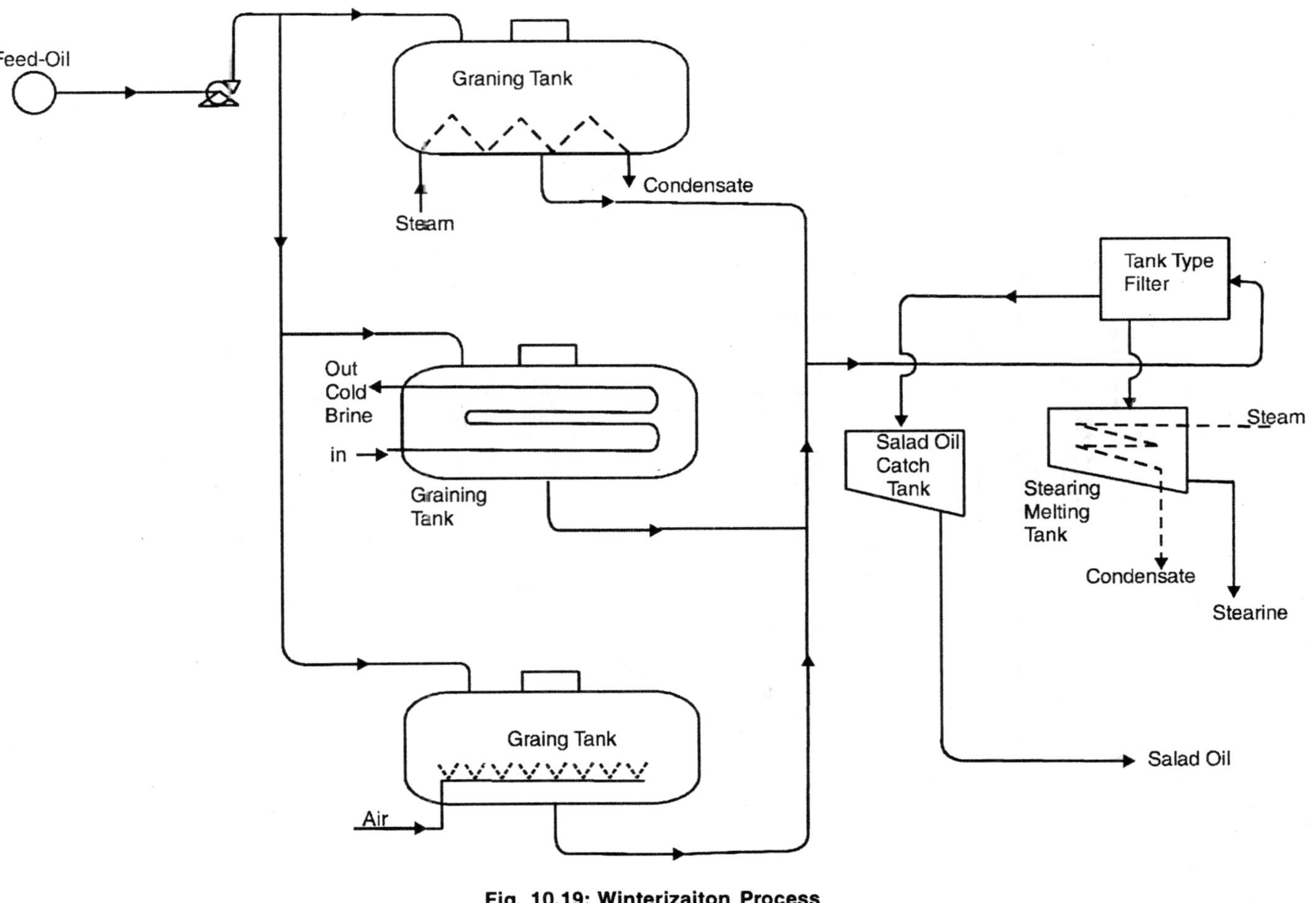

Fig. 10.19: Winterizaiton Process

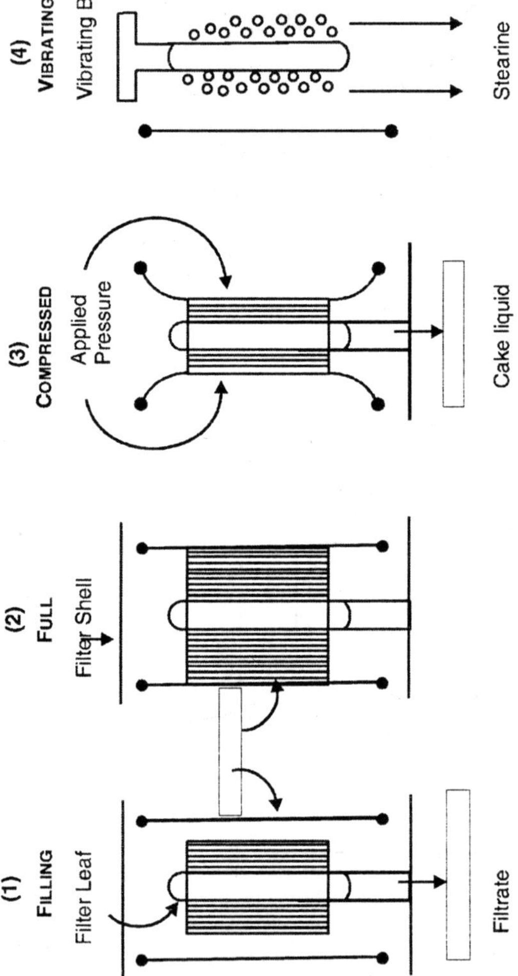

Fig. 10.20 Diaphragm Filter

where it is held for 8–12 hrs. for crystallization to be completed. It continuously undergoes very slight agitation. Crystal inhibitors usually are added to increase crystal size and thus facilitate filtration. One widely used crystal inhibitor is oxystearine which can be used at a maximum of 0.125%. The filtration usually is carried out in a tank type filter.

The removal of stearine from oils during the winterization process has always posed a serious problem. Stearine cakes are relatively dense and non-porous and as a result filtration rate decreases rapidly as the cake thickness increases. The viscosity of the cold oil contributes to poor filtration rate and also makes it difficult to reduce the oil content of the cake.

Capacity of the filtration system is an important factor in winterization. When 15% or more of the feed is removed in the form of stearine, substantial solid retention capacity is required. Recessed plate and frame and pressure leaf filters have been used in winterization since these have the cake holding capacities that the process requires.

A new type of plate and frame or recessed plate filter (21) contains a series of diaphragms which can be hydraulically expanded to press the cake. The pressure reduces the oil content of the cake after the filtration cycle is complete and the frames are full of stearine as illustrated in Fig. 10.20. This type of filters have been used on all types of salad oil stearines. The filtration rates are the same as for the plate and frame process but the oil content of the stearine is greatly reduced. The cake is dry enough so that the presses can be automatically opened and vibrated to be cleaned.

The filtration of stearine from partially hydrogenated soyabean oil is much less difficult than cottonseed oil stearine because of the durable nature of stearine in the former case. A continuous filtration system can be conveniently used. Rotary vacuum filters are extensively used in this application (22). Since the stearine is removed with each revolution of the drum, cake thickness does not built up to the point where it retards filtration rate. The vacuum filter is normally installed in a cold room with a temperature of about 10°C.

Centrifugal separation of the oil and stearine is also commercially practised. It is a basket technique which spins off the oil while the stearine remains in the basket. When the stearine accumulates to a certain point, centrifuging is stopped and the stearine is mechanically scrapped from the basket. It is cyclic operation.

Cottonseed oil can be winterized while in a solvent miscella stage using hexane or acetone. This results in a higher yield and better quality than can be expected from conventional winterization. The chilling and holding time is less, thus making continuous operation practical. However, it is not economical to add solvent to the oil at a refinery just to get the advantage of winterization.

Aqueous detergent fractionation process which has primarily been developed for palm oil fractionation (23) can equally be applied to dewaxing/winterization processes. A process flowsheet is shown in Fig. 10.21. An aqueous solution made up of a wetting agent (sodium lauryl sulphate) and an electrolyte (magnesium sulphate) is mixed into the stream coming from crystalizers (graining tanks). The salad oil is separated in a centrifuge. The heavy aqueous fraction is then heated, which melts the stearine/wax and allows it to be separated in a second centrifuging step. The aqueous solution containing wetting agent and electrolyte can then be recycled after cooling.

Deodorization

In production of edible oils deodorization is the final processing step. It removes the disagreeable flavours and odours that are naturally present or created during the previous steps. Deodorization is primarily a high temperature, high vacuum steam distillation process.

The compounds removed during deodorization have been identified as ketones, aldehydes and free fatty acids. Their concentration is usually quite low, ranging from 0.1 to 1.0% on the weight of the oil. The odour imparted to a hydrogenated oil seems to be the characteristic of the hydrogenation reaction. Deodorization also removes some of the free sterols and tocopherols present in oils.

Steam deodorization is feasible because the flavour and odour compounds that are to be removed have appreciably greater volatility than the triglycerides. Operation at high temperature increases the volatility of these odoriferous compounds and the introduction of stripping steam into the deodorizer greatly increases the rate at which these compounds are volatilized. Reduced pressure operation further aid in removal of odoriferous compounds and protects the oil from atmospheric oxidation.

For most oils of commercial importance, the flavour and odour removal correlates well with the reduction of free fatty acids (FFA) content of the oil. If an oil has an FFA of 0.1%, it will have an odour that will be eliminated when the FFA is reduced to 0.01-0.03% assuming a zero peroxide value.

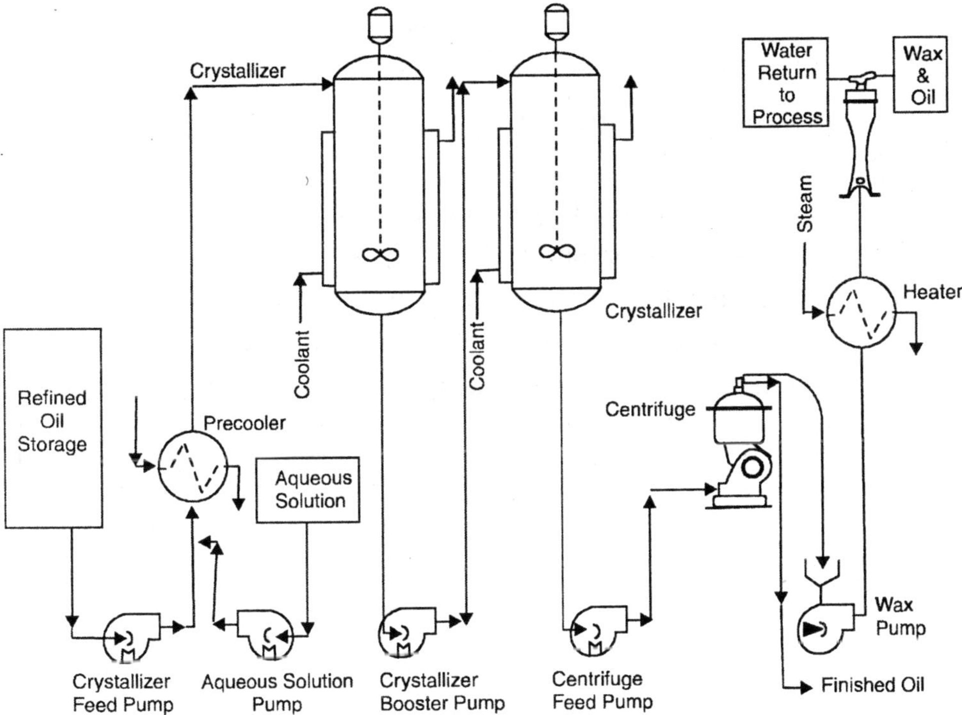

Fig. 10.21: Aqueous Detergent Fractionation in Dewaxing Winterization

Theory of Deodorization

Commercially, the deodorizers are operated as batch, semi-continuous (a series of batch deodorizers within a single vessel) or continuous systems. Deodorizers can also be classified by the method of oil-steam contacting. In one case, steam may be bubbled through a depth of oil while in another, steam may be contacted with a rising or falling film of oil. Bubbling of steam through oil is possible in all modes of operation, viz. batch, semi-continuous and continuous. However, contacting the steam with a film of oil is possible only in continuous mode. In order to understand the effects of operating variables, such as temperature, pressure and stripping steam/oil ratio, the theoretical treatment of deodorization systems are given below:

(i) Batch systems : For a batch deodorizer, the equation indicating the steam requirement based on the FFA removal has been developed (24).

$$S = \frac{PO}{EP_v} \ln \frac{V_1}{V_2} \tag{1}$$

where,

S = moles of stripping steam

O = moles of oil

P_v = Vapour pressure of free fatty acids

P = Total system pressure

V_1 = Initial number of moles of free fatty acids in the oil

V_2 = Final number of moles of free fatty acids in the oil

E = Vaporization efficiency, which is a measure of the completeness with which the steam becomes saturated with volatile substances during its passage through the oil. In continuous or semi-continuous plants with intimate and intense mixing the value of E approaches unity.

Equation (1) is based on the assumption that triglyceride fatty acid solutions behave as ideal solutions. Because of considerable non-ideality, the equation is modified by incorporating an activity coefficient A (25). The modified equation is

$$S = \frac{PO}{EP_v A} \ln \frac{V_1}{V_2} \tag{2}$$

Activity coefficient is usually determined experimentally. At free fatty acid concentration of 0-1%, activity coefficient is reported to be 0.63-0.64 (26).

(ii) Continuous systems: For a continuous deodorizer the equation (2) can be rewritten using the time dependent variables (27)

$$\frac{V_1}{V_2} = 1 + K\frac{P_v}{P} \times \frac{S}{O} \tag{3}$$

where,

V_1 and V_2 are the moles per hour of free acids in the oil entering and leaving the deodorizer.

S = moles per hour of stripping steam

O = moles per hour of oil

and K, P and P_v are same as given in equation (1).

These equations provide some useful relationships in the design of a commercial deodorizer. For a given FFA reduction and temperature, the ratio of the system pressure to the sparge steam is a constant. Also a lower system pressure allows a lower sparge steam rate.

In designing of a deodorization system, the other important considerations are material for construction, contact with oxygen (air) and heat treatment of the oil.

Due to the deleterious effects of copper and iron on oils, deodorizers are fabricated with either T304 or T316 stainless steel at points contacted with oil. Contact with air is another factor which will adversely affect the oil quality at high deodorization temperature. Deodorizers should have proper facilities for effective de-aeration prior to heating and virtual exclusion of air at the temperature of deodorization.

Operation at elevated temperature also promotes thermal decomposition reactions within the oil itself. These are not related to FFA removal but are necessary to provide a stable oil after deodorization. Thermal treatment also heat bleaches certain oils to a much lighter colour. This is particularly noticeable in the case of soyabean oil. These reactions and the heat bleaching are time and temperature dependent. Therefore, commercial deodorization systems should provide a retention period at deodorizing temperatures to allow these reactions and heat bleaching to occur.

Importance of Operating Variables

In deodorization practices understanding of the interdependence of the operating variables are important for a successful and economical operation.

(i) Temperature/time relationship: As vapour pressures of the various volatile components increase rapidly with rise in temperature, deodorization at the highest possible temperature should be carried out for effective removal of odoriferous compounds. The maximum temperature that can be used, however, is limited because of its detrimental effect on oil stability.

Moser *et al.* (28) and Lineberry *et al.* (29) have shown that there exists a definite relationship between temperature and time for deodorization in relation to deodorized oil's flavour and peroxide value. For soyabean oil, these are 1-2 hr. at 210°C, 30 min to 1 hr. at 230°C and 250°C. At 270°C a residence time of 15-20 min gives good result. These time/temperature relationships were fairly independent of pressure in the range of 2 mm to 6 mm Hg absolute. It was also observed that sterols and tocopherols removed at 270°C were almost double that removed at 240°C. More complete removal of tocopherols may adversely affect the keeping quality of deodorized oil.

(ii) Vacuum/stripping steam relationship: Theoretically, a decrease in absolute pressure and/or an increase in stripping steam will increase the removal rate of volatile components. But the operating pressure and stripping steam rates are limited by economical consideration. Lineberry *et al.* (29) found that within the pressure range of 2-6 mm Hg absolute, total distillate removed was independent of pressure but directly proportional to the temperature and quantity of stripping steam. While rise in temperature increases the volatility of odoriferous compounds, increase in stripping steam rate results in increased entrainment. The ratio of percentage stripping steam to absolute pressure (%/mm) is considered as entrainment loss in deodorizer design, and held between 0.5 to 1 in the stripping trays, for example, 3% steam to the stripping tray at 6 mm (ratio = 0.5) or 2% steam at 2 mm (ratio = 1.0).

Deodorization Losses

In the operation of a deodorizer, the process loss includes the FFA content and volatiles present in the oil. Mono- and di-glycerides have appreciable volatilities and are partially distilled. Medium chain triglycerides like coconut, palm-kernel and babassu and more volatile than long chain triglycerides and hence give higher deodorization losses. Tocopherols and unsaponifiables like sterols are also distilled at deodorization temperature, although to a lesser extent than do fatty acids.

Apart from this, an appreciable amount of oil is also lost by way of entrainment in the sparge steam. In addition, because of the hydrolytic action of steam on triglycerides, a small amount of fatty acids is continuously produced which is also lost by distillation.

Deodorization Equipments

For any deodorization equipments, provisions for high vacuum production and high temperature heating are the basic necessities. The other necessary requirements are effective de-aeration of the oil prior to deodorization and cooling of the oil after deodorization is over.

Multi-stage steam ejectors with barometric condensers are routinely used in most commercial deodorizers for obtaining a high operating vacuum. Mechanical pumps are not advisable as they are difficult to operate and maintain. A three stage steam ejector system provides a vacuum of 5–6 mm of Hg absolute, and is found to be quite satisfactory. Four stage steam ejector systems can given a vacuum below 3 mm of Hg absolute, but the insignificant improvement in deodorized oil quality does not justify this additional expenditure. Readers are advised to refer to an excellent article for the detailed discussion on vacuum system (30).

The maximum deodorizing temperature is about 270°C. With a high pressure boiler (about 80 kg/cm^2) this temperature can be attained. Alternatively, a vapour phase, heat transfer medium like Dowtherm vapour is used industrially. This is an eutectic mixture of diphenyl oxide. The vapour phase heating medium operates at 16 psig and can heat the oil to 260°C. For most practices a deodorization temperature of 210–240°C is satisfactory and permits the use of 40–60 kg/cm^2 steam.

(i) Batch deodorization: The conventional batch deodorizer is a vessel in the form of a vertical cylinder with dished or cone head. The vessel is welded and well insulated. They are available in a wide range of capacities, typical being about 12,000 kg. Typical

batch deodorizers are designed to hold 2.4 to 3.0 litres of oil and to have a similar amount of headspace above the surface of the oil. Sufficient headspace is required to minimize the carry over of oil droplets by entrainment with sparge steam. The stripping steam is injected into the bottom of the vessel through a distributor.

With the equipment operating at a high temperature and a 6-8 mm pressure, about 8 hrs time is usually allowed for the complete cycle of charging, de-aerating, heating, deodorizing, cooling and discharging. After deodorization is completed, the oil must be cooled before it is exposed to atmosphere. Hydrogenated oils may be cooled to 60°C but a temperature of 35–40°C is usually preferred for liquid oils. The oil is cooled within deodorizer by circulation of water through cooling coils. Alternatively, hot oil may be pumped through a shell and tube cooler.

(ii) Semi-continuous deodorization: A semi-continuous deodorizer consists principally of a tall cylindrical shell of carbon steel construction in which five trays of T 316 stainless steel are placed. Semi-continuous deodorizers operate on the basis of handling finite batches of oil in a time sequence of de-aerating/heating, holding/steam stripping and cooling. A definite batch size is completely subjected to each condition before proceeding to the next step. Fig. 10.22 shows a flowsheet for a Votator semi-continuous deodorizer with heat recovery system.

Feed stock at about 50°C is pumped to a measuring tank that functions to establish the proper amount of oil to be charged to the deodorizing trays. The top tray takes the batch and is de-aerated by heating with the steam generated from the heat recovery step. Oil then flows to the second tray, where it is heated to the high temperature of deodorization and then flows to the third tray. At this stage the oil is deodorized by stripping steam injected into the third tray. At this stage the oil is deodorized by stripping steam injected into the third tray. An overhead umbrella shapeed baffle, minimizes entrainment losses and exposes the maximum surface of oil to deodorization temperature. The oil is then taken to fourth tray for heat recovery. Here the incoming oil is heated by freshly deodorized oil, resulting 50% heat recovery. Of the various methods available, one is to use the hot deodorized oil to generate steam in a closed-loop thermosyphon, which in turn will heat the incoming oil.

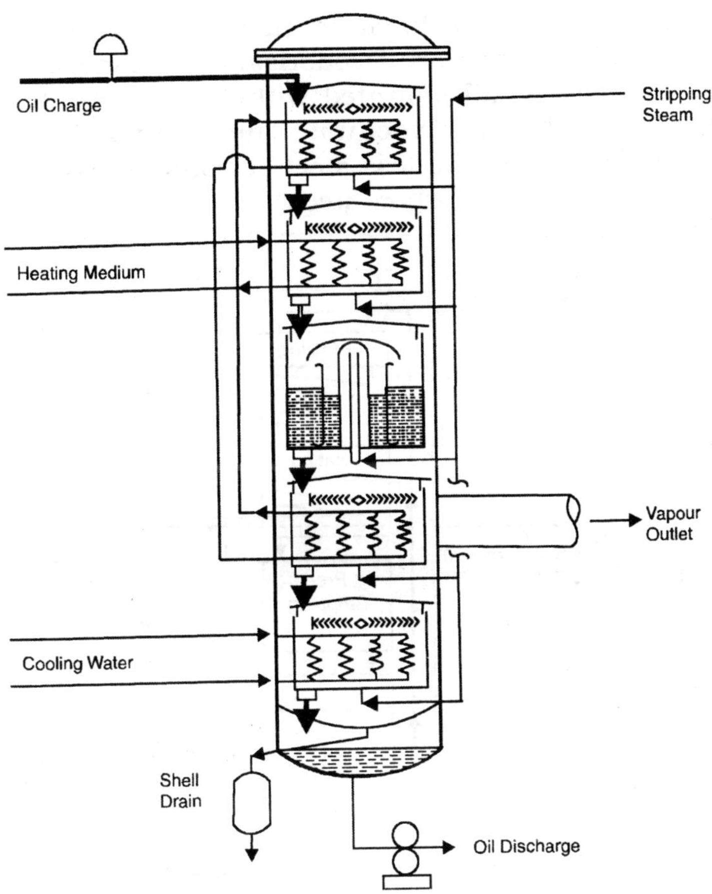

Fig. 10.22: Schematic Diagram of a Semi-continuous Deodorizer

In the fifth tray cooling water is used to cool the oil to about 55°C. Citric acid and other anti-oxidants are added here. The cooled deodorized oil is discharged to a drop tank and from there pumped through a polishing filter.

Stripping steam is introduced in each stage of deodorization. 75% of the total steam is used in third tray for deodorization. Rest 25% is equally distributed to other four trays for mixing.

(iii) Continuous deodorization: In a continuous deodorizer, the oil is passed through the heating, holding and cooling steps of deodorization continuously with controlled holding time for each step.

Fig. 10.23 shows the schematic diagram of a double shell deodorizer by EMI Corporation. Deodorization is carried out in a series of seven vessels, all mounted within a single outer shell which is maintained under vacuum. The internal vessels are fabricated from type 304 stainless steel at points of contact with oil and the out shell is of carbon steel.

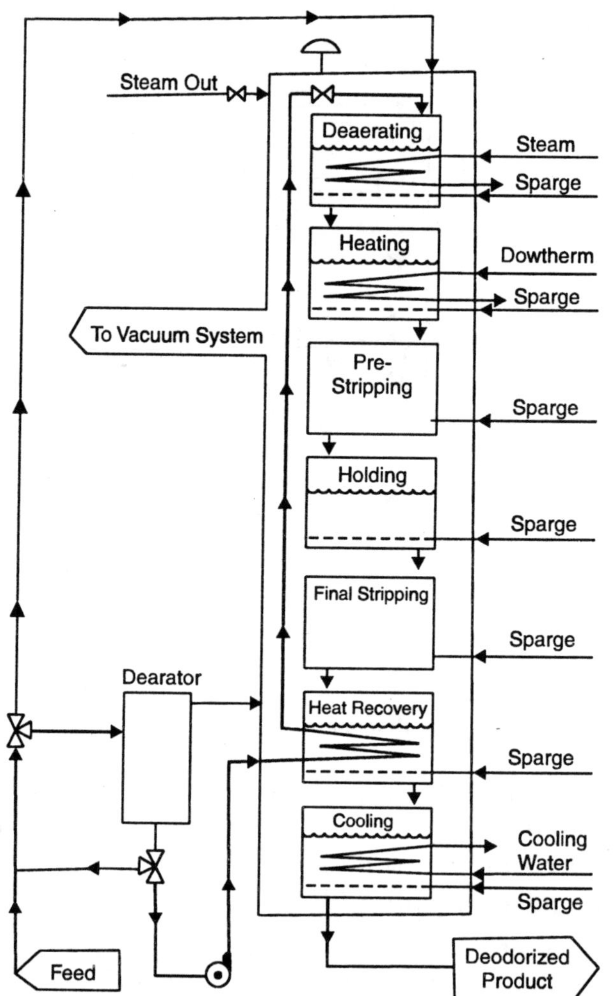

Fig. 10.23: Double-shell Deodorizer Modified for Heat Recovery

Feedstock at about 50°C is pumped to the external de-aerating vessel, which is maintained under the same vacuum as that of the deodorizer. The de-aerated feedstock is pumped through an internal heat recovery tank which is heated by the deodorized oil. The pre-heated and de-aerated oil recovers 50% of the heat normally required for deodorization. The oil then flows to the de-aerating tank. Here, steam-heating is required only when a change is made in feedstock. The oil then flows by gravity to the heating tank.

Oil is now heated with a high temperature heating medium like Dowtherm Vapour or high pressure steam to the deodorization temperature (210-240°C). This tank is sparged with steam to aid in heat transfer and to avoid any local overheating. After reaching the desired deodorization temperature the oil flows by gravity to the third vessel.

In this vessel, also known as pre-stripping tower, the oil flows over a number of stripping trays counter current to the flow of stripping steam injected to the bottom of the vessel. Oil then enters the holding section to allow for the retention time at deodorization temperature for the necessary heat treatment including heat bleaching. It then flows to the final stripping section, where volatile compounds released in the holding vessel are stripped and removed. The oil then enters the heat recovery section to heat the incoming feed. The deodorized oil is then further cooled in the cooling tank to 60°C or lower.

Sparge steam and distillate from each internal section flow to the outer shell and are removed through the vacuum system. Any entrained oil along with the volatile matter condenses on the outer shell and drains to the bottom, from where it can be removed periodically.

Double shell deodorizers are usually suitable for large capacities, ranging from 6–25 tonnes per hour. For lower capacities, single shell deodorizers are more economical to fabricate. They are similar to double shell type except that a carbon steel vapour pipe manifold replaces the carbon steel outer shell. Flow diagram of a single shell deodorizer from EMI Corporation is given in Fig. 10.24.

Deodorization operation build up some carbon deposits because of the high temperature employed in this operation. Sometimes citric acid used in post deodorization step is not completely

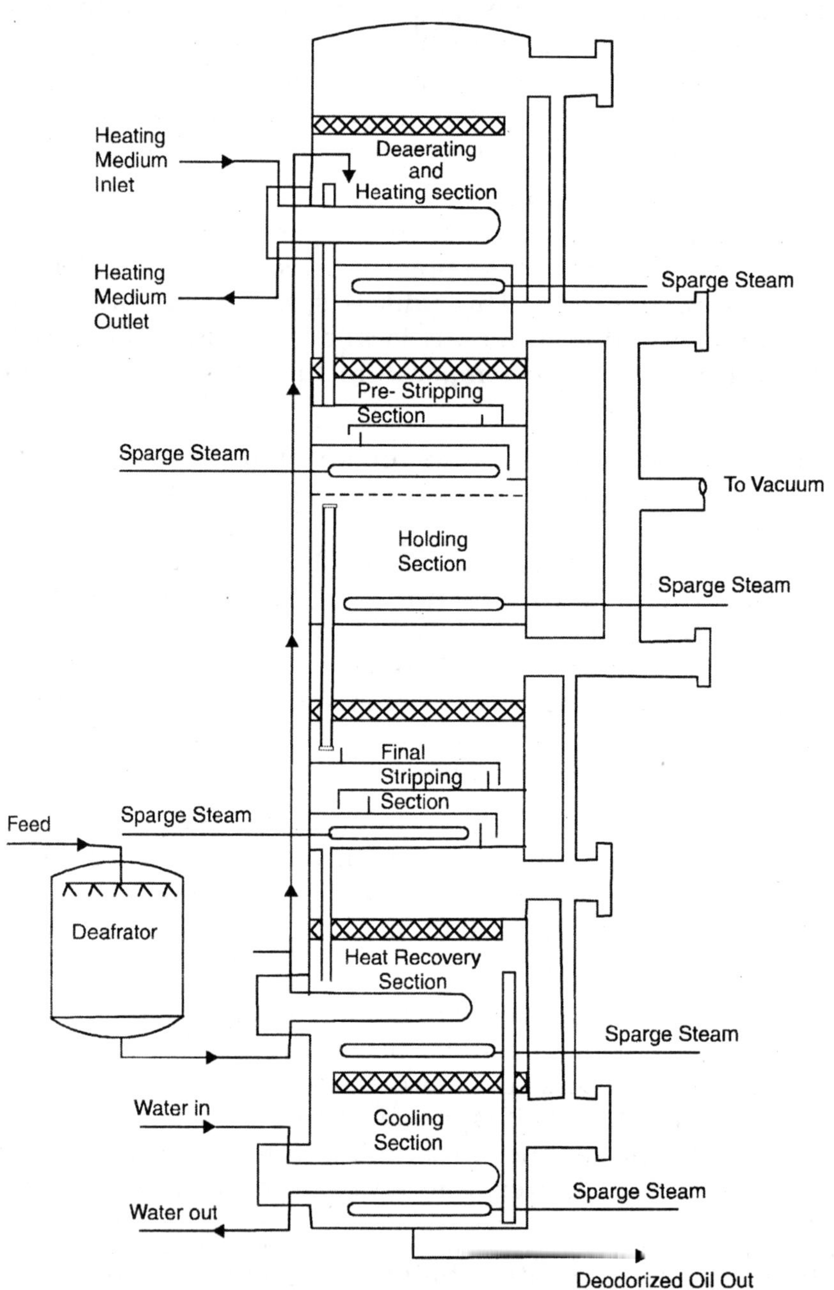

Fig. 10.24: Schematic Diagram of a Single Shell Deodorizer

solubilized. Also the oil may be contaminated with traces of earths from bleaching process. Polishing of the deodorized oil is done to remove these insoluble particles and this is accomplished by filtering the oil through fine filters; 30 μm filters are used for general usage and 10 μm filters for those oils which are to be retailed in transparent bottles. Cartridge filters with low holdups and high throughput rates are used in polishing the oil.

In film type deodorizer the main difference between the two film contacting designs is the manner in which the film is formed. In one, as in the ATT unit, oil is allowed to form a film under the gravity inside a tube and in the other, as in the Campro unit, the oil is forced to rise upward due to the viscous drag of the steam on the oil-steam interface.

Physical Refining

Removal of free fatty acids from the oils by distillation process instead of an alkali treatment is termed as physical refining. In physical refining, de-acidification and deodorization of the oil is achieved by steam distillation under high vacuum. Physical refining is also referred to as steam refining.

To be physically refined, a crude oil should be virtually free from phospholipids, trace metal contaminants and other non-volatile impurities. Only free fatty acids and other volatile impurities are removed in physical refining. It is, therefore, extremely important to have very effective degumming and bleaching processes before physical refining. The most effective pre-treatment for physical refining is the treatment with phosphoric acid and acidic earths (as discussed in the section of degumming). The other important method is as described in Alcon process in which the formation of non-hydratable phosphatides are inhibited during the meal preparation prior to extraction.

In the case of oils with higher free fatty acids content, physical refining has comparatively lower losses than alkali refining. The economy becomes more favourable if physical refining is combined with deodorization. Alkali refining produces a by-product soapstock which requires acidulation for disposal and hence poses an environmental problem. In physical refining, the free fatty acids are recovered as good quality by-products. An excellent description of the relative merits and demerits of both the processes have been given by Noris (31).

Physical refining process can be divided into three component operations all carried out under high vacuum and temperature. First, a part of the free fatty acids is removed by flash evaporation. Secondly, the oil is de-acidified by stripping with steam and finally the oil is deodorized and heat bleached by contacting with multi-stage counter current steam. A graphical description of such a process is given in Fig. 10.25. In order to realize the full potential of a deacidifer-deodoriser, it should operate without substantially greater utilities than a standard deodorizer.

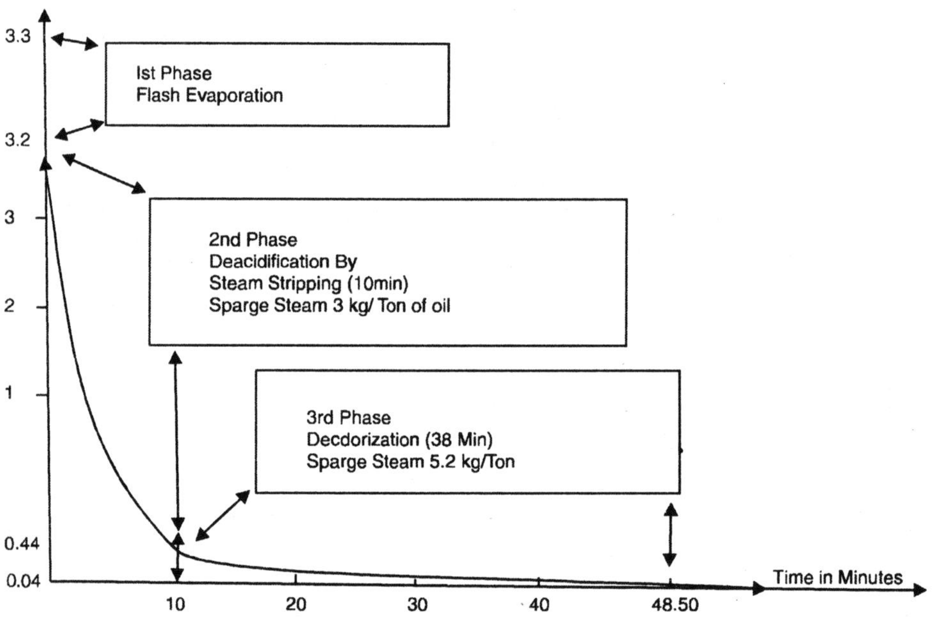

Fig. 10.25: Acidity Variation as Function of Residence Time in Deacidifier and Deodorizer

Because of the corrosive nature of the large quantities of fatty acids at a very high temperature, the material for construction should be of T 316 stainless steel. A deacidifier-deodorizer is usually designed and operated in a manner similar to the single shell deodorizer.

There are two basic arrangements in the operation of a deacidifier-deodoriser (32) and these are shown in Figures 10.26(a) and 10.26(b). In the arrangement described in Fig. 10.26(a) the oil

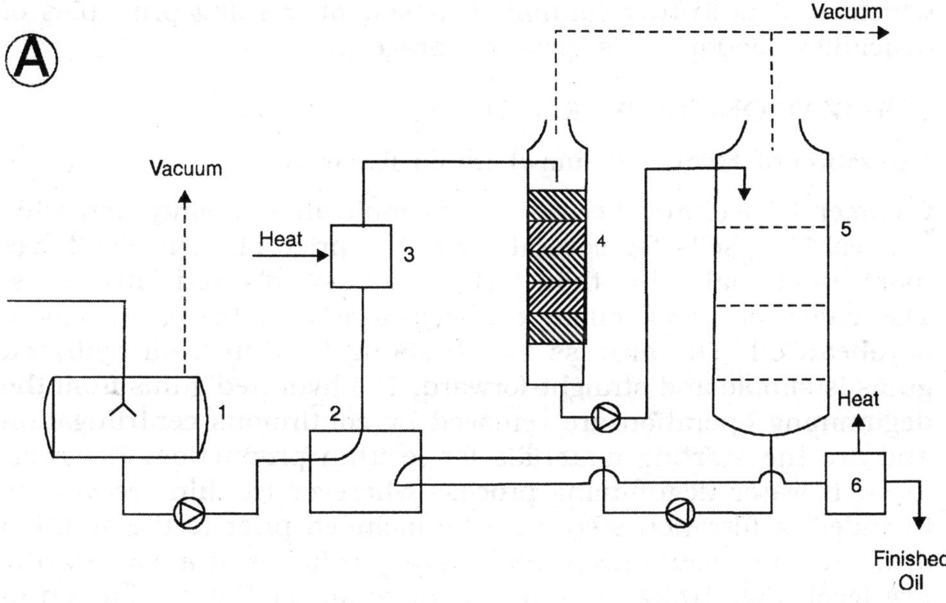

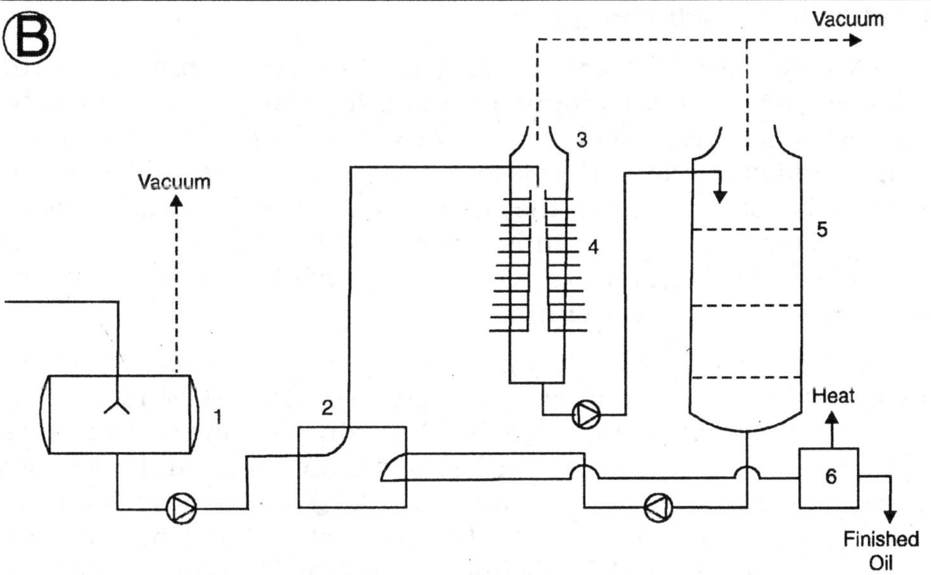

1. Deaerator 2. Heat Exchanger 3. Final Heater 4. Deacidifier 5. Deodorizer 6. Oil Cooler

Fig. 10.26: Operating Principles for Deacidifier Deodorizer

is heated in a spiral or shell and tube heater and the deacidifier may be of packed or falling film type. In the other arrangement, the oil is heated under vacuum at the top section of deacidifier which is a multi-tray column. A review of working principles of deacidifer-deodoriser is given by Stage (33).

UTILIZATION OF BY-PRODUCTS

Utilization of Hydrated Gums/Lecithin Recovery

Commercial lecithin, because of its multi-functionality and wide applicability, sells for several times the price of oil itself. It has more functional value than tocopherols, sterols and fatty acids. The major source of commercial lecithin is from the processing of soyabean oil. The process for obtaining lecithin from hydrated gums is simple and straight-forward. The hydrated gums from the degumming operation are removed by continuous centrifugation and are the starting materials for lecithin production. However, in such water degumming process wherever lecithin recovery is intended, a filtration step must be included prior to the addition of water. This would keep the hexane insolubles at a consistently low level. Fig. 10.27 gives a flow diagram for the production of lecithin from hydrated gums.

A very important part of the lecithin recovery process is the additive step, which maintains consistent quality and uniformity. Fluidizing additives such as oil, fatty acids, or calcium chloride can be added to adjust the viscosity and acetone insolubles. If this process is not carried out, the end product on cooling will be highly plastic solid. Plastic lecithin is usable and functional, but it is not suitable for most end users because it requires heating and mixing to fluidize the crude lecithin.

Colour of the crude lecithin is also to be adjusted before the hydrated lecithins are dried. Hydrogen peroxide (30%) is used up to 1.5% to obtain a single bleached Gardner colour of 14 minus. Benzoyl peroxide up to 1.5% can be used to obtain a double bleached colour body systems. Care must be observed in dose levels as hydrogen peroxide reduces the brown pigments and benzoyl peroxide reduces the red pigments. In bleached products, elevated peroxide value indicates free peroxide. It should be noted that complete elimination of all peroxides at elevated temperature

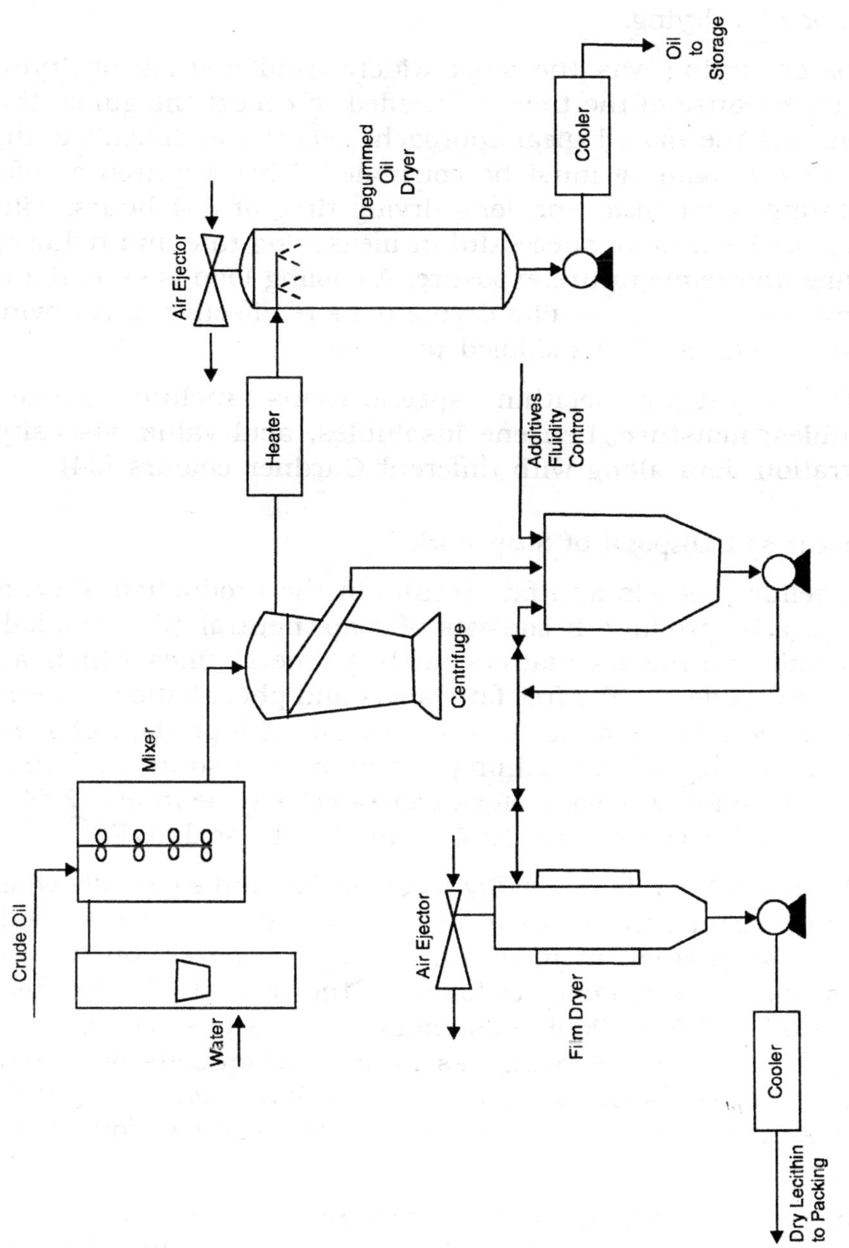

Fig. 10.27: Flow Diagram for Lecithin Production

(70°C) will cause the lecithin to darken rapidly. After the adjustment of fluidity and colour, the wet gums are dried either through batch or film drying.

Batch drying was the most widely used method of drying lecithin. Because of the time lag needed to collect the gums, this system was the most logical approach. Lecithin is difficult to dry because the foaming must be controlled. This required careful monitoring of vacuum and long drying time of 3–4 hours. Film drying has been more successful in measuring rate and reducing the time and temperature exposure. A cooling loop is essential to prevent darkening. Also, film drying is more effective in removing residual acetic acid in modified products.

The soyabean lecithin specifications include acetone insolubles, moisture, benzene insolubles, acid value, viscosity, penetration data along with different Gardner colours (34).

Treatment and Disposal of Soap-stock

Alkali refining of oils and fats results in the production of soap-stock as a by-product. It consists of soap, neutral oil, free alkali, gums and extraneous matters such as metal fines which are present in crude oil. The free fatty acids and phosphatide contents of the crude oil determine the quantity and composition of soap-stock. Efficiency of the refining operation is also an important factor. The amount of soap-stock generated may be from 3-20% of the crude oil and contains 10-40% total fatty acids (TFA).

Soap-stock tends to solidify when cooled and especially when the TFA content exceeds about 30%. It is, therefore, necessary to heat the soap-stock to maintain fluidity. However, on heating soap-stock has a tendency to foam. As the soap-stock contains a good amount of free alkali, on standing it tends to separate in two phases, bottom phase being essentially an aqueous one. Also, because of the presence of water, soap-stock may ferment quickly. As a result the initial processing of the soap-stock is done at the refinery.

In miscella refining, the soap-stock may be added back to the desolventizer/toaster, if the refinery is a part of an extraction plant. In other cases, if the raw soap-stock can be handled promptly, it may be processed directly in a soap kettle. Additional alkali is added to completely saponify it, and then purified to make industrial soaps. In the event of shipment to a distant location, the soap-stock is acidified, made free of mineral acid and the resulting product shipped as acidulated soap-stock (acid oil).

Acidulation of soap-stock is usually carried out in batch process. Soap-stock is charged to a corrosion resistant vessel. In older plants, wooden vats with bronze coils are still used. But in modern installations, monel metal or fiberglass reinforced plastics are preferably used. Sulphuric acid as a 10% solution is added in excess and the mass is boiled with open steam for 2–3 hours. The tank is then settled and the acid layer is drawn off. The acid oil is repeatedly boiled with fresh water and settled thoroughly till free from mineral acid. The acid oil may be stored in tanks or shipped in steel drums.

The waste water containing acid and any unrecovered material is high in biochemical oxygen demand (BOD) and low in pH. Adjustment of the pH and removal of the immiscible materials adds to the cost of waste water treatment. This must be considered while evaluating any soap-stock treatment system.

Several processes have also been developed for continuous acidulation of soap-stock. One of them is a patented process by Blooming and Hutchins (35) and is given in Fig. 10.28. It is especially suitable for those oils which have been thoroughly degummed as even small amounts of gums create emulsions which will not separate.

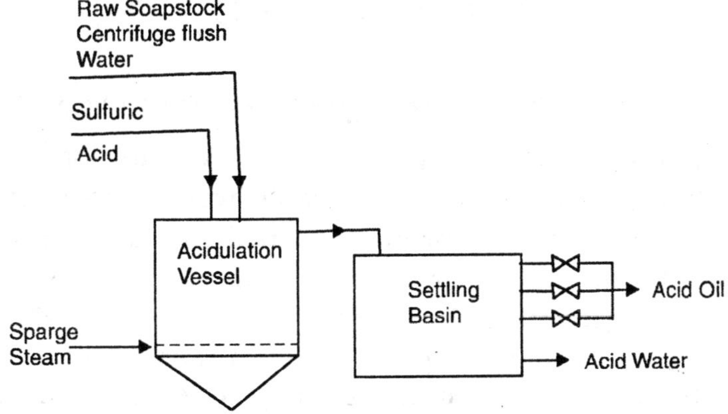

Fig. 10.28: Continuous Acidulation of Degummed Soya Oil Soapstock

A mixture of soap-stock and centrifuge wash water is mixed into an acidulation vessel together with a metered stream of sulphuric acid to bring the pH to 1.5–2.0. The vessel is provided with open steam heating to bring the contents to 90°C. The mixture overflows continuously to a settling basin. In the settling basin the acid oil flows to the top and the acid water settles to the bottom. A series of valves at different levels are provided for draining off the acid oil. The acid water overflows through a stand pipe which maintains the level in the settling basin.

Another process which is offered by De Smet (Fig. 10.29) provides three reactor tanks in parallel. Each tank is charged successively. The sequence of acidulation, water washing and decanting is performed in each vessel. The acid layer from one vessel may be mixed with the soap-stock charged to the next vessel to economise on acid usage.

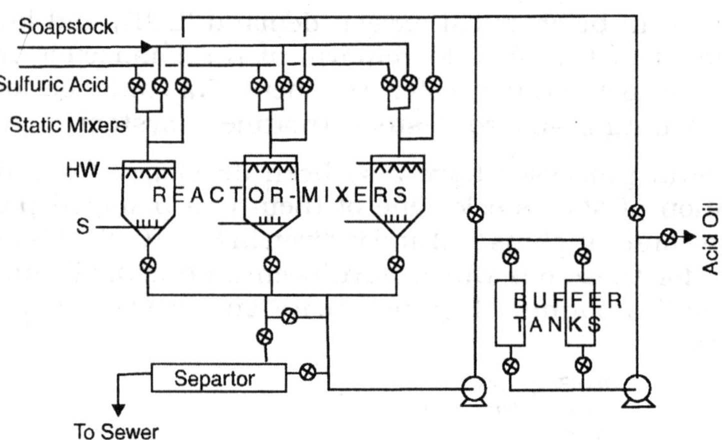

Fig. 10.29: Continuous Soapstock Acidulation (De Smet)

As mentioned earlier, considerable cost in soap-stock processing goes toward an efficient waste water treatment system. Therefore, methods have been developed to eliminate the acidulation step by drying the soap-stock. One such method has been developed using a Votator Turba film processor (3). As shown in Fig 10.30, it consists of acidulation to control pH at 7.0 and evaporation of moisture at 190 mm Hg in a swept surface evaporator. It is stated that foaming can be controlled by close pH monitoring, proper operating pressure and adjustment of heat exchanger conditions. The product is wax like solid at room temperature and is termed as neutralized dried soap-stock (NDSS).

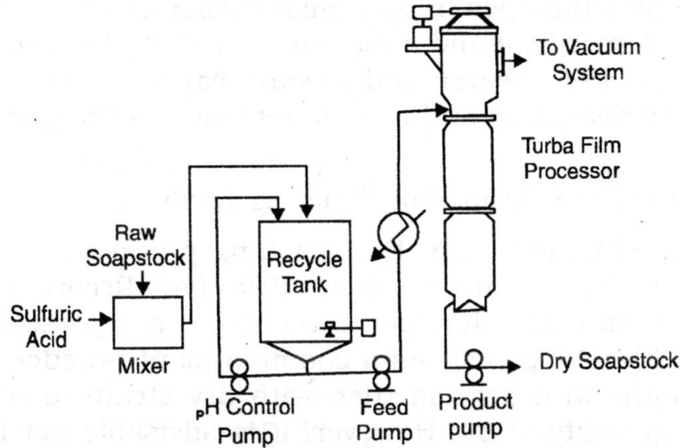

Fig. 10.30: Soapstock Drying Votator

Fig. 10.31 shows a process by Sullivan for desolventizing miscella soap-stock. In this process the soap-stock from soyabean

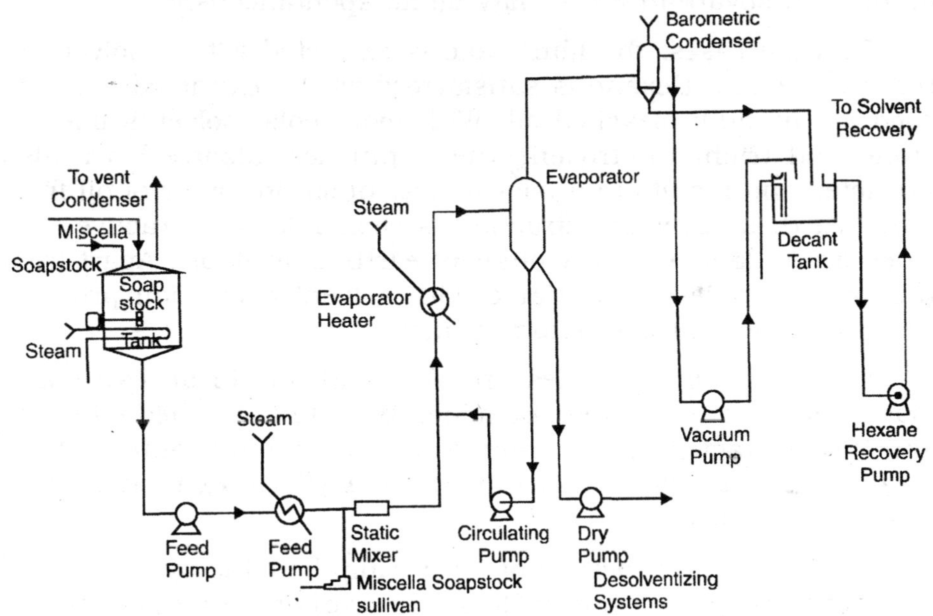

Fig. 10.31: Continuous Acidulation of Degummed Soya Oil Soapstock

oil is heated neutralized with sulphuric acid. It is then fed to a multi-pass forced circulation evaporator. Water and hexane are removed as vapour while the soapmass is recirculated through the evaporator at a high flow rate. Finished soap-stock is reported to contain 2.5% moisture and a maximum of 0.5% hexane. It is also pumpable at 55°C. Water and hexane vapours are condensed, separated by decanting and hexane returned to the plant system.

Recovery of Oil and Disposal of Bleaching Earth

Press cakes obtained from the bleaching process contain, apart from the adhering impurities, about 40% of oil. Recovery of this oil is of great economic value to the refiner as it represents a large part of the bleaching cost. As a common plant practice, the filter press is blown with air and then with dry steam to recover the maximum quantity of oil. However, it is advisable not to recover the oil to a very great extent as the wetting power of steam may desorb some of the impurities and may lower the quality of the recovered oil. Also with reduced oil content in the cake, the oil may oxidize rapidly and in the case of highly unstaurated oils like fish oil or even soyabean oil, it may ignite spontaneously.

In some cases, the filter cake is extracted with a solvent to recover the oils. Hexane is satisfactory as the extracted oil is as good as the bulk bleached oil. With more polar solvents like acetone and trichloro-ethylene the impurities adsorbed are also extracted and results in a poor quality of oil. Recovery of oil from spent earth by solvent extraction is economically attractive only when a very large quantity of spent earth is available. Apart from the safety problems associated with solvent extraction process, the cost of solvent extraction is high.

Where refining facilities are also available in an extraction plant, the spent earth may be added back to the milled oil seeds and then extracted. But this increases the mineral content of the meal and the handling of spent earth in a solvent extraction plant poses a fire hazard.

A rather simple method of recovering the oil is by boiling the spent earth with a solution of 3% sodium carbonate and 3% sodium chloride. The recovered oil is poor quality and may be used only for technical purposes. Because of the alkaline nature of boil, the clay is converted to bentonite, swells and forms a suspension. Dewatering of the suspension by centrifugation and other means is difficult. Lowering the pH to 4.0 or addition of calcium chloride helps in de-aeration.

Disposal of spent earth is yet another problem. Land filling is the most obvious choice. Care must be taken to cover the spent earth as quickly as possible with soil to avoid spontaneous ignition. Other suggested applications are in the manufacture of cement as a source of selenious oxide or mixing with sandy soils in sweet maize corn cultivation.

Recovery of Deodorizer Distillates

The volatile organic materials which are steam distilled during deodorization are termed as deodoriser distillates. Apart from containing fatty acids, neutral oil and volatile odorous compounds, a distillate also contains sterols and tocopherols. Tocopherols are important raw materials for Vitamin E production and sterols are valuable starting compounds for the manufacture of steroid drugs. The amount of sterols and tocopherols recovered are governed by deodorization temperature and sparge steam rate. High sparge steam rate gives a high yield of total deodorizer distillate due to increased entrainment but the per cent content of tocopherols and sterols are correspondingly reduced. High deodorization temperature recovers more tocopherols at the expense of flavour quality of oil.

In earlier practices, the deodorizer distillates were not recovered and were usually discharged into the water bodies like rivers and lakes. With the introduction of strict environmental legislations, closed-circuit water systems were introduced. However, it was soon recognized that water alone was unsatisfactory for this purpose.

A direct contact cooling system for recovery of deodorizer distillate is now-a-days most commonly used. Fig. 10.32 gives a typical flowsheet of such system from EMI Corporation. Steam and distillate vapours coming from the deodorizer at a pressure of 2–5 mm Hg are compressed to 50–60 mm Hg through one or two booster stages. The vapours pass upwards through a packed tower, where they are contacted with liquid distillate cooled to about 60°C by circulating through a heat exchanger. A level controller at the bottom of the tower maintains a constant volume of condensed distillate in the system and the excess is continuously drawn off.

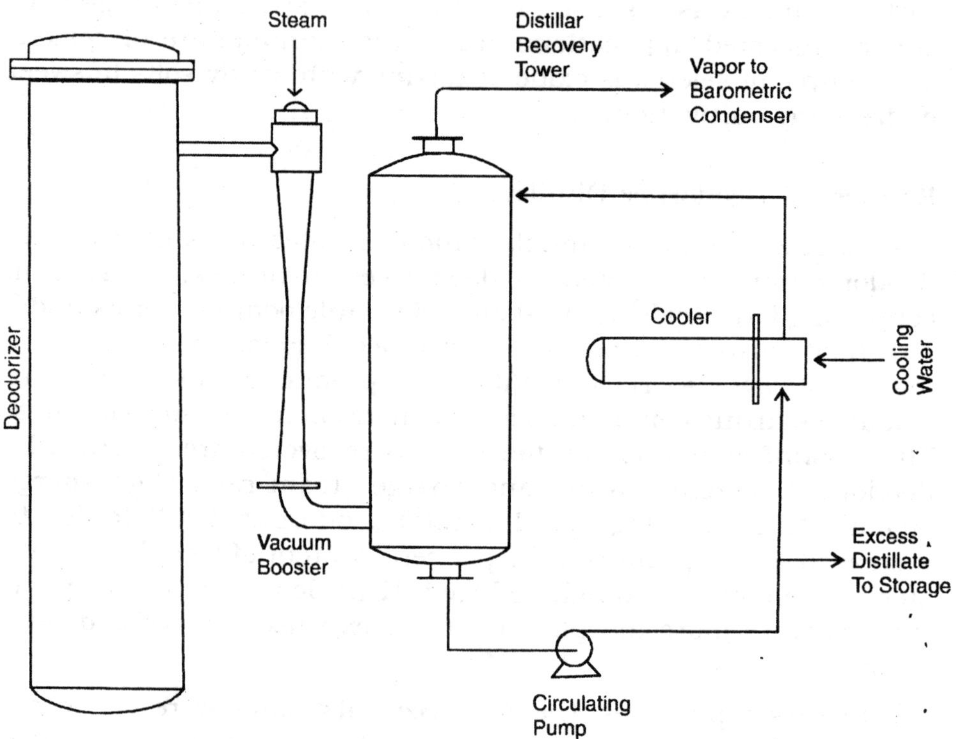

Fig. 10.32: Deodorizer Distillate Revoery System

Packing design in distillate recovery tower is of utmost importance as it must provide minimum pressure drop with good liquid-vapour contact. A tower is either partially packed or the condensate is sprayed as very fine droplets.

Although deodorizer distillate recovery system is primarily a pollution control device, market demand for tocopherols and sterols may sometimes convert it into a profit-making device.

REFERENCES

1. Blak Hendrix, W. *in* Proceedings of the World Conference on Edible Oils & Fats Processing, edited by D.R. Erickson, American Oil Chemists' Society, Champaign, 11 (1990) pp. 94-100.
2. Hvolby, A., *J. Am. Oil Chem. Soc.* **48**: 503 (1971).
3. Hayes, L.P., and H. Wolf, *ibid.* **33**: 440 (1956).
4. Dijkstra, A.J., *in* Proceedings of the World Conference on Oilseed Technology and Utilisation, edited by Thomas H. Applewhite, American Oil Chemists' Society, Champaign, 11 (1993) pp. 138-151.
5. Penk, G., Paper presented at the ISF-AOCS Congress, Tokyo (1988).
6. Farr., W.E., *J. Am. Oil Chem. Soc.* **60**: 571 (1983).
7. Bergmann., L.O. and Johnsson, A., Fett. Seif. Ansch. **66:** 203 (1964).
8. Hoffman, Y., *J. Am. Oil Chem. Soc.* **50**: 260A (1971).
9. Cavanagh, G.C., *ibid.* **53**: 361 (1976).
10. Freundllich, H., Colloid and Capillary Chemistry, Translated from the 3rd German, edited by H.S. Hatfeld, Metheun, London (1926).
11. Brimberg, U.I., *J.Am. Oil Chem. Soc.* **59**: 74 (1992)
12. Khoo, L.E., Morsingh, F. and Liew, K.Y., *ibid.* **56**: 672 (1979)
13. Ney, K.H., *Fett. Seif. Ansch.* **66**: 512 (1964).
14. Pradun., H., Kroll, E. and Werber, O., *ibid.* **70**: 531 (1968).
15. Report of the Oil Colour Committee, April 1950, *J. Am. Oil Chem. Soc.* **27**: 233 (1950).
16. Kogler, H.P., *Fett. Seif. Ansch.* **65**: 834 (1963).
17. Richardson, L.L., *J. Am. Oil Chem. Soc.* **55**: 777 (1978).
18. Rich, A.D., *ibid.* **44**: 298A (1967).
19. idem. *ibid.* **37**: 305 (1960).
20. Hansuddin Siddiqui, M.K., *ibid.* **45**: 885 (1968).
21. Hastert, R.C., *ibid.* **58**: 174 (1981).
22. Neumunz, G.M. *ibid.* **55**: 398A (1978).
23. Braae, B., in International Developments in Palm Oil, edited by Earp D.A. and Newall W. Malaysian Incorporated Society of Planters, Kualalampur, Malaysia (1977) pp 377.
24. Baily, A.E., "Industrial Oil and Fat Products" Second edition, Interscience publishers, New York (1951).
25. Sazabo Sarkadi, D.S., J. Am. Oil Chem. Soc. **35:** 472 (1958).
26. Stage, H., Fett. Seif. Ansch. 104: 445 (1978).

27. Gavin, A.M., *J. Am. Oil Chem. Soc.* **55**: 783 (1978).
28. Moser, H.A., Evan, C.D., Cowan, J.C. and Kowlek, W.F., *ibid.* **42**: 30 (1965).
29. Lineberry, D.D and Dudrow, F.A., US Patent 3,693,322 (1972).
30. Gavin, A.M., *J. Am. Oil Chem. Soc.* **60**: 372A (1983).
31. Frank A. Noris in Bailey's Industrial Oil and Fat Products Vol. 3, edited by Thomas H. Applewhite, John Wiley & Sons (1985) pp. 156-158.
32. Athanassiadis, A., *Fat Sci. Technol.* **90**: 522 (1988).
33. Stage, H., *J. Am. Oil Chem. Soc.* **62**: 306 (1985).
34. National Soyabean Processors' Association's Year Book and Trading Rules, 1981-82.
35. Bloomberg, F.M. and T.W. Hutchins, US Patent 3,425,938 (1969).

CHAPTER 11

Hydrogenation

Hydrogenation is even now regarded from commercial and technical considerations as the most useful process in modifying the chemical (Fatty Acid) composition and properties of fats and oils to cater to edible and industrial requirements.

In the present chapter, some important basic and technological aspects of hydrogenation process and hydrogenated products will be briefly discussed as much extensive information has become available through the pioneering work of distinguished scientists and technologists. Excellent articles, chapters and text books on hydrogenation reaction and process are now available to enrich our knowledge.

BASIC ASPECTS OF HYDROGENATION REACTION

A. Mechanism

Hydrogenation reaction principally consists of addition of hydrogen atoms to the double bonds of various unsaturated fatty acids in fats and oils either wholly or partially. To achieve the addition of hydrogen atoms to the double bonds a heterogeneous or homogeneous catalyst system is needed.

The heterogeneous catalyst system is almost exclusively used in the hydrogenation process. The heterogeneous process consisting of a three-phase system, namely, solid catalyst, liquid oil and gaseous hydrogen involves a complex mechanism of contact. Bailey (1) put forward several mass transfer steps involved in hydrogenation which were later modified by Albright (2) for explaining the simultaneous and also concurrent hydrogenation and isomerisation as:

(1) Transfer of hydrogen from gas to liquid phase;

(2) transfer of hydrogen from liquid phase to catalyst surface;

(3) transfer of unsaturated group (carbon-carbon double bond) from liquid phase to the catalyst surface;

(4) chemical reactions on or near the catalyst surface which include adsorption of reactants, surface reactions, and desorption of product for both hydrogenation and isomerisation; and

(5) transfer of product from catalyst surface to the main body of the liquid.

B. Selectivity

In the hydrogenation reaction two important basic factors, namely, the selective reduction and isomerisation of the unsaturated fatty acid chains accompany the reaction. These two factors largely control the physical and chemical properties of the end products.

Selectivity refers to the preferential hydrogenation of the polyunsaturated acid groups such as linolenic or linoleic acid groups relative to monounsaturated acid groups. The selectivity is largely relative and never absolute which means that in a mixture of linolenic and linoleic acids the hydrogenation of linoleic acid occurs along with linolenic acid at a rate that may be comparatively slow. Similarly, in a mixture of oleic and linoleic acids the hydrogenation of oleic acid always occurs at a comparatively very slow rate. In actual practice, it is possible to vary the ratio of reaction rates, linoleic acid: oleic acid from about 4 :1 in very non-selective hydrogenation to about 50 :1 in very selective hydrogenation (1). Selectivity also refers to the formation of geometric isomers and especially *trans*-isomers. The amount of *trans*-isomers formed is generally in direct proportion to the selectivity of the hydrogenation reaction.

There are concepts of four kinds of selectivity, namely, Selectivity I (S_I), Selectivity II (S_{II}), Isomerisation Selectivity (S_i) and

Triglyceride Selectivity (S_T). Selectivity I (S_I) refers to the preferentially exclusive hydrogenation of linoleic acid to oleic acid in a mixture of linoleic and oleic acids while Selectivity II (S_{II}) relates to the preferential exclusive hydrogenation of linolenate to linoleate in a mixture of these two acids. Isomerisation Selectivity (S_i) refers to the extent of the formation of *trans*-isomer content in the course of selective hydrogenation of the various unsaturated fatty acids in a mixture. Triglyceride Selectivity (S_T) is a relatively new concept. It is concerned with the extent of hydrogenation of the various molecules in the bulk oil in a random manner. The more the hydrogenation of the triglyceride molecules of the bulk oil occurs, the more is the triglyceride selectivity. One can expect that during the hydro-genation of an oil the stearic acids formed can, in turn, distribute randomly in the triglyceride molecules forming tristearins (S_3). On the other hand, all the stearic acid groups can remain confined to tristearin molecules only. Perfect triglyceride selectivity means $S_3 = s^3$ and lowest possible triglyceride selectivity means $S_3 = s$ and high triglyceride selectivity $S_T = (s - s^3)/(s - s^3) = 1$, where S stands for the stearic acid (2a).

The individual Selectivity S_I, S_{II} and S_T can be estimated, but more important is the term Selectivity Ratio SR. The calculation of individual selectivity is based on the first-order irreversible schematic equations for the hydrogenation of unsaturated fatty acids as follows :

$$\text{Linolenate} \longrightarrow \underset{\text{(isomers)}}{\text{Linoleate}} \longrightarrow \underset{\text{(isomers)}}{\text{Oleate}} \longrightarrow \text{Stearate}$$

The rate equation for conversion of Linolenate to Linoleate can be expressed as (2a):

$$\frac{dLn}{dt} = -K_{Ln}C_{Ln} \tag{1}$$

The rate equation for conversion of Linoleate to Oleate is as follows:

$$\frac{dL}{dt} = -K_L C_L + K_{Ln} C_{Ln} \tag{2}$$

Integrating equation (1) and (2) at time t (starting from O)

$$C_{Ln.t} = C_{LnO}\, e^{-K_{Ln.t}} \tag{3}$$

$$C_{Lt} = C_{Ln_o} \frac{K_{Ln}}{K_L - K_{Ln}} (e^{-K_{Ln.t-e}} - {}^{K_{Lt}}) + C_{Lo} e^{-K_{Lt}}$$

For the hydrogenation of Linoleate and Oleate system the reactions are represented as follows :

$$\text{Linoleate} \longrightarrow \underset{\text{(isomers)}}{\text{Oleate}} \longrightarrow \text{Stearate}$$

The rate equation can be expressed as :

$$-\frac{dL}{dt} = K_1L$$

$$\frac{dS}{dt} = K_2O$$

$$\frac{dO}{dt} = K_1L - K_2O$$

The two selectivity ratios can be defined as SR_I and S_I (Linoleic Selectivity) $= \frac{K_L}{K_O}$

SR_{II} or S_{II} (Linolenic Selectivity) $= \frac{K_{Ln}}{K_L}$

$$S_i = \frac{\text{number of } trans \text{ double bonds formed}}{\text{number of hydrogenated double bonds}}$$

A high S_{II} means that high yields of dienoic are obtained. A high S_I means that high yields of monoene are obtained. A high S_i means that sizeable *cis* to *trans* isomerisation occurs. A high S_T means sizeable number of triglyceride molecules are hydrogenated.

Isomerisation

During the hydrogenation reaction there is invariably the formation of isomers of the unsaturated fatty acids whose properties are different from the naturally occurring forms. These isomers comprise both positional and geometric types. The most widely accepted explanations are given by Horiuti-Polanyi Mechanism and the work of Coenen and coworkers (3).

Hydrogen atoms and unsaturated molecules get bonded to the nickel in an activated state. The hydrogenation goes through a half-hydrogenated step, which can follow four different pathways :

(a) The half-hydrogenated state takes a second hydrogen from the surface and the desorption results in the completely hydrogenated molecule.

(b) The hydrogen Ha returns to the nickel surface and the original molecule is desorbed.

(c) The hydrogen H_b returns to the nickel surface and *cis-trans* isomerisation is observed in the desorbed molecule.

(d) The hydrogen H_c or H_d returns to the nickel surface and a shift of the double bond is observed in the desorbed molecule.

A high concentration of dissolved hydrogen or a high coverage of the catalyst surface with hydrogen results in rapid hydrogenation of double bonds and low selectivity. With low hydrogen coverage, the isomerisation mechanisms will be favoured over the completion of the hydrogenation. Coenen (3) postulated that either monoenic or polyenic acids are initially bonded with one double bond. With low hydrogen coverage adsorbed linolenic or linoleic acid may split off a hydrogen atom from its reactive methylene group and become bonded to the catalyst surface with free carbon atoms. It has been proposed that this permits bondings of second double bond to the catalyst surface and increases the bonding strength of linolenic or linoleic acid over that of oleic acid: there is monopolization of the catalyst surface by the polyenoic acids and large reactivity differences therefore result. Relative reactivities of oleic, linoleic and linolenic acids are 1, 20 and 40 respectively.

EFFECTS OF PROCESS FACTORS

The factors that control hydrogenation reaction rate, selectivity and isomerisation are temperature, pressure, agitation and catalyst concentration. Each factor individually and in combination with others controls the hydrogenation reaction characteristics. The effects of the different process factors on hydrogenation can be understood from the table below :

Factors	*Hydrogen concentration on catalyst surface*	*Reaction rate*	*Selectivity*			
			S_I	S_{II}	S_i	S_T
Temperature increase	Decrease	Increase	+	+	+	+
Pressure increase	Increase	Increase	—	—	—	—
Agitation increase	Increase	Increase	—	—	—	—
Catalyst concentration increase	Decrease	Increase	+	+	+	+

The solubility of hydrogen in the oil and its availability on the catalyst surface and within the catalyst pore control the rate and selectivity including isomerisation of hydrogenation reaction. The rate of hydrogen uptake by the oil is given by the equation $r_H = K_L$ a $(C_{HO} - C_H)$ where K_L = mass transfer coefficient in the stagnant liquid film surrounding the bubbles (Fig. 11.1).

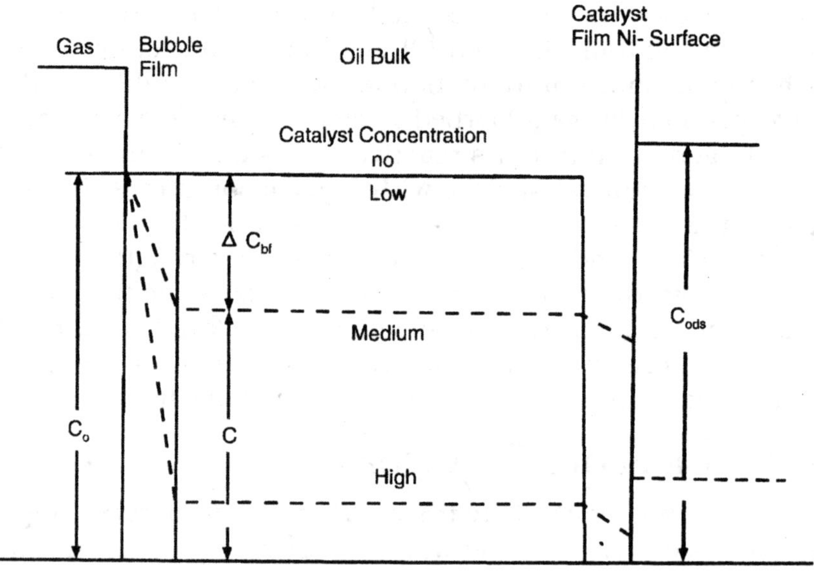

Fig. 11.1: Concentration of Hydrogen at Different Stages in the Mass Transport of Gas Bubbles to Catalyst Pores.

a = interface area separating the oil from the gas in the bubbles and the head space. This 'a' depends mainly on reactor design.

C_{HO} = hydrogen concentration in the gas-oil interface (i.e., the maximum solubility of hydrogen in the oil).

C_H = hydrogen concentration in the bulk oil.

By increasing the dose of catalyst all hydrogen that enters the oil is almost immediately consumed. The hydrogen concentration in the bulk oil (CH) would then drop to zero. The equation becomes:

$$r_{H_{max}} = K_L . a \, C_{HO}$$

The rate of hydrogen uptake reaches a maximum beyond which it is not possible to increase the observed rate of the reaction by adding more (or a more active) catalyst.

The solubility of hydrogen in the oil significantly increases with temperature up to a certain range beyond which the effect is very limited. As a result of this both the selectivity and the isomer formation (*trans*-) increase and the rate of reaction increases but the effect is less than in many reactions.

The solubility of hydrogen in vegetable oils increases remarkably with increase in pressure. In fact, at the hydrogenation temperatures, if the pressure is doubled the amount of hydrogen in the oil doubles. Consequently, the rate of hydrogenation also doubles. Because of the increased concentration of hydrogen on the catalyst surface, the selectivity sharply decreases and the rate of isomer formation also becomes somewhat less.

Agitation influences the supply of hydrogen to the catalyst surface. The greater is the agitation the more is the supply of hydrogen to the catalyst surface resulting in the increased rate of hydrogenation and in the decreased selectivity. The isomer formation is also decreased with increased agitation. An increase in catalyst concentration within a certain level increases the rate, selectivity and isomerisation reaction. When the catalyst concentration is used at a very high level the rate is not increased proportionately and so also the selectivity and isomer formation. In actual hydrogenation practice the agitation is kept sufficiently high so as to eliminate the resistance to mass transfer and then only the rate of hydrogenation varies directly with the temperature, the absolute pressure and the catalyst concentration.

Figures 11.2-11.4 show the rate of hydrogenation under different process parameters.

INFLUENCE OF CATALYST NATURE, CATALYST STRUCTURE AND OTHER ADDITIVES AND POISONS ON RATE, SELECTIVITY AND ISOMERISATION

Transitional metals having electron deficiency in their d-orbitals make their surface susceptible to nucleophilic attacks from the hydrogen molecules. They actually act as hydrogenation catalysts but considerable differences are exhibited among the various metal catalysts in respect of their behaviour towards rate of hydrogenation, selectivity and isomerisation characteristics.

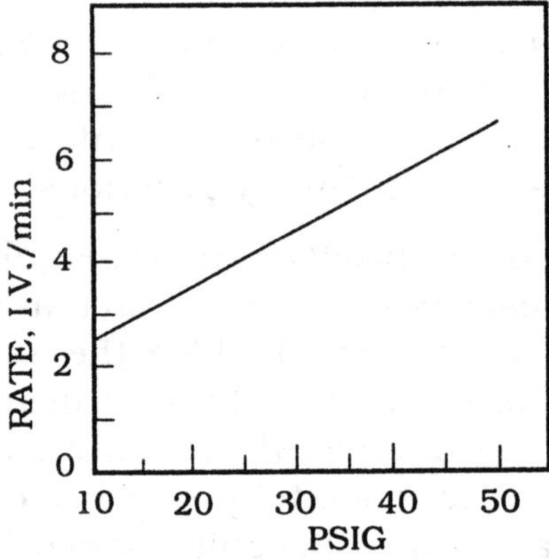

Fig. 11.2: Effect of Reaction Pressure on Hydrogenation at Constant Temperature, Catalyst Concentration and Agitation.

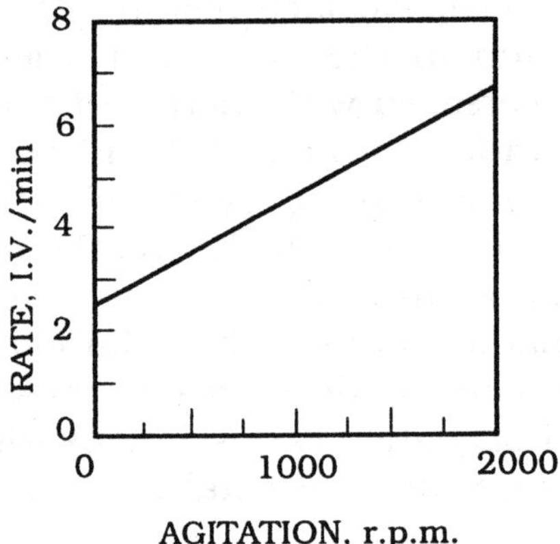

Fig. 11.3: Effect of Agitation on Rate of Hydrogenation

The selectivity varies with the type of metal catalysts used. Thus, for example, the linolenic selectivity varies from 2-3 for nickel catalysts and copper based catalysts have a considerably higher selectivity ratio of around 6. Palladium catalysts hydrogenate oils much more selectively than the nickel catalyst under comparable conditions of hydrogenation. *Trans* isomers formation is, however, less with Pt, Pd and copper catalysts and

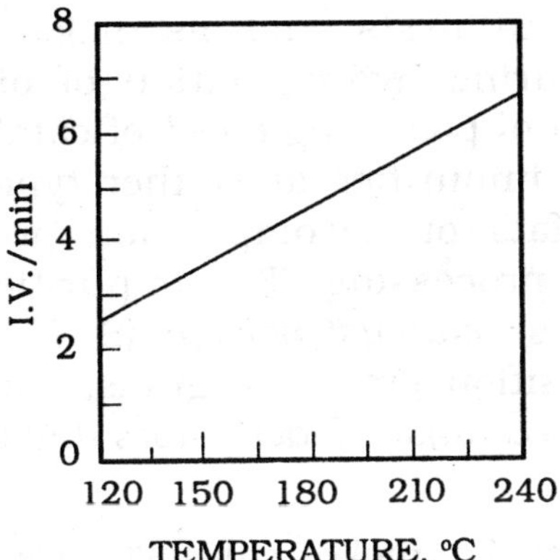

Fig. 11.4: Effect of Temperature on Rate of Hydrogenation

more with nickel catalyst. Homogeneous catalysts are more selective than heterogeneous catalysts. On the other hand, some of the homogeneous catalysts reduce, *trans*-isomer formation. Till now hydrogenation with nickel catalysts has been the choice of industry, largely due to the availability, low cost and inert nature of the metal to the oil.

Coenen (3) explains selectivity by using the pore structure/ size of the catalyst as key parameter. Dimensions of a triglycerides molecule are in the range of 15-20 Å; hydrogen molecules only about 2–4 Å. The conventional nickel catalyst spreads out on a very porous support (usually of kieselguhr-silica-alumina works) in which pore diameters may range from below 20 to several hundreds of angstrom units. Coenen distinguished three pore sizes—narrow with a diameter less than 20 Å, medium with a diameter between 20 Å and 100 Å, and wide with a diameter of 100 Å or more. Pores considerably slow down the transport of molecules causing long residence times inside the pores. The selectivity largely depends on the residence time in the catalyst pores. A long residence time means more possibilities of conjugation, isomerisation and hydrogenation.

The acid character of the support material is an important factor. The acid sites, especially the Bronsted acid sites, are known to favour the isomerisation reactions. Achaya et al (4) reported that the presence of potassium soaps appear to poison acidic sites leading to less hydrogen dissociation and to suppression of the *cis-trans* monoene isomerisation.

Hydrogenation catalysts such as nickel catalysts get their activity reduced during hydrogenation of oils. This has been attributed to a kind of poisoning effect of catalysts due to certain impurities (5). The impurities are either typical constituents of natural oils and fats or decomposition products or may be introduced during processing. The impurities can be grouped into three categories: *catalyst poisons* (S, N, P, Cl, As); *catalyst inhibitors* (decomposition products, gums, alkali soaps, oxidised fatty acids); and *catalyst deactivators* (acids, water, carbon monoxide).

Effective poisons for nickel catalysts are generally found in compounds containing elements of the groups V, VI and VII of the Periodic Table of Elements which possess lone electron pairs e.g. sulphur or phosphorous. The interaction is being accomplished by electron donation to the unoccupied d-orbitals of the catalyst metal.

Nickel Catalysts on Various Carriers (6)

In a study (6) metal oxides (TiO_2, MoO_3, WO_3, ZnO) and alkaline earth oxides (MgO, CaO) were used as carriers instead of kieselguhr when preparing Ni catalysts. Only when TiO_2 was used as the carrier, the soyabean oil hydrogenation tests showed outstanding results : the most selective one among the prepared TiO_2 supported catalysts gave hardly 4.0% solid material at 10 C in the hydrogenated triglyceride, i.e., the lowest solid content reached with the catalysts tested in this study (Table 11.1). Furthermore, only 0.4% conjugated diene was found in the final product.

Table 11.1: Results Obtained with the Most Selective Ni/TiO_2 Catalyst

	(Ni/kieselguhr) (23.5% Nickel)	*Ni/TiO_2*
Activity (min)	64	53.5
C18:0	9	6
C18:1	44.6	48.3
C18:2	32	30.8
C18:3	1.8	1
% trans	15.3	18.5
% Solid at 10 C	13.3	4

Nickel Catalyst Treated with Ammonia or Amines (6)

The two Ni on kieselguhr reference catalysts without and with ZrO_2 were treated with various nitrogen compounds including ammonia, urea and aliphatic amines containing one or four nitrogen atoms (diethylamine, triethylamine, tributylamine, n-butylamine, fatty amines, hexamethylene tetramine).

When the starting Ni catalyst was treated with an aliphatic amine, a low melting point amine was preferred to avoid the presence of heavy components in the reaction mixture. As a total result, a significant decrease of the content in solid matter at 10C was observed with all these basic nitrogen compounds treated catalysts. This phenomenon is mainly due to restricted formation of stearic acid from oleic acid as no real improvement of the retention of the *cis* configuration was observed, in addition, no large drop in catalyst activity occurred. The best result was obtained with the Ni-ZrO_2/kieselguhr catalyst (ref 24/9) treated with a coco primary amine; only 4.8% solid matter at 10C and 0.2% conjugated diene were found in the hydrogenated soyabean oil (Table 11.2).

Table 11.2: Results Obtained with a Ni-ZrO_2/Kieselguhr Catalyst Treated with a Coco Primary Amine

	Ni with ZrO_2 (22.7% Nickel)	*Ni with ZrO_2 treated with coco primary amine*
Activity (min)	67.5	75
C18:0	8.1	5.7
C18:1	45.3	50
C18:2	31.5	30.5
C18:3	1.7	1
% trans	18.1	18
% Solid at 10 C	12.4	4.8

[a]N/100 Ni(at/at):9.8

The amount of nitrogen fixed on the catalyst surface influences the selectivity but has no marked effect on the activity. A definite correlation between the ratio N/100 Ni(at/at) and the amount of solid material at 10C or the stearic formation is observed with the

Ni/kieselguhr catalyst : the higher the N/100 Ni ratio the lower the percentage solid at 10 C and the stearic acid formation (Table 11.3).

Table 11.3: Influence of the N/100 Ni Ratio on Activity and Selectivity of an Amine-Treated Ni/Kieselguhr Catalyst

N/100 Ni (at/at)[a]	*Activity (min)*	*C18:0*	*% trans*	*% Solid at 10C*
0	64	9	15.3	13.3
10	62	7.8	14.5	9.1
20	64	6.8	14.5	6.3
30	63	6.7	16	4.9

[a]Catalyst Ni/Kieselguhr treated with coco primary amine.

The chemical nature of the nitrogen compound used is important. By treating the same Ni/Kieselguhr catalyst with various nitrogen compounds to obtain an identical N/100 Ni ratio of 20, significant differences of the solid content in the hydrogenated soyabean oil were observed. As shown in Table 11.4 ammonia and primary amines lead to lower solid contents than tertiary amines.

Table 11.4: Results Obtained with a Ni/Kieselguhr Catalyst Treated with Various Basic Nitrogen Compounds

Basic nitrogen compound	*N/100Ni*	*Activity (min)*	*C18:0*	*C18:1*	*C18:2*	*C18:3*	*% trans*	*% Solid at 10C*
None	0	67.5	8.1	45.3	31.5	1.7	18.1	12.4
$(C_2H_5)^3N$	20	62.0	8.1	46.1	31.5	1.3	16.0	10.3
$(C_4H_9)^3N$	20	64.0	8.0	46.3	31.3	1.7	16.2	8.9
$(C_2H_5)^2NH$	20	60.0	7.7	46.5	31.7	1.7	15.4	7.5
$C_4H_9NH_2$	20	61.0	7.0	48.0	31.0	1.3	17.2	7.3
NH_3	20	64.0	7.0	48.0	31.5	1.1	15.8	5.1

Nickel Catalysts in Presence of a Feed Doped with Basic Nitrogen Compounds

Hydrogenation tests on soyabean oil doped with various fatty amines have been run in the presence of Ni/kieselguhr catalysts. The catalyst (22.7% Ni and 11.8% kieselguhr promoted with 1.4%

ZrO_2) gave outstanding results with a soyabean oil feed doped with lauric primary amine (Table 11.5). Its activity is slightly decreased but its selectivity greatly improved as the hydrogenated soyabean oil only contains 4.2% solid matter at 10C and 0.1% dienes.

***Table 11.5*: Results Cbtained with the Ni-ZrO_2/Kieselguhr Catalyst by Adding Lauric Primary Amine to the Feed (N/100 Ni = 10)**

	Ref 24/9	*Ref 24/9 + lauryl amine*
Activity (min)	67.5	77.0
C18:0	8.1	5.7
C18:1	45.3	49.8
C18:2	31.5	30.0
C18:3	1.7	1.0
% trans	18.1	17.0
% Solid at 10C	12.4	4.2

Table 11.6 compares the results obtained by both treatments with basic nitrogen chemicals, i.e., the introduction of nitrogen derivatives into the feed and treatment of the catalyst itself before the hydrogenation test. A slight activity decrease but a significant selectivity improvement were shown with both methods. However, the treatment of the catalyst itself led to a slightly higher selectivity.

Characteristics and Control of Hydrogenated Fats and Hydrogenation Process

Hydrogenation reaction leads to both chemical and physical changes of fats and oils. The major chemical change lies in the fatty acid composition of fats/oils. As a consequence of the altered composition, the properties of the hydrogenated products become different. Thus depending on the mode of change in the fatty acid composition the properties of a hydrogenated fat product may be different. It is clearly understood that the selective and non-selective hydrogenations will result in a different kind of fatty acid composition including the *trans*-isomer formation and therefore will result in different physical properties (Fig. 11.5).

Table 11.6: Influence of the Method used to Treat the Catalyst with Basic Nitrogen-containing Compounds

Catalyst	*Basic nitrogen treatment*	*Activity (min)*	*C18:0*	*C18:1*	*C18:2*	*C18:3*	*% trans*	*% Solid at 10C*
Ni-kieselguhr (23.5% Ni + 12% kieselguhr + 64.5% fat)	No	64.0	9.0	44.6	32.0	1.8	15.3	13.3
	Amines added to the feed[a]	83.5	6.0	49.7	30.0	1.4	16.0	5.7
	Catalyst treatment with amines[a]	76.5	6.0	49.1	31.0	1.0	16.5	5.2
Ni-kieselguhr (22.7% Ni + 11.8% kieselguhr + 64.1% fat) promoted with 1.4% ZrO_2	No	67.5	8.1	45.3	31.5	1.7	18.1	12.4
	Amines added to the feed[a]	83.0	6.7	50.7	29.8	1.4	16.1	7.1
	Catalyst treatment with amines[a]	75.0	5.7	50.0	30.5	1.0	18.0	4.8

[a]Coco primary amines have been used to obtain a N/100 Ni ratio of 10.

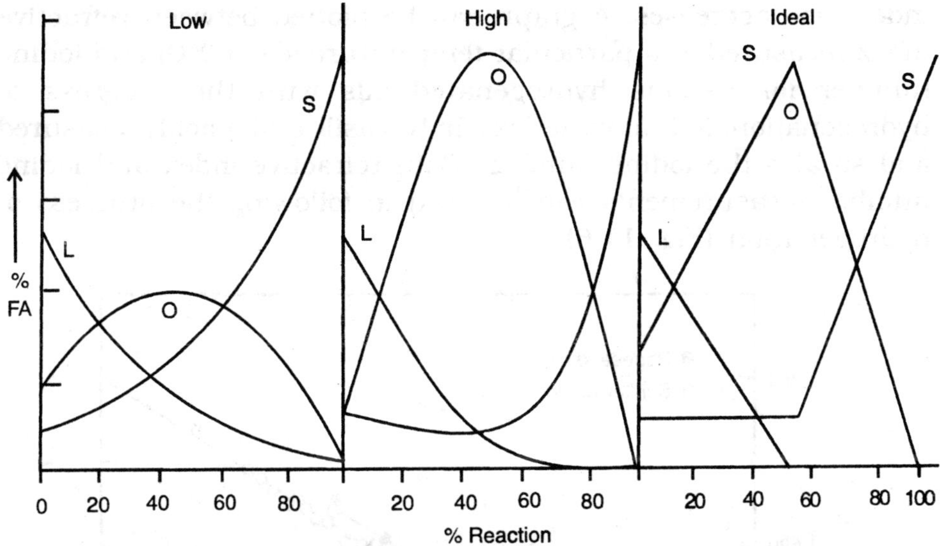

Fig. 11.5: Effect of Selectivity on Fatty Acid Composition

Selectivity or non-selectivity kind of hydrogenation can be ascertained by following the fatty acid composition of the oil or fat by gas-liquid chromatographic separation of the derived methyl esters of the periodically analysed hydrogenated products. *trans*-isomer content also can be determined by infra-red spectroscopic technique. It is known that *trans*-isomers have distinct absorbance at 10.30–10.36 micron and Allen's method is unique in very genuinely and reliably estimating the *trans*-isomer percentage in the hydrogenated fats (7).

The content of *trans*-isomers has an effect on the physical properties of hydrogenated oils (8). The relationship between *trans* content and melting point and SFI has been reported (8).

Hydrogenation vastly improves the colour and odour characteristics of oils and fats and also improves the shelf-life or stability characteristics of fats/oils.

Changes in the Properties

The refractive index of oils undergoes a characteristic change, and a correlation between iodine number and refractive index is often made. As the iodine number of an oil decreases, its refractive index also decreases. A graph can be plotted between refractive index measured at a particular temperature (e.g. 60°C) and iodine number for various hydrogenated oils with the progress of hydrogenation. Refractive index can be easily and quickly measured and so also the iodine number. Both refractive index and iodine number measurements can be used in following the progress of hydrogenation (Fig. 11.6).

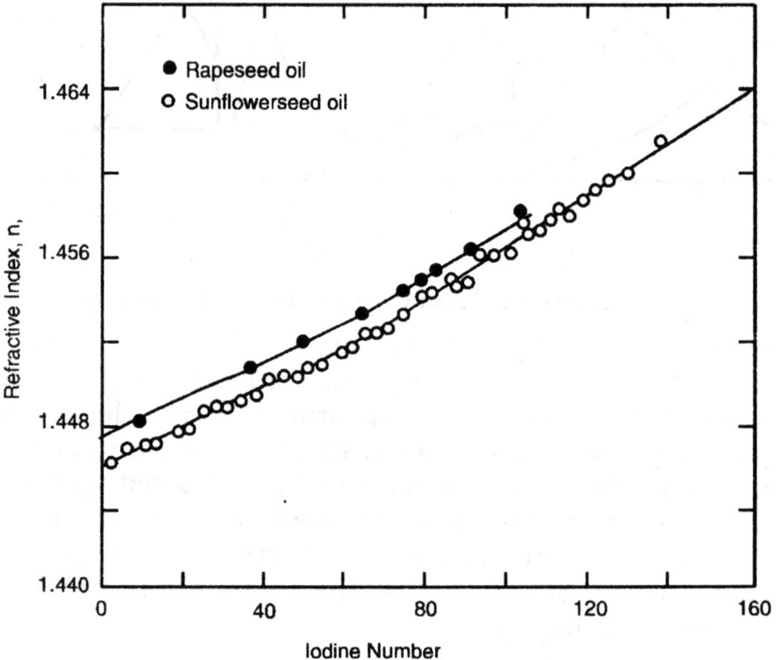

Fig. 11.6: Iodine Value Versus Refractive Index during Hydrogenation of Different Oils

Slip melting point of oils/fats changes after hydrogenation and it varies according to the degree of hydrogenation and to the primary nature of selective and non-selective hydrogenation. Generally, the greater is the selectivity the more is the slip melting point of the hydrogenated fat product. The slip melting point increases when the iodine number decreases; and depending on the content of saturated fatty acids and their nature the slip point of oils/fats also shows variation. It may be noted that vegetable oils having predominantly C_{18} unsaturated fatty acids show almost the same range of slip melting point. On the other hand, oils containing significant amount of $C_{16:0}$ acid and C_{18} unsaturated acids have relatively lower slip points (e.g. cottonseed, palm, tallow) after hydrogenation to nearly complete saturation.

Hydrogenated fats vary in their solid fat content depending on whether selectively or non-selectively hydrogenated. Generally, the selectively hydrogenated fats show steeper SFC or SFI curve than the non-selective ones. The SFC or SFI data at different temperatures are obtained by NMR or dilatometry. In fact, these two techniques can be adopted to evaluate the consistency of a hydrogenated fat over a wide temperature range (Fig. 11.7).

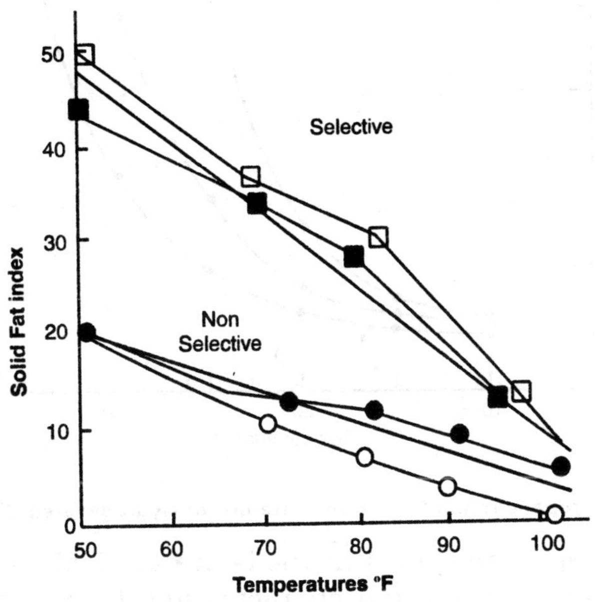

Fig. 11.7: Physical Selectivity

SFC or SFI plot can be made against iodine number also to indicate a variation in the SFI of an oil hydrogenated to different iodine numbers.

Micropenetration is an important test method for measuring consistency of hydrogenated fat products. The consistency of the hydrogenated fat products depends on both selectivity and degree of hydrogenation or iodine number. Micropenetration study can be an elegant method of following the course of hydrogenation and the consistency of hydrogenated fat products at different temperatures. Figure can be plotted with micropenetration, mm/10 at varying temperatures (Fig. 11.8).

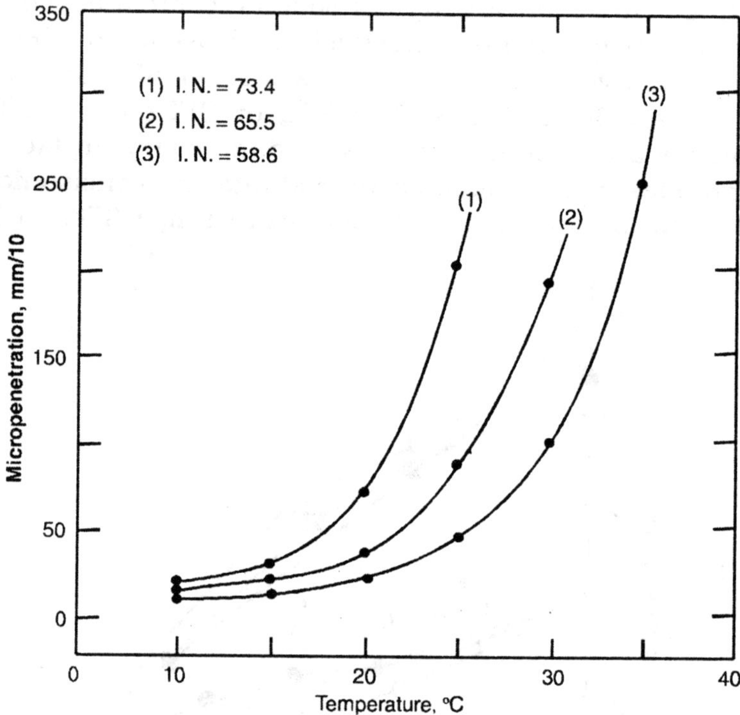

Fig. 11.8: Micropenetration Versus Temperature of Hydrogenated Cottonseed Oil

A number of analytical methods are available for controlling hydrogenation reaction and characterising hydrogenated fats in terms of their compositions and properties.

Hydrogenation Practice

Hydrogenation is invariably carried out on properly degummed, de-acidified and bleached oils and fats. In commercial practice, the preparation of catalyst and the generation of hydrogen gas are important. The choice of the hydrogenation equipment and its materials of construction and design constitute a major aspect.

Production of Catalyst

Nickel catalyst is almost, exclusively used throughout the world in hydrogenation. The nickel catalyst, however, is required to fulfil some essential requirements which are:

(i) Easy separation of catalyst from the hydrogenated product and easy filterability:

The particles or crystallites of nickel (50–100 Å size) are distributed throughout a porous siliceous material with 10 μm (100 Å) particles. The total surface area ranges from 200 to 600 m^2/g,of which 20-30% is catalytically active. Such supported catalysts filter rapidly and by proper manifestation the filtration can be accomplished at the normal rate. The supported catalyst has the desired mechanical properties to permit its repeated filtration on 20-40 occasions depending upon its pattern of use. A nickel catalyst without originally having any support is added with a filter aid (namely hyflosupercel) to facilitate filtration after the hydrogenation reaction.

(ii) Adequate activity of a catalyst during hydrogenation:

Catalyst activity can be best assessed by the rate of drop of iodine value per unit time under a given set of hydrogenation conditions and with varying levels of the catalyst. Oils or fats having almost identical fatty acid compositions especially having identical properties of the unsaturated fatty acid profiles and total saturated acid content should be taken for catalyst activity assessment and evaluation. Different nickel catalysts are available in commerce in the form of proprietory brands. Depending on their methods of preparation, the catalysts may contain varying activity of nickel and, therefore, show different hydrogenation rates. A comparison of activity between the different catalysts is made with respect to the weight of catalyst used and the drop in iodine value observed with time.

(iii) Shelf life of catalyst:

The shelf life of the catalyst (i.e., durability) is very important. It is extremely desirable that the catalyst retains its activity fairly well for a particular length of time. Some highly active catalysts may lose their activity at a relatively much earlier period by picking up poisons from the oils/fats and other sources. Some catalysts may be so designed that poisons cannot get access through the pores to the active sites within.

Production of nickel catalysts: The most important nickel catalysts which have found application in the hydrogenation of oils/fats are nickel-kieselguhr catalyst, nickel from nickel formate (Rufert nickel), and Raney nickel.

Rufert nickel catalyst: Nickel catalyst of Rufert type can be produced by Dry method and by Wet method.

Dry method: The dry method involves the precipitation of nickel from a nickel salt solution at alkaline pH in the presence of kieselguhr or sodium silicate. The catalyst that is eventually obtained after drying, grinding and heating in an atmosphere of hydrogen at high temperature is really a good quality catalyst with high specific surface.

In actual practice, the solution of the nickel salt, is treated with a solution of soda ash or bicarbonate or sodium hydroxide. The mixture is boiled for some time, and then an appropriate quantity of kieselguhr is added. After continued boiling, the mass is filtered hot and washed carefully to remove sulphate, chloride or nitrate ion by hot water as completely as possible.

The washed cake material is then dried (from 60% to 5% water), grinded and is reduced in hydrogen atmosphere at 420°–500°C in a roaster. About 60% of the nickel present is reduced to metal and the catalyst is immediately suspended directly in an appropriate oil (hardened oil of m.pt 60°C) in order to prevent oxidation of the metal which is often pyrophoric. The catalyst is eventually converted to flakes by passage over a chilling drum.

Wet method: The method consists of preparing a nickel catalyst by decomposing a nickel salt of an organic acid (monobasic or dibasic) in the liquid phase as suspending the catalyst in oil or in hardened fat. The nickel salt that is almost exclusively used is the nickel formate which begins to decompose at about 150°C (300°F) with the evolution of a large amount of gas :

$$Ni(OOCH)_2 . 2H_2O = Ni + 2CO_2 + H_2 + 2H_2O$$

At above 200°C decomposition becomes rapid, and by raising the temperature carefully to about 240°C decomposition takes place quantitatively. The oil gets partly hydrogenated by the hydrogen gas which is evolved. The nickel is produced in a very finely divided state and is dispersed in the oil.

In actual practice, the catalyst is prepared in a simple reactor provided with agitator, with arrangement for heating to a high temperature, fitted with a cooling coil to prevent decomposition reaction from becoming too vigorous and usually a vent pipe. Generally, 4 parts of oil (usually a hardened oil) and 1 part of nickel formate (crystals) are used and taken in the reaction vessel. There is actually no rigid oil and nickel formate ratio. The temperature is raised steadily to around 150°–180°C while stirring to remove water of crystallisation. After continuous stirring steadily for an hour additional heat is supplied until about 245°C and then heat is provided to achieve a steady temperature of 250°C either in presence of hydrogen that is passed through the reactor as a slow current through the oil or under vacuum. The gaseous products of decomposition are removed. The catalyst gradually becomes active and hydrogenation of the oil begins. At the end of the reduction period the oil gets completely hardened.

Often kieselguhr equivalent to one to four times the weight of the nickel is added while cooling the catalyst suspended in the hardened oil to around 90°C. The catalyst is discharged from the reactor and flaked. Sometime the catalyst is filtered hot to get rid of the oil that has undergone a kind of thermal decomposition. Fresh oil is instantly added in adequate amount to get a thick paste or slurry. The oil is usually the one that is to be hydrogenated.

In order that the catalyst can be stored, shifted and handled better, the catalyst is suspended in a highly hardened oil that facilitates its casting into blocks, granulated or formed into flakes over a chill roll. The nickel content in the catalyst varies from 7% to about 25% depending on the quantity of nickel formate employed.

The advantage of this catalyst, called Rufert nickel catalyst, lies in the relatively simple method of preparation. Its catalyst properties are very good as indicated by the rate of iodine value drop per minute and the selectivity including *trans* isomer formation achieved. A disadvantage of the catalyst is that it is sometimes difficult to remove completely the finely divided nickel particles of colloidal dimensions by filtration, and hence an after-treatment is necessary.

Raney nickel catalyst: Raney nickel catalyst is a highly active catalyst for hydrogenation of fats and oils. It is, however, low in selectivity for which it has limited range for hydrogenation to make edible fat products. Raney nickel catalyst is prepared by dissolving an appreciable amount of nickel/aluminium alloy (Ni:Al - 1:1 w/w) powder in strong caustic soda solution (20%) in an open digestor. The alloy powder is added slowly to control the rate of foaming and the heat of dissolution. The hydrogen liberated during the reaction is carefully disposed of. The alloy and alkali solution after being combined and heated to 245–250° F and held at this temperature for 2-3 hours. After this period, the catalyst consisting of metallic nickel is obtained in the form of a sludge. The aluminium is in solution as sodium aluminate. The sludge is repeatedly washed with water to remove completely alkali and sodium aluminate. The catalyst after draining off the last wash water is covered with oil. The catalyst sludge is eventually dried by heating under vacuum in another tank, and kieselguhr is next added to the oil/nickel sludge.

Catalyst suppliers normally offer three main types of nickel catalysts with : (1) high S_I - selectivity (ii) high activity and mere good S_I - selectivity and (iii) high S_{II} – selectivity, although low *trans* isomerisation catalyst is produced by one supplier.

Production of Hydrogen Gas

Hydrogen gas can be produced by several methods but the selection of methods depends principally on available raw materials, cost of production and the environment. The methods are : (1) Electrolytic, (2) Steam-iron, (3) Hydrocarbon Reforming, (4) Partial Oxidation of Hydrocarbons, (5) Methanol Decomposition.

Among these processes, upto now the electrolytic process is adopted in many countries.

Table 11.7: Below Lists the Different Types of Nickel Catalysts

Type of catalyst	Suppliers		
	Harshaw	*Sud Chemie AG*	*Unichema*
S_I - selective	Nysel DM_3	KE-NF 20	Pricat 9906
Highly active	Ressan 22	G-50	Pricat 9900
S_i - selective	Nysel SP7 (20% Ni catalyst in flakes; treated with ca. 1% sulphur) to promote *trans*	G-111	Pricat 9908
Low *trans*	Nysel HK-12	G-70	

1. ELLECTROLYTIC HYDROGEN (2a)

In the electrolytic method, an electrolyte, such as sodium hydroxide (23° Be') at 60°C or potassium hydroxide (31° Be') at 80°C, is charged into a cell and an electric current is passed through the solution. The electric current causes the decomposition of water into hydrogen and oxygen. The gases are collected over their respective electrodes and piped to a low pressure storage tank.

In decomposing water into hydrogen and oxygen, a basic minimum of 2400 Ah flowing for 1 hr are needed to produce 1 m^3 dry hydrogen at 0°C and 760 mm Hg or 2180 Ampere hour are needed to produce the hydrogen contained in 1 m^3 of moisture saturated gas at 20°C and 760 mm Hg. To overcome the resistance offered by the electrodes and electrolyte, a final voltage across each cell is required to be 1.8– 2.6 V as the practical decomposition voltage. Further, this voltage range depends on cell design, type and concentration of electrolyte, temperature and pressure.

The energy requirement to produce the 1 m^3 saturated hydrogen gas at 20°C and 760 mm Hg is : $1.8 \times 2.18 = 3.92$ KWh per m^3 H_2 and $2.6 \times 1.18 = 5.68$ KWh per m^3 H_2. Two types of plants are in use for the production of hydrogen. One plant uses a battery of simple cells of the Knowles (1) type (Unipolar) and connected in series, and the bipolar cells of the 'filter press' model, resembling a plate and frame press in appearance and construction. The two cells are shown in 11.9 & 11.10 Figs.

In the unipolar electrolyser, the electrodes are of a single polarity, alternating anode and cathode, with electrodes of like polarity connected in parallel (Fig. 11.9). Each cell is a separate unit at Ca. 2 V. Unipolar equipment consists of a number of individual cells connected by bus-bars. H_2 and O_2 are collected in separate compartments, separated by asbestos diaphragms. Unipolar equipment has the benefits of low cost, relative simplicity and ease of maintenance.

In the bipolar electrolyser, each electrode has a dual polarity and the electrodes are in series so that the entire cell bank operates at relatively high voltage (Fig. 11.10). Bipolar designs tend to be more compact, although considerable space is required around the generator for maintenance purposes.

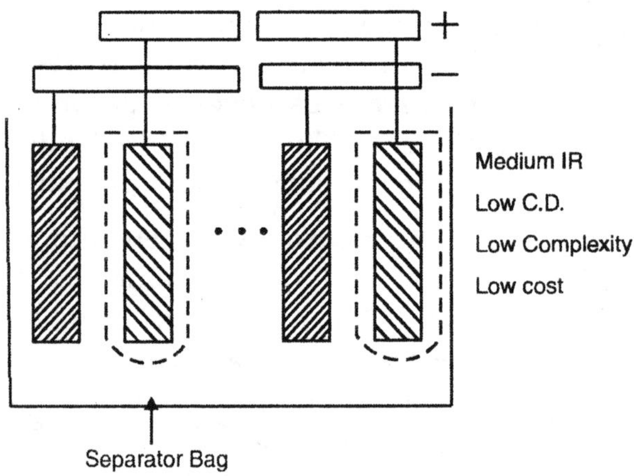

Fig. 11.9: Compact Bipolar Electrodes

Water Requirements and Purity

The electrolyte used is not consumed in the process. Only water is required as the material. The quantity of water needed is calculated as follows:

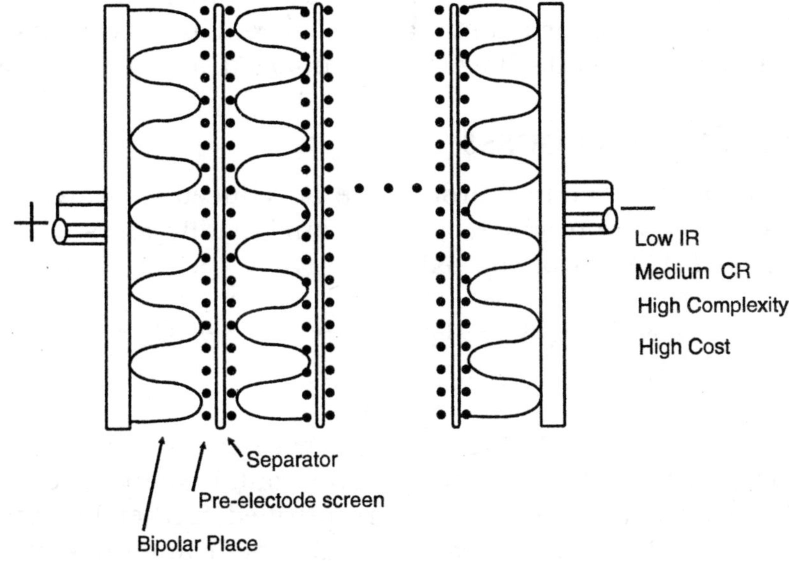

Fig. 11.10: Unipolar Cell Design

M^3 of hydrogen liberated by 1000 Ampere hours.. 0.419 Cum. (Cubic metre)

Litres of water decomposed per 1000 Ampere hours .. 0.336 litre

Hence, 1 Cum. of hydrogen requires $\frac{0.336}{0.419}$ = 0.80 litre of water

This figure in actual practice normally lies in the range of 0.9 – 1.0 litre/Cm^3 allowing for losses due to evaporation and spillage etc. This would work out to 6–7 gallons of water per 1000 cu.ft. of hydrogen gas produced. Equipment is usually provided for automatic addition of this make-up water.

The feed water should be of a high standard of purity. Any presence of mineral salts would increase the power consumption and aggravate the problem of corrosion of the electrolysing equipment.

The de-aerated water should be free from chloride, sulphate and carbon dioxide. On drying, the residue from feed water should not exceed 10 mg/litre. The normal conductivity of water recommended for electrolysis is 2 micro mhos/Cm^3. In any case, it should not exceed 5 micro mhos/Cm^3 at 20°C.

Ideal is double distilled water usually provided by a steam heated still. In modern installations, mixed bed type of water demineralisation plants have been employed for this purpose.

2. STEAM IRON PROCESS

In the steam-iron method, iron oxide is reduced by water gas or producer gas to iron which, with steam, readily gives hydrogen with the reformation of the oxide :

$$Fe_3O_4 + 4(CO + H_2) \longrightarrow 3Fe + 4CO_2 + 4H_2$$

$$3Fe + 4H_2O = Fe_3O_4 + 4H_2$$

The hydrogen is cooled and scrubbed with caustic soda solution to remove carbon dioxide. The gas still contains small quantities of carbon monoxide and is therefore purified further by catalytic conversion. Other impurities are H_2S, small amounts of more complex sulphur compounds, nitrogen and possibly some oxygen. The hydrogen purity is 99.5%

The steamiron process is cheap to operate where electricity is comparatively expensive. On the other hand, capital costs are high for small hydrogenation plants.

3. HYDROCARBON REFORMING PROCESS

Catalytic steam reforming of low molecular hydrocarbons (natural gas, LPG or naptha) is an elegant process for manufacturing high purity hydrogen gas.

The process consists of (i) desulphurisation of the feed by contacting with activated carbon, (ii) steam reforming with an appropriate catalyst (nickel) at high temperature (800-900°C) in heat exchangers at 15-20 bar to form hydrogen and oxides of carbon, (iii) purification of hydrogen by removing the CO by water-gas shift reaction at ca. 400°C with iron oxide — chromium oxide catalyst to CO_2 and additional hydrogen, (iv) removing CO_2 by passing through an absorber to remove CO_2 containing methanolomine and (v) methanating unconverted CO to CH_4 over a nickel catalyst.

The hydrogen content is now 99.5%–99.9% v/v range with CO level at 10 ppm.

4. PARTIAL OXIDATION OF HYDROCARBONS

Natural gas, refinery gas, or other hydrocarbon mixtures are used as feedstocks. Liquid hydrocarbons such as gas oil, diesel oil and even heavy fuel oil are also processed. The hydrocarbon is partially oxidised to oxides of carbon, hydrogen and water.

Natural gas is preheated to ca. 700°C and is reacted with oxygen (also preheated) in a refractory lined pressure vessel. The operating pressure is 14–34 Kg/sq. cm absolute. The feed rate of hydrocarbon is 1000-3000 vol/unit vol. of reactor space. Oxygencarbon ratio is 1.0–1.2 to hold carbon formation below 0.0005 kg/cu m of gas.

The reaction products are removed from the reaction zone and the effluent gases are quenched rapidly to reduce the temperature from 1425°C to 815°C or 530°C.

The partial combustion is moderated by the presence of nitrogen, steam or carbon dioxide. The quench is carried out in waste-heat boiler.

In any case, the gases are scrubbed to remove carbon particles and saturated, and then passed to a shift-converter, where the carbon monoxide reacts with steam to produce carbon dioxide and hydrogen. The hydrogen may be purified by conventional means.

5. METHANOL DECOMPOSITION PROCESS

Methanol can be decomposed catalytically to hydrogen and carbon monoxide which is further converted to carbon dioxide and hydrogen. The reactions are:

$$CH_3OH \longrightarrow 2H_2 + CO - \text{Heat}$$

$$CO + H_2O \longrightarrow H_2 + CO_2 + \text{Heat}$$

$$CH_3OH + H_2O \longrightarrow 3H_2 + CO_2 - \text{Heat}$$

In actual practice, a mixture of methanol and demineralised water is preheated and the vapour mixture is introduced into the reactor where the decomposition and the shift reaction take place in the catalyst filled tubes. The surplus of unreacted methanol and water is condensed and recycled to the feed tank. The raw gas containing 75% H_2 and 25% CO_2 and less than 0.5% CO is purified further. Removal of CO_2 is achieved by CO_2 absorption or CO_2

absorption followed by methanolation, and methanation, and pressure saving adsorption (PSA). The raw gas from the shift reactor is passed through a molecular sieve adsorbent reducing CO and CO_2 to below 1 ppm, methane and other impurities to below detectable levels and water to a dew point less than –50°C. The sweep gas released from the desorption of the molecular sieves is used for firing the process oil heater, thus making the process self-supporting with fuel.

For all systems the purified hydrogen is delivered at a pressure of 18-20 bar (250-280 psig). The systems ensure a hydrogen purity between 99.5 vol. % and 99.999 vol. %. The utility consumption per Nm^3 hydrogen produced by the methanol decomposition process is shown below:

	PSA purification	*Other means of purification*
Methanol, Kg	0.50	0.66
Demineralised water Kg	0.37	0.34
Fuel, K.Cal	0	9.00
Electricity, KWh	0.03	0.18
Cooling water, Kg	20	35

Energy requirements other than feed methanol (fuel and power) are very low, which is an advantage in times of increasing energy costs. Furthermore, the hydrogen is produced at a pressure of 18–20 bar(g) which is common for most hydrogenation purposes, thus avoiding further compression. The process is extremely flexible and has the advantage of consumption of electricity. The initial cost is competitive and there is no need for housing or structural supports. Moreover, the site preparation and installation are reduced to a minimum. Furthermore, supervision and maintenance are hardly required, thus reducing manpower costs. Many developing countries have a shortage of electrical energy. The methanol decomposition process is a better choice than the other processes.

Hydrogenation Plant

In view of the low volume production of hydrogenated fats, the hydrogenation plants are almost exclusively batch in operation. The continuous plants are yet to become popular.

In the hydrogenation plant the main consideration is mixing of the oil and hydrogen for which the recirculation or dead-end systems are operated in commercial practice.

In the recirculation system hydrogen gas is continuously recycled through the oil phase in the reactor (Fig. 11.11). In the dead-end system the reactor is supplied with as much hydrogen as is adsorbed and the dispersion of hydrogen is achieved by mechanical agitation. The dead-end hydrogenation has certain advantages over the recirculation system and these are:

(i) Prevention of oxidation and hydrolysis of the oil through deaeration and dehydration of the oil.

(ii) More positive control of hydrogenation reaction and consequent improved product uniformity since the entire reaction is carried out at a definite and constant temperature, and the amount of hydrogen absorbed by the oils is easily ascertained from the pressure drop on the hydrogen supply tank.

(iii) Greater latitude in selectivity and decreased dependence on the temperature as a factor determining selectivity as the operating pressure can be widely varied.

(iv) Simpler, less expensive, and more easily maintained equipment.

Dead-end hydrogenators of 5–30 tonnes capacity are closed, vertical, cylindrical steel vessels, provided with arrangement for full vacuum and 100–150 pound of working pressure. There are coils within the vessel for heating by steam and cooling by water. A vertical shaft carrying a motor-driven agitator is inserted through a stuffing box or a seal in the top head (Fig. 11.12).

The autoclaves are provided with highly efficient agitators like the turbine-type, with either two or more impellers or paddle-type impellers. In actual operation, the dead-end converter is kept under vacuum while the charge of oil and catalyst is being pumped in as well as during the heating period to deaerate and dry the oil before it reaches a high temperature and also to prevent reaction during heating. When steam in the heating coils has carried the batch to the required operating temperature, the vacuum line is closed and hydrogen is admitted to the vessel and allowed to build up to the desired pressure.

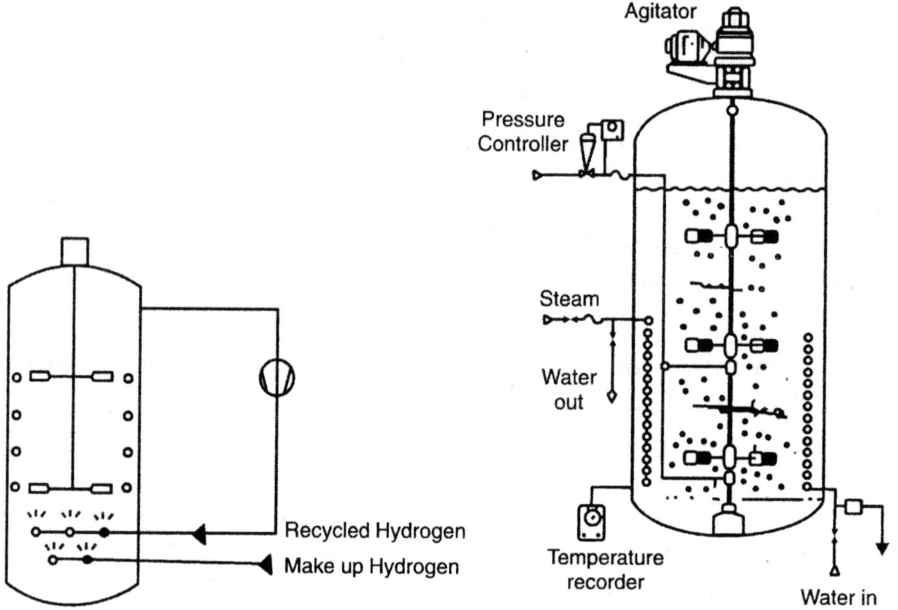

Fig. 11.11: Recirculation System

Fig. 11.12: Dead-end System

At the end of hydrogenation period the flow of hydrogen is stopped; the hydrogen in the head space along with accumulated impurities is blown off to the atmosphere; vacuum is again placed on the vessel, and the charge is cooled to about 160–190°F by water in the coils with the agitator running. Then the vacuum is broken with in the first cloudy oil through the press being diverted to the converter. A vent line to the atmosphere should be installed between double block valves in the hydrogen line to the converter, and this should be opened, except when the converter is under pressure, to prevent accidental leakage of hydrogen into the vessel. Hydrogen loss in the dead-end system does not exceed 5 % of the hydrogen reacted.

Comparison between Dead-end and Recirculation Batch Systems

1. Considerable heat loss takes place when the hot recirculation gases are cooled and then reheated to reaction temperature during the recirculation process.

2. In the dead-end hydrogenator the agitator is more efficient for converting energy to high speed agitation than is possible in the recirculation system.
3. The dead-end hydrogenation system can be operated over a wider range of pressures than the recirculation process.

Continuous hydrogenation

Almost from the beginnings of fat hydrogenation, efforts have been made to develop a continuous hydrogenation process (9–11). The commercial success was rather poor. There are a few processes today which are of some limited importance and they use a suspended nickel catalyst. All attempts to use fixed-bed catalysts failed.

The main advantages of the continuous system are :

(i) Energy saving by use of counter current heat exchange.

(ii) Constant energy and cooling water consumption.

(iii) Constant input of raw materials, catalyst and hydrogen as well as constant discharge of hardened fat.

(iv) Constant quality of finished product by continuous process control.

(v) Minimal manpower as the process is suitable for high grade automation.

The two main disadvantages are : (i) the lack of flexibility in case of frequent product alterations, and (ii) the reduced S_I selectivity to some extent in case of partial hydrogenation.

The three continuous-type hydrogenation equipments are : (i) Lurgi process, (ii) Buss cascade process and (iii) King Dravo process.

The Lurgi process of continuous hydrogenation is outlined in Fig. 11.13. The raw materials mixed with the catalyst flow cocurrent with hydrogen through the tall reactor D_1. Unreacted hydrogen is recycled into the oilcatalyst stream. Most of the installed capacities are used for the hydrogenation of fatty acids.

The Buss AG, Basel, has developed a continuous hydrogenation process basing on the loop reactor. According to Fig. 11.11

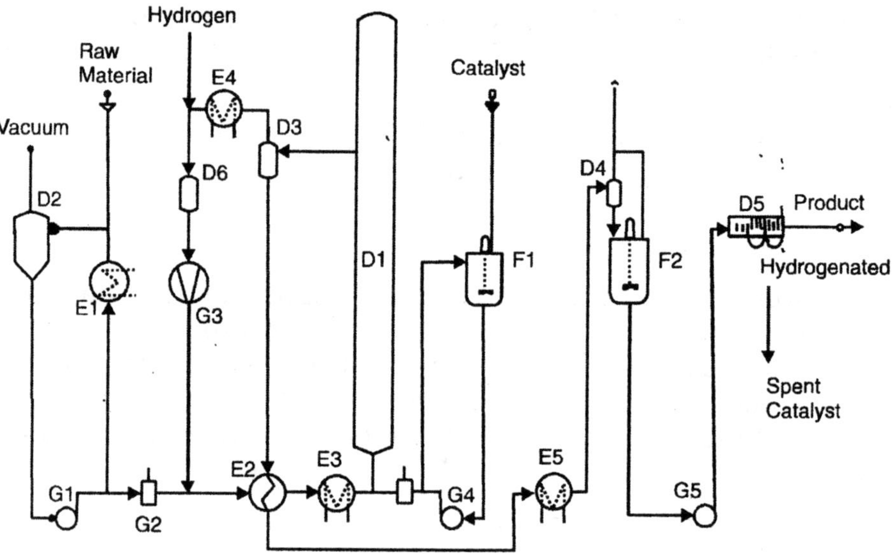

Fig. 11.13: Process Diagram of LURGI Process for Continuous Hydrogenation

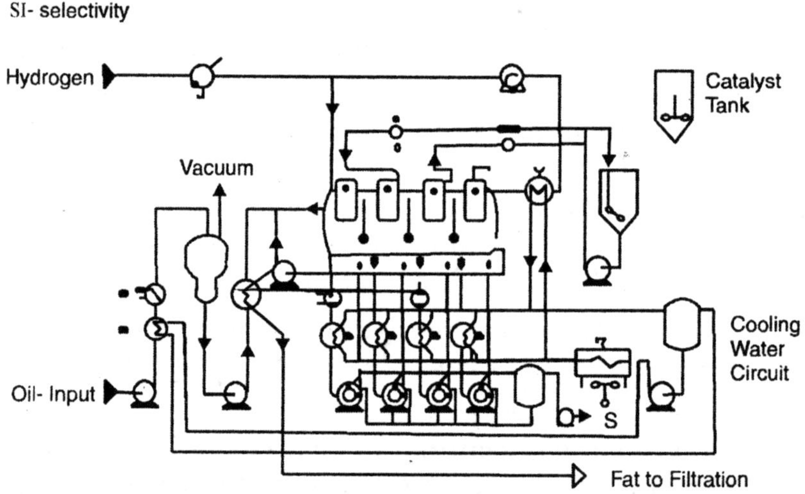

Fig. 11.14: Flow Sheet of a Continous BUSS Hydrogenation Unit with Four Loops

interconnected loop reactors with a joint head space are combined in a cascade in order to avoid substantial backflow and outtage of product and to get homogeneous hydrogenation with reasonable selectivity. The more cascades the steaeper is the residence-time diagram. Instead of one compact processing unit Buss also provides combinations of two to four batch-type loop reactors to a process cascade.

The Dravo process is reported to be in use in USA with good success (12–14). Figure 11.15 demonstrates the principle of the process : The raw materials—catalyst mixture is deaerated and dried under vacuum, heated by heat exchange and preheated to reaction temperature and fed into the reactor consisting of a long pipe. During the passage the oilcatalyst suspension is supplied with proportionate amounts of hydrogen in different sections of the pipe. Under well controlled conditions a definite end product can be obtained. The process is said to have good S_I-selectivity.

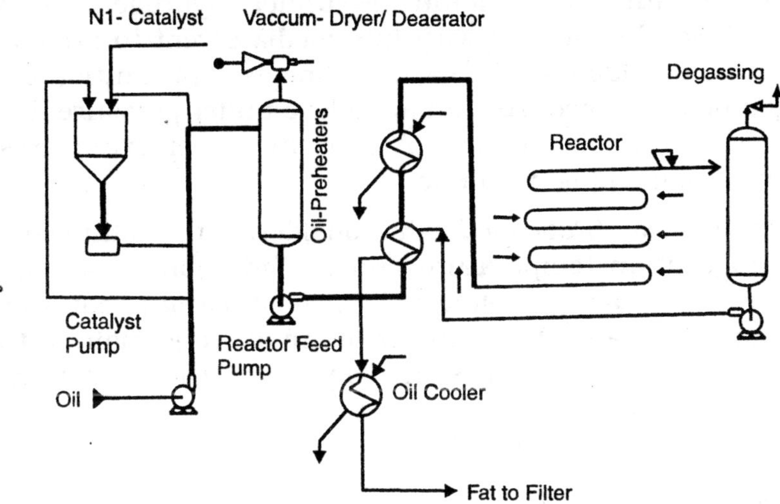

Fig. 11.15: Flow Sheet of a Continuous Hydrogenation According to the KING DRAVO Process

Factories with a great variety of products requiring partial and selective hydrogenation may preferably use batch autoclaves as the latest developments in energy saving techniques make batch operation more attractive. Therefore, Buss has added two suitable vessels, one ahead of and the other after the reactor and two appropriate heat exchangers.

Some Important Design Aspects in Hydrogenation Plant (2a)

In designing a hydrogenation autoclave, the following considerations should be taken into account :

(1) Vacuum, (2) Oil depth and Autoclave Diameter Ratio, (3) Gas Space, (4) Temperature Control, (5) Stirrer and (6) Material of construction.

Vacuum : Vacuum is extremely important in a hydrogenation autoclave for deaerating and drying the oil before it is heated to the operating temperature. Usually a single stage steam ejector is adequate since this is also the safest way of handling hydrogen withdrawn from an autoclave. It is easy to place two block valves on the hydrogen feed to the autoclave with a vent to atmosphere cock between them. The vent should be open except when it has entered the autoclave should be under hydrogen pressure. Similarly, two block valves with intermediate vent to atmosphere cock can be provided on the vacuum line as a precaution against hydrogen being sucked from an autoclave under pressure through a faulty vacuum control valve. A capacity to achieve a pressure of about 50 mm Hg is adequate.

Autoclave diameter/Ratio oil depth: Considering the volume of oil at the hydrogenation temperature and the additional volume from the stirrer and heating/cooling coils, it is important that a proper ratio of oil depth and the diameter of autoclave is maintained for efficient mixing of oil, catalyst and hydrogen gas. Generally, the oil head is greater than 1.5 times the diameter of the autoclave vessel and is not less than 1.2 times the diameter.

Gas Space: In an autoclave, an allowance for the optimum gas space is required which varies depending on the extent of iodine value drop expected. For an oil undergoing iodine value drop of 50–60 units, a gas space of only 23 % total internal volume or roughly 30 % oil volume is adequate and satisfactory. A gas space equal to some 30 % total volume (nearly 43 % oil volume) is quite sensible.

Temperature Control: Temperature control is extremely important during hydrogenation reaction. The reaction being exothermic, there is invariably generation of heat resulting in an increase of temperature of the oil of 1.6–1.7°C for each unit decrease in iodine value. The equivalent heat evolved is 0.942–1.100 KCal/Kg (1.6–1.98 BTu/1b) per unit iodine value drop. In order to cope with the heat evolved it becomes necessary to make provision for controlling the heat and consequent temperature rise.

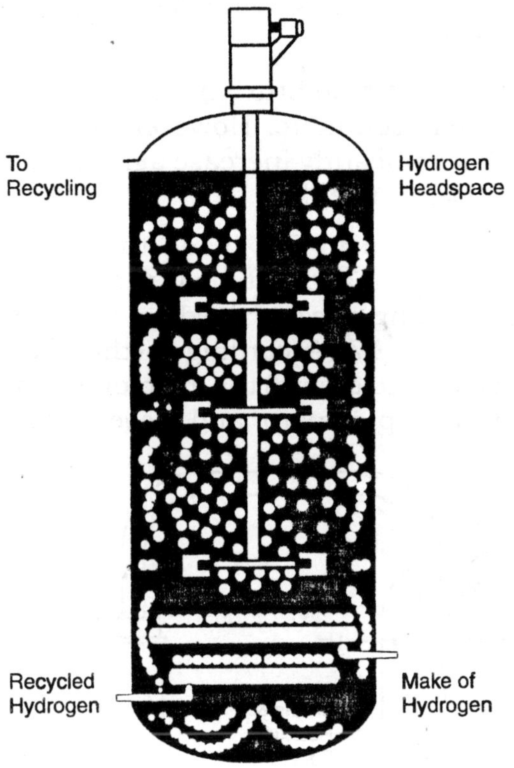

Fig. 11.16: Flat Blade Turbine

The size of the temperature control equipment for an autoclave needs proper estimation. The temperature control will save the oil from getting damaged due to overheating. Normally a coil of proper size is fitted to regulate heating and cooling. Since there is a stirrer, buffle, support on, it becomes often difficult to put the proper size of the coil. With an overall rate of one unit of iodine value/min, a coil of 4.5 m^2 surface will give a good measure of control.

Stirrers : Mass transfer resistances should be eliminated to facilitate the mixing of hydrogen gas,oil and catalyst intimately thereby to promote the diffusion of hydrogen gas through the bulk oil phase in the catalyst pore, adsorption of the unsaturated fatty acid chains of triglyceride molecules and desorption of the partially or completely hydrogenated molecules. Therefore, a number of mixing devices have been developed and used in the hardening plants. Some of the mixing (stirring) techniques are discussed and briefly described below.

Uptill now the flat blade turbine agitator (Fig. 11.16) appears to be the accepted technique. However, several unconventional approaches that significantly increase agitation have been recently proposed or are being used.

Of the alternatives to flat blade turbine mixing, the most commercially successful at this time is the loop reactor. It employs an injector type mixing nozzle based on the venturi principle, as depicted in Figure 11.17. According to the design of the system (Buss Ltd. of Basel, Switzerland) the injector mixing nozzle produces a jet of liquid that picks up and accelerates a gas from its

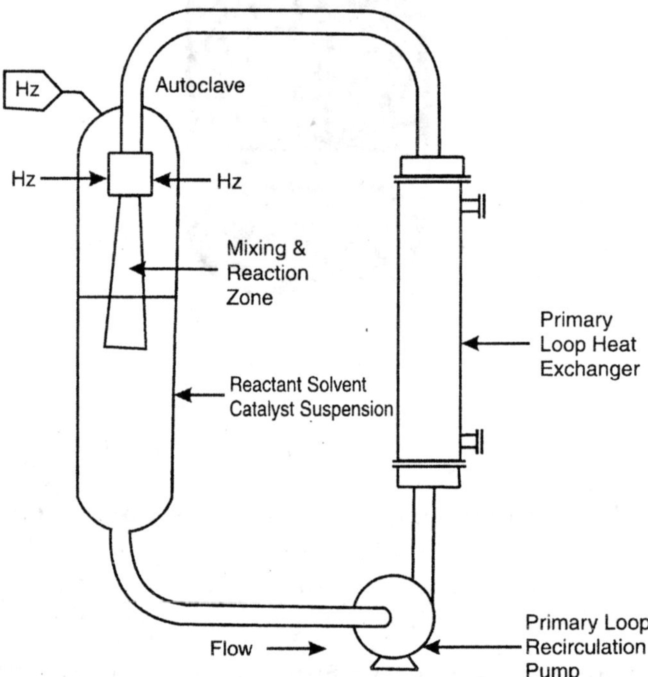

Fig. 11.17: Loop Hydrogenation Reactor

surroundings. The result is an intimate mixture of the oil/catalyst slurry and the hydrogen gas is drawn into it. The velocity of the mixer is reduced and the pressure increased in a second nozzle with an extension tube, called a diffuser. The shear forces developed as this happens disperse the gas very finely in the liquid and at the same time break down the laminar boundary layers of the gas bubbles and the catalyst particles. The hydrogen concentration is increased at the catalyst surface.

Another efficient mixing device is the helical impellar. In the conventional stirred tank reactor, the impellar is a downpumping helical screw within a shroud or draft tube. Because of the buffle arrangements in the inlet of the draft tube the liquid flowing into the top of the draft tube forms a number of vortices. These vortices feed gas from the gas phase above the liquid into the helix and down through the draft tube. Even gas that does escape into the gas phase is recirculated back into the liquid via the cortex ingestion path. The inventors claim that this results in a more even distribution of hydrogen throughout the reactor.

A comparison between the conventional flat blade turbine and the helical impeller, reveals that the greater agitation of the helical impeller results in increased reaction rate along with less S_I selectivity and less *trans*-isomer make.

Another interesting mixing device is the ultrasonic hydrogenation equipment (Fig. 11.18). The oil/catalyst mixture enters the bottom of the converter and hydrogen is introduced into the slurry directly below the base of the processing tank. Ultrasonic agitation increases the reaction rate 50 % and has no significant effect on linolenic (S_{II})selectivity and *trans* isomer formation. A 17-fold increase in reaction rate when comparing ultrasonic with conventional stirring is reported.

Material of construction: The hydrogenation autoclaves are made generally of stainless steel type 316 as the hydrogenation reaction is carried at temperatures above 150°C. Type 304 is satisfactorily used in welding during fabrication.

Utilities in a hydrogenation plant and actual cost of hydrogenation: In order to arrive at the cost of hydrogenation of an oil or fat, it is very necessary to know the utilities and materials required per tonne

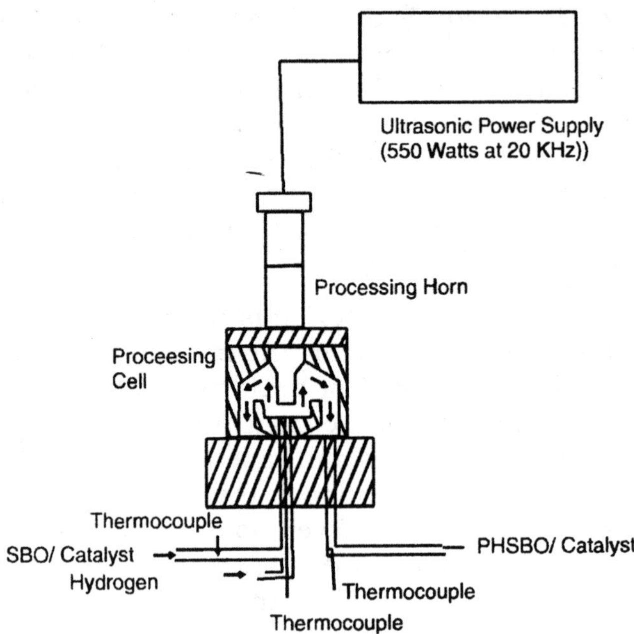

Fig. 11.18: Apparatus for Ultrasonic Hydrogenation of Soybean Oil

of oil/fat. The utilities and the materials are steam, power, cooling water, hydrogen and nickel catalyst. Table below gives an idea about the utilities and materials required for the hydrogenation of 1 tonne of oil/fat with an initial iodine value of 140 and reduction to the iodine number of 65-70.

Utilities and Materials Required per Tonne

Steam 30 kg/cm^2 or 60 psig	..	100 kg
Cooling water at 20°C, m^3	..	4
Power, KWh	..	13
Hydrogen (99.5 % +), m^3 STP	..	72
Nickel catalyst, Kg	..	0.2–0.5
Manpower, per shift	..	1

Cost of Hydrogenation (50 Tonnes/day : Iodine Value Drop by Nearly 70 Units)

1.	Cost of hydrogen gas (2000 cft/tonne oil) for 6 cft H_2, 1 KWh electricity required whose cost is Rs 2/-		Rs. 0.340 lakh
2.	Cost of catalyst (nickel) Rs. 200/- per kg 1.5 kg/tonne oil (51 tonne, 76.5 kg catalyst required)		Rs. 0.153 lakh
3.	Post refining Rs. 1000/- tonne		Rs. 0.510 lakh
	Total production cost	.. ..	Rs. 1.003 lakhs
	Production cost/tonne of hydrogenated fat	.. ..	Rs. 2,000.00

A typical operating cycle for hydrogenating an oil, such as cottonseed, to shortening consistency would be as follows :

(i)	For charging the converter with oil and catalyst slurry	..	1/4 hr.
(ii)	For deaerating, heating to 275°F, and drying the oil under vacuum	..	1/1/4 hrs.
(iii)	Normal reaction time @ 325°F	..	2 hrs.
(iv)	Time required for laboratory for making the final tests	..	1/2 hr.
(v)	For cooling the oil to 150°F	..	3/4 hr.
(vi)	For circulating cloudy oil and for the actual filtration	..	1/1/4 hrs.
	Total time		6 hrs.

Hydrogen Quality and Hydrogen Requirement

Typical Analysis

Hydrogen –	99.95%
Inert –	Nil
Oxygen –	0.05%
Carbon dioxide	Nil
Carbon monoxide	Nil

Hydrogen Required to Drop Iodine Value by One Unit for Metric Tonne (1000 kg) Oil

1 kg-mole iodine is equivalent to 1000 × 22.4 litres of H_2 at a M^3. For 254 kg Iodine Value, volume of hydrogen is

$$\frac{22.4 \times 1000}{100 \times 100 \times 100} \, m^3$$

$$\frac{22.4 \times 1000 \times 1000}{254 \times 100 \times 100 \times 100} \, m^3 = 0.8835 \, m^3$$

Taking 5 percent loss, H_2 required at NTP = 0.9277 m^3

Safety, Energy Saving

Safety is a very important aspect in a hydrogenation plant. Two basic considerations must be taken into account. First is the avoidance of the formation of flammable mixtures at any time in the plant, and the second is avoidance of the source of possible ignition.

The formation of the flammable mixture is the pocketing of hydrogen released by leakage or faulty operation. The best defence is adequate ventilisation because of the very low density of hydrogen. In fact, hydrogenation plants are built completely in the open to get maximum ventilation, and the controls and instrumentation are housed.

To avoid possible ignition of pocketed flammable mixtures, a number of prohibitory measures are recommended. There should be prohibition of smoking and carrying of matches and lighters in the hazardous zone. Electrical equipments should be taken care of. The wirings, electrical equipments like electrical meters, explosion-proof agitator motors and switches, lighting and instruments should be such that they cannot ignite gas in a way that allows flame to propagate. Welding must not be allowed at times when gas may be present. In the hydrogenation process, there are four areas where the process engineer can design a plant such that the plant manager has to spend as little as possible on the utilities. These are energy required for product heating to start the hydrogenation reaction, energy released by the exothermic reaction, hydrogen consumption as near as possible to the stoichiometric quantity required and energy required for the dispersion of hydrogen gas in the oil. Almost all plants built today include heat recovery equipments making hydrogenation at least a non-user of energy. With an oil/oil heat exchanger, it is possible to heat the bleached oil with the hydrogenated oil. This heating can be sufficient to start the hydrogenation reaction without additional steam heating (Fig. 11.19).

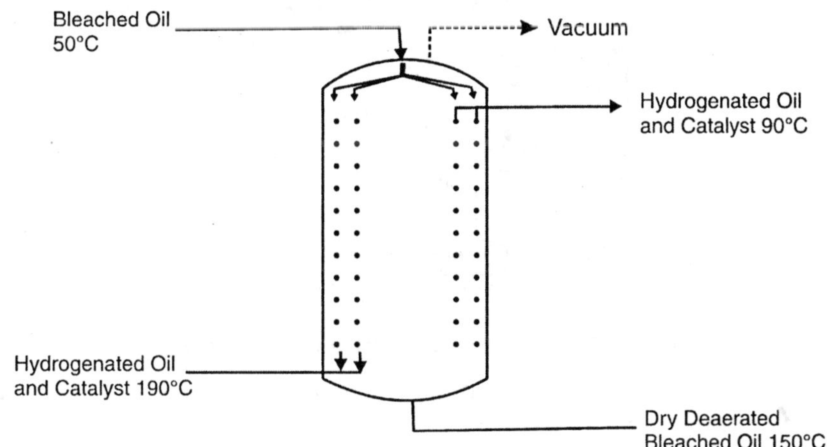

Fig. 11.19: Oil/oil-catalyst Heat Exchange, Drier Aerator

The two standard types of heat exchanger equipment are flow-through tubular and drop-tank static. Since flow-through tubular has a smaller space requirement, it has been more popular for vetrofitting operating plants (Fig. 11.20). Hydrogenated oil at 200°C from the converter is cooled to 100°C by heat exchanging with oil from storage at 60°C. The cold oil from storage is heated to 160°C in the exchanger and is accumulated in a hot oil tank prior to transfer to the converter.

Post-treatments of Hydrogenated Fats

During the process of hydrogenation there is a slight increase in the percentage of free fatty acids and the hydrogenated fat product also acquires a "hardening flavour". This is due to the formation of certain aldehydes from the oxidative breakdown of the *trans* isomers when PUFA (linoleic rich or higher) oils are hydrogenated. Very often off-flavour is generated from the unsaponifiable matters of oils parti-cularly with soyabean and rapeseed as well as marine oils. The hydrogenated products need to be free from nickel to ppm level (1.5 ppm).

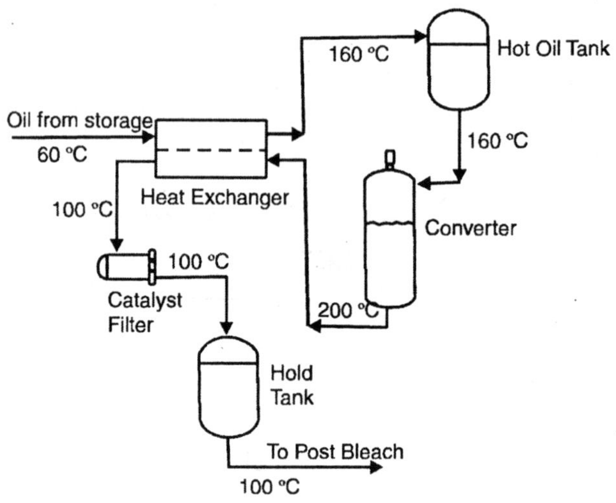

Fig. 11.20: Converter Heat Exchanger Flowsheet

In view of the above considerations, the product requires quite extensive purifications/post-treatments. The product is subjected to neutralisation, bleaching and deodorisation.Normally, neutralisation is done with dilute caustic alkali followed by hot water washings, and bleaching is effected with the usual acid activated clay (0.25–0.5 %) and activated carbon (40 % of the activated clay). Many processes avoid activated carbon to bleach the oil which is very effective in removing the chlorophyll pigments. The filtered oil is treated with citric acid or phosphoric acid to reduce nickel metal content to well under 1 ppm.

The hydrogenated product before subjecting to post-treatment is discharged from the autoclave after being cooled to 60°–90°C in the autoclave depending on the nature of the product and is filtered over a plate-and-frame filter press to remove the bulk amount of nickel catalyst used.

Post-neutralisation with alkali and bleaching remove some flavouring substances but many of the flavouring components still remain in the product and can be eliminated or reduced only by the process of deodorisation. The product is deodorised in any standard deodorisation plant at 180°C–230°C under a high vacuum of 740–755 mm Hg. After deodorisation, the product is cooled under vacuum to about 60°–80°C after which it is discharged into a storage tank. Sometimes, the deodorised fat is filtered before storage.

Plant Requirements

Plant capacities are determined on the basis of production targets. However, actual items of plant do not vary a great deal. The essential requirements for a factory with an output of 25 tonnes/ day are as follows:

1. Oil receiving and Storage : 25 Tonnes/Day

(i) 2 × 200 tonnes capacity storage tanks for raw oil

(ii) 2 × 50 tonnes capacity storage tanks for raw sesame and cottonseed oils

(iii) 1 weigh tank

(iv) 1 weigh scale

(v) 2 underground tanks

(vi) 2 pumps

2. Pre-refining

(i) 1 Neutraliser, 10-14 tonnes capacity

(ii) 1 vacuum bleacher, 5 tonnes capacity

(iii) Caustic soda dissolving tank

(iv) Storage tanks for caustic and hot water

(v) Tanks for soap-stock and emulsion

(vi) Certrifuge

(vii) Filter-press

(viii) 2 storage tanks for neutralised oil

(ix) 2 storage tanks for bleached oil

(x) Pumps

3. Hydrogenation (Hardening)

(i) Autoclave, 7–14 tonnes capacity
(ii) Catalyst reducer
(iii) Catalyst feed tank
(iv) Vent tank
(v) Oil and catalyst pump
(vi) Catalyst mixer
(vii) Filter-press
(viii) 2 cloudy oil tank
(ix) 2 storage tanks for hardened oil
(x) Pumps

4. Post-refining : 25 Tonnes/Day

(i) 1 Neutraliser, 10 tonnes capacity
(ii) 1 vacuum-bleacher, 5 tonnes capacity
(iii) Storage tank for caustic
(iv) Tank for emulsion
(v) Filter-press
(vi) 2 storage tanks for post-neutralised oil
(vii) 2 storage tanks for post-bleached oil
(viii) Pumps

5. Deodorising

(i) Deodoriser, 10 tonnes capacity
(ii) Vacuum pump and steam ejector
(iii) 1 vacuum cooler for deodorised product
(iv) 2 storage tanks for deodorised product
(v) 1 storage tank for deodorised sesame oil
(vi) Pumps

6. Packaging and Finishing

(i) 2 Blending and tempering towers

(ii) Filling and seaming units

(iii) Cooling rooms with refrigeration unit, or chilling roll or votator

(iv) Warehouse

7. Acid Oil (soap-stock treatment)

(i) 4 wooden vats

(ii) Storage tanks for concentrated sulphuric acid

iii) Storage tank for dilute sulphuric acid

(iv) Pumps

8. Hydrogen Production : Electrolytic Process

(i) Electrolytic cells

(ii) Rectifier

(iii) Water distillation shell

(iv) Low pressure hydrogen storage tank

(v) Hydrogen compressor

(vi) High pressure hydrogen storage tank(s)

9. General Services

(i) Boiler plant

(ii) Vacuum equipment for bleachers, autoclave and deodoriser

(iii) Air compressor.

Some Edible Applications of Hydrogenated Fats

Oils and fats, when hydrogenated under different conditions, are converted to products of selective and non-selective compositions and consequent thermal properties, consistency, stability and plasticity. The hydrogenated products find extensive edible utilisations. Some of the major products for utilisation are vanaspati (shortening), hard butters (confectionery fats) and margarines.

Vanaspati is a hydrogenated fat product produced by selective hydrogenation of a number of vegetable oils like soyabean, rice bran, palm, sunflower, mustard/rape, sesame, cottonseed, etc. Generally, hydrogenation is carried out at 160°–200°C with 0.1–0.125 % nickel catalyst employing 30–45 psig hydrogen gas pressure. Vanaspati is used in India as a substitute for butter fat (ghee). The product is also used when votated as a bakery shortening.

Hard Butter Substitutes

Vegetable oils (e.g. rice bran, cottonseed, mowrah, palmolein individually or as blend) can be highly selectively hydrogenated under *trans*-promoting condition with the use of *trans*-promoting catalysts like spent nickel catalyst or sulphur containing nickel catalyst to get hardness and melting character like those of cocoa butter and a good compatibility with cocoa butter and a good resistance to blooming (15,16). The hydrogenation is also often carried out in a two-step process : the reaction is first carried out with the conventional nickel catalyst and next with the isomerisation catalyst using sulphur-poisoned nickel catalyst.

Margarine Fats by Hydrogenation

Vegetable oils like soyabean, sunflower, cottonseed, rice bran, sesame, rape, palmolein, etc. are hydrogenated with varying degrees of sele-ctivity and *trans*-isomer formation so as to obtain hydrogenated products with a varying range of melting points and SFC/SFI profiles to meet the melting and plasticity character of table margarines and bakery margarines.

Some typical hydrogenated fat stocks for use in margarine formulations obtained under different hydrogenation reaction conditions are indicated below. Generally, the blending of two or more stocks hydrogenated to different degrees is done for getting the desired characteristics in margarine fats. Normally, a hydrogenated fat with a relatively high iodine number (I.V. 80-90) is blended with a selectively lower iodine number (55-65) containing hydrogenated products. The following table shows analytical data for a margarine fat made from 80 parts of a softer component and 20 parts of a harder component (17).

Characteristics	*Margarine Fat*	*Softer Component*	*Harder Component*
Iodine number	79	86	55
Melting point (FAC)	100°F	86°F	111°F
Solid fat index at 10°C	27	18	66
21.1°C	16	7	59
26.7°C	12	2	57
33.3°C	3.5	0	43
37.8°C	0	0	27

SFI values for soyabean partially hydrogenated stocks are indicated below which can be blended appropriately to get satisfactory margarine fats (18).

Temperature °C	*1*	*2*	*3*	*4*	*5*
10.0°	58–61	40–43	36–38	18–22	4 max.
21.1°	42–46	27–29	19–21	5–8	2 max.
33.3°	21 max.	9–11	2.0 max.	3.5 max.	–
Final I.V.	64–68	70–72	73–76	85 max.	110

Characteristics of soyabean oil hydrogenated under different conditions for use in margarine (19)

Hydrogenation conditions	*1*	*2*	*3*	*4*
Initial temperature (°F)	300	300	300	300
Hydrogenation temperature (°F)	330	350	425	425
Pressure (psig)	15	15	15	5
Nickel (%)	0.02	0.02	0.02	0.02
Iodine value	80-82	106-108	73-76	64-68
SFI at 50°F)	19-21	4 max.	36-38	58-61
70°F	11-13	2 max.	19-21	42-46
92°F	0	0	2 max.	2 max.

Nutrition of Hydrogenated Fats

Hydrogenated fats find a number of edible uses depending on their melting point, and SFC characteristics. Hydrogenated fats with melting points of 31°-37°C are directly consumed as a frying fat in many food preparations by a large segment of population of the world. Hydrogenated fats of similar range of melting points find extensive use in table margarines between 20 and 40% along with liquid oils for consumption as butter substitute.

A great deal of concern is expressed as to the nutritional quality of hydrogenated fats in view of their fairly high *trans*isomeric fatty acids content.

The metabolism of *trans* fatty acids per se and their effects on other metabolic processes have been reviewed. *Trans* fatty acids are transported similar to their *cis* counterparts. They are oxidised in the body by β-oxidation. After β-oxidation ^{3}*trans* enoyal CoA is formed which is converted to ^{2}*trans* enoyal CoA.

Nutrition Board in the U.S.A and British Nutrition Foundation concluded that the effect of *trans* fatty acids on the cholesterol level was neutral. Heckers *et al* (20) determined levels of 14:1 t, 16:1 t and 18:1 t in myocardium, aorta and jejunum of subjects with advanced atherosclerosis and found them to be similar to those of controls. Thomas et al (21) assayed *trans* fat levels adipose tissue of heart patients and controls in 10 different areas of England. Average levels of *trans* fat were 5.38% in samples and 5.25% in controls. *Trans* fatty acids might be regarded as saturated fatty acids, and indeed some of their effects may be regarded in this way (22). We can regard monounsaturated *trans* fatty acids at the level of current intake to be safe components of the diet. The effects of *trans*, *trans* linoleic acid (23) may present a problem but probably not at the current levels of ingestion.

Recent Advances in Fat Hydrogenation (24)

Work continues on improving the hydrogenation process. One aspect has been researched to decrease the amount of *trans* isomers produced. Associate Professor Peter N. Pintauro and his graduate students at Tulane University, for instance, have developed a low-temperature electrocatalytic process for hydrogenating edible oils that results in less *trans* isomer formation. In this electrocatalytic reaction scheme, atomic hydrogen is produced on an active Raney Nickel powder cathode surface by the electrochemical reduction of water molecules from the electrolytic solution. Adsorbed hydrogen reacts with the oil's triglycerides to form the hydrogenated product. The operation does not require high temperatures or pressures. The electro-hydrogenated oil has higher stearic acid content and lower *trans* isomers than oil produced via traditional hydrogenation.

Conjugated Hydrogenation Process

Conjugated hydrogenation is a process in which the hydrogen which is generated *in situ* by catalytic decomposition of certain hydrogen donors saturates the unsaturated fatty acid chains. The hydrogen donors are the primary and secondary alcohols and compounds like hydrogenated napthalenes and also cyclohexanes. Metal catalysts that favour dehydrogenation are the usual hydrogenation catalysts like metallic nickel and copper. The dehydrogenation reaction occurs at high temperatures (200°–300°C) and the same metal catalyst effects the hydrogenation of the double bonds.

Primary alcohols show relatively slower rate and greater selectivity than the secondary alcohols due primarily to their differences in the rate of dehydrogenation and consequent availability of hydrogen molecules in the reaction system. The variables like the amount of hydrogen donor, reaction temperature and catalyst amount influence the hydrogenation rate and selectivity.

The process was studied in Russia, India and Japan, indicating the special features of the process. Unfortunately, however, the process does not assure any commercial importance till now, perhaps, due to the problem of recovery of unreacted solvent, decomposition products of the hydrogen donors and hydrogen gas.

The data on the hydrogenation of some vegetable oils and fats indicated that the conjugated hydrogenation process may be helpful in making a specific fat-based products (25-27).

Miscella Hydrogenation

Hydrogenation of vegetable oils in the miscella phase, i.e., in the form of solution in solvents can be conducted like that of the oils without a solvent. The nature of solvent, hydrogen partial pressure and temperature influence the rate, selectivity and *cis-trans* isomerisation. The miscella hydrogenation is almost comparable to the non-solvent hydrogenation. The polar solvent shows a different kind of rate and selectivity compared to the use of a non-polar solvent.

Miscella hydrogenation once received some importance in view of the development of miscella refining process. In fact, before the energy crisis due to the scarcity and soaring price of petroleum hydrocarbons the miscella refining followed by miscella hydrogenation did receive some commercial importance.

Albright *et al* (28) investigated the solvent hydrogenation more systematically than others and compared the process with non-solvent hydrogenation. In actual operation refined and bleached cottonseed oil was dissolved in a solvent (e.g. hexane, isopropyl alcohol,or di-isopropyl ether) and was hydrogenated in a dead-end hydrogenator at temperatures from 115° to 145°C at hydrogen partial pressures from 44 to 74 p.s.i.a. and at catalyst concentrations varying from 0.05 to 0.4%. Ni and also at high rates of agitation to eliminate mass transfer resistances. A series of hydrogenation runs under identical conditions were also made without the use of solvent. The rate decreased in the following order : non-solvent, hexane, isopropyl alcohol, and di-isopropyl ether runs. Selectivity and isomerisation were low in all cases and essentially identical for solvent and non-solvent runs. The rate of hydrogenation increased in all cases with higher catalyst concentrations. For the isopropyl runs, the reaction rate was maximum as a function of temperature at about 135°C. In the case of other solvents, the rate of hydrogenation increased with increased temperature in the range from 115° to 145°C, but rate increases of the solvent runs were less than those of the non-solvent runs.

By using a Raney Ni catalyst Kaufmann (29) succeeded in hydrogenating various crude oils in hexane at 35°C. A continuous system employing counter flow between the oil:hexane and hydrogen was used. A very low concentration of *trans* isomers was found in the product. A study (30) of the hydrogenation of methyl oleate with Raney Ni catalyst indicated that hexane was capable of and superior to other solvents in causing a reduction in *trans* isomer and positional isomers.

Analytical data and fatty acid composition of peanut and sesame oil hydrogenated in the form of miscella in n-hexane are indicated below (31).

Concentration of oil in n-hexane	:	60% (w/w)
Catalyst	:	0.3% Ni
Temperature	:	140°C
Total pressure	:	100 p.s.i.g
Agitation	:	200 shakes/min.

Oil	*Period of hydrogen-ation in mins.*	*Iodine Value*	*Slip Point °Csatd.*	*Fatty acid (% w/w) composition*			*transisomer (%w/w) satd.*
				Oleic	*Linoleic*		
Peanut	0	90.5	—	20.5	54.0	25.5	—
"	60	73.4	32.5	27.4	61.3	11.3	15.0
"	120	55.8	37.5	36.5	63.5	—	21.4
Sesame	0	110.7	—	16.5	40.8	42.7	—
"	45	43.4	34.5	24.9	66.3	8.8	21.1
"	60	65.3	39.0	28.9	68.2	2.9	25.4

In view of the trend in the low *trans*-isomer formation, the miscella hydrogenation with n-hexane as solvent may receive importance in producing nutritionally superior hydrogenated fat products for use in margarines and as shortenings including vanaspati.

Hydrogenolysis Process

Hydrogenolysis is a process for conversion of a fatty acid or its ester into an alcohol. The process also involves hydrogenation reaction. There are two ways of carrying out the reaction. The one that has been commercially adopted involves catalytic hydrogenation at high temperature and high pressure. The other process is the chemical Bouveault-Blanc sodium reduction method in which the hydrogen generated by reaction between sodium and an alcohol reduces the carboxyl group of the acid or ester molecule.

The reactions are reported as follows:

Catalytic hydrogenation

$$R\,COOH + 2H_2 \xrightarrow[\text{Catalyst}]{\text{Metal}} R\,CH_2OH + H_2O$$

$$R\,COO'R + 2H_2 \xrightarrow[\text{Catalyst}]{\text{Metal}} R\,CH_2OH + {}'ROH$$

(Monoesters or Triglycerides)

Bouveault Blanc

$$4Na + 4C_2H_5OH \longrightarrow 4C_2H_5ONa + 2H_2$$

$$R\,COOH + 2H_2 \longrightarrow RCH_2OH + H_2O$$

$$C_2H_5ONa + H_2O \longrightarrow C_2H_5OH + NaOH$$

Bouveault-Blanc process involving the use of higher alcohols (e.g. methyl cyclohexanol or methyl amyl alcohol in place of ethanol) and an inert solvent (toluene and xylene) is in use for the production of fatty alcohols on a commercial scale. This process has three distinct advantages over the catalytic hydrogenation : (i) The unsaturation of the fatty acid chain is not affected and hence this process is suitable for making unsaturated fatty alcohols. (ii) The process is carried out at atmospheric pressure and at relatively low temperature as against high pressure (100 atmosphere and above) and high temperature (250°-300°C) employed in the catalytic process. (iii) The quality of glycerol is better in the Bouveault-Blanc process as it has been observed that glycerol undergoes chemical transformations when triglycerides are converted to fatty alcohols by the catalytic hydrogenation process because of high temperature employed.

The catalytic hydrogenation process has been extensively investigated with respect to the choice of fatty material, metal catalyst selection, material of construction and process parameters etc.

As already stated the fatty material is either a fatty acid or its ester. The direct hydrogenation of fatty acids is not much used on a large commercial scale because of the disadvantages that the equipment has to be made of corrosion resistant material and the catalyst used has to be acid-resistant. The quality of alcohol is inferior since there is a higher percentage of hydrocarbon formation.

Lurgi (32) has developed a hydrogenation process which allows the use of fatty acids as feed stock without previous esterification in a separate unit. The fatty alcohol initially formed is recirculated to form alcohol esters of fatty acids (wax esters)*in situ*. The hydrogenation of the wax esters occurs simultaneously in the same reactor. Therefore, the process is not a direct hydrogenation of a fatty acid but the hydrogenolysis of an ester.

Low molecular weight esters of fatty acids, preferably methyl esters, have become the primary feedstock. Today the bulk of natural fatty alcohols is manufactured via the methyl ester route. The fatty alcohols produced from the methyl esters of fatty acids are extremely light coloured, very low in unsaponifiable matter and the catalyst consumption is much less and so also is the reduction of corrosion of the hydrogenation autoclave.

Methyl esters of fatty acids can be made either by esterification of fatty acids or *trans* esterification of fat triglycerides. The predominant process for the manufacture of methyl esters is the *trans*-esterification of fats and oils with methanol.

The metal catalysts used in the hydrogenation reaction include the usual nickel catalyst, but more preferable is the use of the copper chromite catalyst or copper chromite promoted with cadmium, cerium, and so on. The fatty acid salts of a number of metals, including copper, zinc, lead, manganese, cobalt, and mercury, are claimed as catalysts (33).

In general, with the use of copper chromite catalyst the unsaturated chains of the fatty acids get reduced and result in the production of saturated fatty alcohols. If a special catalyst containing zinc is used, the unsaturated esters remain unreduced and are converted into unsaturated fatty alcohols. Copper-cadmium combination catalyst also selectively makes unsaturated fatty alcohols (34, 35).

High-pressure Hydrogenation (36)

High pressure hydrogenation of methyl esters of fatty acid is carried out by suspension process, and fixed bed process. In the suspension process, methyl ester of fatty acid and hydrogen are preheated separately and fed into the reactor. The catalyst (2%

copper chromite), suspended in the ester, is added by a motoring pump. The vertical reactor is tubular and contains no packing. The hydrogenation is accomplished at 250–300 bar and 250°-300°C. Approximately 20 moles of hydrogen per mole of ester are fed into the reactor to serve agitation and reduction of ester bond. The inlet temperature depends upon the degree of unsaturation of the fatty carbon chain. The hydrogenation being exothermic, the reaction of higher unsaturated esters is started at the lower end of the above-mentioned temperature range to avoid undesired side reactions like formation of hydrocarbons.

The reaction mixture is cooled and separated into liquid and gaseous phases. The gas, mainly hydrogen, is recycled and the fatty alcohol-methanol mixture is expanded into a methanol stripping unit. After removal of methanol the catalyst is filtered off. The crude fatty alcohol still contains 2-5% ester which can be removed by addition of alkali to form soap which remains in the bottom of the subsequent distillation. Due to overhydrogenation of the fatty alcohol, up to 2-3% hydrocarbons are formed, lowering the overall yield significantly. Even with recycling, the catalyst consumption is in the range of 0.5-0.7%.

With the use of copper chromite catalyst only saturated fatty alcohols are produced.

In the fixed bed process, the massive catalyst (pelletised or supported type) is packed in the reactor. The ester is preheated and vaporised in 20-50 moles of hydrogen and passed over the fixed bed catalyst. The hydrogenation is accomplished at a pressure of 200-300 bar and a temperature of 200°-250°C. The methyl ester is mixed with fresh and recycled high pressure hydrogen and heated up by heat interchange with the reaction mixture leaving the reactor. The feed is brought up to process temperature by a peak heater and enters one or several reactors from the top. The mainly vaporised material passes over the fixed catalyst bed.

After leaving the reactor, the reaction mixture is cooled and is separated into liquid and gaseous phases in a separator. The gaseous phase consisting mainly of excess hydrogen is recycled and the liquid product stream is expanded into a flash tank in which methanol is stripped from the fatty alcohol. The fatty alcohols do not require further purification.

A comparison of the suspension process and fixed bed process reveals that the capital investment required for the suspension process is somewhat lower than that for the fixed bed process. Costs of additional equipment for separation of the suspended catalyst, distillation of the crude fatty alcohol, and rework of the unreacted methyl ester is more than compensated by larger reaction results, gas recirculation pumps and pipe work in the high pressure part of the fixed bed plant.

As far as the consumption of raw material is concerned, the fixed bed process has an advantage because conversion of the ester into fatty alcohol is almost quantitative with a very low formation of hydrocarbons. The catalyst consumption of the suspension process, which is in the range of 0.4-0.7%, is considerably higher than in the fixed bed process (0.2-0.3%). Energy consumption is estimated to be in the same order of magnitude, the suspension process requiring high temperature, and the fixed bed process having a large power source installed for bigger recirculation pumps and compressors.

The quality of the crude fatty alcohols from the fixed bed process is better with a lower hydrocarbon content and lower saponification values. The fatty alcohol of the suspension process can be brought up to the same quality level which, however, requires an additional purification step.

Production of Amines by Hydrogenation Process (37)

Fatty amines (primary, secondary and tertiary) are commercially produced by catalytic hydrogenation of fatty nitriles.

In the hydrogenation of a nitrile for producing a primary amine in high yield, it is necessary to add ammonia, alkali or metal soaps to suppress the formation of secondary amine. Ammonia acts as the best suppressant. By lowering the temperature and pressure and by using more selective catalyst, the production of a 99% plus yield of primary amine can be achieved. The temperature varies from 120°-180°C and the pressure varies from 20 bar to over 100 bar. The effects of ammonia pressure, hydrogen pressure, temperature, catalyst type and catalyst concentration have been systematically investigated by Engelhard (Research reports from Engelhard dc Meera, Netherlands). It appears from their work that the product specifications of a primary amine can be varied by changing the reaction conditions. For 99% yield of primary amine, the temperature must be lowered and the partial pressure of ammonia has to be increased.

The production of a secondary amine from a fatty nitrile requires high temperature during the catalytic hydrogenation and the avoidance of ammonia in the system. In fact, the byproduct ammonia has to be removed efficiently as the presence of ammonia in the reactor suppresses the formation of secondary amine.

There are many methods of removal of ammonia like (i) intermediate venting of the reactor during the reaction, (ii) continuous venting and (iii) gas circulation through an absorption system.

Engelhard has suggested that the process should be performed in two steps, the first step (hydrogenation) at approximately 150°C and 25 bar and the second step (de-amination) at 220°C and 0-5 bar. Tertiary amines (symmetrical) can be formed directly from nitriles by catalytic hydrogenation and liberation of ammonia or from the mixtures of secondary amines and nitriles.

Hydrogenation of Fatty Acids

Hydrogenation of fatty acids is carried out in the same kind of autoclave built of 316 stainless steel or other corrosion resistant alloys. As fatty acid hydrogenation leads to formation of water at the operating temperatures, it becomes necessary to be extremely careful in conducting the hydrogenation reaction.

The pressure build up is much more and the autoclave is designed for working pressures of 200-300 pounds/sq. inch, almost twice the strength of the autoclaves used in the hydrogenation of fats and oils. Hydrogenation of fatty acids requires very high amount of nickel catalyst, nearly ten times the normal requirement to hydrogenate natural triglycerides. The temperature at which hydrogenation is carried out is very important. It is normally much lower than that used for the triglycerides. The temperature below 150°C is preferred. The pressure is kept much higher to promote the rate of hydrogenation.

Fatty acids are always less selectively hydrogenated than tri-glycerides because of the relatively much greater adsorption affinity on the catalyst surface than the active methylene groups of the fatty acids in the esterified forms.

Commercial stearic acid is major product of fatty acid hydrogenation industry and the free fatty acid source is the Fatty Acid Distillates (FAD) from the physical refining plants.

REFERENCES

1. Bailey, A.E., JAOCS, **26**, 644 (1949)
2. Albright, L.F., JAOCS, **40**, 16 (1963)

2(a). H.B.W. patterson, Hydrogenation of Fats and Oils, Applied Science Publishers,London, 1983, p.22- 25, 76- 7, 102- 103.

3. Coenen, I.W.E., JOTAI, July-Dec., pp. 16-18 (1969)
4. Achaya, K.T. and D.S. Raju, *J. Sc. Ind. Res.* **29**, 68 (1970)
5. Klimmek, H., JAOCS, **61**, 201 (1984)
6. E.D. De Hault and A. Demoulin, JAOCS, **61**, 195-200 (1984)
7. Allen, R.R., JAOCS, **46**, 552 (1969)
8. Stingley, D.V. and Wyobel, R.J., *ibid*, **38**, 201-305 (1961)
9. Lush, J., *Soc. Chem. Ind.* **42**, 219T (1923)
10. Bolten, J., *Soc. Chem. Ind.* **46**, 444T (1927)
11. Leuteritz, G. *Fette Seifen Anstrichmittel*, **71**, 441 (1969)
12. U.S. Patent 3,792,067 (1974)
13. DE 2207909 C 2 Aug. 8, 1983
14. Hastert, R.C., JAOCS, **58**,169 (1981)
15. S. Minowa, T. Tanaka and N. Yasuda, U.S. Patent 4183971, Jan. 15, 1980
16. S. Minowa, Y. Toyoshima, N. Yasuda and T. Tanik, U.S. Patent 4,108,879, August 22, 1978
17. Bailey's Industrial Oil and Fat Products, Vol. 2, 4th edition, Edited by D. Swern, p.52, Wiley-Interscience Publication (1982)
18. Bailey's Industrial Oil and Fat Products, Vol. 3, Edited by T.H. Applewhite, p.70, Wiley-Interscience Publication (1985)
19. E.G. Latondress, *J. Am. Oil Chemists' Soc.*, **58**, 185-187 (1981)
20. Heckers, H., M. Korner, T.W.L. Tuschen, and F.W. Melcher, *Atherosclerosis* **28**, 389 (1977)

21. Thomas, L.H., P.R. Jones, J.A. Winter and H. Smith, *Am. J. Clin. Nutr.* **34,** 877 (1981)
22. Hontsmiiler, U.M.T. *Fette Seifen Anstrichmittel,* **80**, 162 (1978)
23. Kinsella, J.E., G. Bruckner, J. Mai and J. Shemp, *Am. J. Clin. Nutr.* **34**, 2307 (1981)
24. INFORM, 5 (No. 6), 670, (June 1984)
25. Basu, H.N. and Chakrabarty, M.M., JAOCS, **43**, 119 (1966)
26. Chakrabarty, M.M., Bhattacharyya D. and Basu, A., IXth ISF Congress, Rotterdom, III-1 (1968)
27. M.M. Chakrabarty, D. Bhattacharyya and A. Basu, Industries *Chemique Belge-T,* **33**, 883-886 (1968)
28. Albright, L.F., Wei, C.H. and Woods, J.M., *J. Am. Oil Chemists' Soc.*, **37**, 315-320 (1960)
29. Kaufmann, H.P., U.S. Patent, 2,852,541 (1958)
30. Cousins, E.R. and R.O.Feuge, *J. Am. Oil Chemists' Soc.*, **37**, 435-438 (1960)
31. Chakrabarty, M.M. and D. Bhattacharyya, *Fette Seifen Anstrichmittel,* **70**, 716-718 (1968)
32. Buchold, *H. Chem. Eng.* **90**, 42 (1983)
33. Schrauth, W. and T. Bottler, U.S. Pat. 2,023,383 (1935)
34. Pentula, A.J. and K.T. Achaya (unpublished results)
35. Richter, J.D. and P.J. Vandenburg, *J. Am. Oil Chemists' Soc.* **46**, 155-156 (1969)
36. Kreutzer, U.R., JAOCS, **61**, 345-347 (1984)
37. Billenstein, S. and G. Blaschke, *ibid,* 353-354 (1984)

CHAPTER 12

Fractionation

Fats, oils, fatty acids and other fatty derivatives are modified by a number of fractionation techniques into new kinds of products whose compositions and properties become different from the original (unfractionated) ones, thereby offering newer and better avenues of the fractionated products in edible and industrial utilisations. The changes in the composition occur in the fatty acids, triglycerides, phospholipids, and other nonlipids depending on the nature of the fatty materials undergoing the fractionation by various techniques. In fact, on realisation of the importance of correlation between the composition and properties of fatty materials, the fractionation techniques have been the subject of extensive investigations. Modifications of fats, oils, fatty acids and other fatty materials by fractionation technologies appear to be accepted as the convenient and commercially viable approaches.

FRACTIONATION TECHNIQUES

Fractionation techniques include almost exclusively fractional crystallisation although super critical extraction and liquid-liquid extraction are also used.

FRACTIONAL CRYSTALLISATION

Fractionation by crystallisation process can be operated in three different ways, namely, (1) Dry fractionation, (2) Solvent fractionation, and (3) Detergent fractionation in aqueous medium.

The fractionation process is chosen depending on (i) the molecular nature and composition of the fatty materials to be fractionated, (ii) the degree of fractionation and purity of the fractions needed and (iii) the cost involved in carrying out a fractionation.

In view of the operating and production cost for recovery of solvent and the associated environmental pollution problem, the solvent process of fractionation, in spite of high separation efficiency, is receiving less importance. The process is, however, adopted only when the products of fractionation are highly value added. Multistage dry fractionation process is gaining more and more importance over the solvent process in producing valuable fractionated products. Only in the case of fatty materials like phospholipids, the solvent fractionation process appears to be the only way.

The dry fractionation process works well on fats and oils having large amounts of both solid and liquid fractions. The dry fractionation, or the winterisation technique as it is called, is applied for removing high melting low solid material from a large quantity of liquid fraction. The dry fractionation process also works very well on the separation of a small amount of a liquid from a large amount of solid, depending on the choice of the pressing technique adopted.

Fractionation of fatty materials like fats and oils (natural triglycerides), fatty acids, monohydric alcohol esters of fatty acids etc. can be achieved by aqueous detergent crystallisation process at low temperatures. The process has considerable promise and has been adopted on commercial scale in fractionating many fats, oils and fatty acids with enormous success and economic advantages.

However, the aqueous detergent process also has certain limitations. It fails to work when applied to fatty materials having a large amount of solid at temperatures at which the crystallisation is carried out in presence of an aqueous detergent solution.

The process as a whole is more advantageous than the dry process as the separation of fractions is better and more complete and the time required is also much less. The overall output is much more and the energy cost is also less.

Certain basic aspects and factors are involved which are always to be attended to for achieving efficient fractionation irrespective of the fractionation process. The factors are:

(i) Rate of cooling of molten and homogeneous fat phase so as to obtain the right size, shape and number of the crystals for achieving efficient separation of the liquid fraction.

(ii) Proper stabilisation or tempering of the fat phase so as to obtain the most stable polymorphic form or the crystalline phase that will result in the formation of large crystals.

(iii) Proper selection of supercooling point so as to get the right temperature for nucleation of crystals.

In any kind of crystallisation process there are three steps involved in sequence:

(i) Crystallisation by cooling of a liquid oil or a melted fat to the state of nucleation.

(ii) Continued crystallisation to proper growth of crystals in respect of size, shape and crystals number.

(iii) Separation of solid crystals and liquid fractions and purification of the two fractions, if needed.

(1) FRACTIONAL CRYSTALLISATION BY DRY PROCESS

Dry fractionation consists of crystallisation of fats, oils or fatty acids at low temperatures without any additives. The success of dry fractionation depends on the efficiency of separation of the crystallised mass into the solid (stearin) and liquid (olein) fractions. The efficient separation can be carried out by (i) filtration under vacuum on a specially designed filter, (ii) filtration on a filter press equipped with membrane plates, i.e., a membrane filter and or (iii) hydraulic pressing.

Tirtiaux (1) has developed a filtration unit that has been commercially adopted in the fractionation of palm oil into palm olein and palmstearin. The Florentine filter used by Tirtiaux operates in a unique manner. The filtration takes place horizontally and continuously on endlessly rotating, stainless steel, perforated belt under a slight vacuum. The filter is self-cleaning and the filtration area is enclosed and air conditioned. The Florentine filter is shown in Fig. 12.1.

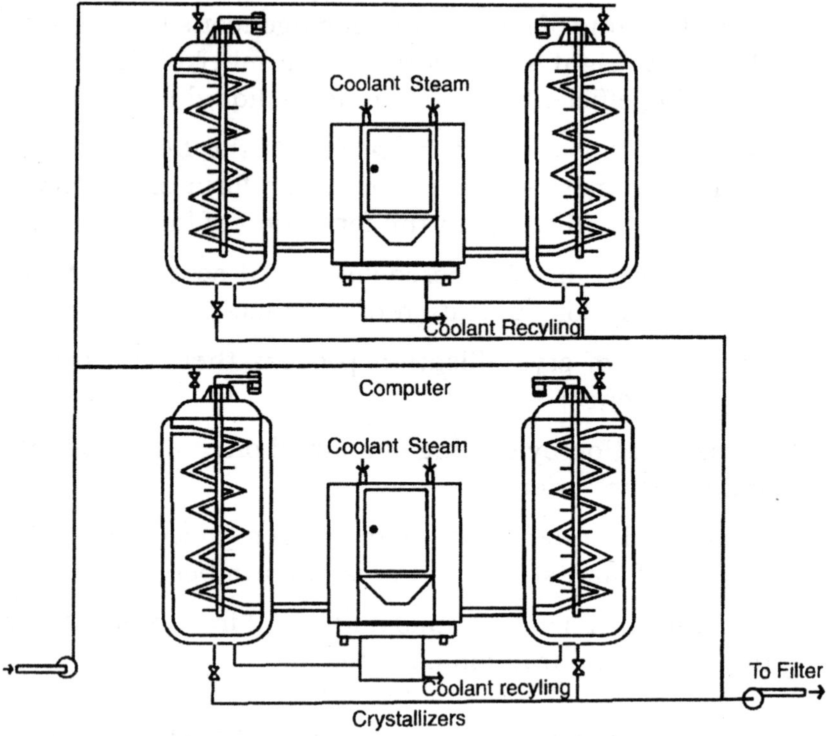

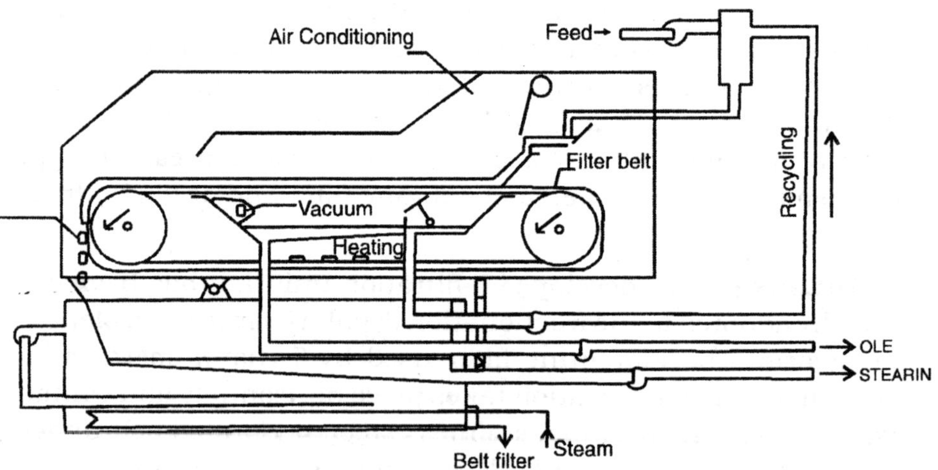

Fig. 12.1: Tirtiaux : Fractination

The oil is crystallised relatively slowly to control the latent heat of crystallisation and to avoid supercooling. The oil is fed into agitated crystalliser tanks fitted with coils and jackets in which the crystals are melted and then allowed to form and grow. The oil mixture is maintained at the fractionation temperature until olein is separated from stearin.

The commercially available processes also include Bernardini (C.M.B.) (2) and De Smet (3) for the fast dry process.

In the Bernardini process, palm oil is cooled in a batch crystalliser fitted with three thermo regulated water tanks at different temperatures. The water from each tank is made to circulate in sequence for a preset period of time. In a semi-continuous version, palm oil is agitated for a pre-established time into four crystallisers thermoregulated from 35°C down to 16°C. The filter is a radial filter where separate filter elements radiate out from a slowly rotating horizontal shaft dipped in the filter chamber in which the palm oil slurry is kept at a constant level. Vacuum is applied to the elements of the filter which become submerged in the liquid phase and olein is sucked leaving a coating of stearin on the surface which is expelled from the filter leaves by compressed air.

De Smet process involves a rapid cooling of the melt to produce crystals of stearin. A preliminary cooling of oil is achieved to reach the temperature at which nuclei are formed. The oil is then cooled to 19°C in a crystalliser provided with a very large cooling surface in comparison with the volume of oil treated. This ensures total exchange with a very small differential temperature between oil and cooling medium.

Filtration is achieved by means of a Stockdale type rotary filter in which the cooled palm oil is fed into a bottom tray in contact with the filter cloth. A vacuum sucks in the olein and the stearin sticks to the filter cloth to be cleaned subsequently by a scraper.

Membrane filters are now available commercially for separating crystals (stearin fraction) from liquid (olein fraction) part. In fact, membrane filters are now replacing the traditional filters used in a number of dry fractionation technique. The efficiency of separation depends on the formation of small, hard and well defined (formed) crystals. The filtration is done on a filter press equipped

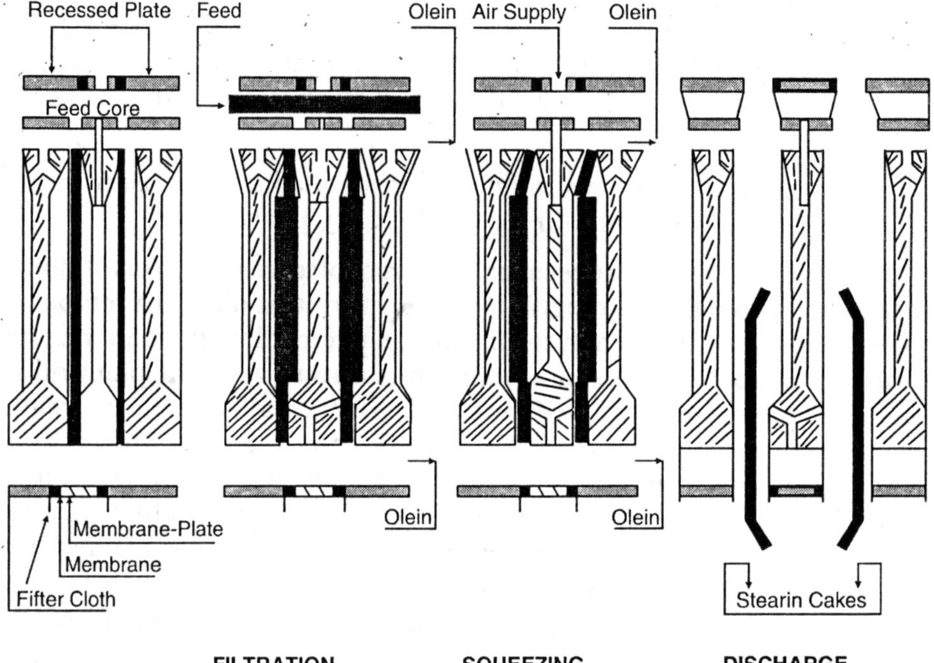

Fig. 12.2: Filters with Membrane Plates.

with membrane plates (Fig. 12.2). The membrane filters are automatic and the filter feed pressure is kept rather low (1 to 1.5 bar); as a result, proper filtration is short and the principal aim at the end of such a step is that each chamber should be full of crystals.

Each chamber is equipped with filter cloths and with one or two membranes made of flexible material (rubber or polypropylene), behind which a high pressure (4 to 6 bar) may slowly be applied by means of a gas or a liquid, but generally with compressed air. At this stage, the interstitial liquid (olein) between the entrapped crystals is expelled first. As the pressure reaches higher values, the olein holding to the crystals by a skin effect will be squeezed out leaving a hard dry cake. The next step is to prepare the filter for cake discharge, firstly by expelling the content of the feed core and then by blowing the olein wetted surfaces and liners. The automatic discharge of the stearin cakes is then started. It should be designed to avoid any disorderly fall-off and unnecessary filling.

Depending upon the type of feed stock fractionated a rinsing operation of the cloths will be effected every 10 to 50 cycles. This rinsing is generally done by circulating hot olein through the filter. The fact that the membrane filter operates with rather high pressure makes it best suited for the separation of small, hard crystals, especially when olein percentage is of great importance.

The characteristics of oleins and stearins obtained from single stage fractionation of palm oil by the above three processes are presented in Tables 12.1, 12.2 and 12.3.

The increasing demand for super oleins and palm mid-fractions has led to the two stage dry fractionation. The physico-chemical characteristics of super oleins and palm mid-fractions by double stage fractionation have been shown in Table 12.4.

Table 12.1: **Tirtiaux Oleins and Stearins from Single Stage Fractionation of Palm Oil**

	RBD (IV 51) Indonesia		*Neutralised (52-53) Malaysia*			
Characteristics	*Stearins*	*Oleins*	*Stearin*	*Olein*	*Stearin*	*Olein*
1	*2*	*3*	*4*	*5*	*6*	*7*
IV (GLC)	39.4–40.4	57.6–58.2	28.6	57.2	40.1	57.2
Yield (%)	30–33	67–70	32	68	28	72
Slip melting point (°C)	50.7–51.4	19.9–23.0	50.7	22.2	49.7	21.5
SFC (%) by pulsed NMR						
5C	82.1–83.0	55.9–56.0	80.3	55.7	76.9	58.4
10C	77.2–78.8	35.2–38.9	76.0	40.0	70.1	34.2
15C	68.7–70.5	15.7–19.8	66.3	19.3	61.2	19.7
20C	58.0–59.9	0–1.8	54.7	0.2	48.7	0.5
25C	46.1–48.9	—	43.1	—	38.0	—

Table 12.1. Contd...

Table 12. 1. Contd...

1	2	3	4	5	6	7
30C	35.0–38.1	—	31.9	—	28.0	—
35C	24.2–28.4	—	23.2	—	19.7	—
40C	18.3–21.1	—	16.9	—	14.3	—
45C	12.3–15.4	—	11.0	—	9.3	—
50C	5.9–7.9	—	4.8	—	3.2	—
55C	—	—	—	—	—	—
Fatty acids (% wt as Me)						
C 16:0	54.6–55.5	39.4–40.2	55.5	39.5	55.2	39.0
C 18:0	4.9	3.8–4.0	4.9	4.6	4.9	4.4
C 18:1	29.9–30.7	41.7–42.3	29.1	41.3	29.9	41.7
C 18:2	7.5–7.6	11.8–11.9	7.2	11.7	7.8	11.6
Diglyceride (% wt.)	4.3–4.5	6.5–6.7	4.7	6.6	4.6	5.8
Triglyceride (% wt.)	95.5–95.7	93.3–93.5	95.3	93.4	95.4	94.2
Triglyceride composition (area %) by HPLC						
S_3	23.3–26.5	0–0.6	21.9	0.4	20.2	0.2
S_2U	44.5–48.1	47.8–48.8	47.0	48.1	45.6	48.7
SU_2	25.3–25.4	44.1–45.4	27.1	44.7	29.7	44.3
U_3	3.3–3.6	6.5–6.8	4.0	6.8	4.5	6.8
Cloud point (C)	—	6.8–9.5	—	8.7	—	8.5
Cold test at 22 C (days)	—	2.5–7 (Minimum)	—	5	—	2

Table 12.2: C.M.B. Oleins and Stearins from Single Stage Fractionation

	RBD Palm Oil					
	Indonesia (IV 51)		*Malaysia (IV 52–53)*			
			Continuous		*Batch*	
Character-istics	*Stearin*	*Olein*	*Stearin*	*Olein*	*Stearin*	*Olein*
1	*2*	*3*	*4*	*5*	*6*	*7*
IV(GLC)	44.3	57.1	46.7	58.0	47.8	58.6
Yield (%)	37	63	40	60	40	60
Slip melting point (C)	48.6	24.5	48.2	24	47.7	23
SFC(%) by pulsed NMR						
5C	75.9	52.8	75.3	54.7	74.2	48.3
10C	68.6	36.9	66.7	37.0	65.8	36.4
15C	59.2	19.9	55.5	17.4	53.9	18.6
20C	47.6	3.0	44.0	3.0	42.2	2.2
25C	36.0	—	32.4	—	30.0	—
30C	26.1	—	23.4	—	21.4	—
35C	18.5	—	16.2	—	14.3	—
40C	13.1	—	9.5	—	8.3	—
45C	8.0	—	5.7	—	4.3	—
50C	2.1	—	0.8	—	—	—
55C	—	—	—	—	—	—
Fatty acids (% wt. as Me)						
C 16:0	51.4	40.4	49.3	39.5	48.4	39.0
C 18:0	4.7	4.3	4.9	4.3	4.9	4.4
C 18:1	32.8	41.1	33.8	41.5	34.5	41.7
C 18:2	8.7	11.9	9.6	12.2	9.9	12.4
Diglyceride (% wt.)	5.1	6.2	5.2	6.9	5.5	6.8
Triglyceride (% wt.)	9.9	93.8	94.8	93.1	94.5	93.2
Triglyceride composition (area %) by HPLC						
S_3	19.5	0.4	17.3	0.5	16.0	0.4
S_2U	46.4	50.0	45.9	48.2	45.8	48.0
SU_2	30.0	42.8	31.6	44.3	33.0	44.5
U_3	4.1	6.8	5.2	7.0	5.2	7.1
Cloud point(C)	—	10	—	10.1	—	9.6
Cold test at 22 C (days)	—	1	—	1	—	2

***Table 12.3*: De Smet Oleins and Stearins from Single Stage Fractionation of Palm Oil**

	RBD Palm Oil (IV 52–53) Malaysia	
Characteristics	*Stearins*	*Oleins*
IV(GLC)	44.9–48.7	57.6–58.4
Yield %	37–40	60–63
Slip melting point (C)	45.9–48.8	23.3–24.2
SFC (%) by pulsed NMR		
5C	72.4–77.2	54.1–55.7
10C	61.9–69.6	36.8–39.0
15C	50.7–58.9	17.2–20.0
20C	38.6–47.5	2.5–9
25C	27.0–36.2	
30C	18.1–26.4	
35C	12.4–18.9	
40C	6.2–12.9	
45C	2.0–8.5	
50C	0–2.2	
55C	–	
Fatty acids (% wt. as Me)		
C 16:0	47.4–50.6	39.2–39.8
C 18:0	5.0–5.2	4.4–4.5
C 18:1	32.9–35.3	41.4–41.7
C 18:2	9.0–10.0	12.1–12.4
Diglyceride (% wt.)	5.3–5.7	6.8–7.0
Triglyceride (% wt.)	94.3–94.7	93–93.2
Triglyceride composition (area %) by HPLC		
S_3	13.8–19.3	0.6–0.7
S_2U	46.0–46.2	49.7–48.1
SU_2	30.0–34.6	44.1–44.6
U_3	4.7–5.4	6.7–7.3
Cloud point (C)	–	10.2–10.3
Cloud test 22 C (days)	–	1.2–1.5

Table 12.4: Physico-Chemical Characteristics of Super Oleins and Palm Mid Fractions (4)

Physico-Chemical characteristics	*Super oleins*	*Palm mids*
Iodine value (GLC)	60.7–62.9	35.4–49.9
Slip melting point (°C)	15.0–18.2	28.6–33.8
Cloud point (°C)	4.8–5.7	2.1–6.7
Cloud test at 16°C (hrs)	40–54	62.1–92.7
		4.7–31.5
SFC (%) by pulsed NMR		
5C	39.6–57.0	71.1–93.7
10C	10.3–33.5	60.3–88.3
15C	0.3–7.9	45.8–80.9
20C	–	26.2–69.4
		4.0–31.3 (25°C)
		0.4–11.5 (30°C)
		0.3–5.0 (35°C)
		0-0.6 (40°C)
Fatty acids (% w/w as Me)		
$C_{16:0}$	34.9–37.3	
$C_{18:0}$	3.7–4.1	
$C_{18:1}$	43.8–46.9	
$C_{18:2}$	12.1–13.4	
Diglyceride (% wt.)	6.3–9.4	2.5–5.6
Triglyceride (% wt.)	90.6–93.7	
Triglyceride composition (area %) by HPLC		
		PPS 0.2–1.0
S_3 PPP	0–2	PPP 1.5–4.8
		MPP 0.3–0.9
		SOS 0.6–3.1
S_2U POS	3.0–4.8	POS 7.5–14.1
POP	21.3–24.8	POP 43.1–69.2
		MOP 1.1–1.3

Table 12.4 .Contd.

Table 12.4: Contd.

Physico-Chemical characteristics	*Super oleins*	*Palm mids*
PLP	11.0–11.9	PLP 7.5–9.1
MLP	0.4–0.6	
SU_2 SOO	3.4–4.0	SOO 0.2–1.8
POO	30.1–33.8	POO 3.4–18.8
PLO	13.2–14.7	PLO 1.0–8.6
PLL	2.7–4.1	PLL 0.1–2.3
U_3 OOO	4.8–6.1	OOO 0.6–2.6
OOL	2.2–2.8	OOL 0.6–1.3
LLO	0.5–0.6	LLO 0.2–0.4
Triglyceride composition (% wt.) by GLC		
C_{46}		0.5–1.1
C_{48}		3.5–6.5
C_{50}		50.6–70.5
C_{52}		19.7–36.3
C_{54}		4.0–8.6
C_{56}		0.2–0.5
$C_{50}/C_{48} + C_{54}$		4.2–8.2

HYDRAULIC PRESSING

The dry fractionation has been done by applying pressure on the crystallised mass in a hydraulic press till the separation of stearin and olein fractions is achieved. Fat or fatty acid samples are at first completely melted and taken into small shallow pans and are solidified slowly to a particular temperature in a refrigerated/air conditioned room. The choice of crystallisation temperature and the time of solidification depend on the nature of solid content at low temperatures and on the polymorphic behaviour of the fatty material. The solid mass or cells from the different pores are wrapped in press cloths and subjected to pressure in a hydraulic press. The pressure applied is generally very moderate and it may vary from 0.5 ton/sq. in. to 1 ton/sq. in. and the time for pressing also varies depending on the degree of fractionation desired and the temperature at which the pressing is done.

Cake/solid (stearin) fraction from the pressing is melted, resolidified at room temperature or at lower temperature and again pressed in steam heated presses. The stearins or stearic acid and the oleins or oleic acids, depending on whether fats/triglycerides (oils/fats) or fatty acids have been fractionated, are used as such or following further bleaching.

Some examples of fractionation of fats are detailed below :

1. Fractionation of Sal fat (5)

Refined and bleached sal fat (Iodine value 45.5)

Rise in temperature in the cooling curve = 6.6°C).

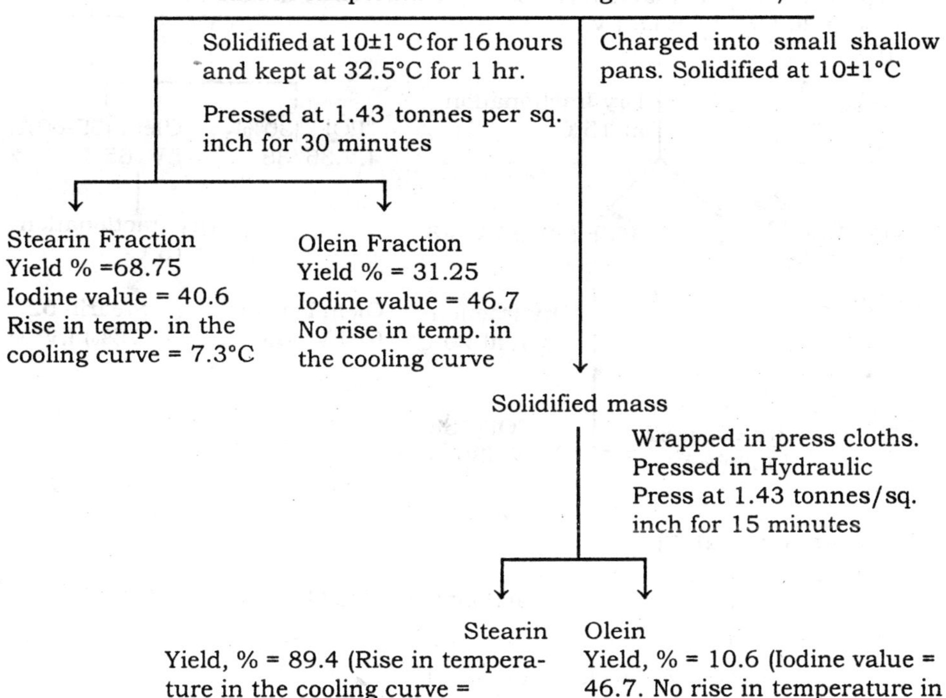

2. Palm Oil fractionation by pressing in a hydraulic press (6)

Palm Oil PO

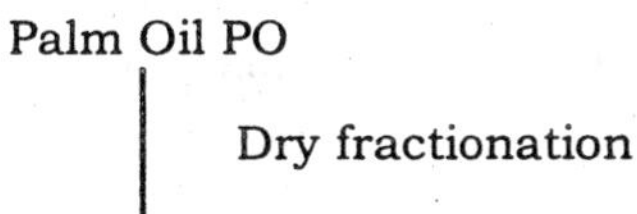

High melting stearin I.V. 28 (20%) | First oleine (80%) 58/59 I.V.

Olein 30%

Palm mid fraction
PMF 48 (50%)

3. Palm Oil 53/54 I.V. (100) (6)

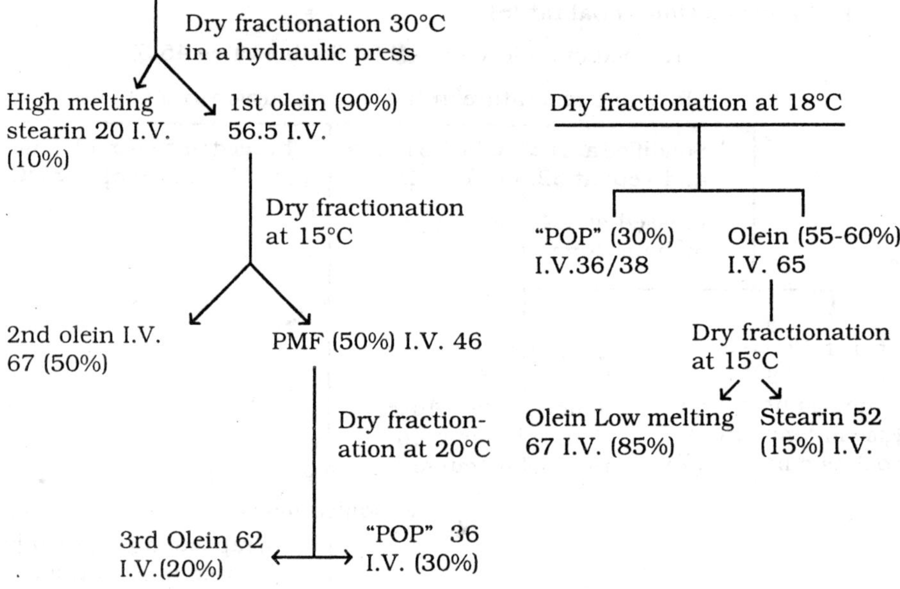

4. Hydrogenated Sal (7)

↓ Dry fractionation 37°C/48 hrs.

Olein 1-85% yield (M.pt. 37°C)
I.V. 33-36 CBS

Stearin 1-15% yield By product

5. Hydrogenated Sal (7)

↓ Hydraulic presses 200-700 psi/20° = 25°C/30-60 minutes

Olein : 2 to 10%
Stearin : 90-98% (compatible with cocoabutter)

FRACTIONATION OF FATTY ACIDS BY DRY FRACTIONATION

The panning and pressing process as applied to fats is also applicable for fractionating fatty acids effectively. In this process mixed fatty acids (fatty acid mixtures) are melted and poured into trays or flat bottom pans and placed in a cold room to crystallise. Normally the pans are placed one above the other on special frames. The melted fatty acids are poured in the upper pan from where the fatty acids overflow in the pans below until all the pans are charged. The pans are then taken in the cold room and kept for a definite period till the fatty acid mixture solidifies. The solidified mass (formed cakes) is wrapped in cloth and placed in a hydraulic press. Pressure is then applied to squeeze out the unsaturated acid oleic acid while the saturated acid (Palmitic or Stearic) remains behind.

The liquid phase is mixed with the feed stock. The solid fraction (single pressed cake) can be remelted, solidified and pressed two or three times in steam heated presses till the desired quality of the stearic acid is obtained.

The double pressed stearic acid with an iodine value of 5-8 can be obtained. The double pressed stearic acid when triple pressed by repeating the solidification and pressing yields stearic acid that has an iodine value of 1 only. This is called the triple pressed grade of stearic acid. The yield of such a high quality of stearic acid is 30 per cent. The yield of oleic acid is also about 30% with a titre of 6-7°C.

(2) SOLVENT FRACTIONATION OF FATS

Fractionation of fats, fatty acids and other lipid derivatives like phospholipids from solvents have been extensively studied. The choice of a solvent system is quite critical and the factors like vapour pressure, latent heat, specific heat, toxicity and cost are important considerations. The types of glycerides present and the fractionated product characteristics desired also influence the selection of a solvent.

The solvent can be polar or non-polar or a combination of both. The polar solvents are mostly acetone, 2-nitropropane as well as isopropanol. The non-polar solvent includes normal hexane. The mixed solvents include isopropanol and acetone, hexane and acetone or any other combination. The polar solvent containing a definite amount of water shows a typical effect on fractionation of fats or fatty acids. Non-polar solvent needs relatively lower temperature than a polar solvent because of higher solubility of triglycerides in a non-polar solvent.

The solvent fractionation can be single stage or multi-stage depending on the nature of the fat and the products to be produced.

The solvent fractionation was extensively investigated by Hilditch *et al* (8) on various fats and oils including palm oil, animal fat (9,10) and some seed oils (11-16, 16a) wherein the solvent like acetone was mostly used at different low temperatures.

The solvent fractionation was primarily aimed at producing cocoabutter substitutes (CBS) from vegetable fats as such or when suitably modified at inter-esterification reaction. Sal stearin has been obtained with thermal properties comparable to those of cocoabutter by fractionating sal fat from acetone (17-20) and from hexane by two stage crystallisation (21). The properties of cocoa butter, sal fat, sal stearin and sal mid (20) are presented in the Table 12.5.

***Table 12.5* : The Physical Properties of Cocoabutter, Sal Fat, Sal Stearin and Sal Mid**

Properties	*Cocoabutter*	*Sal fat*	*Sal Stearin**	*Sal mid***
Yield %		–	66.7	45.0
S.V.	196.0	189.0	—	—
I.V.	37.6	38.9	29.1	31.3
Melting point °C	34.5	36.5	37.2	35.5
Super cooling Temp. °C	20.8	22.2	24.0	22.0
Rise in temperature on solidification, °C	8.0	6.7	8.7	9.0
Solid content index at 30°C	61.9	59.5	65.0	68.3
33°C	16.9	39.5		45.7
36°C	0.0	1.1	6.4	2.0
38°C	0.0	0.0	2.6	0.0

* Sal stearin obtained by crystallisation from 7 volume acetone at 10°C for 3 hours under slow agitation.

** Sal mid obtained by first cooling to 16°C till turbidity from 10 volume acetone and temperature raised and kept at 21°-22°C for 6 hours and filtered (25% solids topped). The filtrate cooled to 5°-6°C and kept for 3½ hours to get 45% mid fraction.

Mowrah (*Madhuca latifolia*) fat has been crystallised from acetone in making cocoabutter substitute (22). Kokum fat has been inter-esterified with mowrah in the ratio of 3:2 and then fractionated from acetone (23). The mid fraction had similar physico-chemical properties as those of cocoabutter shown in paranthesis : I.V. 37.7 (37.5) M.P. 37.5°C (34.5°C), super-cooling temperature 24.1°C (20.8°C), ΔT 7.2°C (8°C), solid content index at 30°C, 65.2 (61.9), 35°C, 38.2 (2.3), 37°C 11.8 (0).

Bringi and Padley (24) fractionated at 15°C a 20 per cent solution of crude mango kernel fat in acetone and obtained a yield of 22 per cent of crude, slightly off white grey coloured stearin with the following characteristics:

	Mango fat	*Stearin*	*Olein*
I.V.	46.3	27.3	51.9
Diglyceride, wt.%	8.4	0.8	8.5
FFA %	25.9	0.48	31.6
Combined fatty acids, wt. % :			
C_{14}	0.1	0.2	0.3
C_{16}	7.6	4.5	9.0
C_{18}	43.7	58.8	36.9
C_{20}	2.3	3.5	3.1
$C_{18:1}$	41.8	32.1	44.4
$C_{18:2}$	4.5	0.9	6.3
Jensen temperature (max) °C	–	37.1	–
Jensen temperature (min) °C	–	28.2	–
Minutes to maximum	–	49.0	–
D_t values:			
D_{25}		2055	
D_{30}		1880	
D_{35}		495	
D_{40}		95	

Acetone fraction (25) (1:7 w/v; 7°C) of a sample of refined fat from Alphonso variety (I.V. 46) gave 60 per cent yield of stearin with D_{20} 2300, D_{30} 2260 and D_{35} of about 1800 similar to that of shea or sal stearin. A D_{35} of about 1500 was obtained with some other commercial samples as well indicating that the fat is a close match to shea or sal as a confectionery raw material.

The stearin product produced chocolate compositions of excellent characteristics:

Ingredients, wt.%	*Chocolate* *dark*	*type* *milk*
Cocoa mass	40.0	10.0
Cocoa butter	6.5	17.5
Mango stearin	5.0	5.0
Full cream milk powder	–	22
Sugar	48.1	45.1
Lecithin	0.4	0.4
	100.0	100.0
Fat content, wt.% of total	33.5	33.5
Cocoabutter, wt.% of fat	85	69
Mango component	15	16
Milk fat	–	15

Baliga and Shitole (26) also fractionated the fat from acetone using 7 ml/g (10°C/four hour) and obtained stearin (I.V. 34.7; slip point 38.6; T max 36.1; (T 9.3) (Fraction 1) in 57.2 per cent yield. Two-stage fractionation was done by first topping at 26°C followed by further cooling at 5°C. The fraction II was obtained in 43.1 per cent yield (I.V. 35.8; slip point 37.2; T max 34, ΔT 8.7).

Fraction I and II and original mango fat were mixed with equal amounts of pre-refined and bleached cocoabutter yielding products I, II and III. The properties were as follows:

Characteristics	Cocoa butter	Mango fat	Product I	Product II	Product III
I.V.	34.5	48.7	34.4	37.1	41.4
Slip point (°C)	35.5	35.7	37.3	36.3	36.6
Sap value	196.5	191.0	193.6	194.0	193.3
Solidification test					
T max	29.0	30.6	33.3	31.8	29.2
ΔT	6.5	9.4	7.9	7.7	6.0
SFI :					
25°C	83.5	54.0	89.0	80.0	64.5
30°C	64.0	44.0	78.0	64.0	44.0

The properties of products I and II are closer to those of cocoabutter, and the study further indicates that the fractions are compatible with cocoabutter.

Ghosh Chaudhari *et al* (27) also fractionated the fat from acetone (1:7 w/v; 50°C) to obtain fraction I in 77 per cent yield (I.V. 44). Straight fractionation at 15°C gave 50 per cent yield of stearin (Fraction II, I.V. 37.4). A mid fraction III was obtained by topping at 20°C followed by cooling at 10°C and by collecting the solids. The characteristics are :

Characteristics	Mango fat	Fraction I	Fraction II	Fraction III
Slip point, °C	37.5	34.5	36.7	32.8
Solidification tests (17°–bath)				
T max	26.6	29.3	33.0	27.8
ΔT	5.1	6.4	8.7	6.5
SFI °C				
20	55.0	64.7	83.5	57.2
30	53.7	48.0	74.3	26.7
35	0	10.2	31.4	traces
Fatty acid composition				
16:0	5.5	5.2	6.7	6.2
18:0	41.1	47.7	52.7	41.2
20:0	1.6	2.1	0.9	2.4
22:0	0.2	0.2	0.3	0.6
18:1	46.8	42.4	38.3	47.5
18:2	4.2	2.0	1.1	1.8
18:3	0.6	0.4	–	0.3

Fractionation of the fat from n-hexane, which is the extraction solvent, has been studied by Mondal (28) under different fat and solvent ratios, temperature and time. The different fractions have been characterised in terms of yield, slip point, and dilatations. The characteristics of some selected fractions are shown in Tables 12.6 and 12.7.

***Table 12.6*: Yield, Slip Point and Fatty Acid Composition of the Fractionated Mango Products from Hexane**

Sample	*Fractionation condition*	*Yield %*	*Slip point °C*	*Major fatty acids (% w/w)*				
				$C_{16:0}$	$C_{18:0}$	$C_{20:0}$	$C_{18:1}$	$C_{18:3}$
Mango oil	–		31.5	5.7	48.8	1.2	41.4	2.9
Product 1	10°C/6 hr.	43.7	34.0	7.0	53.9	1.6	35.2	2.3
Product 2	(i) 10°C/6 hr.	46.0	33.8	9.0	47.2	1.8	39.5	2.5
	(ii) 5°C/2 hr.							
	(iii) 1°C/4 hr.							
Product 3	(i) 10°C/4 hr.	47.5	33.1	7.6	42.0	2.5	42.8	5.1
	(ii) 0°C/8 hr.							

Table 12.7: Dilatation Values of the Fractionated Mango Products from Hexane

Sample	D_{20}	D_{30}	D_{35}
Oil	1372	843	0
Product 1	1899	1386	1010
Product 2	1928	1643	1044
Product 3	1956	1615	1185
Cocoabutter	1720	815	0

D_{40} nil except product 4 with a value of 40.

It appears from the results that high quality confectionery fats useful for preparation of cocoabutter can be produced from mango kernel fat by one stage fractionation or in two stages at first at 10°C to remove a portion and then at combined temperature of –5°C and 0°C for stipulated period. Mondal (28) also fractionated mango kernel fat from isopropanol (fat: solvent ratio 1:3 at 10° and 20°C for periods ranging from one to three hours and fat: solvent ratio 1:4 and 1:1 at 15°C for two and one hours respectively) and recorded yields and the thermal properties like slip point, solidification and dilatation as well as the fatty acids and carbon number triglyceride composition of the products. Some of the fractions exhibited satisfactory properties for use as cocoabutter substitute (Tables 12.8 to 12.11).

Table 12.8: Yield of Mango Stearins Obtained by Fractionating Mango Fat from Isopropanol

Fractionation temperature and time	*Fat : Isopropanol*	*Stearin % w/v*
20 ± 1°C, 2 hr	1:3	47 P_2
10 ± 1°C, 1 hr	1:3	55.4 P_3
10 ± 1°C, 3 hr	1:3	68 P_4

Table 12.9: Fatty Acid and Carbon Number Triglyceride Composition of Fractionated Products Obtained from Isopropanol

Fat/products	*Fatty acid (% w/w)*					*Carbon number of glyceride composition*			
	16:0	*18:0*	*20:0*	*18:1*	*18:2*	C_{50}	C_{52}	C_{54}	C_{56}
Mango fat	5.7	48.8	1.2	41.4	2.9	1.5	22.3	75.5	0.7
P_2	3.7	53.3	1.1	39.9	2.0	–	–	–	–
P_3	4.0	54.2	1.2	38.3	2.3	1.8	24.7	69.2	4.3
P_4	9.1	45.8	3.2	37.5	4.9	1.8	24.8	73.4	–

Table 12.10: Solidification Behaviour at 17°C Bath of Isopropanol Fractionated Products

	Mango fat	P_2	P_3	P_4
Supercooling point °C	21.4	24.0	23.7	22.1
Time (min) for attaining minimum T	17	13	13	17
Solidification point °C	25.6	28.9	29.3	27.4
Time (min) for attaining maximum T	25	21	22	24
Rise in T °C	4.2	4.9	5.6	5.3

Table 12.11: Slip Point and Dilatation of Isopropanol Fractionated Products

	Slip Point °C	Dilatation (D) in $mm^3/25g$ D_{20}	D_{30}	D_{35}
Mango fat	31.5	1372	1043	840
P_2	35	1696	1045	699
P_3	32.5	1731	779	171
P_4	31.5	1670	572	37

Animal fat like tallow was crystallised from acetone in preparing confectionery fat by Luddy and coworkers (29) and by Kozempel *et al* (30). Tallows on fractionation (31, 32) yield a number of edible fat products which are included in the Table 12.12.

Table 12.12: Fractionation of Tallow from Acetone (31, 32)

Solvent : Tallow (Acetone)	*Temp.(°C)*	*Feed*	*Fraction*	*Yield (%)*	*Product Application*
8.7 : 1	19.4	Edible	Olein 1	75	
		Tallow	Stearin 1	25	
12 : 1	6.1	Olein 1	Olein 2	61	
			Stearin 2	14	Cocoabutter replacer CBR
10 : 1	25	Tallow (Beef)	Olein 1		Shortenings, margarines and hard stock
			Stearin 1	7.5	
10 : 1	2	Olein 1	Olein 2	60	Bread shortening salad oil
20 : 1	15	Stearin 2	Stearin 2		
			Olein 3		
			Stearin 3	7.5	Shortenings, margarines
15 : 1	2	Olein 3	Olein 4	5	Shortenings, margarines
			Stearin 4	20	Cocabutter extender

Palm oil has been extensively fractionated from polar and non-polar solvents with a view to making CBS. Palm oil contains 30% of POP glycerides which can be increased to 70–75% concentration by a suitable fractionation procedure.

A two step fractionation has been done to produce palm mid fraction (PMF) with a high content of POP glycerides. The PMF fraction has been examined for properties quite extensively (33). Two stage crystallisation of palm oil yields 20–25% of PMF fraction being composed essentially of monounsaturated glycerides with a melting range of 30°–40°C. This fraction is used as cocoabutter equivalent type fat (34).

Fractionation of palm oil from acetone is preferred to hexane as the crystallisation from hexane results in the concentration of partial glycerides in the desired crystallised fraction, but from acetone miscella the partial glycerides are collected in the olein at the expense of cloud point (35). Palm oil has been fractionated to prepare mid fractions by two stage fractionation from acetone having slip point and SFI characteristics quite close to cocoabutter (36).

Khatoon and Bhattacharyya (37) studied extensively the fractionation of palm oil from hexane and acetone. Conditions have been established that provide a way of producing palm mid from two solvents. Palm mids obtained from acetone are generally superior to that obtained from hexane with reference to slip point, SFI as well as fatty acid and triglyceride profiles. Depending on palm mid yields, the composition and properties have a great deal of correlation (Table 12.13).

Some of the commercially adopted solvent fractionation processes include the Emersol process, Bernardini CMB process, Zondek process and other proprietary processes. All these processes are described in detail in textbooks.

The solvent crystallisation process is losing importance in view of the high cost of production in the solvent process. However, some of the old solvent crystallisation processes are still in commercial practice.

The Armour-Texaco process employs acetone and the Emersol process employs methanol for solubilising the fatty acid mixture and achieving selective crystallisation. The fatty acid mixture is first

Table 12.13 : Properties and Compositions of Palm Mids obtained by Fractionating Palm Oil from Hexane (3 Vol.) and Acetone (7 Vol.)

Solvent & Fractionation (2 stage)	*Condition*	*Yield of Palm Mid Fractions (% wt.)*	*Slip melting point (°C)*	*SFI at °C*					
				15	*20*	*25*	*30*	*35*	*40*
1st stage	2nd stage								
Hexane, 20± 1 °C/2.5 hr	–4± 1 °C/3.5 hr.	30.0 (PMF3)	42.0	64.8	56.0	43.8	23.4	15.0	7.8
Acetone, 30± 1 °C/7.0 hr	8± 1 °C/2.5 hr.	33.5 (PMF4)	44.0	70.6	65.4	56.5	36.3	19.4	11.4
Acetone, 10± 1 °C/2.0 hr	5± 1 °C/5.0 hr.	39.4 (PMF12)	30.2	70.4	66.1	51.9	16.2	1.4	–

Fatty Acid and Carbon Number Triglyceride Compositions

Palm Mid	*Major Fatty Acids (% wt.)*				*Carbon Number Triglycerides (% wt.)*				
Fractions	$C_{16:0}$	$C_{18:0}$	$C_{18:1}$	$C_{18:2}$	C_{46}	C_{48}	C_{50}	C_{52}	C_{54}
PMF_3	45.8	8.2	37.4	6.7	–	5.9	78.9	15.1	–
PMF_4	59.2	5.4	29.0	5.0	1.26	12.4	52.7	29.8	3.8
PMF_{12}	43.7	5.7	37.4	10.1	–	1.5	65.9	31.0	–

dissolved in the solvent and the solution is gradually cooled to the required temperature by passing through the cooled crystalliser. The slurry leaves the crystalliser and is pumped to the separator, usually a filter. The solid fraction is desolventised in a stripping unit to get the stearic (solid) acid rich fraction. The solution (filtrate) from the filter is also desolventised in another separator to yield the oleic (liquid) acid fraction. The solvent is recycled and the saturated and unsaturated acids are utilised either directly or after further purification by distillation.

The Solexol process that uses propane under pressure to separate the fatty acids on the basis of the relative solubilities of the acids in propane is hardly used. Even the only plant in the U.S.A. was dismantled long ago in the 1960's.

The solvent fractionation process can be a multi-stage process depending on whether the solid or liquid fraction requires refractionation. The multi-stage process would need only an additional crystalliser, separator (filter) and stripper.

A comparison of the oil loss and the utilities involved between the non-solvent process and solvent process (Table 12.14) indicates that the solvent process incurs much less oil loss, achieves much better fractionation but involves more energy as indicated by steam consumption and significant amount of solvent loss. The non-solvent process is also usually slow. The choice of the non-solvent or solvent process obviously is guided by the economy considerations, which favour the non-solvent process.

(3) DETERGENT FRACTIONATION (LANZA PROCESS)

The first detergent fractionation plant was set in Malaysia for fractionating palm oil. The process consists of chilling palm oil in the presence of aqueous solution of detergent and magnesium or sodium sulphate at 17°–20°C. When the cooled or partially crystallised oil is mixed with a detergent solution, the crystals (stearin) are wetted by the detergent and pass into suspension in the aqueous phase. The mixture is centrifuged into a liquid oil phase (olein) and a water phase (stearin). The olein mixture is washed with water to remove excess detergent and then dried. The stearin is wetted and then centrifuged to separate it from the detergent solution. It is subsequently washed and dried.

***Table 12.14:* A Comparison between Non-solvent and Solvent Process (38)**

		Non-Solvent Process	Solvent Process (Hexane)
(A)	Cloud point (°C) Oil		
	Cottonseed	+ 4	+ 1
	Sunflower	+ 4	+ 1
(B)	Oil Loss Kg/1000 kg of Oil		
	Cottonseed	35	14
	Sunflower	20	11
(C)	Utilities for a plant of 24 metric tonnes/24 hrs.		
	Steam (kg.)	100	300
	Power (Kwh)	15	15
	Water (M^3)	20	30
	Solvent (kg)	0	7
	Time (worker-hours)	2.5	1
	Time (hours)	30	8

The yield of stearin fraction called palm stearin is 17 –20% while that of the liquid fraction called palmolein is 77–80%.

The physical properties and compositions like fatty acids and triglycerides are as follows:

Composition	*Palmstearins*	*Palmoleins*
Yield %	17–23	77–83
Iodine Value (GLC)	24.9–34.6	57.8–58.0
Physical Properties		
Slip melting point °C	53.8–55.3	19.9–21.7
SFC (%) by	–	8.7–10.9
10C	81.9–92.1	32.7–34.7
15C	76.2–90.8	9.2–14.6
20C	68.6–86.6	0.6
25C	57.3–78.6	
30C	46.6– 68.2	
35C	37.8–57.0	
40C	29.9–46.0	
45C	23.0–35.8	
50C	13.0–25.0	
55C	0.4–9.3	

Table (Contd.)

Table (Contd.)

Composition	*Palmstearins*	*Palmoleins*
Fatty acids (% w/w)		
16:0	60.1–68.8	39.4–40.0
18:0	4.8–5.1	4.3–4.4
18:1	14.1–36.0	40.8–42.3
18:2	4.5–6.6	11.3–12.3
FFA % (as Palmitic)	2.2–2.3	3.5–4.3
Diglyceride % (wt.)	2.8–3.9	5.2–6.5
Triglyceride % wt.		
Triglyceride % HPLC	96.1–96.4	93.4–94.8
	32.9–45.1	6.1–0.3
	41.4–45.5	50.1–50.9
	9.2–19.0	43.2–43.7
	0.8–2.6	5.7–0.3

Rossell (39) reported the yields of stearins and oleins and their iodine values obtained by fractionating by Lipofrac process at different temperatures. According to the author, the original I.V. and FFA of palm kernel oil influences their yields per cent and their I.V.

DETERGENT (HYDROPHILISATION) PROCESS OF FRACTIONATION OF FATTY ACIDS

The separation of fatty acids based upon the hydrophilisation process has been developing in preference to the solvent process. The principle is the same as already pointed out for the separation of triglycerides.

In this process, the molten fatty acids mixture is gradually crystallised by slow cooling and at a particular low temperature. When the proper suspension as fatty acid crystals in the liquid phase is obtained, an aqueous solution of a surfactant is added. Thereby the solid fraction gets transferred in the aqueous phase as a suspension. The suspension containing the saturated acid is separated by centrifugation. The solid fraction is melted and centrifuged to get the saturated acid fraction in the molten (liquid) phase. The aqueous phase obtained contains the detergent, which is recycled back to the process. The liquid phase is again centrifuged when a liquid phase consisting of the unsaturated fatty acid is produced. The saturated acid fraction shows relatively higher iodine value in comparison with that obtained from the solvent process.

Typical results of fractionation of fatty acids by Lipofrac process are shown in Table 12.15. Lipofrac process is influenced by fractionation temperature as can be seen from Table 12.16.

The utilities consumption in Lipofrac process per tonne of fatty acids are shown in Tables 12.17 and 12.18.

***Table 12.15*: Fractionation of Fatty Acids by Lipofrac Process (40, 41).**

	Fractionation Temp.°C	*Iodine value*	*Melting Point (°C)*	*Yield (%)*
Tallow Fatty Acid	12	53	42	
Olein	12	96	–	53
Stearin	12	22	51	47
Palm oil Fatty Acid		57	41	
Olein	15	103	10	44
Stearin		23	49	56
Rice bran oil Fatty Acid				
Olein	15	–	–	63
Stearin				37

***Table 12.16*: Effect of Temperature in Fractionation of Fatty Acids by Lipofrac Process (RiceBran Oil Fatty Acid) (41)**

Fractionation	*Yield (%)*		*Major Fatty Acids (% w/w)*		
Temp. (°C)/Time(hr.)	*Stearin*	*Olein*	*16:0*	*18:1*	*18:2*
Rice bran fatty acid			22.9	45.2	30.3
Rice bran fatty acid at 15°C/1 hr.	37	63	S 51.5	31.7	15.0
			O (10.5)	(53.6)	(33.3)
20°C/1 hr.	36	64	S 48.3	36.1	14.2
			O (13.0)	(52.8)	(32.0)
25°C/1 hr.	35	65	S 45.0	35.3	18.6
			O (13.3)	(51.8)	(33.6)
30°C/1 hr.	32	68	S 32.6	40.3	25.2
			O (18.0)	(48.6)	(31.8)

S – Stearin
O – Olein

Table 12.17: Utilities Consumption in Lipofrac Process per tonne of Fatty Acids (40, 42).

	(41)	(42)
Steam (120°C)/5 bar	170 kg	325 kg
Process water 90°C	0.2 m^3	1500 kg
Cooling water (recirculation)	1 m^3	33 m^3
Electrical energy	34 KW	60 KW
Sodium decyl sulphate	2-5 kg	2-3 kg
Sodium sulphate	1-2 kg	3-4 kg
Instrument air	5 nm^3/hr	–

Table 12.18: A Comparison of Main Characteristics of Fractionation Process (4) (Dry vs Detergent)

	Process		
	Fast dry	*Slow dry*	*Lipofrac/Detergent*
Palm oil ⟵	Crude, semi and fully refined ⟶		Crude
Olein yield %	60-63	67-72	77-83
Cloud point °C	9.6-10.2	8.5-9.5	9.8-10.9
Cold test 22°C (days)	1-2	2-5	3-4
SFC at 20°C	1-3	0.02-1.8	0.6
Stearin Type	Soft	Medium Hard	Hard
Investment Cost	Lower		Higher
Operating Cost	Lower		Higher
Additives	No		Yes
Effluent treatment	No		Yes
Possibility of separation at T°C	Possible	Easy	Difficult

A comparison of fractionation process displayed in Table 12.19 for fatty acids points out their relative merits and demerits.

WINTERISATION : A FRACTIONATION TECHNIQUE

Winterisation is a kind of fractionation carried out on oils, fats and fatty acids to remove from them small quantities of solids which contribute to some unusual features. The winterisation technique is carried out in a manner similar to dry fractionation, solvent fractionation as well as detergent fractionation as applied to fats and fatty acids.

***Table 12.19*: A Comparison of Different Fractionation Processes (40)**

	Fractionation Processes		
	Dry (Pressing)	*Lipofrac*	*Solvent*
Merits	• Simple	• Works on most kinds of fatty acids	• Works on most kinds of fatty acids
	• Low Investment costs	• Low operating costs	• Highly automated
	• Good stearin quality	• Good olein quality	• Good stearin quality
	• High operation costs	• High investment costs for small capacities	• High investment costs
Demerits	• Restricted to mainly hard fats		• High operation costs (energy solvent less)
	• Messy		

The winterisation process is adopted or applied to remove small quantities of solids which tend to separate out from oils or fats at ambient low temperatures or when kept at refrigerated temperature. The solids are the high melting triglycerides (S_3 or S_2U) or the waxes (long chain fatty alcohols and their esters with fatty acids). The removal of these two kinds of solids from oils (e.g. cottonseed, groundnut, sunflower, corn, rice bran) enables the oils to remain clear at refrigerated temperatures. The destearised and dewaxed oils behave like salad oil. Thomas (43) has described in detail the winterisation process.

In commercial practice, the winterisation process is conducted in vessels equipped with slow stirring as well as cooling arrangements internally or externally. The fatty material is first melted completely and then cooled in a controlled manner with slow stirring to the stage of large and easily filterable or separable solid fat crystals.

After complete crystallisation, the crystals and liquid oil are separated usually by filtration. A conventional winterisation plant consists of a prechiller, crystalliser tanks, a refrigeration unit, and a filtration equipment (or filter).

In the prechiller, cooling rate and agitation are controlled to promote the formation of nuclei to facilitate the subsequent growth of large crystals and to the desirable β' form. Depending on the type of oil, the cooling rates may range from 1 to 17°C/hour.

The crystalliser tanks, normally tall and narrow, maintained at refrigeration temperature are used. At this stage the oil remains in a quiescent, slowly stirred or intermittently stirred state for the period of time required to reach saturation. Typically this may require 24-36 hours.

DEOILING AND FRACTIONATION OF PHOSPHOLIPIDS BY SOLVENT

Phospholipid, an important by-product of vegetable oil refinery, contains other constituents like oils, water and other polar/resinous materials in varying proportions. Among the vegetable oils, soyabean oil has the highest (2-3 per cent) phospholipid content commercially known as soyaphospholipids.

In the food, pharmaceutical and industrial products formulations the soyaphospholipids find extensive uses in view of having, in good proportions, the two important phospholipids, namely, phosphatidyl ethanolamine (Cephalin) and phosphatidyl choline (Lecithin) which impart excellent surface activity to soyaphospholipids.

Soyaphospholipids contain nearly 40 per cent oil by weight and the occurrence of oil makes them more susceptible to colour and odour problems (hour shelf life). For many specific food uses, it becomes necessary to deoil soyaphospholipids. Deoiling can be achieved very efficiently by treating the soyaphospholipid sample containing soyabean oil with acetone at low temperature (4°C). The phospholipids are insoluble in acetone while the soyabean oil dissolves in acetone. The acetone solution can be removed by centrifugation or by proper setting in a settler, decanting the acetone solution and leaving in the settler the phospholipid fraction. This fraction can be desolventised to recover total soyaphospholipid free of soya oil.

Deoiled soyaphospholipid obtained as above is often fractionated into an alcohol soluble and an alcohol insoluble fractions under a proper set of conditions (40°C/1 hr). The alcohol (ethanol) soluble part is the Lecithin fraction and the alcohol insoluble portion is the Cephalin or phosphatidyl ethanolamine part.

The fatty acid compositions of the original soyaphospholipid and the fractionated soyaphospholipids like alcohol soluble and alcohol insoluble phospholipids are presented below (44):

Major fatty acids (% w/w)	*Original soya phospholipid*	*Alcohol soluble phospholipid*	*Alcohol insoluble phospholipid*
16:0	22.4	18.7	29.7
18:0	3.5	2.8	6.0
18:1	13.5	13.4	16.9
18:2	546	59.3	40.4
18:3	5.1	5.3	2.6

SUPERCRITICAL EXTRACTION TECHNOLOGY FOR EFFECTING FRACTIONATION OF FATS AND FATTY DERIVATIVES

Supercritical extraction technology involving the use of supercritical carbon dioxide is a promising process for a number of fats and fatty derivatives. It is known that supercritical carbon dioxide dissolves triglycerides and other non-polar non-glycerides but cannot solubilise the polar lipids like phospholipids. This difference in solubilisation of non-polar lipids and insolubilisation of polar lipids in supercritical CO_2 has been taken advantage of for removing (deoiling) the oil from commercial soyaphospholipids containing oil.

A number of patents have been taken on the supercritical CO_2 extraction of soyalecithin containing soyabean oil (45-47).

For dewaxing of oil/vegetable like rice bran oil adoption of supercritical CO_2 extraction technology is feasible (48).

Efforts are made to extract triglycerides from crude sheanut oil (49). The resulting 'shea butter' is a raw material in the production of fractions as substitutes for cocoabutter in chocolate. Polyisoprenoid gum along with other contaminants could be successfully removed from the crude oil using supercritical CO_2 thus eliminating the degumming step.

Butter fat has been fractionated by supercritical CO_2 into triglyceride having short chain fatty acids and long chain fatty acids (50). Low cholesterol butter fat has been made by supercritical CO_2 extraction technology (49).

Supercritical CO_2 extraction technology is gaining importance in the recovery of essential fatty acids like EPA and DHA from fish oils in purified form (49).

Supercritical CO_2 has promise for extracting wax esters from jojoba oil (49).

FRACTIONATION OF FATS AND FATTY ACIDS BY LIQUID-LIQUID EXTRACTION

Fractionation of fats/oils and/or fatty acids can be achieved by taking advantage of solubility differences of triglycerides or fatty acids in a mono or binary solvent systems which are immisible in nature. in one of the solvents it is possible that some particular triglyceride molecules or fatty acids have either preferential solubility or very little solubility. The solvents are polar or non-polar in nature as usual.

A single solvent is used in which the fatty part is not wholly soluble and therefore forms a biphasic system of immisible layers. In the solvent phase, the desirable or the undesirable components of the fatty material can be distributed/segregated, depending on the molecular nature of fatty material. The extract phase and raffinate phase are obtained from which the solvent can be stripped off to get thc fractionated products.

A classical example of fractionation of fat by liquid-liquid extraction is that of soyabean oil which when mixed with four volumes of furfural at about 27°C fractionates soyabean oil into glycerides rich in more unsaturated fatty acids (51). Similarly linseed oil has been fractionated to produce a 75% extraction fraction of 196.3 iodine number (52).

The liquid-liquid extraction process is helpful in producing high concentrate (90%) of linoleic acid from linoleic acid rich oil (53). Liquid-liquid extraction with alcohol fractionates mixtures of mono-, di- and triglycerides (54). The process is also suitable for the recovery of sterols, vitamins, or other unsaponifiables.

REFERENCES

1. Tirtiaux, F. *Oleagineux*, **31**, 279-285 (1976)
2. Bernardini, M. and R. Sussoli, Palm Oil Technology in the Eighties, PORIM Kuala Lumpur (1981)
3. Athanassiadis, A, AOCS Meeting, St. Louis, Missouri (1978)
4. Deffense, E., *JAOCS*, **62**, 376-385 (1985)
5. Chakrabarty, M.M., D.K. Bhattacharyya, T.K. Raha, and P. Ghosh Choudhury, Report Cacao fabrick de Zean, Holland (1978)
6. H. Traitler and A. Dieffenbacher, *JAOCS*, **62**, 418 (1985)
7. R.D. Shinoy and A. Ganapathy, US. Pat. 4,179,456 (1979); U.S. Pat. 4,179,455 (1979)
8. Hilditch, T.P., Meara, M.L. and Reels, O.A., *J. Soc. Chem. Ind.*, **66**, 264 (1947).
9. Hilditch, T.P. and Shrivastava, R.K.; *J. Am. Oil Chemists' Soc.*, **26**, 1 (1949)
10. Gupta, S., Hilditch, T.P. and Meara, M.L., *J. Chem. Soc.* 3145 (1950)
11. Hilditch, T.P. and Maddison, L., *J. Soc. Chem. Ind.* **54**, 262 (1940)
12. *ibid*, **60**, 258 (1941)
13. *ibid*, **67**, 253-257 (1958)
14. Gunstone, F.D., Hilditch, T.P. and Riley, J.P., *J. Soc. Chem. Ind.* **66**, 293 (1947)
15. Dunn, H.C., Hilditch, T.P. and Riley, J.P., *J. Soc. Chem. Ind.*, **67**, 199 (1948)
16. Hilditch, T.P., Laurant, P.A. and Meara, M.L. *ibid*, **66**, 19 (1947)

16a. Crossley, A. and Hilditch, T.P., *J. Soc. Food and Agri.*, **4**, 38-45 (1953)

17. Ghosh Choudhuri, P., Ph.D. Thesis, Calcutta University (1980)
18. Bringi, N.V. (Hindustan Lever Ltd.) Indian Patent, 129348 (1970)
19. Baliga, B.P. and Shitole, A.D., *J. Am. Oil Chemists' Soc.*, **58**, 110-114 (1980)
20. Kane, J.G., Bhambhani, T.R. and Shitole, A.D., *J. Oil Technol. Assoc. India*, **4**, 3 (1972)

21. Chopde, P; Ph. D. Thesis, Bombay University (1961)

22. S. Adhikari, Ph.D. Thesis, Calcutta University (1980)

23. Shitole, A.D., Subrahmanyam, V.V.R. and Kane, J.G., *J. Oil Technol. Assn. India*, **2**, 27 (1970)

24. Bringi, N.V. and F.B. Padley, Ind. Pat. 148928 assigned to Hindustan Lever Ltd., Bombay, U.S. Pat. 4,060,646 (1977)

25. Mehta, D.T., G. Ramachandran and N.V. Bringi, Hindustan Lever Ltd., Bombay (Personal Communication)

26. Baliga, B.P. and A.D. Shitole, *J. Am. Oil Chem. Soc.* **58**, 110 (1981)

27. Ghosh Chaudhuri, P., M.M. Chakrabarty and D.K. Bhattacharyya, Fette Seifen Anstrichmittel (in press)

28. K. Mondal, Ph.D. Thesis, Calcutta University (1984)

29. Luddy, F.C., Hampson, J.W., Herb, S.F. and Rothbart, H.L., U.S. Patent 3,944,585 (1976)

30. Kozempel, M.F.; Graig, J.C., Hoilend, W.K., Jr. Elias S. and Acelo, N.C., *J.Am. Oil Chemists' Soc.*, **59**, 921-25 (1981)

31. Craig, J.G., M.F.Kozempel and S. Elias, U.S. Patent 4,127,547 (1978)

32. Luddy, F.E., *et al* U.S. Patent 4,130,572 (1978) & U.S. Patent 4,205,006 (1980)

33. Kheiri, M.S.A., PORIM, Occasional Paper No. 4, Aug. 1982

34. Malaysian Oil Palm Growers Council; Malaysian Palm Oil, Technical Bulletin, No. 3, 1981; Review of Current Technology

35. Young, F.V.K., *J. Am. Oil Chemists' Soc.*, **62**, 372 (1985)

36. Kheiri, M.S.A. Personal Communication, PORIM.

37. S. Khatoon and D.K. Bhattacharyya, *Oleagineux*, **41**, 519-523 (1986)

38. E. Bernardini, The New Oil and Fat Technology 2nd rev. ed. Technological Review, 564 (1973)

39. Rossell, J.B., *J. Am. Oil Chemists' Soc.*, **62**, 387 (1985)

40. G. Haraldsson, *ibid*, **61**, 222 (1984)

41. B.K. Mukhopadhyay and D.K. Bhattacharyya, *J. Oil Technol. Assoc. India*, 85-86 (1987)

42. O.P. Narula, in Treatise on Fats, Fatty Acids & Oleochemicals, First volume p. J-216, edited by O.P. Narula, Industrial Consultants (India), H-89, Karam Pura, New Delhi-110 015, 1994

43. A.E.Thomas III; in Bailey's Industrial Oil and Fat Products, Volume 3, edited by T.H. Applewhite, pp. 6-8, Wiley Interscience, 1985

44. M. Ghosh, CSIR Project Report, Calcutta University (1994)

45. Heigel, W. and R. Huschens, DE 3,011,185 (1981)
46. Stahl, E. and K. Quirin, DE 3,329,249 (1985)
47. Peter, S. *et al.*, DE 3,630,676 (1988)
48. V.V. Rao, in Proceedings of Seminar on Rice Bran Oil Status and Prospects, Oil Technologists' Association of India (Southern Zone), R.R.L., Hyderabad, August 13, 1983, p. 47
49. INFORM AOCS, Vol. I, Number 9, pp. 810-816 (1990)
50. W.Kaufmann, G. Biernoth, E. Frede, W. Merk, D. Precht and H. Timmen, *Milchwissensehaft*, **37**, 92-96 (1982)
51. S.E. Freeman, U.S. Pats. 2,200,390-1 (1940); 2,278,309 and 2,291,461 (1942); 2,313,636 and 2,315,512 (1943)
52. R.L. Kenyon, S.W. Gloyer, and C.C. Georgian, *Ind. Eng. Chem.* **40**, 1162-1170 (1948)
53. V.E. Sohns, R.E. Beal, and E.C. Griffin, Jr., *J. Am. Oil Chemists' Soc.*, **40** 169-172 (1963)
54. R. Feuge and A.T. Gros, *ibid*, **27**, 117-122 (1950)

CHAPTER 13

Interesterification

Interesterification reactions of oils, fats and other fatty derivatives are well established and recognised for their commercial and technical applications in producing a variety of edible and industrial products. Basically there are three kinds of interesterification reactions, namely, ester-ester interchange, acidolysis and alcoholysis. Each reaction, extensively reported in text books and elsewhere, will be dealt with covering basic and technological aspects.

ESTER-ESTER INTERCHANGE

Ester-ester interchange involves the exchange of acyl radical (s) of one ester molecule with the acyl radical (s) of another ester molecule. Obviously, therefore, the ester molecule can be polyesters (e.g. glycerides), monoesters, and also polyester-monoester combinations.

Since the report of Friedel and Crafts (1), the ester-ester interesterification reaction had been extended to polyesters such as natural triglycerides. Many workers have studied the ester-ester interesterification reaction with respect to the nature of catalyst systems, the reaction mechanism, the condition of the reaction, and the applicability of the reactions in modifying fats to make specific edible and industrial fat products. All these aspects are well described in standard text books (2-4), in review articles (5-6), in patents (7-17), and also in the papers published by various workers (18-30).

The ester-ester interchange reaction of triglycerides proceeds in a random manner as indicated by the positional distribution of fatty acids in the glyceride molecules after the reaction, and also by the formation of theoretically possible glycerides that one can achieve when fatty acids are esterifed with glycerol to form triglycerides. This random interesterification reaction is a liquid phase one. A completely random arrangement of triglycerides at reaction equilibrium and the distribution of the fatty acids in 1, 2 and 3 positions of glycerol in the same amount in relation to their original quantity have been well documented.

The glyceride composition of randomly interesterified fats can be calculated with the following statistical formula if the fatty acid compositions are known :

When A, B, C etc. are the molar percentage of fatty acids a, b, c etc.

$$\text{Glyceride a-a-a} = 1 \times \frac{A.A.A.}{10,000} \text{ mole \%}$$

$$\text{Glyceride a-a-b} = 3 \times \frac{A.A.B.}{10,000} \text{ mole \%}$$

$$\text{Glyceride a-b-c} = 6 \times \frac{A.B.C.}{10,000} \text{ mole \%}$$

Most of the vegetable seed oils get their fatty acid radicals distributed in glyceride moieties almost evenly and contain little or no trisaturated glyceride. In fats, there is no occurrence of trisaturated glyceride if the amount of saturated fatty acids does not exceed 60% of total. But by random distribution of all acid radicals, an oil or a fat containing 66% and 33% saturated fatty acids may be made to contain nearly 30% and 3.7%, respectively, of fully saturated glycerides.

Though randomisation of fats served very useful purpose in modifying fats, it was Eckey (11, 12, 20) who first showed that interesterification reaction can be given a direction from its randomly distributed pattern if one or more component of the product of reaction is removed from the ambit of the reaction by way of crystallisation of the trisaturated and/or monounsaturated-disaturated glycerides. Such withdrawal of component or components disturbs the equilibrium and more trisaturates and/or monounsaturated-disaturates are formed. Since the melting slip point of a fat is very much dependent on its trisaturates content the melting slip point of the fat rises with the result that plastic product with desired consistency may be obtained and these may find use as shortening and margarines fat base. It may be mentioned here that simultaneous to the rise in concentration of saturates there is an obvious increase in the concentration of unsaturates. This segregation of saturated and unsaturated glycerides at the cost of mixed saturated-unsaturated glycerides imparts a gradual melting character to the fats and therefore long plastic range fats are produced. Hence, the directed interesterification process has become immensely useful for the preparation of shortenings from natural fats of medium consistency like palm oil or tallow (17, 43). It is generally accepted that the directed interesterification as a process, is not suitable for the preparation of hard butter and margarine, but Eckey (16) produced a superior hard butter by the directed rearrangement of oils (like coconut oil) with or without subsequent crystallisation and pressing.

A good number of catalyst systems have been examined so far and these include alkoxides (20, 31, 32) such as sodium methoxide, sodium ethoxide, sodium tertiary butoxide; metals and metallic alloys (33-35) like sodium, potassium, sodium-potassium alloy; metallic hydroxides (32, 35) such as stannous hydroxide, sodium hydroxide, potassium hydroxide, lithium hydroxide, sodium hydroxide-glycerol (36), sodium hydroxide-glycerol-water (37), metal soaps (4, 38, 39) like sodium stearates, lithium aluminium stearate, sodiumtitanium stearate; metallic hydrides and amides (39) like sodium hydride, sodamide etc.

MECHANISMS

Weiss *et al* (41) suggested that the interesterification with the methoxide catalyst takes place through the initial formation of an enolate ion. They obtained evidence in favour of their suggested mechanism by studying the infra-red spectra of the reaction mixture.

SOME CASE STUDIES

Van Loon (9) heated a mixture of beef stearin and soyabean oil (1 : 4 W/W) for 16 hours at about 275°C and observed a lowering of melting point from 40.2°C to 30.6°C. Norman (7) heated a mixture of tristearin with triacetin in the molar ratio of 2:1 without any catalyst and found that about 50% interchange takes place after 72 hours. On reacting a mixture of tristearin and coconut oil (1:4) in presence of 1% napthalene β-sulphonic acid at about 250°C for 2.5 hours at atmospheric pressure the author observed a drop in melting point from 57.3°C to 34.5°C; but the same oil mixture when reacted in the presence of 0.1 to 0.2% sodium ethylate as catalyst at 140°C for 0.5 hour under reduced pressure the drop in melting point was from 57.7°C to 31.4°C.

Grun (42) heated equimolecular quantities of tristearin and tricaprylin using tin as catalyst for 3 hours at 230°C and found that the product after reaction contained 20% tristearin, 0.09% tricaprylin and 71% capro-stearo glycerides. Similar results were also obtained with mixtures of tristearin and triolein, tricaproin and tributyrin etc.

Gooding (8) reported that when 75:25 mixture of highly hydrogenated palm kernel oil and highly hydrogenated palm oil were interesterified the drop in melting point was from 50.2°C to 40.2°C.

Eckey (10) noted a lowering of melting point from 58.2°C to 41.1°C when 60% coconut oil was interesterified with 40% of hydrogenated cottonseed oil. Norris and Mattil (18) interesterified synthetic triolein with tripalmitin in proportions of 50:50 (W/W) using stannous hydroxide as catalyst, at a temperature of 225°C and found that the reaction proceeded to equilibrium with the random distribution of acyl radicals. They also studied the interesterification of lard and hydrogenated lard and obtained similar results.

Baur (21) reacted triacetin with cottonseed, soyabean, rapeseed and coconut oils using sodium methoxide or sodium as catalyst allowing a reaction time of 30 minutes at temperatures high enough to keep the components in liquid state. After removing the excess triacetin the diaceto-glycerides were hydrogenated and their properties were studied.

Norris and Mattil (19) and Eckey (11) interesterified several oils and reported various degrees of increment in their melting points. It had been shown that animal fats (tallow, lard) and palm oils were less affected by interesterification than the seed oils as far as the rise in melting point was concerned (11,19). The micro-penetration studies revealed that a good deal of structural change accompanied the ester-ester interesterification reaction (4).

The chemical changes brought about by the interesterification reaction had been studied with the object of modifying fats and oils so as to adapt them to various edible purposes such as margarine oil, confectioner's fats, bakery fats and other plastic fat products mainly from lard. The correlation between chemical change and physical properties such as X-ray diffraction pattern, cooling characteristics, solid fat index and melting point were also studied by many workers (4, 8, 9, 22).

Kaufmann and Grothuse (44) studied the effect of a number of catalysts on the interesterification of oils having unsaturated fatty acid glycerides and reported that there is no conjugation or *cis-trans* isomerisation of the glycerides during the interesterification reaction.

Some of the pioneering works by Van Loon (9), Gooding (8), Norris and Mattil (19), Eckey (11, 12, 20), Baur (21) and others have really led to the development of interesterification process technology based on their fundamental works in modifying the chemical compositions and physical properties of a variety of oils and fats including animal fats.

Chakrabarty and co-workers (26, 30) estimated the trisaturated glycerides of the directed interesterified fat products of some vegetable oils and they also established that a minimum of saturated acids (C_{16} and above) must be present for the production of plastic fats.

Adhikari *et al* (45, 46) studied the interesterification of palm, mowrah, sal fat and cottonseed with rice bran oil as binary blends and obtained a large variety of plastic fat products. They also reported (47) that the stability of the interesterified fats compared favourably with that of the hydrogenated fat after the addition of antioxidants. The same group of workers fractionated the high slip point interesterified fat products obtained from rice bran oil mixed with mowrah, palm, sal and cottonseed oil and showed that suitable fractions could be obtained which may be used as vanaspati and margarine fats with high essential fatty acid contents and without *trans* unsaturated fatty acids. The products of fractionation may also be used as highly stable deep frying fats and as bakery fats (48). Adhikari and coworkers (46) also prepared polyunsaturated fatty acid rich margarine fats, vanaspati, edible plastic shortening agents from sal oil mixed with cottonseed, peanut and sunflower oil by properly manipulating the interesterification reaction conditions like time and temperature.

Majumdar and Bhattacharyya (49, 50) studied the co-randomisation of palm stearin and vegetable oils like soyabean, rapeseed, rice bran, mustard, cottonseed, in making vanaspati of unhydrogenated type for use as substitute of hydrogenated vanaspati (Table 1).

Khatoon and Bhattacharyya (51) could prepare vanaspati like fat and margarine fat base from sal fat after interesterification in the form of blend with soyabean oil and also with rapeseed oil (Table 13.1).

Bhattacharyya and Ghosh (52) reported the random interesterification of palmolein using sodium hydroxide-glycerol with or without water as catalyst. The parameters such as the amount of catalyst, reaction temperature and time were investigated. Tables 13.2 & 13.3 reveal the interesterified product characteristics from palmolein and from the blends of minor fats like sal or mango with rice bran (53).

Eckey (20) first observed that the interesterification reaction can be achieved in presence of solvent. Since then, a number of polar and non-polar solvent systems individually and in combination have been investigated. Separator *et al* (54) studied the simultaneous interesterification and fractionation of fatty oils in presence of a solvent like methyl cellosolve and with sodium methoxide as catalyst. Millar (55) reported a similar study with propane as solvent where the saturated glycerides were separated from the unsaturates by a countercurrent fractionation.

Table 13.1 : Properties of some Co-randomised Fat Products

Co-randomised Products	Slip point °C	Solid Fat Index (ml/kg)					
		15	20	25	30	35	40
Sal + Cottonseed (3:7 W/W)	35.0	10.9	8.9	5.4	3.5	1.4	–
Sal + Sunflower (1:1 W/W)	38.0	16.3	13.3	9.2	6.4	4.6	3.0
Sal + Rice bran (1:1 W/W)	35.5	15.0	11.7	7.4	5.0	2.4	0.5
Sal + Soyabean (6:4 W/W)	36.5	20.1	14.8	10.1	8.0	5.4	3.8
		Dilation mm3/25 g at °C					
		15	20	25	30	35	40
Palm + Soyabean (8 : 2 W/W)	35.0	646	497	306	304	214	123
Palm + Rice bran (8 : 2 W/W)	37.5	557	481	369	250	227	123
Palm + Rapeseed (8 : 2 W/W)	37.0	510	374	330	221	148	66
Palm + Mustard (8 : 2 W/W)	35.5	507	407	277	202	116	0
Palmstearin + Rapeseed (7 : 3 W/W)	37.5	651	458	349	267	197	98
Palmstearin + Rice bran (6 : 4 W/W)	38.5	541	450	345	283	237	148
(1 : 1 W/W)	36.5	422	315	239	197	134	67
Palmstearin + Mustard (7 : 3 W/W)	38.0	707	568	426	349	215	34
Palmstearin + Soyabean (7:3 W/W)	37.5	616	487	376	313	266	143

Table 13.2 : Interesterification Reaction of Palmolein: Effect of Concentration of Catalyst (Sodium Hydroxide and Glycerol with or without Water) and Effect of Reaction Temperature

Product Number	Catalyst (% W/W on oil charge)			Temperature	Slip point
	NaOH	Glycerol	Water	°C	°C
I	0.6	0.6	0.9	160	37.4
II	0.4	0.4	0.7	160	38.2
III	0.3	0.3	0.5	160	38.4
IV	0.3	0.3	0	160	38.6
V	0.2	0.2	0.5	160	36.5
VI	0.2	0.2	0	160	39.1
VII	0.2	0.2	0	140	29.5
VIII	0.2	0.2	0	120	27.0

NaOH : Sodium hydroxide

Table 13.3 : Chemical Interesterification of Palmolein and of Some Minor Oils (Catalyst : 0.2% NaOH + 0.2% Glycerol each on Weight of Oil)

Product Number	Products	Slip point °C	Solid Fat Content (SFC by NMR) °C		
			10	20	30
1.	Palmolein	39.1	45.1	22.4	9.2
2.	Sal (80) + RBO (20)	38.7	49.8	17.4	4.7
3.	Sal (70) + RBO (30)	36.3	36.7	10.5	3.8
4.	Mango (80) + RBO (20)	40.5	44.5	23.9	10.1
5.	Mango (70) + RBO (30)	38.2	27.6	17.3	7.0
6.	Mango (60) + RBO (40)	36.2	23.5	9.7	4.4

RBO : Rice Bran Oil

Chakrabarty and Bhattacharyya (56) reported valuable information on the study of solvent interesterification of various oils like cottonseed, peanut and combination of oils like cottonseed-peanut, cotton seed-sesame and cottonseed-sunflower with special reference to the influence of catalyst (Sodium methoxide, NaOMe), oil concentration of miscella (oil solution in solvent) and the effect of temperature on the reaction. The influence of the nature of solvent was also studied by the authors.

Kaufmann and Schnurbusch (57) studied the effect of catalyst concentration and concentration of oil in miscella on the interesterification of oils and reported that the equilibrium of the randomisation reaction depend on the amount of catalyst. Artman *et al* (58) investigated the interesterification and simultaneous selective separation of the linoleic acid rich glycerides. They used catalysts like sodium, potassium, methylsulphinyl carbanion, and sodium and potassium tertiary butoxide and solvent systems like dimethyl formamide, dimethyl cyanamide and 3,3 dimethyl amino-propionitrile. The resultant mixtures were counter-currently separated with solvents like pentane, hexane, heptane and octane, and the authors claimed a patent.

Chakrabarty *et al* (59) studied the interesterification of partially hydrogenated cottonseed oil in solvent and examined the distribution of fatty acids, glyceride composition changes and the attendant thermal properties such as slip point and solid fat index (S.F.I.). Further studies were carried out on the solvent interesterification of some vegetable oils individually and in various combinations. A study included the interesterification reaction with polar and non-polar solvent systems (60). The work carried out included the interesterification reaction of sunflower oil in combination with some selective natural and hydrogenated fats in the solvent medium like n-hexane. The study indicated the suitability of the solvent interesterification process in making plastic fats under certain manipulated conditions of reaction.

Bhattacharyya and Chakrabarty (61) also studied the miscella interesterification of palm oil and its fraction (S) individually and in the form of blends with selective vegetable oils and could obtain interesting physical properties alterations.

TRIGLYCERIDE-MONOESTER INTERCHANGE

Monoester-polyester interchange reaction involving fatty acid esters of monohydric alcohols like methanol, ethanol, propanol etc. and polyhydric alcohol like glycerol had also been the subject of considerable investigations in making useful edible and industrial fat-based products.

Primarily, the reaction helped in introducing selective and more desirable fatty acids in the natural triglycerides (fats/oils) employing the usual interesterification catalysts and the other operating parameters like temperature, appropriately maintained inert atmosphere, catalyst concentration and the molar proportions of the reactants.

The reaction between a monoester and a polyester interchange also proceeds in a random manner like the interchange of polyester molecules.

Depending on the modified composition of fatty acids and triglycerides induced in the original fats/oils, there occur, as a result, remarkable alterations in their thermal properties.

The modified products may be useful straight away or may be modified further by fractionation into more useful triglyceride concentrates.

One of the early examples of a natural triglycerides oil and ethyl ester of a fatty acid interchange was reported by Normann (62) who reacted ethyl stearate and coconut oil with tin as a catalyst and obtained more than 50% interchange. Konen *et al* (63) reacted dry and neutral triacetin with methyl elaeostearate in a molar ratio of 1:3 in the presence of 0.05% of anhydrous sodium methoxide under reduced pressure at 60-100°C to remove methyl acetate and to obtain excellent quality trielaeostearin. Koslowsky (64) could interesterify palm oil (triglycerides) with ethyl oleate using 0.1-0.5% sodium methoxide at around 60°C for 30 min. The distillation of the saturated ethyl esters yielded a liquid triglyceride oil that was suitable for use as a salad oil.

By interchange of palm oil with ethyl stearate a hard fraction could be obtained. The process has gone into commercial use by a reputed company at Haifa, Israel.

Bhattacharyya and Banerjee (65,66) could modify the composition and properties of some natural vegetable butters (mohua, mango) by interchange with methyl ester of the desired fatty acid.

ANALYTICAL METHODOLOGIES FOR INTERESTERIFICATION PROCESS AND INTERESTERIFIED PRODUCTS

In interesterification reactions of natural oils and fats involving ester-ester interchange there is no change in their fatty acid compositions but there are alterations in their triglyceride compositions. It is because of changes in the triglyceride compositions that oils and fats undergo substantial changes in their physical properties like slip melting point, dilatation, consistency, solidification behaviour, crystalline phases and crystal structures etc.

During interesterification reactions the positional distribution of fatty acids changes are indicated in the text earlier. In the natural oils and fats the saturated acids occupy the 1- and 3- positions whereas the 2- position is occupied by the unsaturated fatty acids. In the interesterified fat products the distribution becomes much different. As a result of the interchange reaction, the saturated acids occupy the 2- position and the unsaturated acids occupy the 1- and 3- positions.

The difference in the positional distribution of the saturated and unsaturated fatty acids between the uninteresterified and interesterified oils and fats can be utilised in characterising interesterification reaction (67).

The triglyceride compositions of the interesterified products on the basis of carbon number by GLC (52) enables detection of the completeness of interesterification reaction and the alterations in the triglyceride profiles from those of the original profiles and content (Table 13.4).

Pancreatic lipase hydrolysis of the uninteresterified and the interesterified fats yields their 2- monoglycerides, which, on subsequent isolation by preparative TLC and derivatisations to the methyl esters of the fatty acids occurring in the 2- positions, provide conclusive evidence of the alterations in the distribution pattern of the saturated and unsaturated fatty acids in the 2- positions as well as in the 1- and 3- positions between the uninteresterified and interesterified products. This distribution pattern difference indicates the interesterification reaction taking place according to randomisation principle (67, 68). Silver nitrate silica gel-TLC is an elegant technique for detection of intereste-rification reaction. Freeman (69), Barrett *et al* (70) and Chakrabarty and Bhattacharyya (71) could distinguish interesterified fats from those of the corresponding uninteresterified fats on the basis of the number of component glycerides which were separated on the basis of unsaturations.

***Table 13.4* : Carbon Number Glyceride Composition (% w/w) of RBD Palmolein and its Interesterified Products**

Sample number	*Diglyceride % (w/w)*	*Carbon number triglycerides (% w/w)*				
		C46	*C48*	*C50*	*C52*	*C54*
RBD Palmolein	0.3	6.6	9.6	40.0	39.7	3.6
I	0.9	5.5	18.1	37.8	33.8	3.9
II	0.6	4.3	18.3	40.6	35.0	1.2
III	0.2	3.4	15.7	34.8	33.5	12.4
IV	0.5	4.7	16.0	31.8	34.6	12.4
V	0.6	3.1	14.8	32.9	36.7	11.9
VI	0.8	4.0	15.4	31.4	36.7	12.7
VII	0.1	4.0	11.2	37.5	39.3	7.9
VIII	0.1	3.8	10.8	36.6	41.2	7.5

The control of interesterification and the characteristics of the interesterified fat products can be easily made by the physical methods like slip melting point, dilatometry and cooling curve.

The changes in the slip melting point of the fats, individual and in mixture, after interesterification can be seen from Tables 13.1 & 13.2.

The dilatometric study of fats before and after intereste-rification enables determination of the solid fat index of the two kinds of fats.

There is invariably a difference in the SFI or SFC values between the uninteresterified and interesterified fats and these values depend on the nature of the individual fat and the fat mixtures (see Tables 13.1 & 13.2).

Cooling curve determination provides an approach for following the completeness of interesterification reaction and for identifying interesterified fats which hardly exhibit sharp supercooling and solidification points (72, 73).

SOME IMPORTANT COMMERCIAL APPLICATIONS OF INTERESTERIFIED PRODUCTS

Interesterification process involving ester-ester interchange aims primarily at producing edible plastic fat products from diverse oils and fats of plant, animal and marine sources. Some of the products are described below:

1. *Shortenings of unhydrogenated type with zero trans-isomer content (74)* : A unique example is that of randomised lard which is a better shortening than natural lard. Natural lard has 16:0 acid in the 2-position of its GS_2U glycerides and crystallises in the β- phase (4.6 Å). Randomised lard has its 16:0 acid drastically lowered in the 2-position and as a result crystallises in β' (4.2 Å) phase.

The changes in the proportion of 16:0 acid in the 2- position and in the crystalline phase produce a smooth textured lard.

Shortenings simulating hydrogenated products but without any *trans*-fatty acid content can be produced by co-randomisation process involving many kinds of fat blends (Tables 13.1 & 13.2).

2. *Margarine fats* : Interesterified fats find extensive use in margarines. High stability margarines and high PUFA with low to zero trans-acid content margarines for better nutrition can be produced using interesterified blends of appropriate oils and fats, which give the desired composition and good physical characteristics like texture, spreadability, eating qualities as well as shelf life and flavour score.

The interesterified fats, for example, are made from a mixture of highly hydrogenated liquid oils with unhydrogenated liquid oils in suitable proportions so as to obtain the desired melting point and SFC values at different temperatures required for good quality margarines (74,76).

3. *Improved quality salad oils and stable frying oils (48,75)*: Interesterification and fractionation of individual oils and fats or appropriate mixtures of oils and fats often enable production of excellent quality salad oils and highly stable frying oils.

4.*Confectionery fats* : Interesterified fats are generally not very appropriate raw materials for making good quality cocoa butter substitutes. However, fat products simulating hard butters can be made from certain selective fats. Thus, hydrogenated palm kernel oil (m. point 46°C) on randomisation yields a satisfactory hard butter (m. point 35°C) as reported by Sreenivasan (74), and this randomised product can be blended with hydrogenated palm kernel oil to make a whole series of hard butters with highly desirable melting qualities.

In general, hard butters have been prepared from lauric acid containing oils, e.g. from palm kernel oil alone or admixed with coconut oil by hydrogenation followed by randomisation and fractionation (77), from fat blends of partial or fully hydrogenated lauric fats with fully hardened cottonseed oil (78, 79) or fully hydrogenated rapeseed, herring etc. oil (80); and from a mixture of lauric (e.g. palm kernel oil) and non-lauric fats (e.g. palm oil) by randomisation and hydrogenation before or after randomisation (81).

Interesterification of completely hydrogenated cottonseed oil and triolein or olive oil in appropriate proportion followed by 2-stage fractionation from acetone yields a cocoa butter substitute having satisfactory thermal properties close to those of cocoa butter (82). Fractionation of the interesterified blends of mowrah fat and kokum fat gives good cocoa butter substitutes (83, 84).

STABILITY AND FRYING

The culinary (frying) characteristics of some interesterified fat products were studied by Bhattacharyya *et al* (85) who observed that the interesterified fats in spite of having more PUFA had frying characteristics comparable to the hydrogenated fat product vanaspati as indicated by changes taking place in physico-chemical characteristics during frying (Tables 13.5 & 13.6).

The stability of the interesterified fat products as examined by Adhikari *et al* (47) indicates that the interesterified fat products have stability characteristics similar to the hydrogenated fat products (Table 13.7). The concept that the interesterified fats are more unstable than the hydrogenated fats having identical melting points is not tenable. The use of antioxidants plays a vital role in protecting the interesterified fats from autoxidation reactions and associated physico-chemical changes.

***Table 13.5* : Frying Characteristics of Randomised Palm (180° ± 2°, 3Min/Batch)**

	Frying Number			
	0	*1st-4th*	*5th-7th*	*8th-10th*
% FFA	0.15	0.19-0.24	0.26-0.27	0.28-0.30
PV	0.75	1.15-2.09	0.12-1.85	1.66-1.13
Y + 5R (1/4" Cell)	1.9	3.0	3.0-6.0	8.0-15.0
% Diene (Conjugation)	0.17	0.19-0.28	0.32-0.44	0.45-0.55
% Linoleic	11.30	11.3-10.9	10.6-9.0	8.7-8.2
Off-flavour score	–	–	+ to +++	++ to +++

- No off flavour
+ Mild off flavour
++ Rancid tendency
+++ Rancid

***Table 13.6* : Frying Characteristics of Randomised Palmstearin and Rice Bran Product (180°C ± 2°C, 3 Min/Batch)**

	Frying Number			
	0	*1st-4th*	*5th-7th*	*8th-10th*
% FFA	0.17	0.19-0.03	0.34-0.38	0.40-0.47
PV	0.27	0.47-0.74	0.86-1.05	0.79-0.68
Y + 5R (1/4" Cell)	3.0	4.5-9.0	13.5-16.0	20.5-26.2
% Diene (Conjugation)	0.36	0.48-0.86	0.86-1.18	1.19-1.50
% Linoleic	21.0	21.0-17.9	16.3-13.3	12.2-11.1
Off flavour score	–	–	+	++ to +++

- No off flavour
+ Mild off flavour
++ Rancid tendency
+++ Rancid

Table 13.7 : Stability of Interesterified Fats and Vanaspati

	Vanaspati		*Randomised Palm*		*Interesterified Palmolein*		*Co-randomised Palmstearin + RBO*	
	Peroxide Values (ml/kg)							
AOM (hrs)	*Control **	*TBHQ (0.01%)*	*Control **	*TBHQ (0.01%)*	*Control **	*TBHQ (0.01%)*	*Control **	*TBHQ (0.01%)*
0	2.3	2.3	2.2	2.2	2.6	2.6	2.0	2.0
1	2.6	2.0	3.9	2.6	4.0	3.5	5.4	2.8
2	3.7	2.2	4.7	2.7	6.0	4.0	5.6	3.3
3	3.7	2.2	5.1	3.5	7.3	4.0	6.6	4.7
4	5.6	2.8	6.4	4.6	8.0	4.4	9.1	5.2
5	7.9	6.5	6.8	4.8	9.2	5.3	9.6	5.6

* Without antioxidant.

NUTRITIONAL QUALITY/CHARACTERISTICS OF INTERESTERIFIED FATS

Study on the nutritional value of interesterified fats appears to be very meagre. Alfin-Slater *et al* (86) were first to investigate the effects on growth, reproduction and lactation, longevities, tissue cholesterol levels and histological examination of tissues of rats fed on fat like interesterified blend of 82.5% unhydrogenated and 17.5% completely hydrogenated cottonseed oil. They observed that the interesterified product was utilised by the animal in the same way as cottonseed oil. Inclusion of the interesterified fats in atherogenic diets showed less atheroselerotic lessions in the coronary arteries and aortas of animals than those fed cholesterol containing diets with butter oil or conventional margarine oil of the all hydrogenated type. The conclusion was that the interesterified fats were at least nutritionally equal to other similar edible fats of equivalent essential fatty acid content.

Chakrabarty and Bhattacharyya (87) carried out studies of the nutritional quality of an edible plastic fat that was made from cottonseed oil by directed interesterification process. In comparison with the hydrogenated fat product vanaspati it was observed that interesterified product showed less content of total cholesterol, free cholesterol and triglyceride in the serum of rats in comparison with the serum of rats from the hydrogenated product group.

Jakubonski (88) reported high nutritive value of the interesterified product obtained from oil blend of hydrogenated tallow (1 part) and liquid sunflower oil or rapeseed oil (2 parts).

Duterte (89) discussed the nutritional quality of interesterified oil blends of hydrogenated lard and tallow with edible vegetable oils.

Bhattacharyya *et al* (90) investigated the nutritional quality of interesterified fat products prepared from sal fat and rice bran oil or cottonseed oil and also from mahua fat by simple randomisation vis-a-vis hydrogenated product such as vanaspati containing substantial amount of *trans* unsaturated fatty acids (Table 13.8). The rats which ingested the fats showed identical response in their body weight gain, average food intake and food intake and food efficiency ratio as well as the weights of different organs like liver, heart, kidney, spleen and testes. The total lipids of serum and liver of rats fed hydrogenated and interesterified fats were also nearly the same. Serum triglyceride levels were observed to be quite low in the case of interesterified fats fed rats. Total and free cholesterol levels in the serum were also not significantly different. Serum phospholipids were slightly higher in the case of interesterified fats fed rats.

***Table 13.8* : Coefficient of Digestibility of Vanaspati and Interesterified Products**

Products	*Coefficient of digestibility*
Vanaspati (hydrogenated)	93.6
Randomised Palm	96.6
Interesterified Palmolein	96.5
Randomised Palmstearin-Soyabean	96.6
Randomised Palmstearin-Rice bran	96.5
Randomised Palmstearin-Rapeseed	93.4
Randomised Palmstearin-Mustard	96.7
Randomised Mahua	89.8
Randomised Sal fat-Cottonseed(55:45)	93.3
Randomised Sal fat-Rice bran oil (50:50)	89.2
Randomised Mustard seed oil	93.0

Majumdar and Bhattacharyya (91-93) investigated the nutritional quality (Table 13.9) of interesterified fat products made from palm stearin plus liquid oils like soyabean, rapeseed, rice bran and from palmolein alone. They used rats as experimental animals and observed that digestibility is better with the interesterified fats than with the hydrogenated vanaspati and that serum/plasma shows invariably lower triglycerides and cholesterol contents in the case of the interesterified fats.

***Table 13.9* : Lipid Profiles (8) of Plasma and Liver of Rats Fed 20% Fat Diet**

Test fat	*TL* *±SEM*	*PL* *±SEM*	*TG* *±SEM*	*TC* *±SEM*	*CE* *±SEM*
Vanaspati					
Plasma (mg/100 ml)	307.62	61.86	65.20	51.50	22.69
	±16.88	±4.4	±10.70	±4.30	±0.49
Liver (mg/g)	41.13	15.26	0.90	4.19	0.37
	±4.24	±0.42	±0.04	±0.15	±0.06
Palmolein product :					
Plasma (mg/100 ml)	299.20	97.89	37.60	64.4	13.69
	±20.5	±8.90	±7.02	±6.46	±0.75
Liver (mg/g)	43.11	24.49	4.85	4.70	0.35
	±1.29	±2.70	±1.05	±0.38	±0.09
Randomised Palmstearin-Mustard :					
Plasma (mg/100 ml)	356.3	53.0	37.9	61.0	32.7
	±47.6	±3.1	±1.4	±8.3	±1.0
Liver (mg/g)	33.5	17.7	2.2	4.0	0.7
	±2.4	±1.2	±0.6	±0.6	±0.05
Randomised Palmstearin-Rapeseed :					
Plasma (mg/100 ml)	443.3	64.3	33.0	47.3	27.5
	±48.0	±9.8	±6.5	±6.2	±1.8
Liver (mg/g)	49.4	10.3	1.2	3.9	0.6
	±0.7	±0.8	±0.2	±0.5	±0.04
Randomised Palmstearin-Soyabean :					
Plasma (mg/100 ml)	290.1	70.3	52.7	52.8	21.1
	±23.6	±5.3	±5.0	±7.7	±2.0
Liver (mg/g)	42.3	11.1	1.5	2.4	0.41
	±1.1	±1.7	±0.05	±0.4	±0.03
Randomised Palmstearin-Rice bran :					
Plasma (mg/100 ml)	405.2	124.7	81.2	78.2	38.1
	±13.3	±4.2	±6.0	±6.8	±8.3
Liver (mg/g)	42.1	23.4	1.4	5.1	0.33
	±13.3	±1.4	±0.1	±0.3	±0.03

Sarkar and Bhattacharyya (94) reported that the nutritional status of high erucic mustard oil can be significantly improved after randomisation of the oil. The digestibility of mustard oil, animal growth response and food efficiency ratios improve when mustard oil is ingested as randomised oil and the results become comparable with those of low erucic rapeseed oil. The total lipid content and its components such as triglyceride, total and free cholesterol are significantly reduced in serum with dietary randomised mustard oil compared with the original mustard oil. In liver, heart and kidney tissues the cholesterol level is significantly reduced on randomisation. In serum and all tissues the erucic acid is deposited in insignificant amounts in the case of original mustard oil and randomised mustard oil.

INTERESTERIFICATION PROCESS (74)

Interesterification reaction can be carried out in both batch and continuous processes. In general, the interesterification process is a batch process in view of relatively low volume production of an interesterified product.

The batch reactor is similar to the conventional batch type bleaching unit and is equipped with vacuum, stirring, heating and cooling arrangements and with inlets for fat and catalyst. There is also a provision for nitrogen sparging to dry the fat charge further during heating under vacuum. Fig. 13.1 shows the batch reactor.

The purified fat (f.f.a 0.05%) moisture 0.1% and peroxide value (1.0) is charged into the reactor and heated to 120-150°C by steam through steam coil charged into the reactor under vacuum and with stirring along with nitrogen sparging. The temperature at which the reaction is done varies with the nature of the catalyst. With sodium methoxide powder catalyst the reaction temperature is 80-110°C. For sodium ethoxide powder catalyst the temperature required is 120-140°C. Alkali hydroxide (NaOH) plus glycerol requires initially 60°C for dispersion of the catalyst to react with f.f.a (residual) under vacuum and for drying the oil and the temperature is then raised to 140-160°C, preferably 160°C for the interesterification.

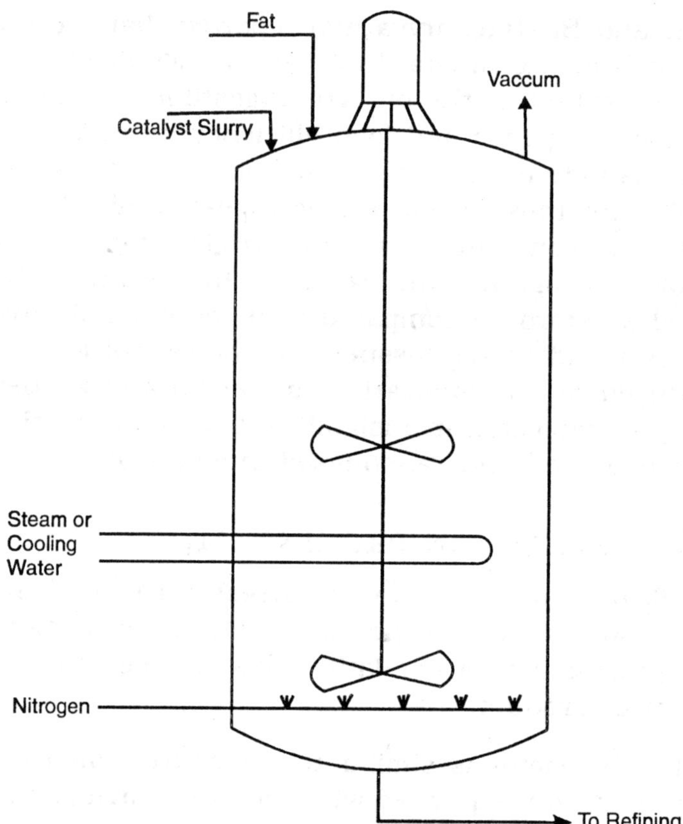

Fig. 13.1: Batch Reactor for Interesterification.

The desired amount of catalyst (0.05-0.4% w/w on oil charge) is added by suction under vacuum. The reaction which is initiated almost instantly as indicated by a sudden characteristic reddish hue colour is conducted to completion for a definite period (30 minutes to 60 minutes). After cooling to the desired temperature, the product is discharged into a vessel where water or preferably an acid (H_3PO_4, acetic, citric etc. of proper concentration) is added to deactivate the catalyst to minimise oil loss.

Next, the product is water washed in a centrifuge, dried and bleached and finally deodorised which is compulsory for the product when sodium methoxide is the catalyst.

A tentative interesterification process cost is given below:

INTERESTERIFICATION PROCESS COST (50 TONNES/DAY) : TENTATIVE (95)

A. Cost of materials	
(i) RBD Oil/Fat @ Rs. 0.30 lakh/tonne	Rs. 15.075 lakhs (Total Oil/Fat input – 50.25 tonnes) 15.1 lakhs
(ii) Catalyst :	
Sodium methoxide (1 kg/tonne) @ Rs. 200/kg	Rs. 0.10 lakh (Total catalyst = 50.25 kg)
B. Post-refining cost : @ Rs. 1000 tonne (includes steam and electricity, chemicals etc.)	Rs. 0.50 lakh
C. Cost due to oil loss : @ Rs. 0.30 lakh/tonne	Rs.0.16 lakh (Oil loss : 275 kg as Methyl Ester Oil loss : 260 kg as soap Total Oil loss : 535 kg)
Total Production cost :	Rs. 15.86 lakhs
D. Sale price of product (Interesterified) Total Product = 50 lakh tonnes Total sale price of product : @ 0.35 lakh/tonne	Rs. 17.5 lakhs
E. Profit (for each 50 tonne product before taxation etc.)	Rs. 1.64 lakh

(In case of sodium ethoxide, the loss of oil can be reduced as the ethyl esters may not have to be removed by deodorisation and oil loss as soap can be reduced by adding H_3PO_4 instead of hot water addition for destroying catalyst).

ACIDOLYSIS

Acidolysis involves the replacement of the acyl group of an ester with an acyl group from a reacting acid as shown :

$$R\,COOR' + R''\,COOH \rightleftharpoons R''\,COOR' + R.COOH$$

The reaction can be conducted without catalysts at high temperatures (250-300°C) and also by using various acidic catalysts (e.g., sulphuric acid) (96) and with certain metallic soaps (92). The reaction can be carried out under reduced pressure. The reaction is reversible and requires an excess of the displacing acid or removal of one of the products during the reaction if a high degree of replacement is desired. When applied to polyesters, the composition of final product is not specific because of the randomness in the exchange of acyl groups between the ester molecule and free acid. Acidolysis reaction involving natural fats and oils offer interesting possibilities for the production of modified fats tailored for specific industrial and edible uses.

In comparison with the ester-ester interchange, acidolysis reaction has been studied to much lesser extent. Schwartz (97) studied the acidolysis reaction of coconut oil with acetic acid with sulphuric acid as catalyst. Barsky (98) studied the acidolysis reaction involving replacement of low molecular weight fatty acids of coconut oil and butter fat by high molecular weight fatty acids derived from cottonseed oil or palm oil. Eckey (99) also succeeded in altering the composition of fats by preferential incorporation of high molecular weight fatty acids by carrying out the reaction under reduced pressure in a vessel equipped with a fractionating column and provision for steam stripping the low-boiling acids displaced. Metallic soaps like aluminium, magnesium, tin or preferably zinc soaps were used as catalysts. These were used in the form of oxides (0.10-0.25%) by weight of the charge. Palm oil enriched with stearic acid could be produced by this technology.

Chakrabarty *et al* (100) studied the acidolysis reactions of several natural oils like cottonseed oil, peanut oil, mahua oil and palm oil with lauric acid by different catalyst systems and in different relative proportions of oil and lauric acid to find out the extent and type of fatty acids displaced from an oil. Chakrabarty *et al* (101) also studied the acidolysis reaction of cottonseed oil with free fatty acids of different chain length to find out the degree of interchange of fatty acids, the mode of distribution of the free acids incorporated in the triglyceride molecules and the glyceride composition of oil after acidolysis.

Kar *et al* (102) studied the acidolysis reaction of sal fat and mohua fat and could modify their composition and properties. The acidolysis reaction has been studied in detail with sal fat and palmitic acid in relation to the operating variables like catalyst system and concentration, reaction temperature, time and molar ratios of reactants that control the reaction. The operating conditions that appeared to be optimum for the reaction between sal fat and palmitic acid are adopted for reaction between sal fat and lauric acid, mahua fat and palmitic acid and mahua fat and lauric acid.

Among the catalyst systems studied, concentrated sulphuric acid (98% w/w, sp. gr. 1.84) has been found to be the most effective catalyst in introducing maximum amount of palmitic acid in sal fat keeping other variables unchanged. The incorporation of palmitic acid in sal fat is also influenced by the amount of sulphuric acid.

The incorporation of palmitic acid in sal fat depends on the temperature employed and molar ratio of reactants used during acidolysis reaction. The highest amount of palmitic acid is introduced in sal fat when the reaction is carried out at a ca 150°C and when the sal fat and palmitic acid used are in 1:3 molar proportion.

In general, acidolysis reaction proceeds very fast and within 30 minutes the introduction of palmitic acid reaches a maximum. Depending on the operating conditions of acidolysis reaction, introduction of palmitic acid takes place to the extent of 2.5% to 39% in a random manner displacing the other component fatty acids of sal fat.

The acidolysed sal fat products with varying percentages of palmitic acid and other constituent fatty acids exhibit alterations in thermal properties in comparison with the original sal fat. The fractions, which are obtained from some of the typical acidolysed fat products by segregating from a solvent like acetone, apparently resemble some edible plastic fats of commerce in physical and chemical characteristics.

Adopting the conditions that favoured maximum introduction of palmitic acid in sal fat, it has been possible to introduce lauric acid in sal fat and mahua fat to 31.6% and 37.4% respectively and 20% additional palmitic acid in mahua fat by acidolysis reaction.

Lauric acid treated sal fat containing 31.6% lauric acid has thermal properties radically different from those of the original sal fat. Lauric acid containing mahua fat on fractionation yields a product of much improved thermal properties than those of original mahua fat. The palmitic acid content in mahua fat has been raised from 20% to 40.1% by acidolysis reaction resulting in much higher slip point. On fractionation, a product has been obtained with somewhat improved thermal properties than those of the original mahua fat indicating possibilities for use as edible plastic fats.

ALCOHOLYSIS

Alcoholysis reaction refers to the exchange of the alcohol or hydroxyl part of an ester with another alcohol or a hydroxy molecule. The ester molecules can be monoesters, diesters and polyesters and the hydroxy molecules can be monohydric alcohols, dihydric alcohols, polyhydric alcohols and even polyhydroxy compounds like carbohydrates. The reactions can take place between (i) monoesters and monohydric alcohols, (ii) monoesters and polyhydric alcohols, (iii) polyesters and monohydric alcohols and (iv) polyesters and polyhydric alcohols.

The above alcoholysis reactions are better known by the hydroxy compounds or alcohols used. Thus the reaction is referred to as methanolysis, ethanolysis, propanolysis, glycerolysis and sucrolysis etc.

Alcoholysis has been extensively investigated embracing basic and technological aspects.

Alcoholysis is a catalytic reaction, occurring with the aid of an acid catalyst or with an alkaline calalyst. The temperature also plays an important part that varies with the nature of the reactants and the nature of the catalyst. Alcoholysis reaction occurs at atmospheric pressure, but often it can be carried out at higher pressures depending on the volatility of alcohol, or under vacuum to remove one of the products.

Among the alkaline catalysts, the alkali metal (sodium or potassium) alkoxides are the most effective. Alkali metal hydroxides such as caustic soda or caustic potash or their carbonates are also effective catalysts. Certain metallic oxides like calcium, magnesium, lead, tin and zinc are used though with less reactivity.

The acid catalysts include sulphuric, sulphonic and hydrochoric acids.

Alcoholysis of polyesters with monohydric alcohols and with polyhydric alcohols have attracted the industries in producing important products of commerce to meet specific industrial and edible applications. In the following sections of this chapter these two kinds of alcoholysis reactions have been dealt with in some details covering basic parameters, technical aspects and scopes of applications.

ALCOHOLYSIS OF POLYESTERS WITH MONOHYDRIC ALCOHOLS

Alcoholysis of polyesters with monohydric alcohols like methanol, ethanol, propanol, butanol, etc. can be accomplished by using catalysts like acids or alkalies.

Alcoholysis of fats like butter fat, cocoa butter, coconut oil, tallow, and oils like linseed, castor etc. could be achieved by treatment with twice the weight of absolute methanol in presence of 1 to 2% of hydrochloric acid or phenolsulphonic acids and/or in presence of a neutral solvent such as diethyl ether, benzene, carbon tetrachloride etc.

Alcoholysis of phospholipids with various monohydric alcohols can be achieved by using hydrochloric acids at 5% on the weight of phospholipids or by the use of 7.5 to 12% of sulphuric acid giving good yields of methyl, ethyl, propyl, butyl, and amyl esters (103).

Alcoholysis of polyesters, like natural triglycerides with monohydric alcohols catalysed by alkalies has been extensively investigated.

Bradshaw and Meuly (104, 105) patented the process of alcoholysis for the production of soaps and glycerol from natural fats.

Wright *et al* (106) extended the method of Bradshaw and Meuly (104) Briefly, the process consists of the following steps:

(i) Neutral and bleached fat and anhydrous methanol (99.7%) in the molar ratio of 1 to 1.6 are heated to about 80°C with 0.1 to 0.5% caustic soda or potash as a catalyst.

(ii) After stirring for 1 hour, by which time the reaction is 98% complete, the whole mass is allowed to stand when the glycerol begins to separate at the bottom of the reactor and is withdrawn. About 90% of the glycerol separates from the mixture and the remainder is present in the upper layer of methyl esters.

(iii) Excess alcohol, alkali, and a small amount of soap remain in the upper layer. The methyl esters are rid of the impurities by repeated washings with a small amount of warm water or by acidifying the soaps followed by repeated washings with warm water.

In the above process, the fat requires pre-neutralisation by alkali refining. The removal of free fatty acids (FFA) is not required if the alcoholysis (107) is carried out under pressure (ca. 90 bar) and at high temperature (240°C). Crude oil/fat, methanol and an alkaline catalyst are preheated to 240°C and are fed into the bottom of the reactor. After flashing off the bulk of the excess alcohol the reaction mixture enters a separator. The upper layer containing the alcohol (methanol) ester and a lower layer containing the glycerine are separated. In a bubble tray column with reflux, pure methanol is topped off and then recycled into the alcoholysis step. The glycerine with a purity of 90% is removed from the ester process, and the methyl ester is transferred to a vacuum distillation unit.

The alcoholysis reaction under atmospheric pressure does not require pressure vessels resulting in lower capital investment; and the excess of methanol can be kept lower than in the pressure process.

The amount of aqueous methanol to be reworked for recycling has a significant impact on the economics because of energy costs. Removal of free fatty acid is mandatory for the process under atmospheric pressure. Alcoholysis under pressure allows the use of inferior grades of fats and oils and acid oils with a high FFA content without previous treatment.

A number of monohydric alcohol esters of fatty acids can be produced from natural polyesters like oils and fats but the actual process may need variations from the usual manner depending on the nature of alcohol. Thus while making ethyl esters of fatty acids from polyesters like the triglycerides of oils/fats it is imperative that the oil is highly purified (FFA below 0.5%) and the total water content in the reaction mixture is well below 0.3% and preferably below 0.1%.

When both FFA and moisture are above the desired limits, the required amount of the alkali will be used up in the formation of soaps, thereby resulting in incomplete ester formation or liberation of glycerol. There will be required more alcohol and alkali to get the expected alcoholysis conversion. With increased alkali, there will be required acidification of the product mixture to get the separation of glycerol.

Wright *et al* (106) had shown the effect of the variables, ethanol, water and alkali, on the yield of glycerol. Thus with dry cottonseed oil, absolute ethanol (99.3%), sodium hydroxide catalyst (1% on wt. of oil) and water not exceeding 0.11%, a 96.6% yield of alcohol esters of cottonseed fatty acids could be obtained.

While investigating the ethanolysis of peanut oil Fenge and Gros (108) found the reaction temperature to be optimum at 50°C, below or above which the yield of glycerol decreased.

In the isopropanolysis of oils, the isopropanol requires to be dried preferably by heating with metallic sodium.

The processes described above are batch in operation, but continuous processes are available for alcoholysis of fats using a variety of different alcohols and for separating and purifying the reaction products (109).

ALCOHOLYSIS OF POLYESTERS WITH POLYHYDRIC ALCOHOLS

The alcoholysis of polyesters with polyhydric alcohols is extremely important as a number of products for edible and industrial uses are produced, depending on the nature of polyesters and polyhydric alcohols.

As usual, among the polyesters, natural oils and fats constitute the important raw materials, and their conversions into a variety of edible and industrial products have been extensively studied. Among the polyhydric alcohols, glycerol, pentaerythritol, polyglycerols, mannitol and sorbitol are employed as well as the polyhydroxy compounds like glucose, sucrose etc.

The alcoholysis reaction is carried out under varying conditions of pressure, temperature, catalyst, and time of reaction. Depending on the nature of polyester, polyalcohol and the alcoholysed product the pressure and temperature of the reaction are chosen. The pressure may be atmospheric, subatmospheric (under vacuum) or above atmospheric, and the temperature may vary between 170°C and 250°C. The use of pressure in glycerolysis reactions assists in increasing the homogeneity of the reaction system (110). The effect of small amounts of water (4-5% on the glycerol) and the use of pressure favour glycerolysis reaction as reported by Dey *et al* (111).

The effective catalysts for the alcoholysis reaction are normally the alkalies and their carbonates and alcoholates. The oxides of certain metals like calcium, magnesium and lead are used although they are relatively less active.

The alcoholysis reaction is referred to by the name of the polyol used. Thus when glycerol is used as a polyol, the alcoholysis reaction is called glycerolysis. Similarly, it is called sucrosolysis when sucrose is used as the polyhydroxy compound.

The glycerolysis reaction has been very extensively investigated. In fact, the reaction is a classical one and a great deal of information is available covering the basic aspects like catalyst selection, temperature of the reaction, pressure maintained during the reaction and time of reaction. When the objective of the glycerolysis reaction is the preparation of monoglycerides, the amount of glycerol in excess over the theoretical is necessary. The excess may vary between 20 and 100% over the theoretical, which is equivalent to 25-40% of the weight of fat glycerol.

The temperature is generally around 225-240°C although variations like 170 to 180°C, 210 to 220°C and 240 to 250°C can be employed. The miscibility of fat with glycerol increases with increased reaction temperature. Sodium hydroxide is the usual alkaline catalyst used at 0.05-0.20% on fat charge but hydrated lime $Ca(OH)_2$ at 0.06-0.10% yields relatively much lighter coloured products.

The time at which the glycerolysis reaction is complete varies from 1-4 hours although 5-6 hours may be needed depending on the temperature and nature of catalyst.

The glycerolysis reaction proceeds in a random manner and leads to an equilibrium composition of monoglycerides, diglycerides and triglycerides. The composition of the various neutral glycerides and free glycerol can be calculated on the basis of the molar proportions of total hydroxyl groups that are esterified and unesterified (112). Thus, if one mole of a neutral fat is reacted with 3 moles of glycerol, 25% of the total hydroxyl groups will be esterified and 75% will be unesterified. At equilibrium, then, the molar percentages of the different neutral glycerides and glycerol will be as follows :

Per cent Triglycerides = (25.25.25) 1/10,000 = 1.56%

Per cent Diglycerides = (25.25.75) 3/10,000 = 14.06%

Per cent Monoglycerides = (25.75.75) 3/10,000 = 42.19%

Per cent Free Glycerol = (75.75.75) 1/10,000 = 42.19%

In actual manufacturing practice for Mono- and Diglycerides, the fat charge properly refined, bleached and deodorised is taken in a special kettle made of stainless steel. To this is added about 25-40% of its own weight of glycerol and 0.1-0.2% of sodium hydroxide or hydrated lime $Ca(OH)_2$ at 0.06-0.10% level as catalyst. The charge is heated by mechanical stirring under vacuum at about 100°C to remove any water present and dissolved air. The blanketing by nitrogen is extremely useful in protection against oxidation and discolouration. The temperature is raised rapidly to 200°C and then upto 230-240°C. The glycerol may be circulated in the charge to direct the reaction of the desired side. The reaction is complete in 4-6 hours and the mass is cooled down to 90-100°C and a concentrated solution of phosphoric acid is added to neutralise the catalyst and the mass is cooled to 60-70°C and is left to settle. After settling, first glycerine and then monoglycerides are discharged. The monoglyceride sample may have a composition like 60-70% monoglycerides, 30-35% diglycerides, and 3-4% glycerol.

Alternatively, monoglyceride preparations may be washed through with difficulty and much unsatisfactorily with strong solutions of sodium chloride, sodium sulphate etc. for the removal of free glycerol.

The monoglyceride content of the technical monoglyceride samples can be enriched by the use of molecular distillation or by extracting out the monoglycerides in 90% ethanol followed by distilling off the ethanol completely.

Monoglycerides can be produced by the glycerolysis reaction of methyl esters of fatty acids (113,114). The methyl esters can be from a vegetable oil or of a fatty acid and 1 mole of methyl ester to 1.5-4.0 moles of glycerol are employed .

The catalyst is as usual potassium hydroxide at 0.1% level. The reaction temperature is 215-220°C. The reactants have been converted by Matsuyama *et al* (113) into 1-monoglycerides assaying 50.6-54.9%. The reaction is carried out under vacuum to remove methanol and about 30 minutes are required for completion of the reaction. The short chain saturated acid esters and unsaturated acid esters give higher monoglyceride yield than the long chain saturated acid esters like palmitate and stearate (114).

Matsuyama *et al* (115) have also prepared 99.0% minimum purity 1-monopalmitin by high vacuum distillations and by crystallisations from a selective solvent of the monopalmitin (60.8%), prepared by glycerolysis of high purity methyl palmitate. The distillation at about 150°C and 0.02 mm removes the excess glycerol and another distillation at ca. 160°C and 0.008 mm gives a distillate, which, after two crystallisations from 1:1 hexane ether, yields the above high purity product.

Alcoholysis of polyesters like natural triglycerides (fats and oils) with polyhydroxy compounds like tetrahydroxy or higher polyhydroxy alcohols to make products for coating, detergent and food applications has also been extensively studied by Burrell (116) and by Wright *et al* (106). According to them, the alcoholysis of glycerides with high boiling alcohols like pentaerythritol, mannitol, sorbitol, and polyglycerols can be carried out at 200-250°C with 1.0-1.5 equivalents of the alcohol along with 0.1-2.0% of a catalyst, such as lead salt or an alkaline compound. Glycerol is removed continuously by stripping with steam under reduced pressure of 40-300 nm. Wright *et al* (106) reacted pentaerythritol with the methyl esters of linseed oil fatty acids at 225-280°C and atmospheric pressure with the use of 1% lead as catalyst, added as napthenate and could get products with acid values of 4-8 after 1-4 hours reaction. Mleziva and Hanzlik (117) could prepare pentaerythritol esters of linseed oil fatty acids by reacting linseed oil (2320 parts) for 4 hours in CO_2 atmosphere at 260°C with pentaerythritol (618 parts) which was added with stirring over a period of 2 hours. The heating was continued for 9 additional hours. The product after 9 hours contained: 27% pentaerythritol monoester, 53% pentaerythritol diester, 15.9% monoglyceride, and 4.1% pentaerythritol polyesters and triglycerides.

Alcoholysis of fats or methyl esters of fatty acids of fats with polyhydroxy compound like sucrose has also been studied in detail for making sucrose esters. The reaction can be carried out in presence of a solvent or without using any solvent. As usual there needed a catalyst which is again a standard alkaline catalyst like sodium methoxide, potassium carbonate or a soap catalyst like lithium oleate or mixtures of sodium, potassium, or lithium soaps as emulsifier catalysts. Solvents include dimethyl formamide, dimethyl sulphoxide, or dimethyl benzylamine. The solvents are however expensive and difficult to remove.

To avoid the use of solvents, Osipon and Rosenblatt (118) developed a 'microemulsion process' for the preparation of sucrose esters. Sucrose, methyl ester of a fatty acid (stearate or palmitate), sodium stearate, and potassium carbonate suspended in propylene glycol were microdispersed until a transparent emulsion was formed. Using a starting molar ratio of 1.5:1:0.9 for sucrose, methyl stearate and sodium stearate, the slow distillation of propylene glycol was continued until reaction completion. The product assayed 85% sucrose monostearate and 15% sucrose distearate.

Feuge *et al* (119) were able to prepare sucrose esters of fatty acids without using any solvent by interesterifying sucrose with esters like methyl 'carbitol' palmitate, monopalmitin, distearin, and 40% technical mono- and diglycerides (from hydrogenated cottonseed oil by glycerolysis) at 170-187°C without extensive decomposition using lithium oleate or mixtures of sodium, potassium or lithium soaps as emulsifier catalysts. With lithium oleate the products were 90% tetra esters or higher.

James (120) described the preparation of sucrose monoesters from non solvent media at 125°C. The process involves the sucrosolysis of tallow methyl esters, with potassium carbonate and an emulsifier consisting of sucrose monoester and tallow fatty acids. The Proctor & Gamble Company (121) developed a non solvent two step process for the preparation of sucrose polyesters(SPE). In the first step, methyl esters and sucrose at 3:1 molar ratio are reacted with potassium soap emulsifier to form low sucrose esters. In the second step, more methyl esters are added and reacted to produce SPE in yields upto 90% based on sucrose.

REFERENCE

1. Friedel, *C. and Crafts*, J.M. Ann., **133**, 207 (1865)
2. Reid, E.E., in P.H. Groggins "*Unit Process in Organic Synthesis*" McGraw Hills Book Co., New York (1955), pp. 508-511
3. Markley, K.S., *Fatty Acids Pt II*, Interscicence Publishers, New York (1961) pp. 885-892
4. Baily, A.E., "*Industrial Oil and Fat Products*" 3rd ed., Interscience Publishers, New York (1964) pp. 182, 198, 735-743, 771-783, 958-972
5. Marcinkiewicz, J. and Niewiadomiski, *H. Rev. Fr. Crops. Gras.* **15** 8-9, 511 (1968)
6. Sreenivasan, B., *J.A.O.C.S.*, **55**, 11, 796 (1978)
7. Olwerke Germania G.m.b.H. and Norman,W., *German Pat.* 417, 215 (Sept. 7, 1925)
8. Gooding, C.M., *U.S., Pat.* 2,309,949 (Feb. 2, 1943)
9. Vanloon, C., (to N.V. Anton Jurgens Margarinefabrieken), *British Pat.* 249916 (Dec. 30, 1924), *Dutch Pat.* 16703 (Aug. 15, 1927), *U.S. Pat.* 1,744,596 (Jan. 21, 1930) *U.S. Pat.* 1,373,513 (Aug. 13, 1932)
10. Eckey, E.W., *U.S. Pat.* 2,378,005 (June 12, 1945)
11. Eckey, E.W., *U.S. Pat.* 2,378,006 (1945)
12. Eckey, E.W., *U.S. Pat.* 2,442,531 (June 1, 1948)

13. Chakrabarty, M.M. and Bhattacharyya, D., *Indian Pat.* 89943
14. Abott, A.D., *U.S. Pat.* 2,442,538 (June 1, 1948)
15. Eckey, E.W., *U.S. Pat.* 2,442,533 (June 1, 1948)
16. Eckey, E.W., *U.S. Pat.* 2,442,536 (June 1, 1948)
17. Eckey, E.W., *U.S. Pat.* 2,442,537 (June 1, 1948)
18. Norris, F.A. and Mattil, K.F., *Oil and Soap,* **33**, 289 (1946)
19. Norris, F.A and Mattil, K.F., *J.A.O.C.S.*, **34**, 274-275 (1947)
20. Eckey, E.W., *Ind. Eng. Chem.* **40**, 1183-1190 (1948)
21. Baur, F.J., *J.A.O.C.S.* **31**, 196 (1954)
22. Wiedermann, L.H., Weiss, T.J., Jacobson, G.A. and Mattil, K.F., *J.A.O.C.S.*, **38**, 389 (1961)
23. Formo, M.W., *J.A.O.C.S.*, **31**, 548 (1954)
24. Cowan, J.C., *J.A.O.C.S.*, **27**, 492 (1950)
25. Kaufmann, H.P. and Grothues, B., *Fette Seifen Anstrichmittel,* **62**, 425-429 (1959)
26. Chakrabarty, M.M., Talapatra, K. and Das, S., *Fette Seifen Anstrichmittel,* **68**, 310 (1966)
27. Spadare, J.J., Lovegren, N.V., Feuge, R.O. and Patton, E.L., *J.A.O.C.S.*, **38**, 461 (1961)
28. Herman, D., and Weihe, J., *Dairy Science,* **44**, 944 (1961)
29. Freeman, I.P., *J.A.O.C.S.*, **45**, 456 (1968)
30. Chakrabarty, M.M. and Talapatra, K., *Ind. J. Appl. Chem.* **29**, 19-21 (1966)
31. Vanderwal, R.J. and Van Akkeren, L.A., *U.S. Pat.* 2,571,315 (1951)
32. Mattil, K.F. and Nelson, D., *U.S. Pat.* 2,625,483 (1953)
33. Dominick, W.E., Nelson, D. and Mattil, K.F., *U.S. Pat.* 2,625,484 (1953)
34. Howley, H.K. and Dobson, R.D., *U.S. Pat.* 2,733,251 (1956)
35. Holman, G.W. and Going, L.H., *U.S. Pat.* 2,875,066 (1959)
36. Durgore, J., Mott, C.W. and Soiden, P., *U.S. Pat.* 3,170,798 (1965)
37. Katsuy Niki, Iliroto Suzuki and Koji Ito, Miyoshi Oil and Fat Co. Ltd., Tokyo, Japan
38. Holman, G.W., Laylock, T.W. and Willie, R.L. *U.S. Pat.* 2,738,278 (1956)
39. Brockaw, G.Y., *British Pat,* 791, 165 (1958)
40. Nelson, D. and Mattil, K.F., *U.S. Pat.* 2,625,486 (1953)
    Nelson, D. and Mattil, K.F., *U.S. Pat.* 2,625,487 (1953)

41. Weiss, T.J., Jacobson, C.A. and Weidermann, L.H., *J.A.O.C.S.*, **38**, 396 (1961)

42. Grun, A., Ztsohr Angew Chem., **38**, S27 (1925)

43. Eckey, E.W., *U.S. Pat.* 2,442,532 (June 1, 1948)

44. Kaufmann, H.P. and Grothues, B., Fette Seifen Anstrichmittel **61**, 425 (1958)

45. Adhikari, S., Dasgupta, J., Bhattacharyya, D.K. and Chakrabarty, M.M. Fette Scifen Anstrichmittel **83**, 262-267 (1981)

46. Adhikari S., Dasgupta, J., Chakrabarty, M.M. and Bhattacharya, D.K., Fette Seifen Anstrichmittel **84**, 146-149 (1982)

47. Adhikari, S., Dasgupta, J. Bhattacharyya, D.K. and Chakrabarty M.M., Fette Seifen Anstrichmittel **83**, 345-347 (1961)

48. Adhikari, S., Dasgupta, J., Chakrabarty M.M. and Bhattacharyya, D.K., Fette Seifen Anstrichmittel **84**, 185-188 (1982)

49. Majumdar, S. and Bhattacharyya, D.K., Oleagineux, **41**, 235-240 (1986)

50. Majumdar, S. and Bhattacharyya, D.K., J. Oil Tech. Assocn. India, **18,** 37 (1986)

51. Khatoon, S. and Bhattacharyya, D.K., J. Oil Tech Assocn. India, **17**, 58 (1985)

52. Bhattacharyya, D.K. and Ghosh, S., 1993 PIPOC Porim International Palm Oil Congress Update and Vision, Chemistry and Technology CT6, 20-25, September, 1993, Kuala Lampur, Malaysia.

53. Ghosh, S. (personal communication), Ph.D (Tech), Thesis, Calcutta University (1995)

54. Separator, A. and Aktiebolaget, S.O., *British Pat.* 682, 797 (Nov. 19, 1932), C.A. **47**, 4635f. (1953)

55. Miller, R., U.S. Pat. 2,688,626 (Sept. 7, 1954), C.A. **49**, 1349e (1955)

56. Chakrabarty, M.M. and Bhattacharyya, D., Fette Seifen Anstrichmittel **74**, 389-393 (1972)

57. Kaufmann, H.P. and Schnurbusch, H., Die Umesterung Von Fettn and Olen, Westdentscher Verlag, Koln and Opladen (1963)

58. Artman, N.R., Carter, R.M. and Whyte, D.D., *U.S. Pat.* 3,376,326 (1968)

59. Chakrabarty, M.M. Gayen, A.K. and Chakrabarty, M.K., Fette Seifen Anstrichmittel **81**, 233-236 (1979)

60. Sen, G., Ph.D. Thesis, Calcutta University (1982)

61. Bhattacharyya, D.K. and Chakrabarti, R., Fat Science, Proceedings 16th ISF Congress, Budapest, pp. 517-528 (1983)

62. Normann, W., *German Pat.* 417,215 (September 7, 1925)

63. Konen, J.C., Clocker, E.T. and Cox, R.P., Oil & Soap, **22**, 57-60 (1945)

64. Koslowsky, L., Oleagineux, **30**, 221-224 (1975)

65. Bhattacharyya, D.K. and Banerjee, K., J. Am. Oil Chem. Soc., **60**, 841 (1983)

66. Bhattacharyya, D.K. and Banerjee, K., J. Oil Technologists' Assocn., India, **13**, 123 (1981)

67. Dasgupta, J., Ph.D. Thesis, Calcutta University (1982)

68. Sil Roy, S. and Bhattacharyya, D.K., J. Am. Oil Chemists' Soc., **70**, 1293 (1993)

69. Freeman, I.P. *ibid*, **45**, 456, (1968)

70. Barrett, C.B., Dallas, M.S.J. and Padley, F.B., J. Am. Oil Chemists' Soc., **40**, 580 (1963)

71. Chakrabarty, M.M. and Bhattacharyya, D.K., J. Chromatography, **31**, 556 (1967)

72. Sil Roy, S. and Bhattacharyya, D.K., J. Am. Oil Chemists' Soc. **70**, 923 (1993)

73. Sil Roy, S. and Bhattacharyya, D.K., J. Oil Tech. Assocn. India, **21**, 31-32 (1989)

74. Sreenivasan, B., *ibid*, **55**, 802 (1978)

75. Babia, A., Oleagineux, **29**, 375 (1974)

76. List, G.R., Othozfer, W.N. and Mounts, T., INFORM, 5, Paper 31c, 1994 AOCS Annual Meeting Abstracts

77. Roylance, A., *U.S. Pat.* 2,928,745 (March 15, 1960)

78. Gooding, C.M. and Cravens, C.A., *U.S. Pat.* 3,085,882 (1963)

79. Bell, R.J., Campbell Jr., R.L., Gibson, P. and Sims, J.F., *U.S. Pat.* 3,396,037 (1968)

80. Kidger, D.P., *U.S. Pat.* 3,361,568 (1968)

81. Ainger, G.E. and Caverly, B.L.; Ger. Offen., 2,200,461, Aug. 10, 1972, Brit.

82. Spadaro, J.J., Lovegren, N.V., Feuge, R.O. and Patton, E.L., J. Am. Oil Chemists' Soc. **38**, 461 (1961)

83. Kane, J.G., Fette Seifen, Anstrichmittel, **68**, 1 (1966)

84. Shitole, A.D., Subrahmanyan, V.V.R. and Kane, J.G., J. Oil Tech., Assn. India, **2**, 27 (1970)

85. Bhattacharyya, D.K., Majumdar, S. and Basu, R., Fat Science Proc. 16th ISF Congress, Budapest, pp. 529 (1983)

86. Alfin-Slater, R.P., Aftergood, L., Hansen, H., Morris, R.S., Melnick, D. and Gooding, C.M., J. Am. Oil Chemists' Soc., **43**, 110-112 (1966)

87. Chakrabarty, M.M. and Bhattacharyya, D.K., Fette Seifen Anstrichmittel, **71**, 37-39 (1969)

88. Jakubonski, A. Rev. Fr. Crops. Gros. **18,** 93-101 (1971)

89. Duterte, R. *ibid*, **19**, 587-597 (1972)

90. Kar, K., Chakrabarty, M.M. and Bhattacharyya, D.K., Fette Seifen Anstrichmittel, **87**, 486-489 (1985)

91. Majumdar, S. and Bhattacharyya, D.K., Oleagineux, **41**, 393-399 (1986)

92. Majumdar, S. and Bhattacharyya, D.K., *ibid*, 29-33 (1988)

93. Majumdar, S. Bhattacharyya, A.C. and Bhattacharyya, D.K., Proceedings of ISF-JOCS Congress, Tokyo, September (1986)

94. Sarkar, S. and Bhattacharyya, D.K., Oleagineux, **47**, 713-714 (1992)

95. Bhattacharyya, D.K., *Soaps, Detergents and Toiletries Review*, p. 15 (June 1993)

96. Markley, K.S., "*Fatty acids, Their Chemistry, Properties, Production and Uses*", 2nd edition, Part-2, Interscience Publishers, Inc., New York, p. 858 (1961)

97. Schwartz, G.L. (to E.I. du Pont de Nemours & Co.), *U.S. Pat.* 1,588,299) (October 20, 1925)

98. Barsky, G. (to Wecoline Products, Inc.) *U.S. Pat.* 2,182,322 (December 5, 1939)

99. Eckey, E.W. (to Procter & Gamble Co.), *U.S. Pat.* 2,378,006 (June 12, 1945)

100. Chakrabarty, M.M. and Talapatra, K.J. Am. Oil Chemists' Soc., **45**, 172 (1968)

101. Chakrabarty, M.M. Bhattacharyya, D and Gayen, A.K., Fette Seifen, Anstrichmittel, **77**, 468 (1975)

102. Kar, K., Chakrabarty, M.M. and Bhattacharyya, D.K. Fette Seifen, Anstrichmittel, **87**, 278-282 (1985)

103. Shinowara, G.Y. and Brown, J.O., Oil & Soap, **15**, 151-152 (1938)

104. Bradshaw, G.B. and Menley, W.G., *U.S. Pats.* 2,271,629 (February 3, 1942) and 2,360,844 (October 24, 1944)

105. Bradshaw, G.B., Soap Sanit. Chemicals, **18**, No. 5, 23-24, 69-70 (1942)

106. Wright, H.J., Segan, J.B., Clark, H.V., Coburn, S.K., Langdon, E.E.; and Duphis, R.N.; Oil & Soap, **21**, 145-148 (1944)

107. Krentza, K.R., J. Am. Oil Chemists' Soc., **61**, 346 (1984)

108. Feuge, R.O. and Gros, A., *ibid*, **26**, 97 (1949)

109. U.S. Pats. 2,383,579; 2,383,580-581; 2,383,596; 2,383,599; 2,383,614; 2,383,632-633, August 28, 1945 assigned to Colgate-Palmolive - Peet Co.

110. Kochar, I.R.K., and Bhatnagar, R.K., Indian Patent to Council of Scientific and Industrial Research, 71,979 (April 7, 1962)

111. Dey, S.K., Kochar, I.R.K., and Bhatnagar, R.K. Indian Oilseeds J., **6**, 215-225 (1962)

112. Bailey's Industrial Oil and Fat Products, edited by Daniel Swern, Interscience Publishers, New York (1964) p. 966

113. Matsuyama, S., Takasago, M. and Hirokawa, K., Kagaku to Kogyo (Osaka), **42**, 239-242 (1968)

114. Schmalinzon, C., Yaron, A. and Letan, A; Riv. Ital. Sostanze Grasse, **48**, 168-169 (1971)

115. Matsuyama, S., Takasago, M., Horikawa, K; Fujiwara, F. and Sugimoto, C., Kagaku to Kogyo (Osaka), **42**, 242-248 (1968)

116. Barrell, H., *Oil Soap*, **21**, 206-211, 1944; *Ind. Eng. Chem.* **37**, 81-89, 1945, *U.S. Pat.* 2,360,394; 1944 (to Heyden Chemical Corp.)

117. Melziva, J. and Hanzlik, V., Fette Seifen, Anstrichmittel, **60**, 197-201 (1958)

118. Osipon, L.I. and Rosenblatt, W., J. Am. Oil Chemists' Soc. **44**, 307-309 (1967)

119. Feuge, R.O., Zeringne, H.J. Jr.; Weiss, T.J. and Brown, M., *ibid*, **47**, 56-60 (1970)

120. James, K. (Tata and Lyle, Ltd.), German Pat. 2,546,716 (April 29, 1976)

121. Rizzi, G.P. and Taylor, H.M., J. Am. Oil Chemists' Soc., **55**, 396-401, 1978; U.S. Patent 3,963,699 (to Procter & Gamble Co. 1, (June 15 , 1976)

CHAPTER 14

Blending

Compositions and properties of individual fats and fat fractions can be significantly modified by blending with other fat(s) or fat-fraction(s) having specific compositions and properties. This kind of blending results invariably in the products displaying compositions and properties functionally more desirable, suited and tailored to making value-added products to cater to edible and industrial consumptions. Blending of fats or fat fractions has been extensively examined and formulated for the purpose of making cocoa butter substitutes, margarines and shortenings. Blending of vegetable oils to obtain nutritionally better oils at relatively cheaper price and fulfil the consumers' acceptability and preference is also recognised. Certain additives are also blended with edible oils, fats and their products to improve their characteristics such as shelf-life, frying character, crystallisation behaviour, nutritional character etc.

BLENDING OF FATS AND FAT FRACTIONS

One of the important applications of fat blending is in the area of confectionery fats and chocolates. In good quality chocolates, the fat phase is invariably the cocoa butter. The unique chemical composition and physical properties make cocoa butter ideally suited in chocolates to impart eating qualities.

The price of cocoa butter is soaring and for the past few decades efforts have been made to develop technology for making cocoa butter substitutes in total or partial replacements of cocoa butter. The natural hard fats, the so called vegetable butters and other natural fats, with compositions and properties close to those of cocoa butter can be blended as such or following fractionations into the appropriate fat fractions. The new blends are evaluated in respect of their compatibility with cocoa butter without affecting its properties.

Sal stearin produced in 76-78 per cent yield from the fractionation of sal fat from acetone at 7°C is harder than cocoa butter and has a sharp melting point of 39°C. This stearin can be used at 20-40 per cent as replacement fat in the manufacture of plain chocolates, for example, 13 per cent sal stearin and 33 per cent cocoa butter (1). Sal stearin can be blended with palm mid fraction (PMF) at 40-60 per cent level and the blend can be used with cocoa butter in plain or milk chocolates.

Baliga and Shitole (2) fractionated mango kernel fat from acetone using 7 ml/g at 10°C for 4 hours and obtained in 57.2% yield mango stearin having 34.7 I.V., 38.6 slip point; T max 36.1; Δ T 9.3 (Fraction I). Two stage fractionation was done by first topping at 26°C followed by further cooling at 5°C. The fraction II was obtained in 43.1% yield having I.V. 35.8; slip point 37.2; T max 34, Δ T 8.7. Fractions I and II were mixed with equal amounts of cocoa butter and the properties obtained were as follows:

Characteristics	*Cocoa butter*	*Product I*	*Product II*
I.V.	34.5	34.4	37.1
Slip point (°C)	35.5	37.3	36.3
Solidification Test			
T max	29.0	33.3	31.8
Δ T	6.5	7.9	7.7
SFI : 25°C	83.5	89.0	80.0
30°C	64.0	78.0	64.0

Properties of products I and II are closer to those of cocoa butter, and the fractions are compatible with cocoa butter.

Ghosh Chaudhuri (3) fractionated mango kernel fat from acetone (1:7 w/v, 15°C) to yield a mango stearin fraction in 50% yield (I.V. 37.4, slip point 36.7°C, T max 33.0, Δ T 8.7, SFI : 83.5 at 20°C, 74.3 at 30°C and 31.4 at 35°C) and blended 87 parts of this stearin fraction with 17 parts of the weight of palm mid fraction. The blend characteristics were :

Characteristics	*Blend*
Slip Point °C	35.6
T max	31.1
Δ T	6.9
SFI °C	
20	68.3
30	48.2
35	23.0
40	0

The characteristics of the blend indicate that cocoa butter substitute of high quality can be made from mango kernel stearin and palm mid fraction.

Traitler and Dieffenbacher (4) observed strong interactions between PMF and sal stearin, shea stearin etc., resulting in maximum depression of the solid contents at 30°C at 30% addition of foreign fats (stearins). The authors examined the physical properties of a blend of 50% PMF with mango stearin and pointed out that CBE's can be formulated according to the desired physical properties.

Khatoon and Bhattacharyya (5) reported that cocoa butter substitutes with satisfactory thermal properties can be produced from the blends of suitable palm mids and sal stearins or sal mids. Some data are shown in Table 14.1.

A report (6) is available on the solidification and S.F.I. characteristics of the blends of sal stearin and mango kernel fat stearin as pointed out in Table 14.2 indicating highly satisfactory CBS products. The thermal properties of two samples of cocoa butter in the form of blends with sal stearin and mango kernel fat stearin are also included in this report.

Table 14.1 : Thermal Properties of Blended Products of Palm Mid Fractions with Sal Stearin or Sal Mids

Thermal Properties		$PMF_2 + S_2(BL_3)$ 15 / 85	$PMF_2 + S_5(BL_7)$ 10 / 90	$PMF_1 + S_6(BL_{11})$ 10 / 90	$PMF_1 + S_3(BL_4)$ 10 / 90	$PMF_2 + S_4(BL_6)$ 15 / 85
Slip point (ºC)		31.0	32.0	32.6	30.0	32.5
Super cooling temp. (ºC) time, min.		21.5º ; 20 min.	22.9º ; 16 min.	23.1º ; 14 min.	21.8º ;17 min.	20.55º; 19 min.
Solidification temp. (ºC) time, min.		28.85º ; 30 min.	28.7º ; 23 min.	29.45º ; 23 min.	28.3º ; 27 min.	27.45º ; 32 min.
Rise		7.3	5.8	6.3	6.5	6.9
T (ºC/min.)		0.73	0.82	0.7	0.65	0.53
SFI at	15 C	71.3	72.3	75.7	69.8	71.8
	20 C	68.0	69.7	74.1	64.7	66.2
	25 C	55.4	59.3	64.5	49.8	53.5
	30 C	20.0	26.1	42.9	16.2	18.4
	35 C	0.2	1.0	2.8	–	0.6

PMF_1 : Palm mid fraction 1 having slip point of 35ºC and SFI as 59.4/15º, 52.0/20º, 39.5/25º, 25.0/30º, 13.6/35º, 8.0/40º.

PMF_2 : Palm mid fraction 2 having slip point of 30.2ºC and SFI as 70.4/15º, 51.9/25º; 16.2/30º, 1.4/35º.

S_2 : Sal stearin of slip point 31.0ºC with SFI as 74.0/15º, 73.0/20º, 64.4/25º, 27.5/30º, 0/35º.

S_5 : Sal stearin of 32.0ºC slip point and SFI as 76.3/15º, 73.8/20º, 66.7/25º, 39.7/30º, 2.0/35º.

S_6 : Sal stearin of 33.6ºC slip point and SFI as 78.1/15º, 77.6/20º, 72.4/25º, 50.1/30º, 3.4/35º.

S_3 : Sal mid of 30.1 slip point and SFI as 72.9/15º, 71.0/20º, 63.7/25º, 28.0/30º, 0.35º.

S_4 : Sal mid of 30.8 slip point and SFI at 74.9/15º, 72.2/20º, 65.1/25º, 32.8/30º, 0.7/35º.

Table 14.2 : Physical Properties of Sal stearin, Mango stearin, Cocoa butter as individual and as blends.

Blended Fat	*Solidification behaviour*					*SFI at °C*				
	Super cooling Temp. °C/time, Min.	*Solidification Temp. °C/ Time, Min.*	*Rise °C*	*T °C/ Min.*	*Slip point °C*	*20*	*25*	*30*	*35*	*40*
Sal stearin (crystallised from 3 vol.Hexane at –3°C/4 hr)	21.4/19	29.4/30	8.0	0.72	32.5	73.8	66.7	34.7	2.0	–
Mango kernel fat stearin 1 (crystallised from 3 vol. Hexane at 8°C/ 18 hr)	23.3/9	33.4/20	10.6	0.96	35.5	77.0	75.0	63.5	54.7	44.2
Mango kernel fat stearin 2 (crystallised from 3 vol. Hexane at 8°C/20 hr)	24.5/12	34.7/21	10.2	1.13	34.6	78.9	71.0	60.6	51.9	40.4
Cocoa butter A (Australian)	21.1/15	27.4/23	6.3	0.79	29.0	79.5 (Tempered at 26°C/40 hrs.)	73.5	48.1	1.8	–
Cocoa butter S (Swedish)	20.8/15	27.8/25	7.0	0.71	30.3	85.4 (Tempered at 26°C/40 hrs.)	82.0	57.7	–	–

Table 14.2 (contd.)

Table 14.2 (contd.)

Blended Fat	*Solidification behaviour*					*SFI at °C*				
	Super cool-ing Temp. °C/time, Min.	*Solidifi-cation Temp. °C/ Time, Min.*	*Rise °C*	*T °C/ Min.*	*Slip point °C*	*20*	*25*	*30*	*35*	*40*
Cocoa butter A + Mango kernel fat stearin (1:1)	21.9/13	30.3/15	8.4	0.70	32.1	79.9	70.9	61.1	39.0	–
						(Tempered at 26°C/40 hrs.)				
Cocoa butter S + Mango stearin 2 (1:1)	22.7/12	29.3/22	7.1	0.71	32.8	84.0	72.8	62.4	41.0	3.3
						(Tempered at 26°C/40 hrs.)				
Sal stearin + Mango stearin 1 (85:15)	21.65/19	30.05/32	8.4	0.64	33.0	70.9	64.4	37.6	6.8	–
						(Tempered at 15°C/16 hrs.)				
Sal stearin + Mango stearin 2 (70:30)		31.6/31	9.8	0.75	33.7	72.8	67.9	44.7	1.7	–
						(Tempered at 15°C/16 hrs.)				
Sal stearin + Cocoa butter A (1:1)	20.1/17	27.2/26	7.1	0.78	31.5	83.0	80.5	64.6	22.8	–
-do- S (1:1)	21.0/15	28.4/27	7.4	0.61	31.5	82.4	76.5	55.5	24.8	2.3

In the formulation of a CBE, the PMF rich in POP is blended with vegetable fats rich in SOS and POS fractions. Illipe and Shea are typically used as sources of SOS and POS. Illipe needs no fractionation and can be directly blended while Shea requires fractionation to provide a source rich in SOS and POS. By carefully blending these sources, it is possible to produce a CBE that may be combined with CB in any ratio.

Kane *et al* (7) blended a harder fraction, also called least soluble fraction (LSF), from mowrah with unfractionated kokum in different proportions and could obtain products with a clear match to cocoa butter. The mowrah hard fraction was obtained by first crystallising the fat from acetone (7 ml/g; 0°C; 3 hrs.) followed by recrystallising the solids at higher temperature (5 ml acetone/g; 12-14°C, yield 20-23%.

The fraction was mixed with kokum at 70:30, 60:40, proportion. The properties are as follows:

	I.V.	*M.P. °C*	*Super cooling °C*	*ΔT*	*Solid content index*		
					30°	*35°*	*37°*
Cocoa butter	37.5	34.5	20.8	8	61.9	2.3	0.0
Kokum	33.5	42	27	11.9	98.5	92.8	79.7
Mowrah hard fraction	40.2	34.5	23	4.2	40	2.2	0.0
Mowrah hard fraction + Kokum (70:30)	38.2	36.3	23.5	5.6	65.3	25.7	1.8
Mowrah hard fraction + Kokum (60:40)	37.5	37	24.9	6.8	80.6	48.0	4.5

At 30 per cent kokum incorporation, the product is closer to the properties of cocoa butter. The melting point and solid fat index at 25°C are higher but this should be advantageous for use in tropical countries in white or dark chocolates.

Dhupa (*Vateria indica*) fat contains nearly 67-70% (PU St & St U St) symmetrical triglycerides and is an important material for cocoa butter substitute. Its stearin fraction is suitable for formulating a CBS in admixture with PMF or mowrah least soluble fraction (LSF).

Phulwara (*Madhuca butyraea*) can be fractionated to enrich POP glycerides to nearly the same level as the PMF. The blend of unfractionated phulwara with other fat rich in St U St can offer CBS.

BLENDING OF FATS IN MARGARINE FORMULATIONS AND IN SHORTENINGS (VANASPATI)

In formulating margarines of various kinds, there is invariably required a variety of fats and oils and in fact, the margarines contain blends of a variety of plastic fats and liquid oils. The guideline in the selection of fat bases is mainly premised on the SFI characteristics at different temperatures. As long as the SFI/ SFC profiles satisfy the spreadability and taste in terms of aroma and other butter features as well as the storage behaviour etc., the manufacturers tend to use the base stocks more or less unaltered in their product formulations.

A great deal is known about soyabean oil based margarine formulations involving hydrogenated soyabean oil of varying iodine value, congeal point and solid fat index and soyabean oil and other liquid oils. The proportions of hydrogenated stock and liquid soyabean oil are variable depending on the types of stock margarines (8).

A blend consisting of 18% soyabean oil, 57% hydrogenated soyabean oil (53% slip melting point 28°C, I.V. 105; plus 4% slip melting point 36°C, I.V. 50), 21% interesterified palm oil, and 4% hydrogenated palm oil (slip melting point 45°C, I.V. 45) resulted in a product having 33.3°C slip melting point which could be used in making a margarine with pleasing texture and ready spread (9).

A margarine fat suitable for preparing a tub margarine can be prepared by blending 80% of an interesterified fat phase (made by randomisation of 20% sunflower oil, 40% hydrogenated soyabean —M.P. 41°C, palm kernel fat 25% and palm fat 15%) with 20% sunflower oil. The D_t values of the fat base are : 650/10°C 355/ 20°C, 240/25°C, 95/30°C and 10/36°C.

A margarine fat (10) can be prepared by randomising a blend of 30% palm oil, 30% palm stearin (M.P. 55°C) and 40% palm kernel oil and blending the randomised product with liquid vegetable oils containing high levels of polyunsaturated fatty acids in an appropriate proportion, preferably 1:1.

The co-randomised blends of palm oil, palm stearin and soyabean oil at 80% level with 25% of blended palm oil, palmolein and hardened palm oil can be utilised in margarines (11).

The slip melting point and NMR data of blends of interesterified palmolein and liquid oils with soyabean, rice bran, sal, palm kernel and coconut point out their use in margarine formulation (12) (Table 14.3).

The blending of interesterified palmolein with selectively hydrogenated liquid oils like hydrogenated cottonseed, soyabean, palm oil, rice bran displays highly satisfactory melting point and NMR data which point out that the blends offer excellent shortenings including vanaspati (12) (Table 14.4).

Table 14.3 : Effect of Addition of Hydrogenated Fats on the Properties of Interesterified Palmolein

Fat blend	*Melting point °C*	*Solid fat content (SFC by NMR) at °C*		
		10	25	35
IPOL	39.1	45.1	16.3	5.4
IPOL + HPO[1] (1:1)	40.4	66.4	31.7	10.2
IPOL + HPO[2] (1:1)	39.4	70.4	36.5	11.4
IPOL + HCSO (1:1)	39.9	68.7	36.0	10.5
IPOL + HSBO (3:7)	36.0	69.7	29.1	3.4
IPOL + HSBO (7:3)	31.1	58.6	30.4	10.7
IPOL + HSBO (8:2)	38.2	54.2	29.1	11.3
IPOL + HRBO (8:2)	38.2	52.8	27.3	10.9

IPOL - Interesterified palmolein
HPO [1,2] - Hydrogenated palm oil
HCSO - Hydrogenated cottonseed oil
HSBO - Hydrogenated soyabean oil
HRBO - Hydrogenated rice bran oil

Table 14.4 : Effect of Addition of Liquid Oils on the Properties of Interesterified Palmolein

Fat/Fat blend	*Melting point °C*	*Solid fat content (SFC by NMR) at °C*				
		10	20	25	30	35
IPOL	39.1	45.1	-	16.3	-	5.6
IPOL + SBO (8:2)	34.4	33.7	17.7	-	8.3	-
IPOL + RBO (8:2)	36.4	33.4	17.9	-	8.5	-
IPOL + Sal (6:4)	33.0	45.4	18.0	-	6.2	-
IPOL + PKO (7:3)	34.4	47.8	16.2	-	6.3	-
IPOL + CNO (7:3)	32.4	53.1	17.1	-	5.8	-

IPOL = Interesterified palmolein
SBO = Soyabean oil
RBO = Rice bran oil
PKO = Palm kernel oil
CNO = Coconut oil

Klimes (13) has reported typical blends for biscuit and cake shortenings comprising palm oil plus hydrogenated soyabean (75/25), hydrogenated soyabean (34-36°C) plus hydrogenated groundnut (30-32°C) plus groundnut oil (50+40+10). Also, the author has given blends of fish oil based, tallow based, soyabean oil based and palm oil based fat blends for puff pastry.

Blends are also available for possible shortenings from hydrogenated soyabean products and from sunflower oil and hydrogenated sunflower (46-48%) (98+2), 96% soyabean oil + 4% palm oil and 80% SBO + 20% palm oil. The SFCS at 20°C varying from 1-4% are required for typical frying fat formulations.

BLENDING OF LIQUID OILS

In many countries vegetable oils are blended by simple mixing in definite proportions to meet the preference of consumers and to provide edible oils of improved shelf life, frying characteristics, nutritional value and above all at relatively cheaper prices. A number of examples can be cited here to indicate the scope of blending of vegetable oils.

Palmolein can be blended with rapeseed, mustard, soyabean, groundnut, sunflower, sesame, rice bran, etc. to meet a frying or cooking oil specification. In a country like Malaysia, palmolein is blended with sesame oil or groundnut oil to meet the preference of many Malaysian people. The amount of sesame oil or groundnut oil is mixed in very small proportion (1%) in palmolein. In India also, a number of combinations of binary blends of vegetable oils with one of the component oils at 20% by weight have been allowed by the Government of India to meet the preference of consumers and also to provide nutritionally balanced edible oils and price-wise cheaper oils to the vast majority of population. Some typical blends are shown in the Table 14.5. As usual, like the individual edible oils, the blended oils should conform to the specifications normally laid down with, perhaps, one or two exceptions when one of the oils in the blend has some special features.

Table 14.5 **: Some Typical Blended Oils for Linolenic/Linoleic Ratio For Better Nutrition**

Blended oils (weight ratio)	*18:2*	*18:3*	*18:3/ 18:2*	*18:1+ (20:1)*	*SAFA*	*22:1*
Rice bran + Soyabean (1:1)	45	5	0.11	33	17	–
Rice bran + Mustard (1:1)	25	5	0.20	33	12	25
Sunflower + Mustard (1:1)	35	5	0.14	24	11	25
Soyabean + Mustard (1:1)	35	10	0.30	20	10	25
Groundnut + Soyabean (1:1)	40	5	0.11	37	18	–
Palmolein + Soyabean (1:1)	34	5	0.11	35	25	–
Palmolein + Mustard (1:1)	14	5	0.35	35	20	25
Sunflower + Soyabean (1:1)	55	5	0.09	20	20	–

SAFA : Saturated fatty acids.

REFERENCE

1. Bringi, N.V., Hindustan Lever Ltd., *Ind. Pat.* 129348 (1970)
2. Baliga, B.P. and A.D. Shitole, J. Am. Oil Chem. Soc., **58**, 110 (1985)
3. P. Ghosh Chaudhuri, Ph.D. Thesis, Calcutta University (1981)
4. H. Traitler and A. Dieffenbacher, J. Am. Oil Chem. Soc. **62**, 419 (1985)
5. S. Khatoon and D.K. Bhattacharyya, Oleagineux, **42**, 337-340 (1987)
6. S. Khatoon, Ph.D. Thesis, Calcutta University (1985)
7. Kane, J.G., A.D. Shitole and V.V.R. Subrahmanyam, J. Oil Technol. Asson. of India, **2**, 27, (1970)
8. E.G. Latondress, J. Am. Oil Chemists' Soc., **58**, 185-187
9. N.J. Carlile and T.J. Van Selm; U.S. Patent 3,939,262 (February 17, 1976)
10. Wiegesd, E., *et al,* Federal Republic of Germany Patent 1910, 862 (1969)
11. Teasdele, B.F., and G.A. Helmel, U.S. Patent 3,174,868 (1965)
12. S. Ghosh, Ph.D. (Tech.) Thesis, Calcutta University (1995)
13. J. Klimes, Proceedings of the World Conference, Edible Fats and Oils Processing, American Oil Chemists' Society, Edited by D. Erickson, J. Am. Oil Chemists Soc. Vol. 208-209 (1990)

Chapter 15

(A) Margarine, Shortenings, Spreads (B) Confectionary Fats

PRODUCTION OF EDIBLE PRODUCTS OF HIGH FAT CONTENTS

(A) MARGARINES, SHORTENINGS, SPREADS

1. Margarines and minimum calorie margarines

(i) Historical background

The scarcity and high price of butter around the time of the Franco-Prusian War (1870-71) was the substantial reason to induce the search for a butter substitute. By an order Emperor Nepolean III appointed Hyppolite Mege-Mouries, a previous doctor's assistant and a famous inventor, to carry out the research work. As a result of his work Mege-Mouries filed a patent on 2nd October, 1869. The famous French Patent 86480 claimed the production of a congealed whitish edible fat product with a gloss of pearls called Margarine (Margaritos—the Greek word for pearls). It consisted of the semi-solid part of low temperature rendered beef and/or mutton tallow fraction called oleo margarine obtained by crystallising, panning and pressing the tallow. The tallow was then mixed with soured skimmed milk. The fat including finely dispersed milk globules gradually formed a coarse water-in-oil type of emulsion. This colloidal mixture was cooled and subsequently intensively kneaded. Some other additives like salt and a mixture of cow udders were mixed. The latter was considered at that time to be an indispensable step on margarine production. The large scale production of butter substitute was started only after the war in 1873.

One aim of the invention was based on the supposed activities of the cow's udder for extra cellular milk fat formation. Mege Mouries hoped to perform the direct transformation of tallow to milk fat by adding pieces of cow's udder to the milk fat mixture. This is of course biochemically impossible. The other aim to imitate physically the customary appearance of butter became a technical success.

Butter is produced from either fresh or fermented (ripened) milk and during the process partially crystallised cream of 40-50% milk fat content which in modern processes increased to 80% butter fat (1). The cooled cream is mechanically beaten (churned) to cause the separation of most of the serum or butter milk (aqueous part of the cream from the fat).

The milk fat (compositionally and rheologically totally different from the tissue fat of the same animal) is first dispersed in the continuous water phase of the milk in the form of tiny globules. This relation does not change when the fat is concentrated during skimming into cream. Only during churning, the nature of the dispersion changes to the opposite form with the inclusion of tiny serum droplets into a continuous phase of butter fat. The butter consists of small partly crystallised oil globules and fat crystals. This conversion of oil-in-water type dispersion into the water-in-oil type is called inversion. Butter has to be kneaded and then rested for a certain period of time to obtain its typical texture. This results in the formation of dense network (matrix) of small solid fat crystals which do not grow beyond a certain size during prolonged storage with temperature fluctuation. This special network provides butter with its plasticity and elasticity.

Although margarine was invented to imitate butter in the beginning, it was a near substitute for it. However, margarine has been improved to obtain a product which is closer to butter now and whose characteristics make it possible to recognise margarine as a product in its own right.

Margarine is a representative of water-in-oil type of emulsion. The fat crystals formed during cooling would stabilise this type of emulsion, but monoacyl glycerols (glycerol monostearate) must be used as emulsifiers to produce the formation of water-in-oil type of emulsions because of their water affinity and fat solubility characteristics. Therefore, margarines are produced using these emulsifiers although other emulsifiers are also used serving special purposes in industrial margarines and shortenings.

The main building blocks of the margarine incorporated food emulsifiers are fatty acids (the long alkyl and alkenyl chain constituting the polar part) and glycerol (the polar part), mostly in the form of a solid or liquid monoacylglycerol (monoglyceride). Modifications are compounds derived from this structure which contain organic acids such as acetic, citric, lactic, diacyl tartaric and inorganic acid derivatives such as phosphoric acid (phosphoacyl glycerol, lecithin). Instead of glycerol other alcohols which might be used as bases are polyglycerols, propylene glycol, sorbitan, sucrose, polyoxyethelenes etc.

Milk and mayonnaise are the most characteristic representatives of oil-in-water emulsion. The protein of milk (casein, lactalbumine, proteins derived from soyabean and other seed oil meals) and the proteins and lecithin of egg yolk (the traditional emulsifier of mayonnaise) all stabilise this type of emulsions. Because of the electrolytic dissociation of their hydrophilic groups in an aqueous milieu, these stabilisers are more than sensitive to increased salt concentration, pH and temperature changes than are the above mentioned less water soluble counterparts. The inversion to emulsion type observed in the traditional preparation during churning of butter is partly explained by (a) the change in the original volume proportion of the two phases by mechanical expulsion of a considerable part of the original continuous phase and (b) the removal (inactivation by aeration) of emulsifiers stabilising the original oil-in-water type dispersion.

Different monoacyl glycerols in margarine processing could be revaluated from this angle. When margarine had to be made by preparing a coarse preemulsion which had to be cooled by ice and mechanically kneaded in a discontinuous operation the preemulsifying action of the monoacyl glycerol was vital. Mostly solid monoglycerol palmitate was required to stabilise the preemulsion until solid fat crystals had formed and also embedding hydrophobic fatty acid residues of the emulsifier. The solidified fat thereby took over the emulsion stabilising role of palmitate. Present day margarine machines produce a very fine dispersion/ emulsion of water in fat phase followed by immediate fat crystallisation in a votator or fat crystalliser. In most cases this process greatly reduces the amount of emulsifier required. Because of the different structure and stability of minimum calorie margarines, the amount of emulsifier used is to be modified or increased.

(ii) Crystallisation of fats

Fats, if left alone without mixing, whether under isothermic or adiabatic conditions, crystallise slowly with an equilibrium being reached only after several hours. On the other hand, if mixed and cooled quickly (shock chilling), a proportion of the triacyl glycerols, because of their limited solubility, will crystallise, but solid phase formed will consist of thinly dispersed and generally microscopic sized crystals. The remaining liquid triacyl glycerols fill up the interstitial space around the crystals which are interconnected by 'bridges' forming a network; also called a solid-liquid matrix. Small crystals can form more contact points than can larger ones. The margarine matrix crystals stabilise the water added to the system in the form of a fine water-in-oil dispersion.

(iii) The mechanism of crystallisation

The driving force of crystallisation is supersaturation which governs both the formation of crystallised nuclei and the growth of crystals on those nuclei (the precipitation of solids on them from the melt). The rates of two processes may differ. If both processes have a high rate at the freezing point, the system cannot be supercooled. Seeding with small crystals and mechanical impact by stirring both induce nuclei formation. If the rate of nuclei formation is higher than the crystal growth, the solidification mass will consist of small crystals. In the opposite case, a smaller amount of larger crystals will be obtained. By appropriate selection of the time, temperature profile (the temperature of crystals and the cooling rate) supercooling and mechanical impact the texture and properties of the crystallised fat can be controlled over wide ranges.

(iv) Polymorphism

The phenomenon of polymorphism affects many triacyl glycerol molecules which means that they tend to crystallise in different forms or modifications. This affects not only the external differences but also internal ones, determined most often by X-ray analysis. In general, the triacyl glycerols may have 3 polymorphic forms α, β' and β forms. In the α modifications the hydrocarbon chains are at right angles to the glycerol plane; in the β modification they are about 67° to this plane. The β' form is a intermediate position between the α and β forms. The α form is thermodynamically least stable. The β' form has medium stability and the β form has the highest stability. On cooling the fats, margarines are in a thermodynamically metastable α form and later they are in the β' form. Some of the metastable α form will be formed if the cooling is done rapidly. One triacyl glycerol may block the other stearically and the β' to β form i.e. (transfer) is often reverted. Simply built symmetric triacyl glycerols tend to be transferred further to the highest melting point β form. The transformation goes easily from the lowest to the highest stability form, but not in the reverse direction. Such an irreversible transition is called the monotropic transition, i.e. going only in one direction. Predominantly the melting point, but also the density, the dilatation and the heat of melting increase proportionately with the stability of the crystal modification.

(v) Solid fat content had its significance to product properties

The solid fat content (SFC) is the percentage of solidified triacyl glycerols in an oil at a given temperature. It can be used as a measure of fat crystallisation. It can be measured by Pulsed Nuclear Magnetic Resonance (NMR) spectroscopy or by the relatively old fashioned melting dilatation. The dilatation, as measured, delivers a solid fat index (SFI), which can be converted to a percentage of solid fat content by using statistically tested conversion tables (Vanden Enden *et al* 1978). The measurements are normally done at a few selected temperatures. The data, obtained after standardised tempering procedures, are treated as equilibrium results. However, in practice this very often appears not to be the case. Therefore, the tempering methods used must be quoted if the data obtained are to be used in practical calculations, for example, blending properties.

DEVELOPMENT OF MARGARINE-CONTRIBUTORY PROCESSES

After the first development of margarine in 1869 efforts were made to improve it (3). By 1917 coconut oil had replaced tallow in the oil phases. From 1923 onwards, Vitamin A was added which enhanced the nutritional value of margarine. From 1934 onwards gradually cottonseed oil and later soyabean oil began to be used. The base oil differs from country to country. In recent years palm oil has become an important raw material for the margarine industry. In India, palm oil, hydrogenated sunflower and soyabean and also coconut oil are being used.

Four significant developments contributed to the rapid worldwide acceptance of margarine. These were (1) advances in oil refining technology, (3) development of the hydrogenation process that allow the use of partially hydrogenated vegetable oils in margarine, (4) development of the brine cooled revolving drum for rapid chilling of the fat emulsion and (5) development in 1937 of closed continuous internal chiller and plasticiser that become known as a Votator or perfector or kombinator. In most of the United States and also in Western Europe except in France, per capita consumption of margarine surpasses that of butter which is a development from 1956–57 onwards. In the developed western countries there has been a definite shift in preference towards margarine as against butter by consumers. This is not due primarily to any nutritional attributes but because of the fact that margarine unlike butter is a product whose physical attributes and chemical composition can be changed at will to meet ever changing consumer demands. Different types of margarines can be produced as has been stated earlier.

FORMULATION OF MARGARINE OIL BLENDS

The consistency, plasticity and/or spreadability of margarine is controlled by the following factors, which are important aspects in determining the final taste and spreading quality of margarine :

(i) Per cent or concentration of the fat phase solids as determined by SFI dilatometric method or the SFC as determined by the NMR technique.

(ii) Melting and solidification of characteristics of the individual triglycerides.

(iii) Crystal behaviour of the individual oil or fat used in the final margarine blend.

(iv) Processing conditions and their effect on crystallisation of fat.

(v) Emulsification of the water phase.

(vi) Post processing storage and handling conditions.

To produce the proper consistency in the table grade industrial or commercial margarine the formula must focus on the following factors :

(i) For each margarine to be produced the desired spreadability at refrigerator temperature and acceptable mouth-feel requires that a specific balance be maintained between the liquid and hard fractions.

(ii) Crystallisation of the high melting triglycerides or the solid phase must be accomplished quickly so that almost all of the crystal nuclei are formed in the desired β' form (1 μ or less).

(iii) The low melting or liquid triglycerides must be entrapped within the crystal lattice of the solidified high melting fat as honey is held in honeycomb.

(iv) The finely formed crystals must form a lattice strong enough to hold the firmly divided water phase droplets which range from 2 to 20 μ (with the majority being 2-5 μ) as a discontinuous phase.

With thc above factors in mind the formulator must compound the particular margarine oil blend desired from the partially hydrogenated oil stock available. Under normal conditions a refinery has 7-8 base stock for the formulator to work with. However, preparation of these stocks differs between different manufacturers and even between the two refineries of the same manufacturer. Generally, a margarine manufacturer would purchase oil blends from several suppliers, but the specification required in each case should be rigidly maintained. It is imperative that detailed specifications including processing conditions should be given to the plant operators to follow. This is a very important point if the product quality as regards its plasticity and spreadability are to be maintained. It is to be emphasised that close control of processing temperature of the emulsion during passage through the votator must be maintained within a narrow operating range. For the formulation if different grades of table grade and bakery margarine, among others the following oil stocks are maintained:

(i) Refined, bleached and deodorised (RBD)—Soyabean, Corn, Sunflower, Safflower or similar oils

(ii) RBD palm oil

(iii) Partially hydrogenated cottonseed oil

(iv) Partially hydrogenated cottonseed flakes

(v) Partially hydrogenated palm oil flakes

The polymorphism and crystal habit characteristics of the oil blend may cause consistency problems in the finished margarine that could require the addition of one of the β' promoter fats or the addition of a crystal modifying additive. Initially the water phase is dispersed in the continuous oil phase by means of a good shear agitator mixer, and finally the emulsification is achieved in the positive displacement pump feeding either the votator or the heat exchanger chilling tube. Once the emulsion gels, fine water droplets are immobilised and stabilised in the plastic fat crystal mass.

The size of droplets in the dispersed water phase plays an important role in the flavour release of the water soluble flavour ingredients. As previously mentioned, the majority of the water droplets range in size are 2-5 μ. However, for a good flavour release some droplets should remain close to the upper limits (20 μ). Usually care is taken that the spread will stay in the mouth, will allow the fat to melt and release the flavour laden water phase. It is very important that temperature cycling be avoided during in-plant storage, shipping or at the retail outlets. Melting and solidification as a continuous process could ultimately result in the shifting of β′ crystals to the larger β type. This condition usually is accompanied by a reduction in its specific area of the crystal mass and leads to oiling out. Usually, as has been stated, in Western Europe and America there are mainly two or three products—(1) standard margarine with 80% minimum fat content, (2) diet margarine containing 40% maximum fat, and (3) spread with 60% fat content.

It has to be remembered that the nutritional quality has to be adequate according to fat content, but low fat margarines have come to be marketed because of the concern for heart patient. It has been recommended by various medical bodies that total diatery fat should comprise not more than 30% of the daily calorie intake and a good portion of it should consist of polyunsaturated fatty acids.

Since vitamin A, D and E are present in significant quantities in standard margarines, the low fat spreads are also fortified with these vitamins upto the same level in order to make them nutritionally equivalent to the standard margarines. In recent years, all over the western developed countries the sale of spreads, i.e. low fat margarines, have gradually increased. For example, in the U.S.A., between 1979 and 1988 among the different margarines and spreads taken together the per cent age of spread in the sale has increased from about 8% to about 25%.

In a typical semi-continuous margarine production oil soluble ingredients such as lecithin, mono- and diglycerides, vitamins and, to some extent, oil soluble flavours are added to the oil blend and kept moderately heated and mildly agitated in order not to adversely affect the oil quality or the colour of the blend. The water phase which can range from 100% fresh skimmed milk or reconstituted nonfat dried milk solids and water pasteurised and stored at 4.4°C until use. Some of that milk may be cultured with special lactic bacteria for proper development of the typical battery flavour when used in the formulation of margarine. A typical milk plant will be made of 70-75% uncultured milk and 25-30% cultured milk to which water soluble ingredients such as salt preservatives and water soluble flavours are added. The oil and water phases are weighed into another tank mounted on a beam scale and provided with mixer of appropriate size, and then transferred to the proper emulsion tank feeding into the proper chilling and packaging equipment. This system works well with plants producing a limited number of varieties which require the use of some oil blends and water phase.

However, if the plant is producing a variety of formulations that differ in both oil phase and water phase as well as in colour, flavour and preservatives, then a different system is to be used. Usually the production department is computerised, and in the plant the computer feeds the necessary commands to the different oils and water phase valves to open and close as needed. In this system the different oil based stocks such as soyabean, cotton-seed, sunflower, palm etc., refined, bleached and deodorised and partially hydrogenated as needed, are stored separately in individual storage tanks. The same is done to the water phase where each type of milk or water is stored in its own tank with its own pump. When a particular formula is worked upon the proper oil valve will open and the oil will be delivered to the weighing scale. At the same time the appropriate amount of lecithin is metered in with the oil from a calibrated positive displacement pump followed by the other oil or oils as needed. In each case the valve will partially close towards the end of addition until the load cell registers the proper weight of that particular oil and so on. The same thing will take place with the water phase; and as a final correction either water or oil is added to the top tank depending on the final calculations made by the stored weights of both oil and water. This is done in order to comply with the formula specifications before dropping to the lower mix tank where the rest of the minor ingredients such as colour, flavour, preservatives, mono- and diglycerides are added from calibrated stainless steel positive displacement pumps. High shear agitation and mixing is provided for the proper time before the mixed emulsion is sent to the appropriate holding tank.

Such a system will work well with margarine formulae from 80-53% fat. However, as the fat content decreases below that point, attention must be given to the proper formation of the water temperature and proper rate of moisture addition. In such cases the above system can be modified to accommodate these new spreads by having two load cells and two weigh tanks of the appropriate sizes—one for weighing the oil phase and the other for weighing the water phase. The oil blend will be dropped into the blend tank and the water phase will be discharged slowly to the lower emulsion plant through a partially open valve, thus allowing for proper mixing and emulsion formation below. The details of procedure vary from plant to plant and from manufacturer to manufacturer. Typically, margarine formulations have between 12-14%, upto 16% of water, 1.6-1.8% salt, 1-2% emulsifier, anti-oxidants, colour and flavour.

CONSISTENCY OF EDIBLE FAT PRODUCTS AND OF HIGH FAT CONTENT

Fats and fatty emulsions have many properties which can be used as quality criteria. One of these properties is the consistency of the product, an aspect based on its rheological behaviour (flow). The processor judges the consistency by changes of hardness during processing, packaging and storage, whereas the consumer judges it in terms of spreadability, workability and oral melting properties of the product during use.

Whatever the criteria for judging, the consistency is connected with a certain change in the shape of the sample in question depending on the conditions used in the processing of fats (in margarines and shortenings) which are generally considered to be plastic/elastic solids or visco elastic liquids. In order to be plastic, fats must have (i) the coexistence of two phases, the solid one embedded in a liquid one; (ii) a proper proportion of the two as expressed by the term solid fat content at a given temperature; and (iii) interconnection/cohesion of the solid fat crystals by bridges (or primary bonds). Of course, all other factors such as polymorphic crystal modification, the complexity or crystal composition, the shape of the individual crystals affect the consistency of a high fatty product. Large needle shaped crystals interlace more easily and these primary bonds result in firm, rigid structures. The same amount of solid fat in a more compact form is bound by less direct secondary bonds which results in a softer structure. The agregation of smaller crystals to clusters with substantial amounts of liquid phase also makes the system softer but more plastic than those described above.

It is generally not well realised that even the water phase makes a distinct contribution to the consistency of an emulsion. The contribution of the water phase compared to that of the solid fat (per cent solids) is negligible at room temperature. At higher temperature (close to melting point) its relative contribution becomes more pronounced in determining the consistency with the disappearance of most of the solid fat crystals. This fact makes the consistency of shortenings, margarines and minimum calorie margarines virtually the same at room temperature. However, the minimum calorie margarine close to its final melting point is much thicker in the mouth than are other margarines and these, in turn, are somewhat thicker than the shortenings based on the same fat blend.

LEGAL ASPECTS

Margarine by definition is a food in the form of plastic, partly solidified or fluid water-in-oil type of emulsion dispersion produced principally from edible fats and oils which are not usually derived from milk fat. Under a proper microscope it can be seen that there are small crystals in the fat phase, mostly larger than 5 μm. These crystals form a stabilising matrix (frame) held together by different forces in which the oil and the finely dispersed water (most of the droplets are less than 3 μm, many of them upto 20 μm in diameter) and some air bubbles are enclosed like fluids in a sponge. In a so called non-refrigerated table margarine at room temperature the oil phase amounts to about 60%, the solid crystals to about 20% and the water phase is about 20% of the total. If heated excessively stabilising fat crystals melt and dissolve in the oil which is the main reason why the whole emulsion dispersion destabilises quickly.

According to the legislation of many countries including India the fat content of margarine is generally regulated to be not less than 80% with water mixture not more than 16%. A part of the water is derived from the added milk of whey. This product is labelled as margarine. In Europe if the fat content is around 40% (not less than 39% and not more than 41%) and the water content not less than 50% the spreadable water-in-oil type emulsion is called minimum calorie margarine (Minarine), half calorie margarine (Halvarin) or just a low calorie spread (Locas). type of margarine is produced principally from edible oils and fats which are not solely derived from milk fats.

The labelling of the margarine packet in all cases should carry "Minarine" (or a similar name) or spread with the stated fat and ingredients content. European health margarines are produced with high content of fat of not less than 80% with an EFA level of minimum amount of polyunsaturated fatty acids (PUFA). These margarines contain generally low amounts of salts. In some countries, particularly USA, any claim or advertised aspects regarding PUFA saturated fatty acid (SAFA) or cholesterol level is to be labelled on each package. In USA all products containing less than 80% fat have to be considered as 'spreads' and are to be labelled as such. In India the standard margarine contain 80% fat. However, recently one concern has marketted 60% fat spread. In some countries margarines or spreads may contain butter fat or vegetable butter fat mixes, also called dietary spreads. All additions whether fat of water phase soluble must be of edible grade and to be approved by appropriate health authorities. This means that they are tested toxicologically conforming to the guidelines developed by FAO or WHO. Different countries prescribe their own legislations for margarine, spreads or similar products. In India the concerned committees are the Bureau of Indian Standards of the Ministry of Industries and the Central Committee for Food Standards of the Ministry of Health and also the Directorate of Vanaspati Vegetable Oils & Fats (DVVOF) of the Ministry of Civil Supplies. All ingredients are required to be stated on the packages.

DIFFERENT TYPES OF MARGARINE PRODUCTS

Different users have different needs and these differences can be used as the criteria for grouping the many types of margarine produced for the 3 main market segments :

(i) Household margarine—also used by restaurants and caterers;

(ii) Margarine produced for semi-industrial users; and

(iii) Health margarine for the special segments for household users. It is possible to characterise different products by ranges of the solid fat content at a few different temperatures.

Both the traditional wrapper packaged block margarine and tub packaged soft margarines try to imitate at high temperature ranges genuine steep melting (large solid fat content, differences) of the butter fat. This type of melting means the total and if possible rapid disappearance of the solids in the mouth. The mouth feeling which is obtained from butter is to be imitated by better quality margarine. All household margarines are to be preferably kept in the refrigerator. The soft and high PUFA (health margarines) and the minimum calorie margarines, are readily spreadable at this low temperature. Margarines such as wrapper packed kitchen and table margarines have to be warmed up to room temperature in order to become reasonably spreadable.

The creaming power is the extent of air retention at the working temperature (20-35°C). Good creaming power requires fat crystals of relatively small surface area which on whipping readily enclose air in a fine dispersion. The product must also have a high work softening when beaten in a mixing device. The main criterion is the yield (specific volume) of the finished cream or cake batter. Some margarines are composed of blends with a special aim besides normal caloric, general nutritional and practical performance aspects :

(1) These special margarines may contain high levels of essential linoleic acid. This is not the only but the most important part of the PUFAs. The daily ingestion of substantial amounts of linoleic acid lowers the blood plasma cholesterol levels in humans. It is a statistically corroborated fact that cholesterol is a risk factor in atherosclerotic heart diseases.

(2) Besides high levels of linoleic and oleic acids definitely stated, low levels of saturated fatty acids are also required. The saturated fatty acids increase the plasma cholesterol level more than the essential fatty acids decrease it.

(3) From lauric acids which contain high levels (over 95% medium chain fatty acids (C_8-C_{10}) fractions are obtained by fractional distillation and are reesterified with glycerols to get

the glycerols containing medium chain fatty acids for preparation of Special Margarine. These margarines are used for small children and aged persons with some gastric disorder.

(4) Some specially formulated margarines also contain high levels of triacyl glycerols with γ-linolenic acid (6,9,12-linolenic acid). This is a precursor of certain prostaglandins.

(5) As of now these margarines are not produced as its claim in effects have not yet been amply corroborated. In Europe and certain other countries margarines. with high levels of PUFA and SAFA are called diet margarines. This is in contrast with the diet margarines of USA which are mostly reduced or minimum calorie spreads. These special margarines are cholesterol free margarines which are composed mostly of edible oils.

(6) Margarines free of *trans* fatty acids : These are supposed to be free from hydrogenated fat which contains *trans* fatty acids.

(7) Sucrose polyesters (SPE) on different fatty bases of the sucrose octaoleate should be considered as no calorie fat substitutes constitute the components of some spreads which are indigestible for humans and these are also not calorific fat substitute. They are being used for clinical investigation for slimming diets and to control plasma cholesterol levels.

(8) Soft margarines composed of high levels of mono- and diglycerides with liquid oil with no solid fats added, the emulsifiers give from the required gelling matrix thereby retaining the oil and the emulsified water phase.

PRODUCT CHARACTERISTICS OF MARGARINE

(a) *Spreadability*:

The penetration is the distance in units of 0.1 mm travelled by standard cone in 5 seconds after it is released on the surface of the product. These values can be converted mathematically to a hardness index or yield value that is independent of the cone weight. Spreadability with a knife can be correlated with penetrometer values as both evaluate the force required to bring about a significant deformation. Table margarine should have good spreadability with a uniform value in the range of 100-800 gms per sq. cm. at 20°C.

(b) *Oil Separation*:

Testing of oil separation tendency is often conducted by placing a margarine sample of defined shape and weight on a wire screen or a filter paper at about 25°C or sometimes higher for a period of 24-48 hours. The oil collected in the filter paper or through this screen is measured. Another test used for determining the structural stability is slump. This involves placing of a standard size cube or margarine between 20-30°C for several hours. Deformation of the cube is graded according to visual reference to a standard slump chart.

(c) *Organoleptic appeal*:

The third characteristic which is important is organoleptic appeal. A high quality table margarine melts quickly and completely with a chilling sensation on the palate. Flavour and salt components of the aqueous phase are immediately perceptible by the taste buds. The factors affecting these qualities are the melting profile of the fat, the nature of emulsion and conditions for storage of the product. Although several parameters and processing conditions influence the organoleptic appeal, it is the actual test on the palate which is the ultimate parameter. A well formulated table margarine should have good oral melt, be spreadable at 10°C and soft at 20-25°C depending on the local climate.

(d) *Additives used in margarine*:

(1) *Preservatives* : 3 types of preservatives are used in margarine : antioxidants, metal scavengers, and antimicrobial agents. The commonly used additives are (i) moderate doses of tocopherol obtained from constituent vegetable oils, (ii) lecithin, (iii) citric acid and (iv) EDTA (ethylene diamine tetra-acetate).

(2) *Flavours* : Many synthetic butter flavours are available for margarines. Diacetyl, which is a normal constituent of butter during the butter fat preparation process, is used as a primary flavouring agent.

(3) *Emulsifiers* : Emulsifier normally used to make margarine is glycerol monostearate (or GMS). Such monoglycerides are prepared by reacting fats with excess glycerols.

(4) *Vitamins and colours* : Margarine is fortified with Vitamins A and D. For colouring, carotenoids, particularly synthetic carotenes, are normally used. Natural extracts, e.g. annatto (Bixin), carrot oil and red palm oil, have also been utilised. The extracts of annatto and turmeric together can impart more typical colour than annatto alone. A typical formulation of margarine may be as follows :

			%
Fat blend	..	..	80.30
Glycerol monostearate		..	1.25
Lecithin	..	..	1.25
Colour and flavour		..	1.25
Milk	..	..	14.35
Salt	..	..	1.60

The oils are introduced in the mixing vessel and heated with agitation to about 35°C. The colouring matter, emulsifiers, vitamins and lecithin are added with agitation until a smooth homogenous oil results. The oil is allowed to cool to a temperature of about 45°C. When the oil is cooling the required quantities of milk or salt are stirred together and then added to the oil.

MANUFACTURE OF MARGARINE

Briefly the process for the manufacture of margarine can be divided into the following steps :

1. First, selecting and preparing a suitable mixture of oils and fats, the so called 'fat blend'.
2. Preparing the milk or some other aqueous liquid which forms the basis for the flavour and aroma of the margarine.
3. Mixing or emulsifying the fat blend and the aqueous liquid in suitable proportions by the so called 'churning' or emulsification process.

4. Cooling and crystallising the fluid emulsion.
5. Working and kneading the solid fat to give the desired plasticity and body to the margarine.
6. Packing the margarine.

FUTURE DEVELOPMENTS POSSIBLE

At the present time in all countries several types of margarines are being marketed including the standard table margarine, bakery margarine, the specialised margarine such as puff pastry type, fluid type margarine etc. Now-a-days special types of margarines are being marketed in developed western countries with fish oils containing fish oil fatty acids such as docoso hexanoeic acid (DHA) and Eicoso Pentaenoeic Acid (EPA) for its supposed medical advantages in the case of people having high cholesterol or tendencies towards rheumatoid artharitis. It is obvious that from time to time the consumer preference would guide the market. It is also possible that in the most developed countries instead of many companies there will be only a few companies offering most of the products demanded by the market.

SHORTENINGS

The term 'shortenings' was coined by the workers in the bakery industry. It refers to all fats and oils which when formulated into the baked goods imparted a 'short' or tender quality. The early products were limited to butter, lard, beef fat or tallow and liquid vegetable oils. In the western countries, for many years animal fats were the principal edible fats used in the shortenings, whereas in India mostly butter and vanaspati have been used for confectioners apart from liquid vegetable oils.

In recent years the techniques of hydrogenation, interesterification or rearrangement, emulsification and fluidisation have made it possible to 'tailor' shortenings for specific bakery products and processes. There are hundreds of products in existence in the developed western countries. Some of these shortenings are classified as anhydrous all purpose shortenings and unemulsified and emulsified oil blends. Bakery shortenings, anhydrous, emulsified and unemulsified are further classified as semi-solid (plastic), fluid and clear liquid. Commercially available clear liquid shortenings are frying media.

In the case of plastic and fluid shortenings two important characteristics are solid content (as measured by SFI or SFC) and polymorphic form. There is also a fluid shortening which is a stable dispersion or solids in the proper polymorphic form in a continuous oil phase which flows and can be pumped over a temperature range of about 25-40°C. The solids are derived from additives made up of stearine and emulsifiers. Stearins are characterised by low iodine values and high melting points. Emulsifiers provide the functional attributes of aeration, starch complexing and dough conditioning or machinability.

For cake application the high ratio fluid cake shortening consists of lactilated monoglycerides and monodiglycerides in a hydrogenated liquid oil like soyabean oil. It provides cakes with improved tenderness moistness and overall quality. It can be measured in bulk and can be used directly by the cake manufacturer. Among the other ingredients of cake formulations are : flour and shortening, sugar, milk, salts, eggs, leavening agents, water. Typical cake formulae are given as illustration:

Ingredients	*Souffle percentage*	*Whitelayer percentage*
Flour	100	100
Sugar	114	110–140
Shortening	70	35–65
Milk solids	4	6–10
Eggs	140	40–70
Salt	3	3–4
Leavening (yeast)	7	6–4
Water	30	90–72

Some further developments have taken place in the fluid shortening. Earlier, the fluid shortenings for yeast raised goods incorporated monodiglycerides as the softening agent in a hydrogenated vegetable oil. It provided increased initial softness and longer shelf life. Later, from 1970 onwards a substantial improvement was achieved by the use of ethoxylated monoglycerides/

mono- and diglycerides in a hydrogenated vegetable oil. This provided excellent dough conditioning. That gives the dough greater tolerance to manufacturing stresses, particularly in machine handling.

After 1977 further refinement led to the optimum fluid shortening for yeast raised goods. Two conditioners e.g. ethoxylated monoglycerides and sodium stearoyl 1-2-lactilate and a mono- and diglyceride softener were incorporated in a hydrogenated vegetable oil. This type of product assures the baker of consistently good quality yeast raised product despite wide variation in flour quality. The shortening levels could also be reduced. The detailed practices are described in the AOCS Proceedings of World Conference, Mastrict. Some of the recent practices are elimination of stearates like stearic acid that are produced by hydrogenation and the practice formulated with noncaloric sweeteners, fibre and non-absorbable fat ingredients like sucrose polyester.

In various countries the feed stocks for shortening manufacture may vary according to the availability of local oils and fats. Palm oil is used for shortening in its refined stage and also stearin and olein fractions as in the hydrogenated product of palm oil and palmolein. In recent years, the availability of palm oil has led to its increased use in many countries besides the palm oil producing countries like Malyasia. Among vegetable oils soyabean oil and rapeseed oil are in use both in the hydrogenated and unhydrogenated forms in the soyabean producing countries and rapeseed oil producing countries respectively. Selectively hydrogenated groundnut oil is also in use as is the hardened cottonseed oil. In recent formulations coconut oil and palm kernel oil are important. In Europe, in particular, some typical shortenings are used which are classified as follows:

1. Cake and Pastry shortenings
2. General purpose shortenings
3. Puff Pastry shortenings
4. Creaming shortenings
5. Pourable shortenings
6. Dry shortenings

Shortenings are also used in biscuits. Some typical products are groundnut oil and soyabean oil based, also fish oil and palm oil based in Europe. Following are some of the typical examples.

Groundnut oil Soya oil base			*Fish oil palm oil based*			*Fish oil based*		
HSBO	34/36	50%	HFO	30/32	65%	HFO	36/38	75%
HGNO	30/32	40%	PO		35%	HSBO	IV 105	25%
GNO		10%						

HSBO = Hydrogenated Soyabean oil
HGNO = Hydrogenated Groundnut oil
GNO = Groundnut oil
HFO = Hydrogenated Fish oil

Shortenings listed in the table are low melting formulations. There are also medium melting products which are blends of palm oil, medium melting hydrogenated soyabean or rapeseed oils and, in Europe and USA, interesterified tallow are also being added.

GENERAL PURPOSE SHORTENINGS

Fats which show both good baking performance and satisfactory frying qualities can be classified as general purpose fats, the constituents of which are hardened vegetable oils like groundnut oil, palm oil, palmolein etc.

PUFF PASTRY FATS

Typical puff pastry fats have a tough waxy consistency over a wide temperature range that permits sufficient stretching and a stable crystal structure that prevents oiling off the softness during roll-out. β' stabilising hard stocks melt from hardened palm oil, hardened animal oils like tallow or fish oils or hardened shea oils to improve waxyness. Following are the examples of soyabean and palm oil:

Soyabean

HSBO	36/38	60%
HPO	42/44	20%
SBO		20%

Palm oil based PO

PO		50%
HPO	42/42	40%
SBO		10%

CREAMING SHORTENINGS

Fats used for soft flexible icing and cream fillings are required to have excellent aeration properties and quick melting characteristics. Therefore, these fat systems should be β′ tending and should have a steep SFC curve. The blends are preferably made from palm oil, hardened palm oil or palmoleins in combination with lauric oils. Typical creaming shortenings show melting points between 30-34°C and solid fat content of about 18-25% at 20°C. A typical all vegetable palm oil based shortenings would consist of :

Palm oil	50%
Coconut oil	20%
HPO 40–42	10%
Liquid oil	20%

Traditional shortenings for the preparation of high standard fillings contain hardened groundnut oil with melting point near 30°C.

POURABLE SHORTENINGS

The term pourable shortenings covers all fluid suspensions which consist of a solid, preferably β tending fat or a high melting emulsifier, dispersed in liquid oil. This type of product offers easy handling and metering and is used for preparation of snacks and fried fat. Typical formulae are given below.

1.	Sunflower oil	96%
	Hydrogenated Sunflower oil (HSFO)	4%
2.	Non-hydrogenated Soyabean Oil (NHSO)	80%
	PO	20%

DRY SHORTENINGS

Dry shortenings are crystallised or spray dried powder fats which are used in blends to prepare dry ready mixed products. Typical fats are:

Hardened soyabean oil with melting range of 36-44°C and Hardened palm oil with melting range of 42-46°C

PROCESSING

The solidification and texturing of shortenings are usually achieved by a shock chilling process in which molten fat is rapidly cooled in a scraped surface tubular heat exchanger. Depending on the properties of the shortenings and requirements of the final products various chillers/crystallisers combination can be used.

AERATED PLASTIC SHORTENINGS

Shortening blends contain fats which are characterised by early or fast crystallisation rates, e.g. lauric fats, hardened soyabean, palm or fish oils, which can be chilled and aerated in a tube chiller. At the start of the process air or nitrogen is incorporated to the extent of 25% into precooled liquid fat in the gas absorber tank and also dispersed by high pressure pump.

PUFF PASTRY FATS

Puff pastry fats as well as puff pastry margarines can be produced in an open drum cooler/vacuum kneader systems. Tube chiller systems have been introduced. Pourable shortenings are prepared often by chilling the liquid mixtures of base stock and solid fat in a tube chiller plant.

DRY SHORTENINGS

For the preparation of dry baking mixes it is customary to blend powdered fats with other main components such as flour, starch, milk components etc. The emulsion shortenings and skimmed milk is spray dried to form powder containing up to approximately 80% fats.

REFERENCES

1. *Bailey's Industrial oil and Fat Products*, Volume 3, Edited by John Weily (1985) page 91-111.

2. Vanden Endew, J.C; Haighton, A.J; Van Putte, K; Vermaas, L.F. and Waddington, D, Jour. Fette Seifen Anstrichm, **80**, pp. 180-186 (1978)

3. Hoffmann, G. The Chemistry and technology of edible oils and fats and their high fat products, 1st edition, Academic Press, London, New York, Toronto, (1989), pp. 279-327

4. Mustafa, in "Edible fats and oils processing (Basic Principle and Modern Practice", Proceeding of the World Conference, AOCS Illinois, 1990, edited by D.R. Erickson, page 217-221

5. Roy, T.K. and Banerjee, S.K., in Treatise on fats, fatty acids and oleochemicals, Section N-1, Narula, Industrial Consultants, New Delhi-10015 (1994)

(B) Confectionery Fats

INTRODUCTION

Confectionery fats are the special fats to which belongs mainly the cocoa butter, obtained from the plant '*Theobroma cocoa*' grown in tropical areas, mainly in Africa. Because of the high cost of cocoa butter it is tried to be replaced by some cocoa butter replacer fats partially or fully. Cocoa butter is a pale yellow solid having a characteristic flavour and composition of about 24-27% palmitic, 32.6-35.4% stearic, 33.8–36.9% oleic and 2.7-4% linoleic acid.

Cocoa butter replacers can be classified into three groups—cocoa butter equivalent, cocoa butter extenders (CBX) and cocoa butter substitutes (CBS). Cocoa butter equivalents have characteristics and properties similar to those of cocoa butter and can be mixed with cocoa butter in any proportion. Cocoa butter extenders do not necessarily have the physico-chemical characteristics similar to those of cocoa butter, and they are used as extenders where partial replacement of cocoa butter is to be done. Cocoa butter substitute fats have their own physico-chemical properties which may not be similar to those of cocoa butter. Cocoa butter substitutes serve as the cheaper source of fats which perform as cocoa butter and which can be processed like cocoa butter. These fats include Sal (*Shorea robusta*), Mowrah (*Madhuca latifolia*), Kokum (*Garcinia indica choisy*), Mango (*Mangifera indica*) etc.[3]

Confectionery fats are generally divided into three groups according to their use—sugar confectionery, chocolate-like substances and chocolate. Sugar confectionery is a stable oil-in-water emulsion where emulsifiers used are glycerol monostearate or soya lecithin.

Chocolate-like substances are the cheaper alternatives to chocolate and are generally the mixture of sucrose, low fat cocoa powder or cocoa liquor, vegetable fats, milk products with milk fat and, if required, lecithin and flavours. Like chocolate the chocolate-like substances are two phase systems, a continuous fat phase and a dispersed phase containing solid materials. The quality of the product depends mainly on the type of fat used. The cheapest products are usually prepared from hydrogenated palm kernel (HPKO) or coconut oil (HCO) type lauric rich fats. The use of these lauric-based fats is advantageous as they need no thermal processing. Hence they are vastly used in small bakeries and confectioneries where expensive plants are not available. For a slightly higher quality product, the fractionated HPKOs, known as vegetable stearins, are used. For high quality products, cocoa liquor, and in the case of milk products, full cream milk powder are required. These fats are generally based on fractionated palm oil, fractionated hydrogenated cottonseed oil and soyabean oil. The quality of chocolate-like substances can be improved up to the point where they are indistinguishable in flavour and texture from chocolate.

Chocolate contains mainly the cocoa butter fat with other fats like milk fat, non-cocoa butter vegetable fats known as cocoa butter equivalents (CBS), lecithin and polyglyceryl polyricinolete. All these fats constitute about 28 to 40% of chocolate. Milk fat used in chocolate serves different purposes. Milk fat modifies texture, makes chocolate softer than the products prepared with cocoa butter alone, inhibits blooming of fats (the whitish appearance due to the appearance of fat crystals on the surface of chocolate) as chocolates made from cocoa butter or with the addition of certain CBE tend to bloom, and milk fat reduces the cost of product as it is cheaper than cocoa butter.

Vegetable fats can be used in chocolate at the level of up to 5%. In the crystallisation of the fat phase during the manufacture of chocolate, they exhibit a number of melting points depending on the thermal history and the type of polymorphic crystalline form produced. Three main types of crystal structure are found in fats:

(i) α-form : This form is of lowest packing density, fastest growth rate, lowest melting point and least stable.

(ii) β'-form : This has more dense packing system than α-form, has a higher melting point and is more stable.

(iii) β-form : This has highest packing density and melting point and is the most stable.

The type of crystal present can be determined by using X-ray diffraction. All fats do not exhibit all three types of crystal structure for example, HPKO are stable in β′ phase. On the other hand, certain other fats including cocoa butter, because of minor variations in crystal structure, exist in more than 3 forms.

In table 15.1, the six different polymorphic forms identified for cocoa butter are given with their respective melting points.

***Table 15.1* : Polymorphic forms in Cocoa Butter**

Willie and Lutton classification	*Crystal type*	*Melting point*
1	β'	17.3
2	α	23.3
3	α and β'	25.5
4	β'	27.5
5	β	33.8
6	β	36.4

Every polymorphic form can be transformed to the next more stable one under proper conditions. In the case of chocolate manufacture, to get the required appearance and eating quality of the products, the cocoa butter crystals must be present in form 5.

Lecithin is used in chocolate as a fat extender and it partially replaces the expensive cocoa butter. It is seen that 0.5 to 0.7% of lecithin can replace the cocoa butter up to 7%, i.e. a reduction based on total fat is 37 to 30%.

Polyglyceryl polyricinoleate serves as a surfactant which greatly reduces the yield value of liquid chocolate and is used up to a level of 0.2%.

In the case of milk chocolate, use of non-cocoa butter vegetable fats causes a lot of problems as most of them are not compatible in properties with milk fat. However, this problem mainly occurs in the case of partial replacers but, generally, true CBEs do not pose any problem.

Some new Indian origin vegetable fats as cocoa butter replacer :

Cocoa butter has unique chemical and physical characteristics which make chocolate universally popular. Cocoa butter becomes hard and brittle at 30°C but melts completely at 35°C. It rapidly transforms into the most stable β polymeric form from unstable α form which allows tempering, demoulding and hardening of bars and this is due to the presence of about 80% symmetrical triglycerides, i.e., POP, POS and SOS.

As mentioned earlier, cocoa butter replacer fats, which replace cocoa butter partially or fully in different chocolate and chocolate products, can be classified into three groups :

(1) Laurics (hydrogenated palm kernel oil or hydrogenated coconut oil)

(2) Hydrogenated fat fractions

(3) Fats with symmetrical disaturated, monounsaturated glycerides (GS_2U), so called cocoa butter equivalents, CBEs.

The first two types are not compatible with cocoa butter and are used in cheaper products. The use of these fats produces blooming and softening of the product. Blooming results from incomplete crystallisation.

The fats called cocoa butter equivalents are totally compatible with cocoa butter and can be mixed with it in any proportion without affecting the physical characteristics of cocoa butter. Palm oil contains a high amount of POP glyceride; some natural butter such as Shea contains SOS in large quantities and illipe is rich in both SOS and POS. Thus a judicious mixture of solvent fractionated palm oil, solvent fractionated Shea and illipe make a blend which matches with cocoa butter. The first commercial CBE was Coberine and now other products similar to it are available. Sal or sal fractions are the main cocoa butter equivalent of Indian origin.

There are some cocoa butter substitutes which include partially hardened and pressed or crystallised sal fat, mowrah fat, mango kernel fat, kokum fat etc.

Sal : Sal contains 69% symmetrical triglycerides, mainly SOS, which resembles shea and illipe and serves as an additional new raw material for CBEs. A comparison of the glyceride composition of sal, shea and illipe is shown below :

	Shea	*Illipe*	*Sal*
SOS	35	40	40–45
POS	–	31	8
AOS	–	0	10–14
SOO	45	–	21–25
AOO	–	–	4–8
Total S_2U	35	71	62–70

A = Arachidic acid

Sal or sal fractions are used only to a limited extent in CBEs. They are now mainly preferred to be termed as cocoa butter substitute, though it is totally compatible with cocoa butter.

Fractionation of sal from acetone and comparative evaluation of their physical properties like melting point, micropenetration, solid content index, plasticity range in admixture with cocoa butter have been studied. It was concluded that sal and sal stearin may be good extenders for cocoa butter. Mid fraction is an excellent substitute for cocoa butter.

The fractionation of sal fat from acetone at 7°C to yield 76-78% stearin fraction containing 75% SOS, which is utilised in a large variety of confectionery products containing various proportions of cocoa butter, has been described in an Indian Patent from Hindusthan Lever Limited. The stearin is harder than cocoa butter and melts at 39°C sharply which is advantageous in hot climates. It can replace about 20-40% of fat in palm chocolates. It can be used in a similar manner in milk chocolate.

As sal stearin contains about 6-8% arachidic acid which gives 14% arachido-oleostearin (AOS) and 4% archiodiolein, it has the advantage of imparting resistance against bloom and heart and gives very good melting characteristics to chocolates. Another advantage is the possibility of formulating CBS with higher proportions of SOO, i.e., 25-40% CB and CBEs containing 80% SOS and 15% SOO (S = stearic/palmitic and O = oleic).

Partially hardened and pressed or crystallised sal fat as CBS : Normally solvent fractionation or pressing is employed for removing SOO fraction and enrich the stearin with SOS. But hydrogenation is a more efficient method for hardening of fat which is seldom used due to *trans*-acid and isooleic acid formation even under absolute selectivity conditions. The glycerides formed may not be fully compatible with cocoa butter and may not be acceptable for high quality chocolates.

Mowrah (Madhuca latifolia) as an ingredient in cocoa butter substitute (CBS) : Mixtures of mowrah stearin and StOSt rich fats have been found to be more suitable as CBS. Blends of kokum and mowrah stearin (30:70) have composition very similar to that of cocoa butter. It is also reported that blending of sal stearin and mowrah stearin obtained from aceton crystallisation in the ratio of 85:15 gives rise to products having satisfactory properties.

Mowrah fat can also be used in small quantities in the manufacture of vanaspati (m.p. 37°C) after partial hydrogenation. Mowrah fat can be interesterified to get higher proportion of polyunsaturated fatty acids (PUFA). Blends of sal and groundnut oil (30:70) or sal and sunflower oil (50:50) or cottonseed oil alone could be interesterified to yield plastic fat of melting point 33–34°C.

Application of mango kernel (Mangifera indica) fat : Mango kernel fat is mainly used for making cocoa butter substitute. Mango kernel fat is much softer than cocoa butter, it must be treated properly to achieve compatible composition and properties like those of cocoa butter. For this reason, the fat is to be fractionated by selective solvent, particularly acetone.

Cocoa butter substitutes from kokum and mowrah fats : The blending of a harder fraction of mowrah with unfractionated kokum gives a closer match to cocoa butter. The hard fraction of mowrah has comparable fatty acids and glyceride distribution (as obtained from silver nitrate—TLC) but is softer than cocoa butter. It is seen that by increasing the amount of kokum the desired hardness and solid fat index can be achieved. With about 30% kokum incorporation the product becomes closer to properties of cocoa butter.

The above mentioned fats of Indian origin, rich in the amounts of 2-oleosymmetrical triglycerides (with palmitic/stearic) perform as cocoa butter replacers as partially or wholly.

REFERENCES

1. *Bailey's Industrial Oil and Fat Products*, Vol. 1, 4th edition (Wiley) 1979
2. '*Fats and Oils Chemistry and Technology*,' edited by Hamilton and Bhati—Applied Science Publishers Ltd., London (1980)
3. Bringi, N.V., editor, '*Non Traditional Oilseeds and Oils in India*'—Oxford and IBH Publishing Co. Pvt. Ltd., New Delhi, Bombay and Calcutta (1987)

Chapter 16

Biotechnology of Fats

INTRODUCTION

During the past few decades microbial enzymes having specific and non-specific reaction activities have shown considerable potentialities in the production, purification and modification of fats and oils to make edible products and oleochemicals. Countries like Japan, U.S.A., France, West Germany, U.K., U.S.S.R. and Holland have made enormous progress in the application of microbial enzymes. Considerable scope exists for adopting the microbial enzyme technology for the fats and oil industries to meet edible and industrial requirements.

Enzyme process, an alternative of conventional process has certain advantages like lower energy consumption, mild reaction condition, high catalytic activity and positional specificity. Most of the enzymatic reactions are carried out at low temperature requiring minimum amount of heat and electrical energy. With minimum or no usage of solvent such processes not only eliminate the need for high price and potentially hazardous solvent but also help to maintain the balanced ecological environment keeping it pollution free.

Regarding the quality of product, enzyme process yields high quality products, and the specificity of enzymes helps to obtain a tailor made product which would be impossible in conventional chemical processes.

BIOEXTRACTION

Biotechnology has been receiving importance in processing oil bearing materials primarily for extraction of vegetable oils. Certain researchers have suggested the use of different enzymes in extraction processes based on the principle that enzymatic degradation of cell wall is possible. Very little is known about the bio-extraction process of oilseeds for recovery of oils and non-oil constituents until now though lot of research work is being done in this aspect.

The cell wall of a seed is generally made of pectic substances, cellulose and hemicellulose. In order to extract the oil from different oil seeds, it is necessary to degrade the cell wall to increase the permeability for oil. Enzymatic degradation opens up the cell walls and also breaks up complex lipoprotein and lipopolysaccharide molecules into simple molecules releasing extra oil for extraction.

Generally, the enzymes used for extraction from oilseeds belong to the carbohydrate class which are cellulase, hemicellulase, pectinase, amylase, β–glucanase etc. Proteolytic enzymes like α-proteinase and enzymes with mixed activity have also been used for good result.

Many microorganisms are unique in having enzyme activities on the carbohydrates including cellulose and hemicellulose as well as proteins. Obviously, such enzyme activities of microorganisms can be useful in treating oilseeds to release oil globules readily by their action on proteins, on cellulose and hemicellulose, preferably on both, in a controlled manner.

Among the microorganisms, *Bacillus subtilis* and *Aspergillus niger*, have the desired enzymes and have been examined in the extraction of oilseeds like soyabean and rapeseed (1). Aqueous enzymatic extraction was first studied on olives (2,3) by Italian and Spanish teams, and later on has been investigated for different oil seeds like coconut (4), avocado (5), palm (6,7) cottonseed (8), groundnut (9), canola (10,11) and also corn germ (12,13). Novo (14) developed an enzymatic process for the extraction of rapeseed oil with the aid of an enzyme preparation SB 311 from a selected strain of *Aspergillus niger*. Enzymatic extraction of mustard seed and rice bran (15) with pectinase and cellulase showed almost 100 per cent recovery of oils. Among the advantages claimed with the use of enzymes in processing of oil seeds are low energy consumption and solvent usage, higher nutritive value of protein and extremely good quality of oil, requiring hardly any refining after extraction. A typical flow diagram of the bioextraction process is shown in Diagram 16.1.

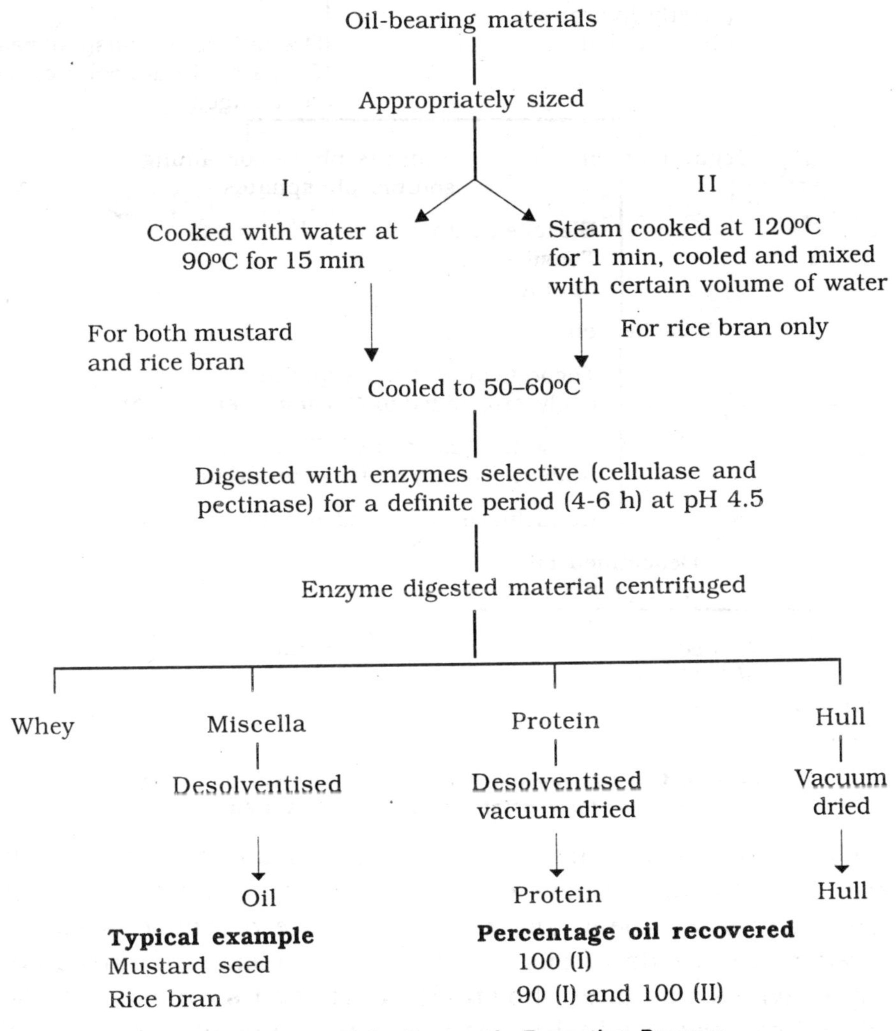

Diagram 16.1 : Enzymatic Extraction Process

The efficiency of aqueous enzymatic oil extraction depends on several factors like the degree of grinding, time of hydrothermal pretreatment and enzymatic conversion, enzyme preparation, temperature, concentration of seed in water, pH of the medium as well as the speed of centrifugation.

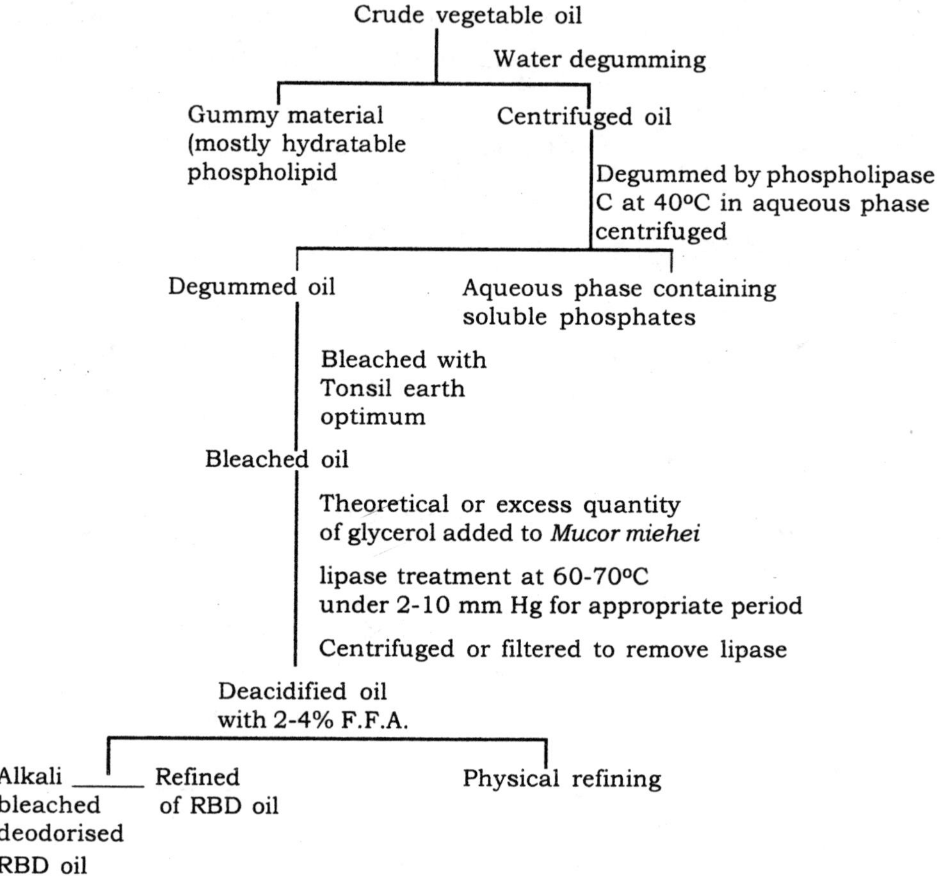

Diagram 16.2 : Enzymatic Degumming and Deacidification Process for Vegetable Oils and Typical Data.

Oil yield during enzymatic extraction also increases in the presence of hexane. Enzyme preparation from *Aspergillus niger* when used for soyabean and rapeseed in the presence of solvent showed better yields of oils (1). Enhanced release of extractable oil by enzymes was later shown (8) in crushed soyabean, cotton-seed and castor bean hydrolysis in the presence of hexane. Similar results are found also in the case of mustard seed and rice bran (15).

Although laboratory experiments are carried out on enzymatic extraction as a pretreatment step before pressing or solvent extraction or as an alternate way of oil production it has not been commercially scaled up as yet. Table 16.1 gives extraction data.

ENZYMATIC DEACIDIFICATION

One of the major steps in the purification of oils/fats is deacidification. Depending on the quantity of free fatty acids (F.F.A.) in an oil/fat the deacidification technology varies. The most important and practised method of refining of oils with low free fatty acid is the classical alkali neutralisation. For fats/oils having high acidity the processes commercially adopted are miscella refining with single or mixed solvent and physical refining. All these processes have distinct advantages and disadvantages.

It is, however, quite possible that a biotechnological process can be developed for deacidifying oils and fats. Enzymes can play an important role in the deacidification of oils/fats by converting the F.F.A. into neutral glycerides, predominantly into triglycerides, under certain conditions.

***Table 16.1*: Comparative Characteristics of Crude Mustard Oil and Rice Bran Oil Obtained from the Enzyme Process and Solvent Process**

Analytical characteristics	*Rice bran oil*		*Mustard oil*	
	Solvent extracted	*Enzyme extracted*	*Solvent extracted*	*Enzyme extracted*
Oil (%, w/w)	20.7	20.6	35.1	35.0
Free fatty acid %	3.5	3.6	1.1	1.3
Saponification value	184.4	188.4	175.0	174.3
Iodine value	89.4	88.9	111.8	113.5
Peroxide value (mequ/1000 gm of oil)	2.0	3.9	1.6	2.0
Unsaponifiable matter (%)	5.0	4.9	1.2	0.9
Acetone insoluble (%)	4.4	4.0	—	—
Lovibond colour (1 cm cell)	25Y + 8.2R + 3B	12Y + 5.3R + 0.2B	78.7Y + 2.2R	45.6Y + 2.2R
Odour (pungency)	—	—	strong	mild
% Protein in meal	18.5	29.4	43.3	55.4

Lipase, the enzyme excreted by microorganisms to degrade lipids, can be produced cheaply on large scale from several microbial species by fermentation. This lipase catalyses the hydrolysis of fats and oils, and as the reaction is reversible it can also catalyse the formation of acyl-glycerols from glycerol and fatty acids.

Utilising this property of lipase having specific and non-specific activities, it may be possible to neutralise the free fatty acids of oils and fats by esterifying them with the hydroxyl groups already in the oils and with the hydroxyls from outside by adding glycerol or monoglycerides.

There are lots of references about using lipase as a catalyst in esterification producing different kinds of esters, but there are very few references about the re-esterification of free fatty acid with glycerol to produce triglyceride. Recently, biorefining processes (16,17) have been reported on a laboratory scale to deacidify oils of high acidity (25-30% F.F.A) by esterification of the F.F.A. enzymatically with added glycerol in theoretical quantity or in excess over theoretical with the aid of 1,3 specific *Mucor miehie* lipase under the optimum conditions of temperature, vacuum, quantity of enzyme and time. Biorefining process investigated for rice bran oils of varying acidity (18) (5-17% F.F.A.) showed that the process is also effective for oils having different acidity, though its effectiveness in reducing the F.F.A. level increases with increase in its acidity. In other words biorefining is more effective for oils particularly of high acidity (F.F.A). Monoglyceride can also be used in the enzymatic deacidification as a source of hydroxyl groups instead of glycerol. It is reported (19) that monoglyceride, either commercial or made from the same oil, when utilised in enzymatic deacidification instead of glycerol gives better results in producing more triglycerides. It is therefore possible to use monoglycerides of the same oil to be biorefined in order to maintain the sequence of its fatty acids in the triglyceride molecules. The time taken for monoglycerides in esterification is however more than glycerol. Tables 16.2-16.4 contain the relevant data.

Biorefined oils having 2-4% F.F.A. can be further alkali refined, bleached and deodorised to extremely good quality edible oils. The refining factor and the other characteristics of the combined and alkali refined oils are comparable with those of the single solvent and double solvent alkali refining and physical refining processes. Biorefining process has also been successfully applied to some dark coloured and high F.F.A. minor oils (20).

Tables 16.5 & 16.6 give a comparative study of refining.

Table 16.2: High Acid Rice Bran Oil Biorefining

Enzymatic deacidification		Characteristics of oil		
		F.F.A.	U.M.	Colour (1 cm cell)
1(a) Crude rice bran oil	—	30.0	4.0	27Y + 4R
(b) Degummed dewaxed and bleached rice bran oil	*Mucor miehei* lipase (10%) 70°C, glycerol (theoretical quantity), 10 mm Hg pressure	3.6	2.0	9Y + 2.2R
(c) Biorefined oil further alkali refined, bleached	—	0.1	2.0	5Y + 0.6R
2(a) Crude rice bran oil	—	10.2	4.4	8.6Y + 3.8R +0.6B
(b) Degummed, dewaxed and bleached rice bran oil	Biorefined at the same condition mentioned above, but at 2 mm Hg pressure, 9 hours	3.0	2.2	3.3Y + 0.3R
(c) Biorefined oil further alkali refined bleached and deodorised	—	0.2	2.0	7.6Y + 1.4R
3(a) Crude mohua oil	—	25.1	3.2	5.0Y + 0.8R (1/4" cell)
(b) Degummed and bleached mohua oil	Biorefined at 60°C/20h/2mm/Hg	3.8	2.0	2.0Y + 0.1R
(c) Biorefined, alkali refined bleached and deodorised oil	—	0.2	1.0	0.6Y + 0.0R

BIOTECHNOLOGY IN MODIFYING COMPOSITION AND PROPERTIES

Biotechnological processes like bio-interesterification have been developed to modify fats/oils, fatty acids and other lipids like phospholipids to produce fat products of desired composition and properties for specific edible and industrial applications.

Table 16.3: Extent of Deacidification of Rice Bran Oil of Varying Acidity with _Mucor Miehei_ Lipase (10% on oil w/w) at 70°C and 2 mm Hg with Stoichiometric Amount of Glycerol

Reaction time (hr)	*Free fatty acid (%) in oil*	
	Initial	*After esterification*
7	30.0	3.6
7	16.9	3.8
7	9.5	3.2
9	8.6	3.0
8	5.4	4.9

Table 16.4: Effect of Monoglyceride on the Deacidification of Rice Bran Oil by _Mucor Miehei_ Lipase (10% on oil w/w) at 2 mm Hg and 70°C

Sl. No.	*Nature of reacting molecule*	*Amount used*	*Reaction time (hr)*	*F.F.A. (%) in oil*		*MG %*	*DG %*	*TG %*
				Initial	*Final*			
1(a)	Glycerol	S.A	9	8.6	3.0	-	5.0	89.7
(b)	Monoglyceride (commercial)	S.A.	12	8,6	2,6	-	0.34	94.4
(c)	Monoglyceride (pure)	S.A	15	8.6	2.2	-	2.7	93.0
2(a)	Glycerol	40% excess of S.A	7	9.5	1.9	-	10.4	85.2
(b)	Rice bran oil monoglyceride (58.9% MG)	40% excess of S.A.	7	9.5	1.9	-	2.0	93.8
3(a)	Glycerol	40% excess of S.A.	7	16.9	2.8	-	-	-
(b)	Rice bran oil monoglyceride (58.9% MG)	40% excess of S.A.	9	16.9	3.0	-	-	-

Fats and oils are modified for use in producing edible fat products by hydrogenation and interesterification process using chemical catalyst systems. For nearly a decade, considerable interest has grown on doing these two processes by biochemical means while biohydrogenation has not yet found to be useful, the bio-interesterifications have proved to be commercially very useful. In fact, interesterification process with enzymes has gone commercial in Unilever, Holland for making shortenings, margarine fats and cocoa butter substitute. The enzymes are microbial lipases and have specific and non-specific reactions like hydrolysis and re-synthesis of triglycerides when reacted with oils and fats. A non-specific or random lipase catalyses the interesterification reactions like those obtained by chemical interesterification. Specific lipases catalyse the reactions of 1- and 3-positions of triglycerides and accordingly give interesterified reaction products which are selective and different from the non-specific lipase catalysed and chemically interesterified fat products.

***Table 16.5*: Comparison of Refining Characteristics of Rice Bran Oil by Enzymatic Deacidification and by other Refining Processes**

Sl. No.	*Oil*	*F.F.A. %*	*R.F.*	*Overall Process loss %*	*U.M. %*	*MG %*	*DG %*	*TG %*	*Lovibond colour (1 cm cell)*
1.	Crude oil	10.2	—	—	4.4	—	—	—	8.6Y + 3.8R +0.6B
2.	Degummed, dewaxed and bleached oil	8.6	—	—	2.2	—	7.2	79.7	5.7Y + 1.1R
3.	Biorefined oil	3.0	—	—	2.2	—	5.0	89.7	3.3Y + 0.3R
4.	Biorefined, alkali refined, bleached and deodorised oil	0.2	1.2	9.6	2.0	—	0.73	96.8	6.0Y + 1.1R
5.	Biorefined, bleached and physical refined oil	0.3	1.1	10.2	3.0	—	2.0	93.8	8.2Y + 1.2R
6.	Alkali refined, bleached and deodorised oil	0.2	1.8	20.4	2.0	—	5.6	92.2	7.6Y + 1.4R
7.	Physical refined oil	0.4	1.3	16.2	3.9	—	4.7	90.7	9.0Y + 1.8R

Table 16.6 **: Biorefining of High F.F.A. Mohua oil and Comparison with other Processes**

Characteristics	*Crude oil*	*Biorefined oil 10% Mucor Miehei 60°C/20 hr/ 2 mm Hg Th. glycerol*	*Biorefined alkali refined bleached and deodorised*	*Biorefined bleached and physical refined oil*	*Physical refined oil*
F.F.A. %	25.1	3.8	0.2	0.3	0.3
U.M. %	3.2	2.0	1.0	2.2	2.6
Triglyceride %	65.6	84.2	98.3	94.7	92.3
Diglyceride %	6.1	9.9	0.48	2.8	4.8
Monoglyceride %	Nil	Nil	Nil	Nil	Nil
Lovibond colour (1/4" cell)	5Y + 0.8R	2Y + 0.1R	0.6Y + 0.0R	2.8Y + 0.8R	6.8Y + 1R
Total process oil loss %	—	—	11.4	7.5	31.8

The interesterification reactions of importance for making specialised fat products include ester-ester interchange, acidolysis, and alcoholysis. A great deal of work has already been done on these aspects (21-31) in producing vanaspati, margarine and high value confectionery fats. These reactions have been investigated for palm oil and its fractions and for some of the minor oils and fats of India with a view to making edible plastic fat products like vanaspati, cocoa butter and low caloric fats.

Tree-borne seed fats like sal, mango, and mowrah have been interesterified as such and with liquid oils like rice bran (a minor oil) and cottonseed by lipozyme or other enzyme to produce vanaspati and margarine (32,33) fats (Tables 16.7, 16.8 & 16.9).

Sal, mango and mowrah have been interesterified by exchange reaction with methyl ester of palmitic acid or stearic acid, as the case may be, with the aid of lipozyme (34). The interesterified product from sal fat, for example, shows properties comparable to those of cocoa butter (Table 16.10).

Enzymatic acidolysis (23) of sal fat, mango kernel fat and mowrah oil with palmitic acid or stearic acid in the presence of lipase also leads to fat products having fatty acid composition closer to that of cocoa butter. Selective fraction of sal fat, when acidolysed with palmitic acid using lipase as such, or following fractionation, yields product of satisfactory fatty acid composition and NMR data similar to those of cocoa butter (31) (vide Tables 16.11 & 16.12).

The bio-interesterification process involving ester-ester interchange has a number of advantages, for example,

(i) low oil loss,

(ii) light coloured products,

(iii) prevention or limiting of the formation of high melting point triglyceride with specific lipases,

(iv) the reaction being slower than chemical process can be controlled in respect of composition and rate,

(v) bio-interesterification reactions not requiring high purity oils/fats,

(vi) low temperature improving the quality and making for substantial energy savings, and

(vii) with specific lipase fatty acid composition in 2-position remaining unchanged.

Flow diagram with certain concrete/typical examples of enzymatic ester-ester interchange and acidolysis reactions are shown in Diagrams 16.3 and 16.4.

Diagram 16.3 : Enzymatic ester-ester interesterification of vegetable oils and fats along with some typical examples

RBD oil(s) of appropriate composition (blends)

Random or regio-specific lipases 40°–70°C, stirred reactor lipase filtered off

Interesterified fat

Bio-hydrogenation has been receiving considerable attention to achieve high selectivity of hydrogenation of polyunsaturated fatty acids without the *trans* or double bond positional isomers.

Table 16.7: Lipase Catalysed Ester-Ester Interchange Reaction

Oil/Fat	*Lipase*	*Temp.°C /Time hr*	*Slip point*	*SFC/SFI/at °C*						
			°C	*10*	*15*	*20*	*25*	*30*	*35*	*40*
Palm oil	—	—	36.5	—	35.5	29.9	21.3	12.0	7.7	0
Palm oil	*Mucor miehei*	70°/1	31.0	—	35.2	25.0	16.4	4.3	0	0
Palm oil + Cottonseed oil (70 + 30)	—	—	36.0	—	23.8	19.5	14.0	9.4	6.3	1.0
-do-	*Mucor miehei*	70°/1	33.0	—	25.9	14.6	13.1	9.7	4.0	0
Palm + Rice bran (70 + 30)	—	—	37.0	—	22.8	17.2	11.9	7.1	4.7	2.1
-do-	*Mucor miehei*	70°/6	31.0	—	26.2	15.6	13.7	7.8	1.4	0
Sal	*Mucor miehei*	60°/1	33.5	—	50.4	41.0	23.8	5.2	0	
Sal + Rice bran (80 + 20)	*Mucor miehei*	60°/1	33.8	41.8 (NMR SFC)	—	13.1	—	5.2		
Mango kernel	*Mucor miehei*	60°/1	35.0	—	18.3	13.9	10.3	7.8	3.9	0

TABLE 16.7 (CONTD.)

TABLE 16.7 (CONTD.)

Mango kernel + Rice bran (80 + 20)	*Mucor miehei*	60°/1	37.6	30.3	—	12.4	—	6.5		
				(NMR SFC)						
Mowrah	*Candida cylindraceal*	40°/48	35.2	—	19.1	15.2	10.4	7.1		
Mowrah + Rice bran (80 + 20)	*Mucor miehei*	60°/5	35.2	22.2	—	8.1	—	4.2		
					(NMR SFC)					
Sal + Cottonseed (90 + 10)	*Mucor miehei*	60°/1	33.5	—	43.1	38.6	28.9	12.6	5.3	0
Palm stearin + Rice bran (30+70)	*Mucor miehei*	60°/1	32.0	—	19.6	16.5	12.9	7.6	2.9	1.4
Palm stearin + Mowrah (70 + 30)	*Mucor miehei*	60°/1	38.0	—	35.9	32.1	25.1	18.3	14.4	7.8

Table 16.8: Characteristics of Oils Interesterified with Candida Cylindracae Lipase for 48 hours at 37°C

Oil	*Slip point °C*	*Solid Fat Index at 25°C ml/g*
Sal	29.5	45.0
Sal : Cottonseed oil 7:3 (w/w)	39.0	11.4
Sal : Soyabean oil 6:4 (w/w)	36.5	5.6

Table 16.9: Characteristics of Oils Interesterified with Mucor Miehei Lipase for 48 hours at 37°C

Oil	*Type*	*Slip point*	*Solid Fat Index at 25°C*
	w/w	*°C*	*ml/g*
Sal : Soyabean	8:2	37.0	8.5
Sal : Soyabean	6:4	40.5	8.9
Sal : Rice bran	6:4	41.0	12.9

Table 16.10: Enzymatic Triglyceride-Methyl Ester Interchange

Fat + Me-ester	*Fatty acid composition % (w/w)*				
	16:0	18:0	18:1	18:2	20:0
Sal + Methyl palmitate	24.3	32.7	37.7	1.2	3.9
Cocoa butter	25.2	35.5	35.2	3.3	—
	NMR data at °C				
	20	25	30	35	37
Sal + Methyl palmitate	90.0	69.0	44.4	12.6	8.7
Cocoa butter	96.6	93.8	71.2	4.8	1.9

Bio-hydrogenation has been investigated mainly with rumen bacteria (*Butyrivibrio fibrisolvens*) and research is going on with anaerobic bacteria from sources other than the rumen. Bio-process works mainly with fatty acids as the enzymes do not react with the fats and oils. The process is therefore highly promising for the hydrogenation of unsaturated fatty acids with mixed bacterial culture.

Table 16.11 : Fatty Acid Composition of Sal, Mango kernel and Mowrah Fats after Enzymatic Acidolysis with Mucor Miehei Lipase

Fat	Reaction time (hr)	Fatty acid composition % (w/w)				
		16:0	18:0	18:1	18:2	20:0
Sal fat	—	4.2	43.8	48.0	0.0	3.8
Sal fat acidolysed with palmitic acid	3.5	25.6	29.3	42.3	0.0	1.6
Mango kernel fat	—	7.0	40.7	47.2	2.4	2.4
Mango kernel fat acidolysed with palmitic acid	4.0	25.7	28.1	42.8	1.8	1.8
Mowrah fat	—	23.3	19.0	46.2	11.2	0.0
Mowrah fat acidolysed with stearic acid	4.0	22.5	34.6	33.9	7.2	0.0

Table 16.12 : Enzymatic Acidolysis of Sal fat / Sal fat fractions

Fat/fat product	Slip point °C	NMR data at			Fatty acid composition, % (w/w)				
		10°	20°	30°	16:0	18:0	18:1	18:2	20:0
Sal	34.5	77.3	62.1	2.2	6.3	44.6	41.6	1.7	5.7
Sal acidolysed with palmitic acid	29.5	76.0	52.8	1.8	14.0	33.8	40.9	4.1	7.0
Stearin fraction from acidolysed sal	31.0	77.1	67.5	3.1	20.9	36.2	33.0	3.0	7.6
Sal olein acidolysed with palmitic acid	29.0	72.1	44.7	3.8	19.4	32.3	41.1	3.0	4.7
Stearin fraction from acidolysed sal olein	29.5	77.5	58.7	2.1	20.7	37.6	41.1	2.4	6.0

Diagram 16.4: Enzymatic Acidolysis Reaction of Fats Along with Data for Typical Examples

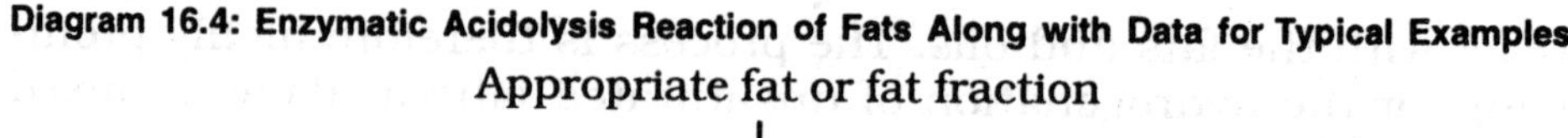

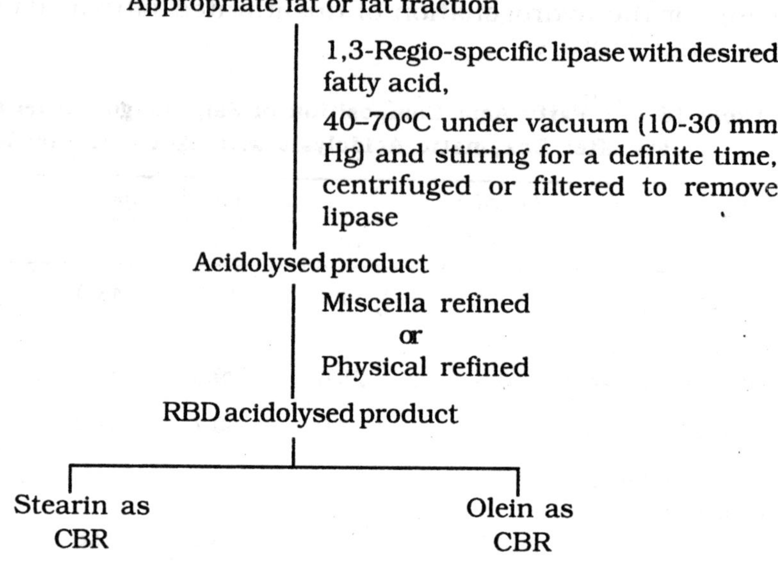

It is too early to claim the viability of the process on an industrial scale, though there will be ample scope for hydrogenating the unsaturated fatty acids of the vast majority of minor oils which can be hydrolysed by bio-hydrolysis process followed by separation of the fatty acids as fatty acid distillates.

BIOTECHNOLOGY IN PRODUCING BASIC OLEOCHEMICALS

Biotechnology has been gaining enormous importance in the production of basic and downstream oleochemicals. The major basic oleochemicals are fatty acids, glycerol, fatty alcohols, fatty amines, methyl esters of fatty acids and dibasic acids. Some of the downstream oleochemicals are esters of fatty acids, fatty acid amides and modified phospholipids. The biotechnological processes developed for oleochemicals are based on traditional reactions like hydrolysis, esterification, *trans*-esterification or alcoholysis hydrogenation, oxidation etc. carried out with the aid of enzymes, principally lipase. The lipases obtained from a variety of microbes have been demonstrated to be random, 1-3 specific and 2-specific in nature, and therefore contribute to making different oleochemicals.

The production of fatty acids and glycerol from a variety of oils and fats by *bio-hydrolysis* process with the aid of microbial lipases has been well documented (35-39). The process is also highly suitable for the production of polyunsaturated fatty acids from linseed oil, tung oil, fish oil, etc., unsaturated hydroxy acids and epoxy acids. The different non-specific lipases and the reaction conditions as well as the different reactors, viz. stirred tank reactor, droplet type, fluid bed and membrane type reactors have now been well established. The downstream separation units like microfilter and ultrafilter have also been developed. The droplet type reactor is successfully employed for the hydrolysis of a fat. Continuous hydrolysis of oil can also occur in a membrane type reactor, which uses a sheet of microporous hydrophobic membrane. Lipases from *Candida rugosa, Rizopus arrhizus, Aspergillus niger* and *Candida cylindraceae* are commercially suitable for almost complete fat splitting. Now, a 1-3 specific lipase like *Mucor miehei* lipase splits the fatty acids from 1- and 3-positions of triglycerides enabling the production of selected fatty acids. The use of specific lipase can yield fatty acids like erucic, stearic, ricinoleic etc. from the specific oils. Solid phase bio-hydrolysis has also been developed for solid fats like tallow, palm stearine and hydrogenated fats.

Bio-hydrolysis process produces fatty acids and glycerol of extremely good quality in relatively greater yields in comparison with the conventional high temperature, high pressure fat splitting process due to the fact that fatty acids and glycerol do not undergo any side reaction. The bio-hydrolysis process, being a process of low energy and low cost has tremendous commercial viability. In Japan, a batch process has been in use and some continuous processes have also been used industrially at least on pilot plant scale. In India also, this process can be utilised in producing fatty acids and glycerol from the vast majority of minor oils/fats (38,40) and also from the refinery by-product "Acid Oils" readily and economically (Table 16.13).

Flow diagram and typical bio-hydrolysis process conditions are shown in Diagram 16.5.

Fatty acid esters, e.g. monoglycerides, monohydric alcohol esters of saturated and unsaturated fatty acids including unsaturated hydroxy acids, and polyolesters including sugar esters of fatty acids that cater to food, pharmaceutical, cosmetics and chemical industries, can be produced by enzymatic interesterification and esterification process (41-43). A number of lipases are now known and so also are the specific reaction conditions for producing a variety of fatty acid based esters (42,44) vide Table 16.14.

Table 16.13: Some Examples of Biohydrolysis in a Batch Stirred Reactor

Oil/fat : Water (w/v)	*Lipase*	*Temp., °C/Time (hr.)*	*Fatty acid yield (% w/w)*
Rice bran : Water (1:2)	*Mucor meihei* SP 398 preparation 375 KLV on oil	37 ± 2°C/30 hr.	82.0
Castor : Water (1:2)	" "	37 ± 2°C/30 hr.	89.7
Palm : Water (1:2)	" "	37 ± 2°C/30 hr.	86.8
Rice bran acid oil (56.2% F.F.A. as oleic)	*Candida cylindracea*	25 ± 2°C/72 hr.	93.8 (as oleic)
Coconut acid oil (69.0% F.F.A. as lauric acid)	" "	40 ± 2°C/24 hr.	82.7 (as lauric)
Mustard (76.2% F.F.A. as oleic)	" "	25 ± 2°C/24 hr.	94.4 (as oleic)

Diagram 16.5 : Biohydrolysis Process for oils and fats along with typical data

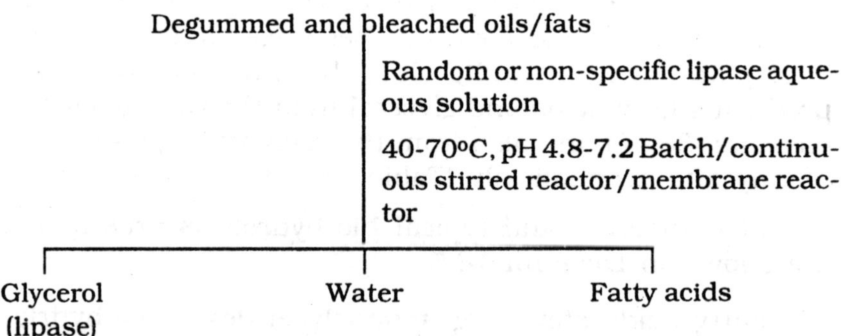

Table 16.14: Enzymatic Synthesis of Esters of Fatty Acids of Monohydric Alcohols

Fatty acid	*Alcohol*	*Lipase*	*Ester conversion%*
Oleic	Methanol	*Pseudomonas spec.*	37.1
		Rhizopus arrhizus	25.9
Oleic	Ethanol	*Pseudomonas spec.*	34.5
		Rhizopus arrhizus	37.6
Oleic	Isopropanol	*Pseudomonas* spec.	64.6
		Mucor spec.	64.9
Oleic	Butanol	*Candida cylindracea*	67.7
		Pseudomonous spec.	73.6
		Rhizopus arrhizus	75.1
		Mucor spec.	64.3
Oleic	Oleyl alcohol	*Candida cylindracea*	84.5
		Rhizopus arrhizus	75.8
Oleic	Erucyl alcohol	*Candida cylindracea*	85.8
		Pseudomonas spec.	84.9
		Rhizopus arrhizus	80.2
		Mucor spec.	80.7
Erucic	Erucyl alcohol	*Humicola lanuginosa*	90.0
		Pseudomonas spec.	84.2
		Rhizopus arrhizus	65.0
		Mucor spec.	40.8
Erucic	Methanol	*Pseudomonas* spec.	36.5
Erucic	Ethanol	-do-	39.2
Erucic	Butanol	-do-	59.2

With specific lipase, α monoglyceride and esters of specific fatty acids may be produced by *bio-alcoholysis* process. In making esters from high F.F.A. oils or acid oils a combination of bio-esterification and bio-alcoholysis provide an elegant approach.

The bio-alcoholysis process with random lipase is more advantageous than the bio-hydrolysis process from commercial point of view as the glycerol can be obtained in highly concentrated form and the esters of fatty acids can be fractionally distilled straightaway to yield high purity individual fatty acid esters when required. The bio-alcoholysis process for making high mono-glyceride from vegetable oils and oil fraction and for making monohydric alcohol esters shown in Diagram 16.6 along with typical conditions adopted (Tables 16.15 and 16.16).

***Table 16.15*: Monoglyceride Preparation by Lipase Catalysed Reaction**

Fat/oil	Lipase	Optimum temp. °C	Monoglyceride yield (max.) (%)
Beef tallow	*P. fluorescens*	42	76
Palm oil	*P. fluorescens*	40	67
Palmolein	-do-	30	71
Palm stearin	-do-	40	86
Coconut oil	-do-	30	77
Rapeseed oil	-do-	5	77
Olive oil	-do-	10	90
Palm oil	*Mucor miehei*	40	57
Palm oil	Sp 398	40	43
Palm oil	*Chromobacterium Viscosum*	40	65
Palm stearin	*Mucor miehei*	40	65
Palm stearin	Sp 398	40	47
Palm stearin	*Chromobacterium viscosum*	40	83

***Table 16.16*: Bioalcoholysis of Vegetable Oils**

Oil	Lipase	Alcohol	Ester yield %
Mustard	*Mucor miehei*	Butyl	74.4
Mustard	-do-	Lauryl	97.1
Castor	-do-	Lauryl	62.1
		Myristyl	65.3

Enzymatic esterification and *enzymatic alcoholysis* in combination provide route for making monohydric alcohol esters from acid oils and high acid natural oils (Tables 16.17 & 16.18).

The synthesis of polyolesters of fatty acids having surface activities can be readily achieved (Table 16.19) by *enzymatic esterification.*

In the preparation of methyl esters of fatty acids by lipase catalysed esterification and alcoholysis process there has been little success so far, but it has been reported (45) recently.

***Table 16.17*: Enzymatic Esterification and Alcoholysis of Minor Oils to Esters of various Monohydric Alcohols using Mucor Miehei Lipase at 60°C for 5 hours**

Minor oils	*Ester yield % (gms of ester per 100 gms of oil)*						
	Isopropyl	*Butyl*	*Octyl*	*Decyl*	*Dodecyl*	*Cetyl*	*Stearyl*
Rice bran acid oil	76.8	89.2	111.9	119.2	127.6	143.5	151.5
Rice bran oil	74.4						
Karanja oil	72.3						
Sal oil	75.8						

***Table 16.18*: Bioesterification and Bio-alcoholysis of Acid Oils of Commerce**

Acid oil	*Lipase*	*Butyl ester yield %*
Rice bran	Mucor miehei	89.2
Coconut	-do-	94.3
Mustard	-do-	99.2

***Table 16.19*: Enzymatic synthesis of Polyol Esters of Fatty Acids**

Polyol	*Fatty Acid*	*Ester %*	*Surface Tension (dyne/cm)*
Polyethylene glycol			
– 200 M	Lauric/ Myristic	98.5/98.1	35.0/34.7
– 300 M	-do-	96.7/95.4	36.7/35.7
– 400 M	-do-	93.9/91.9	36.0/35.7
– 600 M	-do-	91.4/89.9	33.7/36.9

Diagram 16.6 : Bio-alcoholysis process along with typical conditions

Fats and Oils

↓ Glycerols + water (all in appropriate proportions). Stirred at below critical temperature (optimum temperature) until the reaction mixture solidifies

Product
(mainly monoglycerides)

The conversion of some natural oils and fats and fat fraction into their monoglycerides is presented in Table 16.17.

That methyl ester of a fatty acid can be obtained in 100% yield in 4 hours by reacting a fatty acid with dimethyl phosphite in the presence of Lipozyme as a catalyst.

Lipases, particularly 1,3 specific lipase, acylate long chain fatty amines to yield amides. Under optimum conditions one can obtain high yield (80%) of a fatty acid amide. Similarly, one can obtain in high yield (85%) N-acylated amino alcohols by achieving acylation of the amine fraction with the aid of Lipozyme (45,46). The products are reported to have unique surface active properties, providing scope of utilising in shampoos and bubble baths as foam boosters.

BIOTECHNOLOGICAL MODIFICATION OF POLAR LIPIDS (E.G. PHOSPHOLIPIDS)

Phospholipids, e.g. soyaphospholipids, can be modified by lipase catalysed hydrolysis and alcoholysis reactions. Hydrolysis with 1,3 specific and 2 specific lipases yields 1- and 2-lysophospholipids (47). Alcoholysis with 1,3 specific lipase yields 1-lysophospholipid and α monoglyceride with glycerol (48) and long chain alcohol esters of fatty acids when long chain alcohols are used (47).

Dibasic acids constitute an important group of oleochemicals and can be best produced by bio-process instead of the traditional chemical oxidative method. Yeast such as *Candida tropicalis* M 2030 utilises a number of fatty acids and oil fats producing a wide range of dibasic acids that can be used as raw materials for polyester and allied industries (49).

Table 16.20, shows the oxidation of monobasic fatty acids into dibasic acids.

Table 16.20*: Dicarboxylic Acid Production from Fatty Acids and Ester of Fatty Acid by *Candida Tropicalis

Fatty Acid/Ester	*Dicarboxylic acid (g/l)*
Petroselenic acid	53.8
Oleic acid	48.1
Palmitic acid	72.0
Butyl palmitate	57.5

The foregoing examples point out that biotechnology can be very helpful to produce some of the major basic oleochemicals and a host of downstream oleochemicals making use of microbial lipases. The optimisation of the biotechnological processes are required along with special stress laid on reactors and downstream separation equipments. The production of a wide range of lipases and making them available in adequate quantity at reasonable prices must be made to make the bio-processes commercially viable.

REFERENCES

1. Fullbrook, P.D., JAOCS **60**, 476 (1983)
2. Leone, A.M., Almparelli F., Notte E. LA., Lluizzi V.A., and Padula M.,Rev. Ital. Sost. Gr., **54**, 514 (1977)
3. Martinez-Squarez E., Munoz Aranda E., Mendosa, J. Alba, Lauzon Rey A., Grassas Y Aceitas **28**, 31 (1977)
4. McGlone, O.C., Lopez-Munguia Canales and J.V. Carter, J. Food Sci., **51**, 695 (1986)
5. Buenrostro, M. and Lopez Munguia Canales A., Biotech. Lett., **8**, 505 (1986)

6. Bouvier, F. and Entressangles B., Actes du Congress Chevreuller Congress EUROLIPID, Angers, France, 6-9 Juin, Tome 11, 1027 (1989)
7. Cheah, S.C., Augustin M.A. and Ooi L.C.L., Palm Oil Developments, No.10, 12 (1989)
8. Bhatnagar, S. and Johari B.N., Curr. Sci., **56**, 775 (1987)
9. Lauzani, A,, Petrini M.C., Cozzoli O., Gallavresi P., Carola C. and Jacini G., Riv. Ital. Sost. Gr. **52**, 226 (1975)
10. Sosulski, K., Sosulski F.W. and Coxworth E., JAOCS, **65**, 357 (1988)
11. Sosulski, K. and Sosulski F.W., Quality of Oil and Meal from Enzyme-Treated canola seeds, 5th International Congress of Engg. and Food, Cologne, Federal Republic of Germany, May 25-June 3 (1989)
12. Karlovic, Dj., Bocevska M. Pericin D. and Turkulov J., New Technology of Obtaining Highly Valuable Corn Germ Oil, VIII Yugoslav Nutrition Congress, Beograd, Oct. 11-19 (1990)
13. Bocevska, M., Marlovic Dj., Turkulov J. and Pericin J., JAOCS, **70**, 1273 (1993)
14. Sejr Olsen, H., Lecture given at the workshop on "Agriculture Refineries"- A Bridge from Farm to Industry at Hotel Fredensborg, Bornholm, Sep. 16-18 (1987)
15. Sengupta R. and Bhattacharyya D.K., JAOCS, **73**, 687 (1996)
16. Bhattacharyya, S. and Bhattacharyya, D.K., JAOCS **66**, 1469 (1989)
17. Sengupta R. and Bhattacharyya, D.K. *ibid*, **69**, 1146 (1992)
18. Sengupta R. and Bhattacharyya, D.K., JOTAI **28**, 17 (1995)
19. Sengupta R and Bhattacharyya D.K., *ibid*, **28**, 125 (1996)
20. Bhattacharyya, S., Bhattacharyya, D.K., Chakrabarty A.R. and Sengupta R., Fat Sci. Technol. **91**, 27 (1989)
21. Macrae A.R., JAOCS **60**, 291 (1983)
22. Hansen T.T. and Eigtved P., JAOCS Proceedings, World Conference on Emerging Technologies in the Fats and Oils Industry, 365 (1986)
23. Bhattacharyya D.K. and Bhattacharyya S., *ibid* Proceedings, World Conference on Biotechnology for the Fats and Oils Industry, 309 (1988)
24. Bhattacharyya D.K. and Bhattacharyya S., DGF Session DGF-AOCS World Conference on Biotechnology, Hamburg, September 30 to October 1, 1987

25. Chakrabarty M.M., Khatoon S. and Chaudhuri S., JAOCS, Proceedings, World Conference on Biotechnology for the Fats and Oils Industry, 288 (1988)
26. Chakrabarty M.M., Chakrabarty S. and Khatoon S., *ibid*, World Conference on Oilseeds and Edible Oil Proceedings, 234 (1990)
27. Bhattacharyya D.K., Bhattacharyya S., Sarkar S., and Sil S., Proceedings PORIM International Palm Oil Development Conference, Kualalumpur, 1989.
28. Muderthwa J., Pina M., Montet D., Fenillared P. and Graille J., Oleagineux **44**, 3 (1989)
29. Muderthwa J., Pina M., and Montet D., Rev. Fr. Corps Grass, **36**, 11 (1980)
30. Kalo P., Kemppinen A. and Antila M. ISF-JOCS World Congress, 1988 Abstracts 8514, 283 (1988)
31. Bhattacharyya S. and Bhattacharyya D.K. Oleagineux **46**, 509 (1991)
32. Bhattacharyya S. and Bhattacharyya D.K.JOTAI, **24**, 19 (1992)
33. Ghosh S. and Bhattacharyya D.K., Manuscript under communication.
34. Sridhar R., Lakshminarayanan G. and Kaimal T.N.B. JAOCS; **68**, 726 (1991)
35. Linfield W.M., JAOCS, Proceedings of World Conference on Biotechnology in the Fats and Oils Industry, 131 (1988)
36. Ishida S., JAOCS, Proceedings of World Conference on Emerging Technologies in the Fats and Oils Industry, 354 (1988)
37. Khor H.T., Tan N.H. and Chua C.L., JAOCS, **63**, 538 (1986)
38. Bhattacharyya S. and Bhattacharyya, D.K., Oleagineux, **47**, 241 (1992)
39. Taylor F. and O. Brien D.J., Proceedings of Sessions Lectures and Scientific Presentations of ISF-JOCS World Congress 1988 Vol II, 1076 (1989)
40. Ghosh S. and Bhattacharyya, D.K. Manuscript under communication
41. Mc. Neill G.P., Shimizu S. and Yamane T., JAOCS, **68**, 1 (1991)
42. Lazer G., Weiss A. and Schmid R.D. *ibid* Proceedings World Conference on Emerging Technologies in the Fats and Oils Industry, 346 (1986)

43. Ghoshroy S. and Bhattacharyya D.K., JAOCS **69**, 85 (1992)
44. Ghosh S., Bhattacharyya D.K. and Basu S., JOTAI, **25**, 79 (1993)
45. Graille J., Plenary Lecture, International Meeting on Fats and Oils Technology, Symposium and Exhibition Campinas State University, Sao, Paulo, Brazil July 8-11 (1991)
46. S. Basu and D.K. Bhattacharyya, Private communication
47. M Ghosh, C.S.I.R. Report, 1995, New Delhi
48. M. Ghosh (Basu) and D.K. Bhattacharyya, J. Oil Technol. Asson. India, **27**, 175 (1995)
49. Uemara N., Taoka A. and Takagi M., JAOCS, Proceedings of World Conference on Biotechnology for the Fats and Oils Industry 148, (1988)

CHAPTER 17

Oleochemicals

INTRODUCTION

Oleochemicals are chemicals derived from natural oils and fats (natural route) as opposed to chemicals derived from petrochemical feedstock (synthetic route) such as ethylene, propylene, xylenes etc. Oleochemicals refer mainly to fatty acid methyl esters, fatty acids, fatty alcohol, and glycerine as well as their derivatives that are derived from animal and vegetable oils and fats.

World interest in oleochemicals appears to be increasing, particularly in developed nations as a means to develop new uses for what seems to be an ever increasing supply of vegetable oils. In recent years the more publicised ventures are printing inks, diesel fuel extenders and replacements, lubricating oils, alkyl polyglycosides, surfactants and many others. In many cases, oleochemicals are the more "environmentally friendly" alternate raw materials.

RAW MATERIALS

Raw materials for oleochemicals are plentiful. World production of the prime raw materials — tallow, coconut oil, palm oil and palm kernel oil, accounted for about 36% of total world fats and oils production in 1990; by the year 2000 they should account for 42%. Castor oil which consists predominantly of ricinoleic acid is a major source. Recently high erucic rapeseed and crambe oils have been used for oleochemicals.

Southeast Asia is where most of the recent expansion in oleochemicals production facilities has occurred. The major increase in raw material supplies will be in Southeast Asia. Malaysia leads the world in palm oil and palm kernel oil production with Indonesia following in a rapid expansion of its oil palm acres.

Oleochemicals are bound to enlarge their lease and improve because of their source being the natural products, viz. oils and fats. Natural raw materials are interchangeable; no parallel exists in the petrochemical industry. Natural raw materials are replenishable since they can be grown and so their availability is guaranteed whereas synthetic raw materials are obtained from petroleum products which are getting depleted and are not replenishable.

Asian countries comprising Malaysia, Indonesia, Philippines have developed as major centres for production of oleochemicals. In 1989, ASEAN countries produced 326,100 tonnes of oleochemicals or about 11.5% of the world production. In 1995 and 2000, ASEAN is expected to produce respectively about 25% and 35% of the world demand for these green products.

A brief view of Asian oleochemical production capacity and the world production of oils and fats, suitable as raw materials for oleochemicals, is presented in Tables 17.1 and 17.2.

***Table 17.1*: Asian Oleochemical Production Capacities (Thousand tonnes)**

Oleochemicals	*1989*	*1990*	*1995**	*2000**
Fatty acids	172.0	202.0	609.5	800.0
Methyl esters	75.1	126.1	130.0	170.0
Fatty alcohols	47.5	115.0	363.0	770.0
Fatty esters	—	10.0	50.0	80.0
Fatty amines	—	5.0	50.0	100.0
Glycerol	31.5	503.0	116.0	214.0
Total :	326.1	508.4	1318.4	2134.0

Source: Porim Handbook (1993).

Table 17.2: World Production of Oils and Fats Suitable as Raw Materials for Oleochemicals (Million tonnes)

	1960	1970	1980	1989	1990	1995*	2000*
Soyabean	4.0	6.1	12.2	15.0	16.1	18.2	21.0
Tallow	3.6	4.4	6.0	6.6	6.6	7.1	7.0
Coconut	2.1	2.2	3.3	2.8	3.1	3.8	4.2
Palm	1.1	1.7	5.0	10.3	10.8	17.8	22.0
Palm kernel	0.4	0.4	0.7	1.3	1.4	2.3	2.8
Others	20.9	25.3	29.6	41.3	41.2	35.2	50.0
Total :	32.1	40.1	56.8	77.2	79.2	84.4	107.0

* Forecast

Source : Porim Handbook and Papers - Malaysia, AOCS 1990-1993 – Inform.

In addition to the established sources for oleochemical raw materials, there are several new sources with future ramifications for the oleochemical industry — flaxseed and meadowfoam. Oils from these crops can be used for a variety of industrial purposes, such as paints and lubricants. USDA (U.S. Department of Agriculture) scientists are examining another plant *Lesquerella* with potential future uses in resins, waxes, nylons, plastics, high performance lubricants, corrosion inhibitors, coatings, cosmetics and hand soaps. Unichema International of Netherlands is researching the possibilities of *Crambe abyssinica*, which contains a high proportion of erucic acid. Potential uses include polyethylene films, lubricants and as building block for polymers.

Various oleochemical raw materials and the production of their derivatives are briefly presented in the charts (Figs. 17.1-17.6) and Tables 17.3 and 17.4.

Fig. 17.1: Structure of Fatty Acids

Common Fatty Acids

COOH
n

n=1, Lauric; n=2, Myristic; n=3, Palmitic; n=4, Stearic; n=6, Behenic

Saturates Monounsaturates

COOH 18:1, Oleic acid

7-18:1, Elaidic acid

COOH

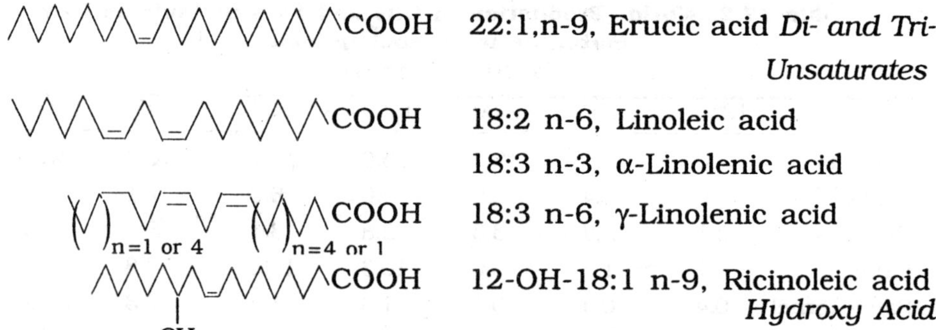

22:1,n-9, Erucic acid *Di- and Tri-Unsaturates*

COOH

18:2 n-6, Linoleic acid

COOH
n=1 or 4
n=4 or 1

18:3 n-3, α-Linolenic acid

18:3 n-6, γ-Linolenic acid

COOH
OH

12-OH-18:1 n-9, Ricinoleic acid *Hydroxy Acid*

Less Common Fatty Acids

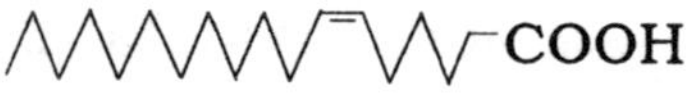

COOH

18:1 n-12, Petroselinic acid *Monounsaturates*

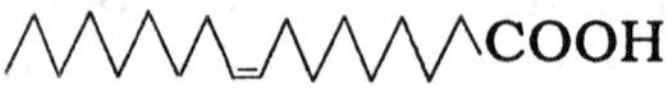

COOH

20:1 n-9, cis-11-Eicosenoic acid

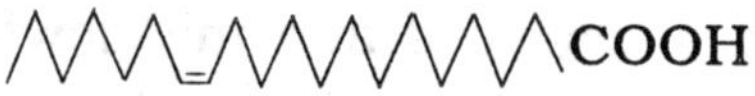

COOH

20:1 n-6, cis-13-Eicosenoic acid

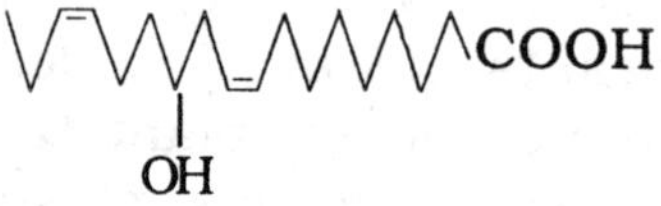

COOH
OH

Densipolic acid *Hydroxy Acids*

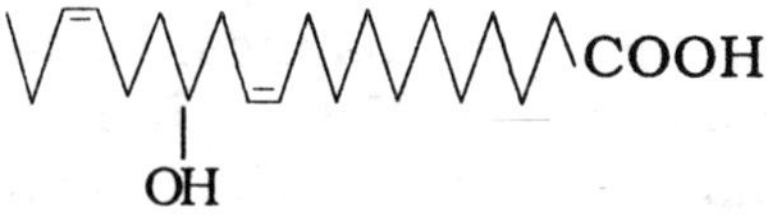

COOH
OH

Auricolic acid

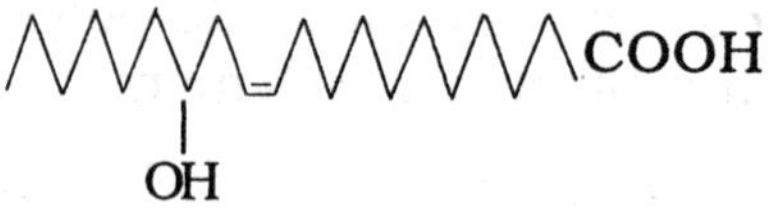

COOH
OH

Lesquerolic acid

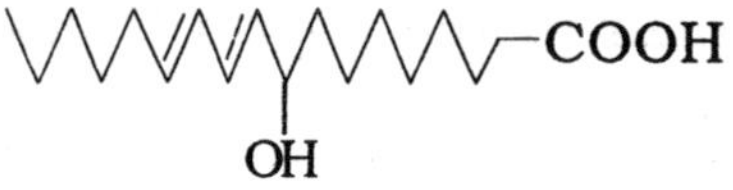

COOH
OH

Dimorphecolic acid

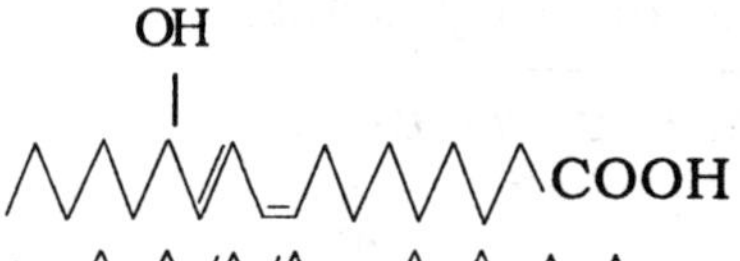

OH
COOH

Coriolic acid

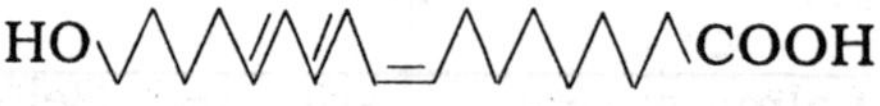

HO
COOH

Kamolenic acid

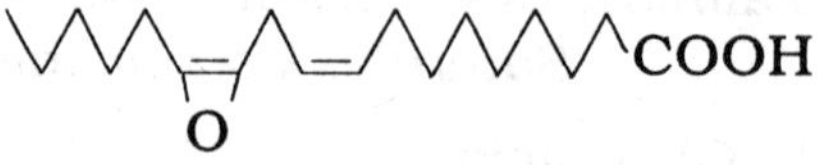

COOH
O

Vernolic acid *Epoxy Acids*

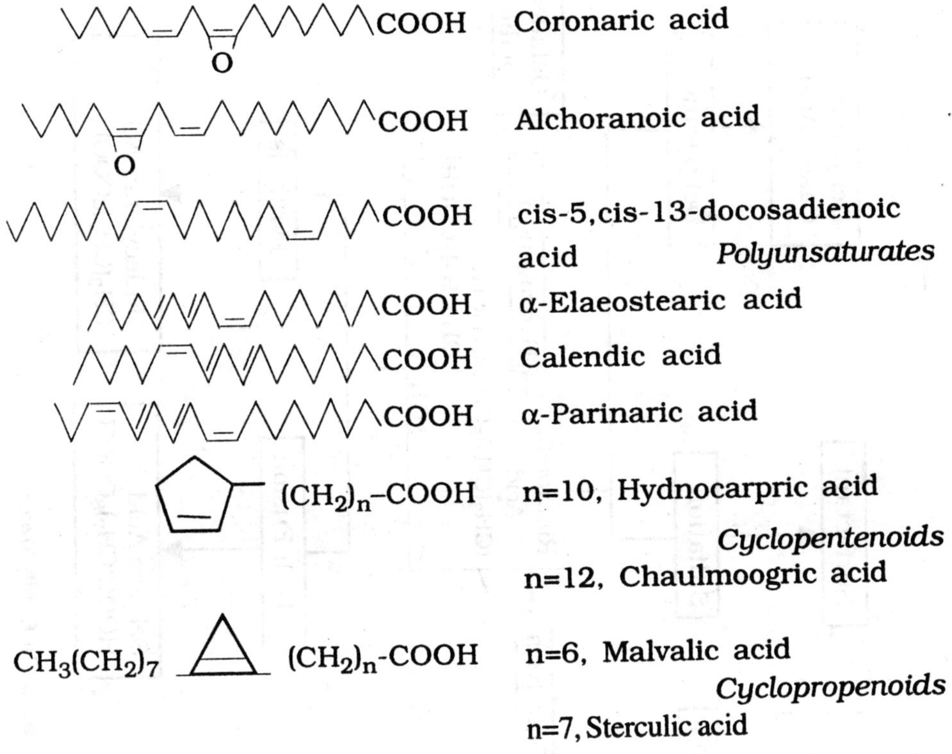

Fig. 17.1 : Structures of Fatty Acids

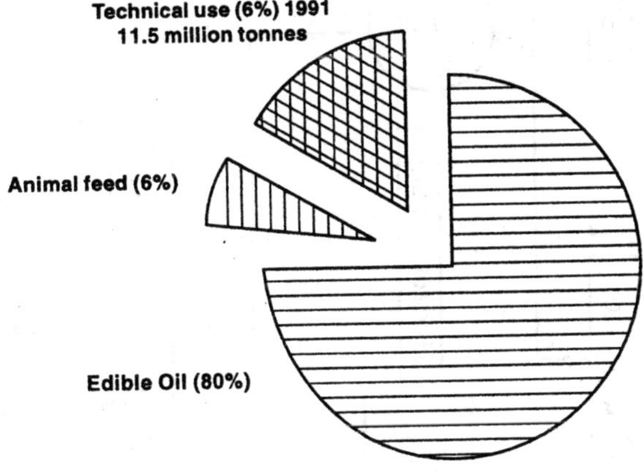

Fig. 17.2 : World Consumption of Oils and Fats (1991)

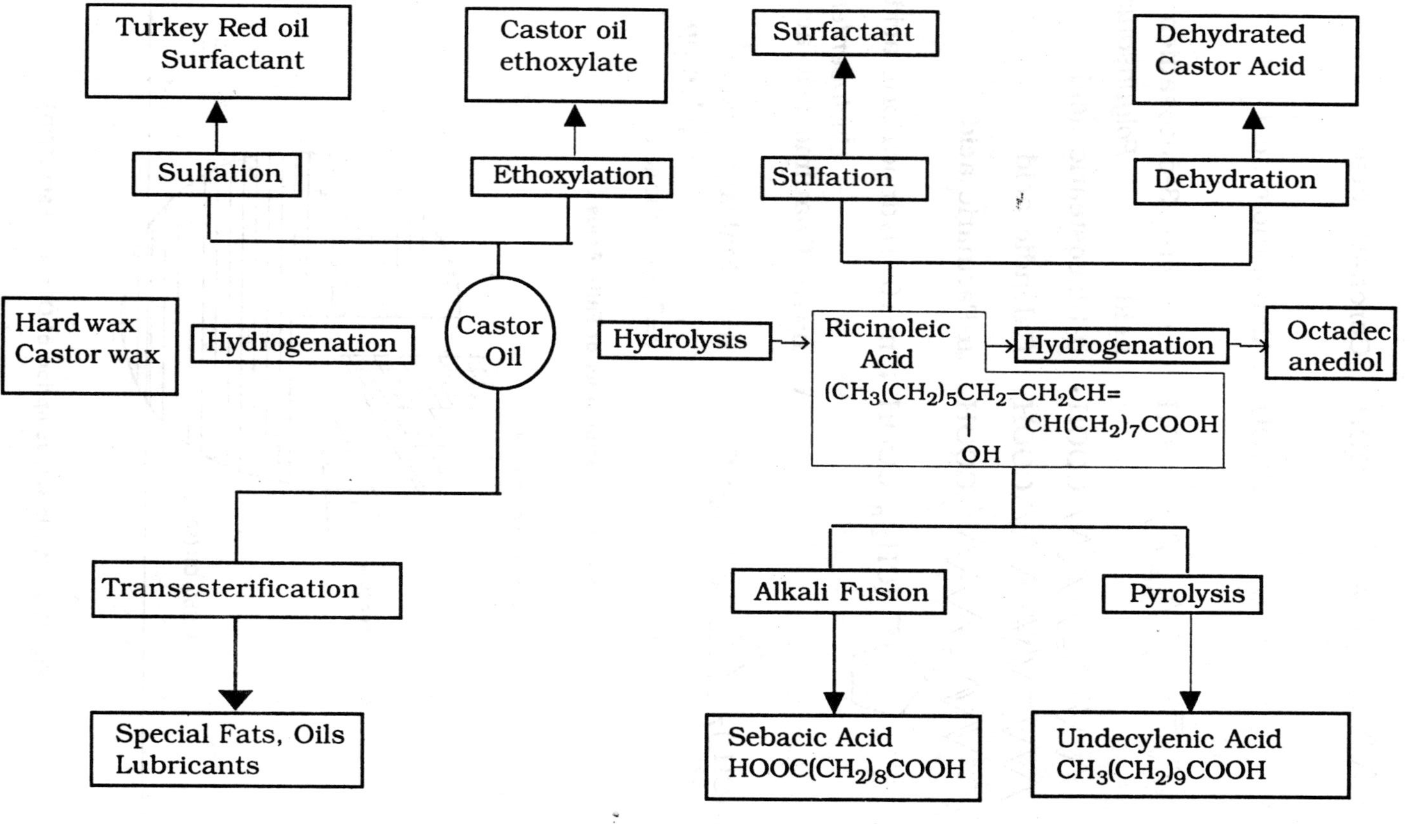

Fig 17.3: Castor Oil Family Tree

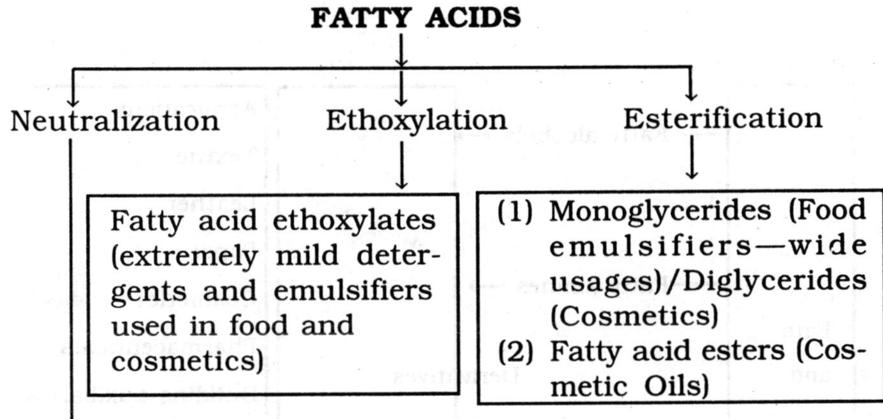

(1) Toilet soaps (sodium soaps)

(2) Liquid soaps (potassium soap)

(3) Calcium soaps (for animal feed as a source of calcium and fat)

(4) Lithium soap (component of heavy duty lubricants)

(5) Other metallic soaps — for mould release agents, lubricants

Fig. 17.4: World Consumption of Oils and Fats (1991)

Source : Malaysian Oil Science and Technology, Vol. 1, No.2, July 1992

***Table 17.3*: The Main Oils and Fats Producing Countries, 1991**

Country	*mmt*	%	*Type*
Asia, Pacific	29.2	35	palm, palm kernel, coconut, groundnut, castor oil
USA, Canada	14.0	17	soyabean, tallow, cotton, groundnut
Europe	16.0	19	rapeseed, olive, tallow, sunflower, butter
Latin America	8.2	10	soyabean, castor oil, fish, sunflower, cotton
Russia (CIS)	5.5	7	sunflower, cotton, castor oil
Africa	2.2	3	groundnut, palm, palm kernel
Others	7.4	9	
Total :	82.5	100	

• m mt = million tonnes

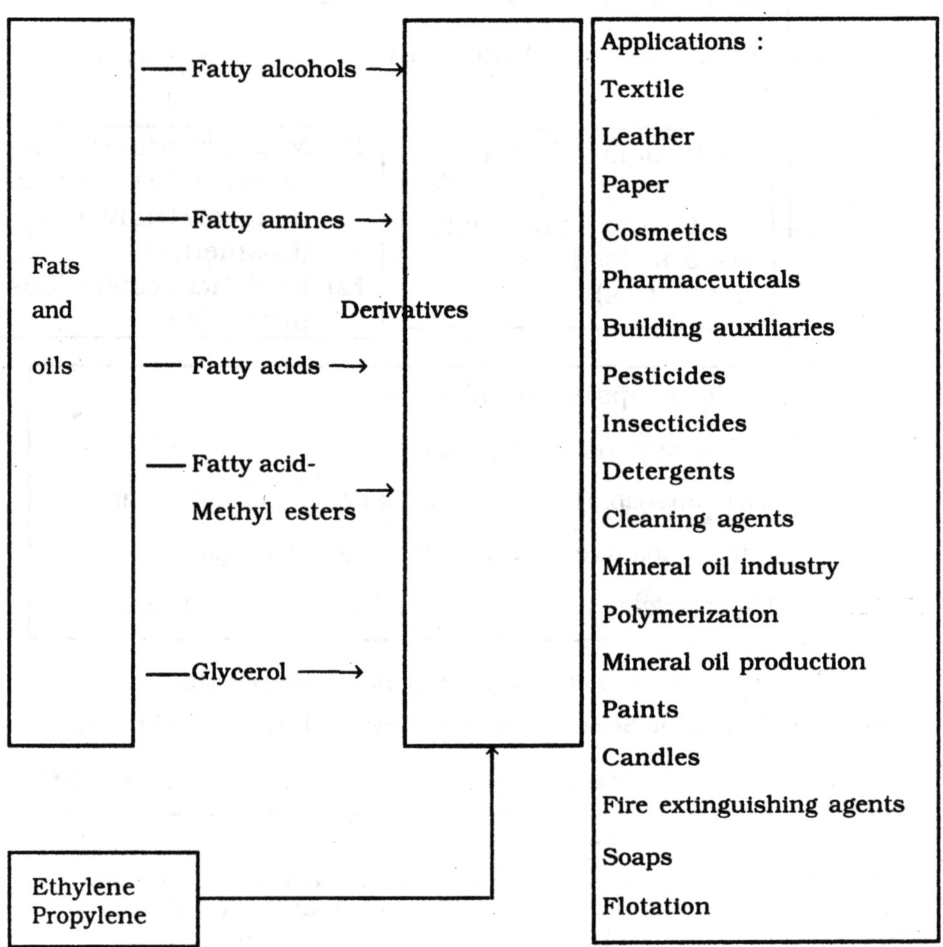

Fig. 17.5: Oleochemicals - Fields of Application

Source : Malaysian Oil Science and Technology, Vol. 3, No.2, July 1994

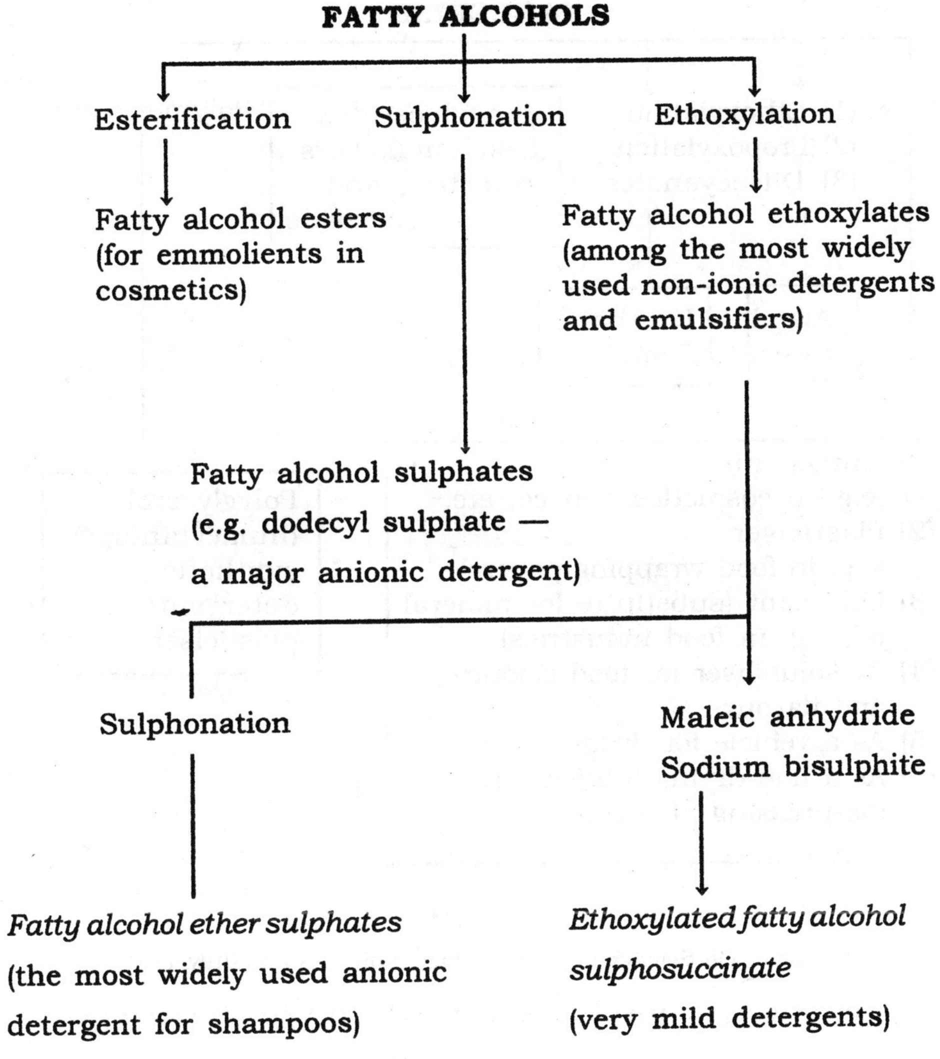

Fig. 17.6: Fatty Alcohols

Source : Malaysian Oil Science and Technology, Vol. 1, No. 2, July 1992

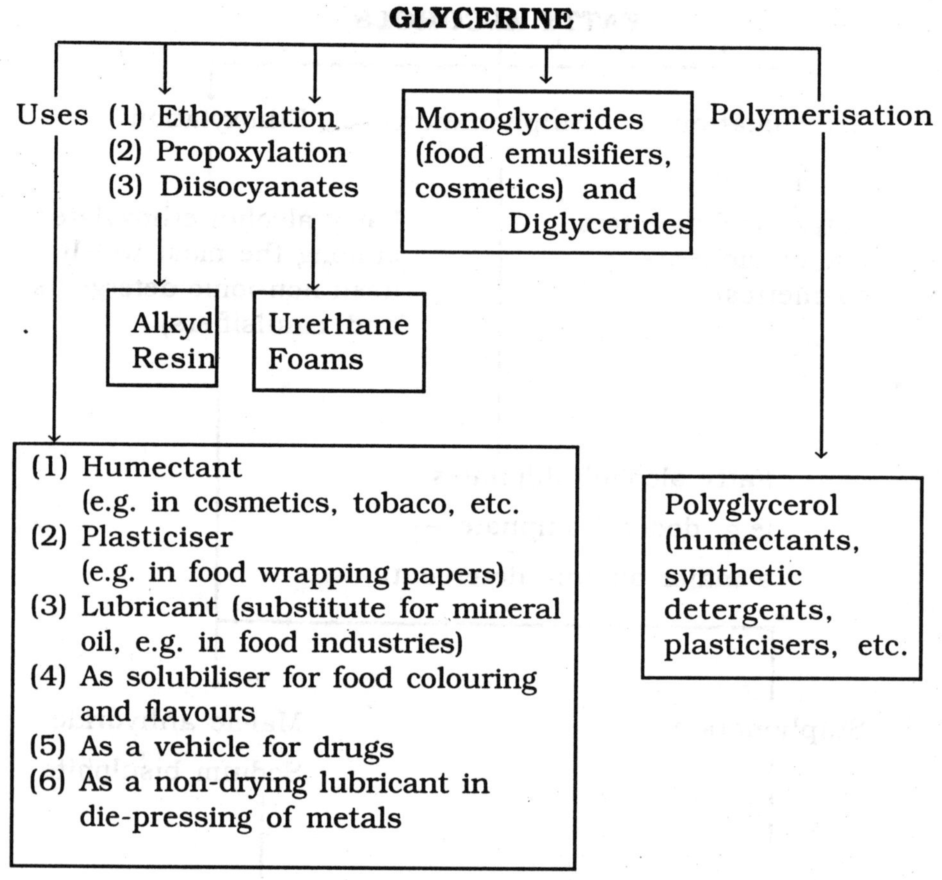

Fig.17.7: Glycerine

Source: Malaysian Oil Science and Technology, Vol. 1, No. 2, July 1992

REACTIONS FOR PREPARATION OF OLEOCHEMICALS

Reactions involving oils, fats and oleochemicals : A very wide spectrum of organic chemical reactions are carried out in the oleochemical industry. An attempt is made herein to illustrate the different pathways to specific oleochemical products.

1. *Hydrolysis (fat splitting)*

The triglycerides of fats and oils can be split under proper conditions by the process of hydrolysis to free fatty acids and glycerol.

$$
\begin{array}{l}
\qquad\quad H \\
\qquad\quad | \\
H - C - OOCR_1 \qquad HOH \qquad\qquad CH_2-OH \qquad R_1COOH \\
\qquad\quad | \qquad\qquad\qquad\qquad\qquad\qquad | \\
H - C - OOCR_2 + HOH \qquad\quad CH-OH + R_2COOH \\
\qquad\quad | \qquad\qquad\qquad \longleftrightarrow \qquad | \\
H - C - OOCR_3 \qquad HOH \qquad\qquad CH_2-OH \qquad R_3COOH \\
\qquad\quad | \\
\qquad\quad H
\end{array}
$$

R_1, R_2 and R_3 may be same or different. It is a stagewise reversible process. This aqueous reaction is favoured and accelerated by high temperature and high pressure. The reaction catalysed by lipolytic enzymes can be carried out at ordinary temperatures but they are ineffective at high pressure and high temperatures.

2. *Hydrogenation*

Basic Reaction $-CH=CH-CH_2+H_2 \longrightarrow CH_2-CH_2$

Hydrogenation of fatty acids or their esters to fatty alcohols is an important industrial process. A number of catalysts are known for the hydrogenation of carboxylic acids and their esters. Copper chromite or its modified catalysts are commonly employed but the reaction requires high pressure (300 atm) and high temperature (250°C). Thus, there is no satisfactory catalyst which can catalyse the hydrogenation of fatty acids in high alcohol selectivity under mild conditions. Hydrogenation of carboxylic acids to alcohols at low temperature and under low pressure was achieved by a new catalyst system, a rhenium-osmium bimetallic catalyst. The most active catalyst was prepared by reduction of the corresponding metal oxides with hydrogen in the presence of succinic acid.

Decanoic acid was hydrogenated to decanol in high conversions at 25-100 atm.

3. *Esterification*

In this process a fatty acid reacts with alcohols, glycerol, glycol etc. to form an ester of the acid. Other polyhydric compounds like sugars and starches can also participate in this reaction with the fatty acid to form ester.

Different triglycerides may be prepared by esterification process. Besides the desirable triglycerides, mono- and diglycerides can also be prepared by restricting the reaction in a proper way.

4. *Interesterification*

A triglyceride and an alcohol react in this process to form an alkyl ester by displacing the glycerol from the fat by using an alkaline catalyst.

$$C_3H_5\,(OOCR)_3 + 3CH_3OH \longrightarrow 3CH_3OCOR + C_3H_5(OH)_3$$

This process of preparation of methyl ester from triglyceride by methyl alcohol is termed as methanolysis.

The reaction of triglycerides with glycerol (glycerolysis) may be employed for the preparation of a mixture of mono- and diglycerides. Different triglycerides may also be prepared by using a very small quantity of alcohol.

5. *Saponification*

Saponification reaction is employed for the preparation of 'Soap' which is produced by the reaction of a fat with an alkali; glycerol is a byproduct of this reaction.

$$C_3H_5(OOCR)_3 + 3NaOH \longrightarrow C_3H_5(OH)_3 + 3NaOOCR$$

6. *Hydrogenation of the carboxyl group*

Fatty acids react with hydrogen in the carbo group to form fatty alcohols at high pressure by using a metal catalyst.

$$RCOOH + 2H_2 \longrightarrow RCH_2OH + H_2O$$

The fatty alcohols undergo similar reactions with alcohols. They produce vinyl ethers at 100-120°C. The selectivity of alcohol was improved by the addition of thiophens as a modifier of the catalyst.

7. *Hydrogenation of the fatty acid chain*

Unsaturated fatty acids are converted to their corresponding saturated fatty acids by hydrogenation using a metal catalyst like nickel, platinum or palladium.

$$\underset{\text{Oleic acid}}{CH_3(CH_2)_7CH = CH\,(CH_2)_7\,COOH} \longrightarrow \underset{\text{Stearic acid}}{CH_3(CH_2)_{16}COOH}$$

8. *Amidation*

Amidation reaction is carried out by reacting fatty acids with ammonia at high temperatures to form amides.

$$RCOOH + NH_3 \longrightarrow RCONH_2 + H_2O$$

$$RCOOH + R'NH_2 \longrightarrow RCONHR' + H_2O$$

9. *Formation of Nitrile*

Nitriles can be formed from amides at slightly higher temperatures by dehydration process :

$$RCONH_2 \longrightarrow RCN + H_2O$$

10. *Amination*

Nitriles can be converted to the corresponding primary or secondary amines by catalytic hydrogenation process.

$$RCN + 2H_2 \longrightarrow RCH_2NH_2$$

$$2RCN + 4H_2 \longrightarrow (RCH_2)_2NH + NH_3$$

Primary amines are formed at low temperature with excess ammonia and secondary amines are formed at high temperatures and removal of ammonia.

11. *Epoxidation*

Epoxide is formed from unsaturated fatty acids by the reaction of epoxidation at the place of double bond unsaturation.

$$-CH = CH^- + CH_3COOH \longrightarrow -\overset{\overset{O}{\diagup\ \diagdown}}{CH - CH^-} + CH_3COOH$$

Peroxyacetic acid is used as oxygen supplier. The *in situ* use of peroxyacetic acid lowers the possibility of hazard in the reaction.

Epoxidised soyabean oil, tallow oil and oleic acid esters are used as important plasticisers and stabilisers for polyvinyl chloride resins.

REFERENCES

1. Wordclmann Bruns, Exotic Oils and Fats in South East Asia, *Malaysian Oil Science & Technology*, Vol. 3, No. 2, 13-21 (July 1994)
2. Kirk-Othmer, *Encyclopedia of Chemical Technology*, Vol. 5, 3rd Edition: Castor oil, pp 1-15
3. Appleqvist, A.I. and R. Ohlson, *Non-Nutritional Uses of Rapeseed Oil and Rapeseed Fatty Acids : Rapeseed*, pp. 274-286
4. Oleochemical growth should continue. INFORM, Vol. 4, No. 2 (February 1993)
5. Oorke, Y.M. Choo and S.H. Ooh, *Natural Fatty Acids for Oleochemicals, Malaysian Oil Science & Technology*, Vol. 3, No. 2, 8-12 (July 1994)
6. PORIM Handbook and papers, AOCS 1990-1993—INFORM
7. An Oleochemical Adventure into the wonders from a Golden Crop, *Malaysian Oil Science and Technology*, Vol. 1, No. 2, 23-25 (July 1992)
8. Formo W. Marvin, Eric Jungermann *et al. Bailey's Industrial Oil and Fat Products*, Vol. 1, 4th edition, (1979)
9. Harwood, J. Henrick, *Oleochemical as a Fuel : Mechanical and Economic Feasibility*, JAOCS, Vol. 61, No. 2, 315-323 (February 1984)

CHAPTER 18

Chemistry and Technology of Soap

INTRODUCTION

Soaps are the sodium, potassium or ammonium salts of fatty acids and they are formed from either fatty acids or fats according to the following reactions:

$$R.COOH + NaOH = R\ COONa + H_2O$$

$$C_3H_5(OOCR)_3 + 3NaOH = 3\ R\ COONa + C_3H_5OH)_3$$

The reaction between a fatty acid and alkali is known as neutralisation and the process is known as neutralisation process. The reaction between a fat and alkali is known as saponification and the process is known as saponification process. Oils and fats consist of several fatty acids in mixtures. The soaps made accordingly consist of sodium, potassium or ammonium salts of a variety of fatty acids.

The soaps of various fatty acids show distinct solubility, hardness, texture, plasticity, colour, lathering etc. These characteristics are important in soap making.

Raw Materials in Soap Making and their Selection

Fats and oils or their fatty acids are the principal raw materials for soap manufacture. The selection of fatty materials for soap manufacture is the most crucial and important part.

In the selection of fats and oils, the classical 'INS' factor is used as a general rule.

INS factor of a fat is the saponification number minus iodine number. INS factor of a mixture of fats is obtained by calculating proportionally from the INS factors of individual fats. Solubility Ratio (SR) factor in a mixture is calculated by dividing the INS factor of a mixture by another similar factor in which the individual INS factors of coconut and palm kernel oils and all liquid oils are taken as zero.

The recommended INS factor and SR factor for toilet soaps are 160-170 and 1.3-1.5 respectively.

***Table 18.1:* Characteristics of Oils and Fats for Soaps[1]**

SI. No.	OIL/FAT	Iodine Value	Saponifi--cation Value	Titre °C	Unsap. matter
1.	COCONUT OIL	10	256	24	0.4
2.	PALM KERNEL	17	250	25	0.5
3.	BEEF TALLOW	40	198	40	0.3
4.	PALM OIL	52	200	44	0.5
5.	PALM STEARIN	40	210	50	0.5
6.	RICE BRAN OIL*	100	186	27	4.2
7.	RICE BRAN OIL* (HYDROGENATED)	55	186	27	4.2
8.	MOWRAH OIL	62	193	40	2.5
9.	GROUNDNUT OIL	90	191	30	0.5
10.	NEEM OIL*	90	191	28	2.5
11.	KARANJA OIL*	90	191	28	2.5
12.	COTTONSEED OIL	110	196	35	1.5
13.	CASTOR OIL	85	182	-	0.5
14.	KUSUM OIL	53	227	45	2.2
15.	SAL OIL*	40	190	52	1.5

Actual values may vary with the source of oil/fat.
*Often oil needs treatment for upgrading quality before actual use.

Hard fats like tallow, palm oil, palm stearin, hydrogenated oils like hydrogenated rice bran, mowrah, sal, are important fatty materials. These fats help to produce hard, slow lathering soaps with a stable foam and a good cleansing action.

The soft oils include rice bran, castor, cottonseed, neem, karanja, olive oil etc. Besides these are used also refinery soapstocks or acid oils, distilled fatty acids or fatty acid distillates. These soft oils contribute the foam and cleansing action. Castor oil does not foam well although its cleaning action is fairly well. Oils of coconut family like coconut, palm kernel, pisa etc. produce hard and brittle soap. The foam is quickly formed with relatively large bubbles.

Raw Materials other than Fatty Materials

Rosin although is not a fat or oil, is an important raw material for soap making. Consisting largely of abietic acid rosin produces soaps which are soft and sticky and give quick foam. It is never used alone as a soapery raw material. In laundry bar soaps and flakes rosin is used at around 20% level, while 25% rosin may be incorporated in toilet soaps as well.

A laundry soap having 42-45% total fatty matter content and *18-22%.* by weight alkaline sodium silicate as a builder, is normally made from a fat charge consisting of 60-70% hard fats and 30-35% soft oils. The ultimate fat charge can have an iodine value of 68-70 and a titre of 38°C. During winter time, however, when ambient temperature falls below 20°C, the hard fat is used at 60% and the soft oil is increased in the fat charge, bringing down the titre to 37°C, as high titre soaps tend to crack during winter months.

For bathing soap (unmilled toilet soap) having 60% total fatty matter and 27% moisture, the iodine value may be maintained at 65 and the titre at 39°C. However, bathing soaps at 60%. TFM (total fatty matter) and 30%. moisture are available in the market with a much higher titre (43-44°C) and a lower iodine value (50) to compensate for the softness introduced by the high moisture content. These very high titre soaps with high moisture are known to shrink fast on the shelf and -tend to develop cracks on usage, particularly during the winter months.

For a milled toilet soap of the premium category, containing 76-80% total fatty matter with 10-14% moisture, and the iodine value of fat charge may have within 55-58 and the titre of mixed fatty acids at 38-38.5°C. Coconut/palm kernel oils -are mostly used in 10 to 20% by weight of the fat charge for premium quality toilet soaps. These oils reduce the -iodine value of the fat charge considerably without causing much increase in the titre value of mixed fatty acids. A titre value in excess of 40°C for milled toilet soaps tends to reduce the plasticity of the soap, thereby presenting problems in processing often causing blisters to appear on the soap cakes on cooling and keeping.

It is customary to bleach a fat charge meant for toilet soap production. In certain cases, individual oils requiring specific bleaching techniques may be bleached separately. Palm oil gets bleached by blowing air through the oil at a temperature of 110°C. However, residual colour may be removed by adsorption bleaching under vacuum at 110°C-120°C, preceded by a pre-treatment with oxalic acid to remove gums 'and muscilage. After bleaching the fat charge is cooled to 80-9C°C and filtered. Activated carbon may sometimes be used in combination with active earth for a better bleaching action.

In addition, chemical bleaching like hydrogen peroxide bleaching and sodium chlorite plus sulphuric acid bleaching techniques are adopted, the latter being" suitable for very dark coloured fats like low grade tallow, animal grease, salseed oil, rice bran oil, etc. However, sodium chlorite being a hazardous and explosive chemical, due care must be exercised in its handling and storage.

Other Raw Materials

(A) Caustic soda (NaOH) is another important raw material used in the saponification process to saponify glycerides or to neutralise fatty acids. Caustic soda for soap making can be purchased as a strong solution containing 45-47% NaOH or in solid form as flakes or as caustic soda cakes in drums.

(B) Common Salt, NaCl : Common salt is used by the soap maker for salting out or graining out of soap by common ion effect.

A high grade salt, containing not more than 0.5% calcium and magnesium salts, is desirable for use in soap making.

(C) Water : Normally large quantities of process water preferably soft or with no excessive hardness in order to avoid loss of fatty matter, is used. Liquid soaps present "clarity" problems unless soft water is used. Presence of excessive amounts of iron in process water is also undesirable as it may reduce the shelf life of the final product.

(D) A number of other ingredients are used in soaps particularly in making toilet soaps.These are perfumes, silicates, colouring agents, opacifiers, antioxidants and emollients.

Physico-Chemical Aspects of Soap[2]

Soap exists in a variety of forms and the application of phase diagrams helps us considerably in understanding just what happens in the kettle and also in subsequent finishing operations such as plodding. The soap boiler recognises such aqueous systems of soap as (i) neat soap, which is the finished kettle soap and

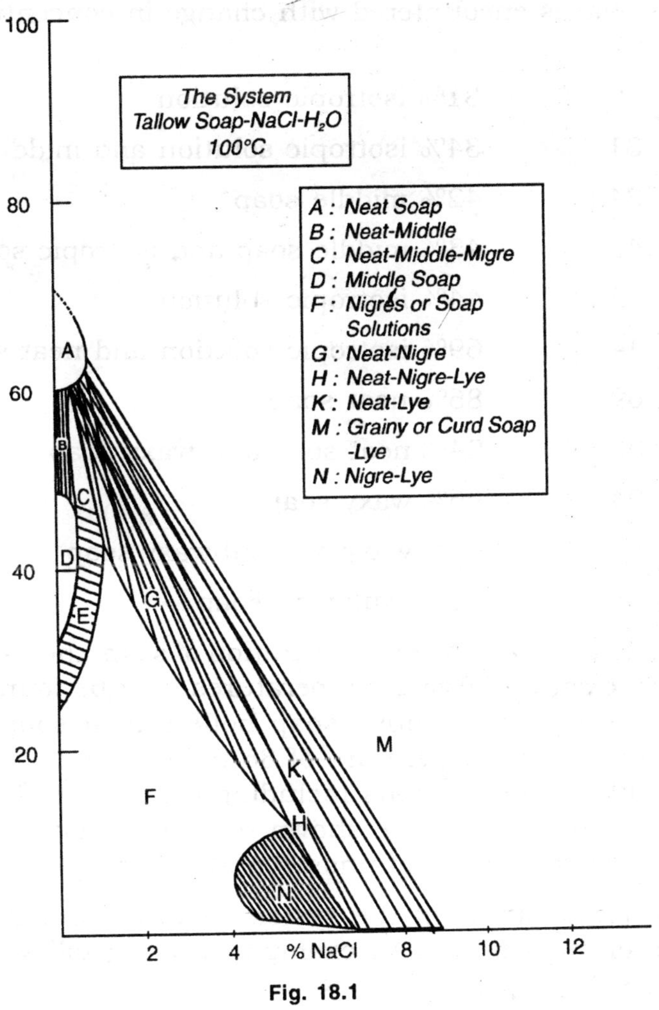

Fig. 18.1

is an anisotropic liquid; (ii) nigre soap, which is an isotropic solution; (iii) soap curd, which is dried soap consisting of fibrous crystals enmeshing a liquid phase; and finally (iv) middle soap, which is a viscous dilute crystalline phase. To these may be added such further phases as superneat soap and kettle wax`(Figure 18.1). Figure 18.1 shows the system sodium oleate and water at varying concentrations and temperatures. The shaded areas represent mixtures of soap phases, determined from the boundary phases connected by each isothermal tie line. The unshaded portion represents single phases. At a temperature of 100⁰ the various phases encountered with change in concentration are as follows :

0.	31% isotropic solution
31.	34% isotropic solution and middle soap
34.	42% middle soap
42.	43% middle soap and isotropic solution
43.	44% isotropic solution
44.	69% isotropic solution and neat soap
69.	86% neat soap
86.	94% neat soap and waxy soap
94.	96% waxy soap
96.	98% waxy and subwaxy soap
98.	100% subwaxy soap

Similarly, at any one concentration of soap, the various phases that may exist at varying temperatures may be found. At a 65% concentration, for example, soap boilers neat soap exists as a single phase at a temperature of 36 to 54°C and again at 130 to 175°C. Other investigations including the work of McBain, Void, Ferguson, and others have shown that anhydrous soaps exist in an even greater variety of phases and polymorphic forms.

The phase study of the system soap-water-salt at 100°C is perhaps of more help3. An example covering tallow is shown in Figure 18.2.

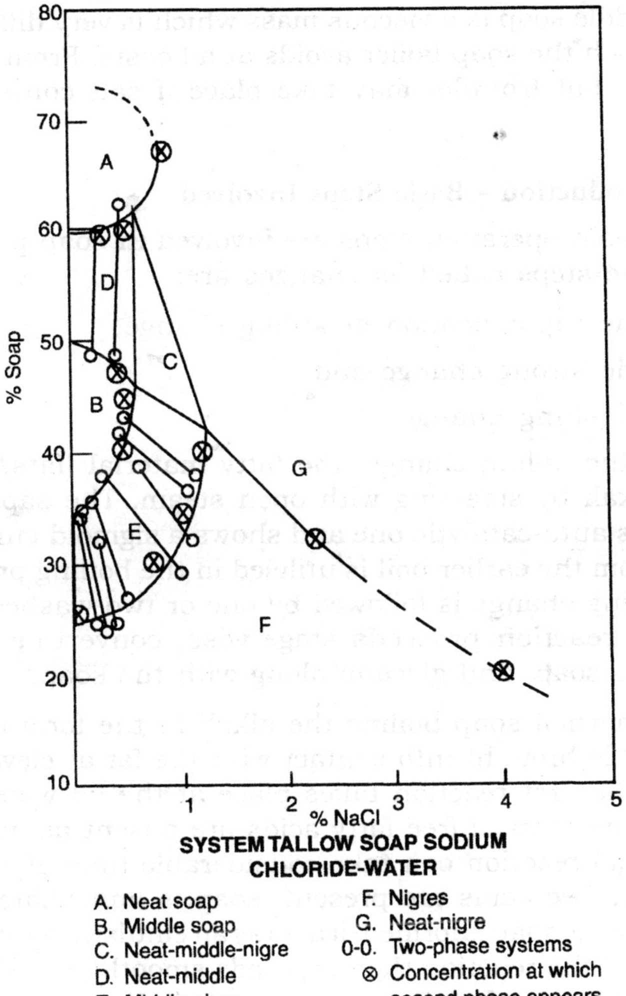

Fig. 18.2: Phase diagram of a Tallow Soap-Sodium Chloride-Water.

Considering the diagram, Fig. 18.2, it is seen that when zero salt is present, we pass through the following phases with increase in soap concentration as follows:

Area F Nigre	25% (approx.) cone. of soap
Area E Middle Nigre	25.35% (approx.) cone. of soap
Area B Middle	35.50% (approx.) cone, of soap
Area D Neat Middle	50.58% (approx.) cone, of soap
Area A Neat	58% +

Middle soap is a viscous mass which is very difficult to handle and which the soap boiler avoids at all costs. From the diagram it is seen that trouble- may take place if salt content falls below $1^1/_4\%$.

Soap Production - Basic Steps Involved

Three main operation steps are involved in soap production from fats. The steps called as changes are:

i. the saponification or killing change,

ii. the strong change and

iii. finishing change.

In the killing change, the fatty material (fats/oils) is boiled with alkali by steaming with open steam. The saponification reaction is auto-catalytic one and shows a sigmoid curve. Generally, niger from the earlier boil is utilised in the boiling process of soap. The killing change is followed by one or two washes. The saponification reaction proceeds stage-wise, converting the triglycerides into soap, and glycerol along with the soap.

In normal soap boiling the alkali in the form of an aqueous solution is brought into contact with the fat at elevated temperatures. The first reaction takes place at the oil water interface to form some soap. If free fatty acids are absent as in a refined fat, this initial reaction can take considerable time. If, as is normally the case, free acids are present, soap is immediately formed. As soon as some soap is formed, it acts as an emulsifying and solubilising agent and the reaction then proceeds smoothly as a homogeneous one. The work of McBain and others indicates that there is no relation between the molecular weight of the glyceride and the rate of saponification, the latter depending more on the ease with which the oil is emulsified after some soap is formed. Such factors are somewhat, although not wholly, dependent on the solubility of the soap, which in turn increases greatly as the molecular weight decreases. Sodium stearate, for example, is more insoluble in water than sodium laurate. The addition of salt or the reduction of water in the lye, which is another way of increasing the electrolyte content, can decrease this solubility This decrease in water solubility of the soap is important and is under the control of the operator by the addition of brine along with the lye. Soap thus partially grained out is improved as an oil emulsifying agent, and for this reason the addition of salt is extremely important in kettle house boiling operation.

Kettle Soap

The main operations in kettle soap boiling may be classed in three parts : (a) the saponification or killing change, followed by one or two washes, (b) the strong change; and (c) finishing operations.

As the oil is being pumped into the kettle, a very weak solution of about 1% caustic alkali is a'dded. The rate of lye addition must be very carefully regulated with the flow of oil. Usually the high fatty acid oils are added to the kettle first. This is to enable the saponification to start quickly. The acids unite with the lye quite readily, forming soap which emulsifies the fresh oil and enables them to come into intimate contact, thus aiding further saponification. If the lye is added too rapidly, the emulsion will be broken and the saponification retarded. This is brought about by the graining -out action that the lye has on the soap because of the common ion effect. (Salt has the same action as will be discussed later on). If the lye solution is too dilute, saponification will also be retarded. After saponification has once begun or, in terms of the soap-boiler, a catch has been made, the heat of theaction speeds it up. This heat is considerable and amounts to about 65 to 75 Cal. per kg. of fat saponified.

The strong change is now given to the soap charge in a manner identical to that for the wash, except that alkali anywhere from 10 to 14% Na_2O may be added. This insures the complete saponification of all fats. The kettle is now grained as before and let stand. The lye is drawn off to storage.

The finishing of the soap consists of one or more brine washes known as "pickling" to remove glycerine and free alkali, followed by a final "pitch", which consists of boiling up on water alone and allowing to settle several days. At this point *the* salt and alkali content must be such that the soap boiler arrives at point H of Figure 18.3 :

This diagram is the same as Fig. 18.2 except that it shows the area G under conditions of higher salt concentration, where lye

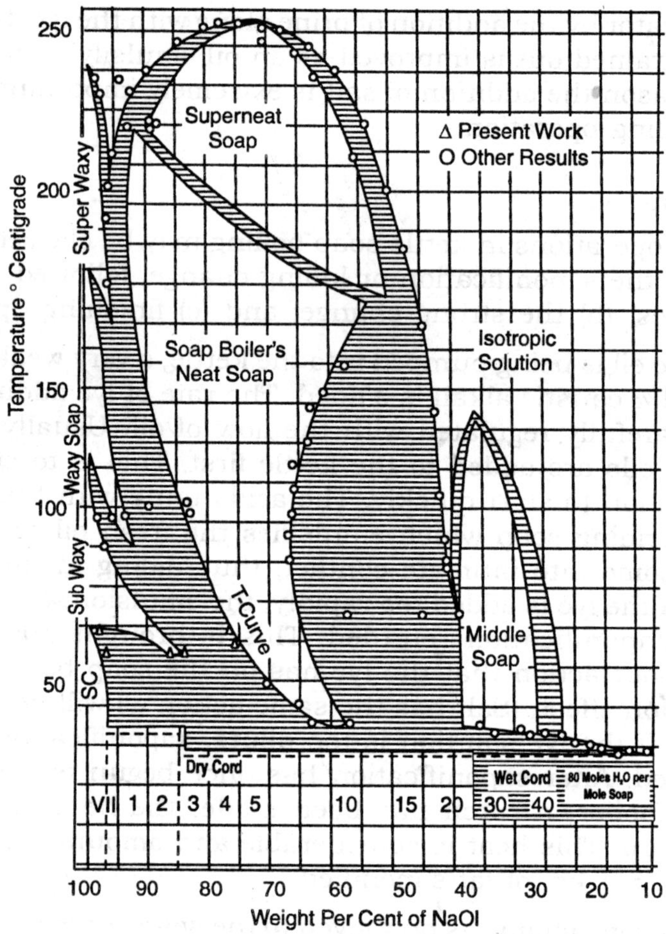

Fig. 18.3: Phase diagram for sodium oleate showing homogeneous fields of each of the different phases and the isothermal tie lines connecting the phases in heterogeneous equlibria.

exists as a separate phase. That is, one can have a large neat soap plus a small nigre plus lye, and one must adjust the water and salt content so as to procure these conditions. It will be noted from the diagram that finished kettle soap will run consistently from 65 to 68% anhydrous soap. The nigres are 15- 0% soap solution and contain the dirt and considerable colouring matter from the oil. They are usually reworked in lower grade soaps. The nigres of such low grade soaps are then accumulated and reworked to yield a nigre of a nigre, which last is discarded.

Although the phase diagram points up a rigid scientific basis for boiling operations, the individual modifications of the process are quite extensive for example, if a 98% or higher saponification is obtained on the killing operation, a comparatively weak alkali wash will suffice for the strong change. Sometimes a nigre is removed after the strong change and a second one on the finish.

Some Soap makers use only salt for graining, some use salt and brine and some only brine. Each manufacturer follows a very definite procedure depending on the product to be made.

Variations in the procedure are dictated by finished specifications, lye space, available glycerine recovery, and other factors. For example, the tabulation below shows typical kettle-house soap specifications on three classes of soap.

Typical specifications for Kettle House Soaps

	Toilet Base	*Laundry Bar Base*	*Industrial Flakes Base*
Real Soap ..	67.0% min.	67.0% min.	65.5% min,
Free Na2O ..	0.03-0.05%	0.3-0.5%	0.2-0.4%
Glycerine ..	0.3% max.	0.2% max.	0.7% max.
NaCl ..	0.40-0.50%	...	...

When rosin is used in the charge, it may be added over the top if small, such as 10% of the total charge. In this case it usually is added with the fats or directly after the saponification. When large amounts of rosin are used, they are generally saponified separately with less expensive soda ash, washed, and grained. The lyes are dropped to the sewer since they contain no glycerine. The soap is then half-grained and pumped on top of the saponified fat charge, where it is strong-charged and finished in the usual manner.

The amount of glycerine recovered is of very great importance, and as little as possible is left in the soap commensurate with economical recovery. It is not uncommon for a well-run soap plant to lose as much as 10% glycerine by having it in their kettles. It is possible to pursue this recovery but the cost of the additional evaporation may exceed the price of the crude obtained. For example, the steam consumption required for evaporation of a 5% glycerine lye is about three times as much as it is for a 12% glycerine lye. Cases have been known where soap washing operation were cut to avoid high costs of recovery. The total volume of lye evaporated bears a very simple relation to the fats saponified. The system of lye handling in the kettle house also determines largely the efficiency of recovery, and it is frequently the practice to use the lye from the strong change on the second wash of the next kettle and the first wash of a third, etc. This is known as the countercurrent system of lye handling.

Cases have been known where such a system will reduce the lye to fat volume from a ratio of 1.0 to 0.7, yielding glycerine lyes as high as 15% while still keeping the known loss of glycerine below 7% In certain types of operation the countercurrent system of lye transfer is not always feasible as, for example, when production is limited and contains various types of soaps.

Generally speaking the yield of kettle soap will be between 98.5 and 99.5% of theory with primary fats and oils as raw materials. With large quantities of foots, yields of both glycerine and soap will be adversely affected. The quality of the finished soap is a joint responsibility of the boiling operations and raw material selection, and each must stand on its own feet.

Other methods of saponification exist which modify regular kettle house practice. In the semi-boiled process for example, the soap is saponified in a crutcher with the requisite amount of lye added. Builders are added and the whole mixed and dropped in frames. The entire glycerine goes with the soap. This type of operation is exceedingly rare.

Cold-made soap is processed on the same principle as semi-boiled soap. For example, coconut oil having a saponi-fication value of 260 would require 8.4 kgs of NaOH for each 45 kgs of oil used. If a *60%* superfatted soap product is desired, a little arithmetic shows that 25.7 kgs of water will be needed for the alkali. By holding out 400 gms of NaOH and replacing it with an equal weight of water *the* desired product is obtained having superfat of 5.3%. The fat is charged to the crutcher and the alkali solution added slowly. When saponification begins, it may be necessary to cool the crutcher with water due to the heat released through the saponification.

Soaps from Fatty Acids

One of the present trends in soap making is the increasing use of fatty acids in place of the fats and oils formerly used. While it is only in recent years that appreciable amounts of fatty acids have been consumed by the soap industry, the idea of soap from fatty acids is not new.

By about 1900 the Twitchell, Autoclave, and Enzyme splitting processes were in use, and fatty acids were relatively plentiful with a consequent increase in their use in the soap industry. However, even these acids were relatively low in quality, and only recent developments in fat splitting and distillation processes have given the high quality fatty acids now used in soap making.

The use of fats and oils was curtailed in soap processes from which glycerine was not recovered. Since glycerine cannot be recovered from potash soaps made by the saponification of fats and oils, this regulation could be complied with only by the use of fatty acids.

Advantages and Disadvantages of Fatty Acids

Some of the advantages of using fatty acids in soap manufacturing are given as follows :

Because the reaction between fatty acid and alkali to form soap is a simple neutralisation, it takes place immediately on contacting the reactants. Complete neutralisation is achieved without the long boiling time or the large excess of alkali needed to completely saponify an oil. This, in itself, gives an appreciable saving in operating time and steam usage.. In addition, the lack of glycerine in fatty acid soaps makes unnecessary brine washing for the recovery of glycerine as is required with soaps made from oils. Also, the alkali content of the final product made from fatty acids can be controlled by a simple titration of the soap and addition of the small amount of alkali or fatty acid required for the adjustment.

The rapid, complete reaction between fatty acid and alkali to form soap allows the use of simpler equipment than is required for the saponification of oils. In place of the large kettles normally used for soap making fatty acids can be neutralised in relatively small crutchers. Crutcher neutralisation also allows incorporation of other ingredients as part of the same operation. The neutralisation can even be carried out continuously, with a large production volume possible from a unit occupying very little factory space.

Preparation of soap from fatty acids permits considerably more variation in moisture content of the soap than can be obtained by kettle saponification of fats. In conventional kettle boiling the final neat soap obtained after settling and separating the nigre layer has a moisture content which must be maintained at a fairly set value if thė excess salt and alkali are in drop out in the nigre as desired. The moisture content is usually in the range of *30%,* water in the neat soap. When neutralising fatty acids in a crutcher, the moisture content can be higher or lower than this value. The actual moisture required is fixed only by the need to maintain the soap in a physical state which will allow it to be handled satisfactorily.

The use of fatty acids often allows upgrading of stocks with a consequent saving in money. Thus a low grade fat stock unsuitable for use in a given product can be split and the resulting acids distilled to give considerable improvement in colour and odour. It will often be found that the split and distilled acids will be cheaper than the high grade fat stocks they replace despite the additional processing costs. An example of this is the use of acidulated foots from alkali refining. Splitting and distillation also make possible more uniform raw materials for soap making than are obtained by the usual refining and bleaching of fat-stocks, and the soaps made from them are accordingly more uniform.

An additional cost advantage in using fatty acids for soap production is that sodium carbonate can be used for the reaction in place of the more expensive sodium hydroxide. This is possible because the fatty acids will react with a fairly weak alkali while glycerides require strong alkali for saponification. In the case of potash soaps the carbonate is more expensive than the hydroxide, and this advantage is not realised.

A final benefit gained in using fatty acids, and one of the most important, is that it permits the precise selection of both the fatty acid and the alkali to give desired properties in the product. Lauric, myristic, palmitic, stearic, and oleic acids, which are the major constituents of the fats and oils commonly used in soap making, are sold commercially in purities of 90% or better. Also singly and doubly distilled acids from individual fats and oils such as coconut, cottonseed, soybean, tallow, etc. are available. Since soaps of the different fatty acids have different solubility, foaming, and detergency properties, it is possible by blending acids to obtain a product which gives best performance for a given use. Likewise, in using fatty acids, the alkali can be varied in a manner not possible in saponifying fats and oils. Sodium, potassium, and amine soaps free of glycerine can be prepared, each type having its own distinctive properties. Mixed alkalies can also be used. It can be seen from the foregoing that fatty acids, far from being a substitute raw material, are necessary for making many soaps which cannot be made directly from oils.

Soap Making Procedures

Processes for making soap from fatty acids vary from the regular kettle boiling technique used with fats and oils to the newer methods for continuous neutralisation.

Different types of crutchers, kneaders or, dough mixers -can be used instead of kettle for neutralisation of fatty acids. High pressure fat splitting and fatty acid distillation are avoided in the case of fatty acid saponification. There are other advantages like greater flexibility in the choice of raw materials, in blending, and in formulations suitable for the production of solid or liquid soaps, paste and cream soaps, waterless hand-cleansers and powdered hand soaps etc.

Soap boiling time: To prepare a batch of full boiled soap, the total time required is about 5-10 days which depends on the number of changes carried out, size of the kettle and whether rosin is a component in the process may be required.

Bleaching of soap: Bleaching is often done to improve upon soaps. Sodium hypochlorite or other chemical agent, dichromate and sulphuric acid are used for bleaching of the soap in the kettle, preferably after the batch has been washed several times. Bleaching is done after water has been added to the curd to form a solution, and after complete bleaching the soap is grained out.

Purification of nigers: Niger contains all the impurities of the soap batch and it must be purified before use in another fresh batch. Purification of niger is carried out by boiling and either pitching or salting out.

Semi-boiled and cold processes: The simplest soap making processes are the semi-boiled and cold processes. In these processes fat is reacted with strong alkali which is very nearly equal to the quantity just required for complete saponification. After the reaction, the whole mass gets solid and no free glycerine or neat and niger phases separate out. These processes have the advantages of requiring simple equipment, less skill of the soap maker and production of a soap with an anhydrous soap content of any desired value over a wide range. But in these processes glycerine cannot be recovered, purification of raw materials does, not occur like -full-boiled process and soap produced is somewhat inferior in quality.

These processes are mainly used for those soaps which are difficult to handle in the full-boiled process like marine or other coconut oil soaps, the soaps which cannot be easily salted out like soft or potash soap, and for the manufacture of cheap and heavily filled soaps made in small plants.

Semi-boiled soaps: Semi-boiled soaps are prepared by simple heating of the fat charge with the required amount of strong caustic soda (25-35° Be') in either a soap kettle or in small batches in a crutcher. After the saponification, builders like sodium silicate or non-soap ingredients are added. If the operation is carried out in a kettle, the batch may be boiled but if it is carried out in *a* crutcher the temperature must be kept little below the boiling point and; thorough mixing is done by mechanical agitation.

Cold made soaps: In the cold made process temperature is not employed and the operation is carried out in crutcher. Mixing with thorough emulsification is done at the ambient temperature and very strong lyes 34-45° Be' are used for saponification. The emulsion of strong lye and fat is the water-in-oil type which differs from the emulsion of soap boiling which is oil-in-water type[4] . Several days are required to complete the process. As there is little scope of exact calculation of the quantity of the fat or alkali required for the reaction, these soaps usually contain excess of either alkali or more commonly fat.

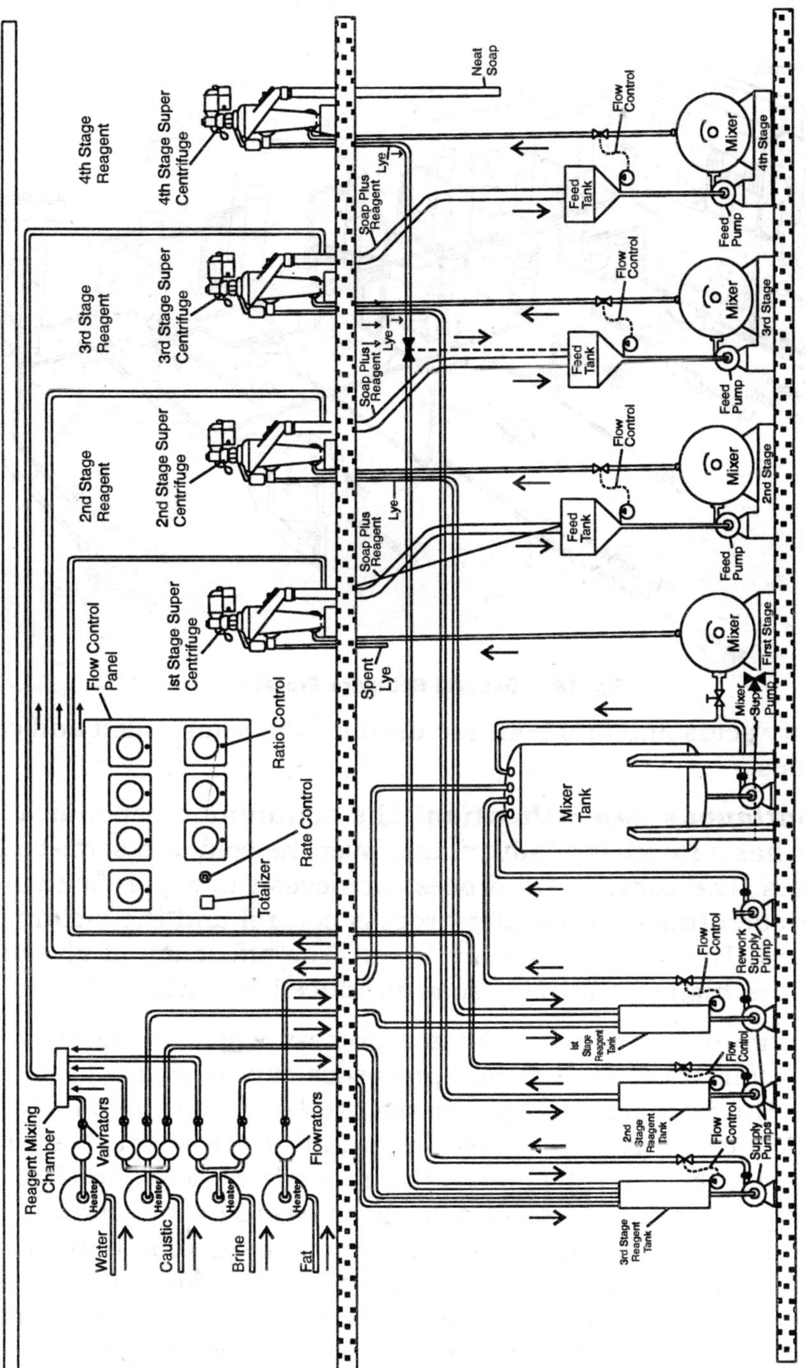

Fig. 18.4: Sharples Centrifugal Soap Process

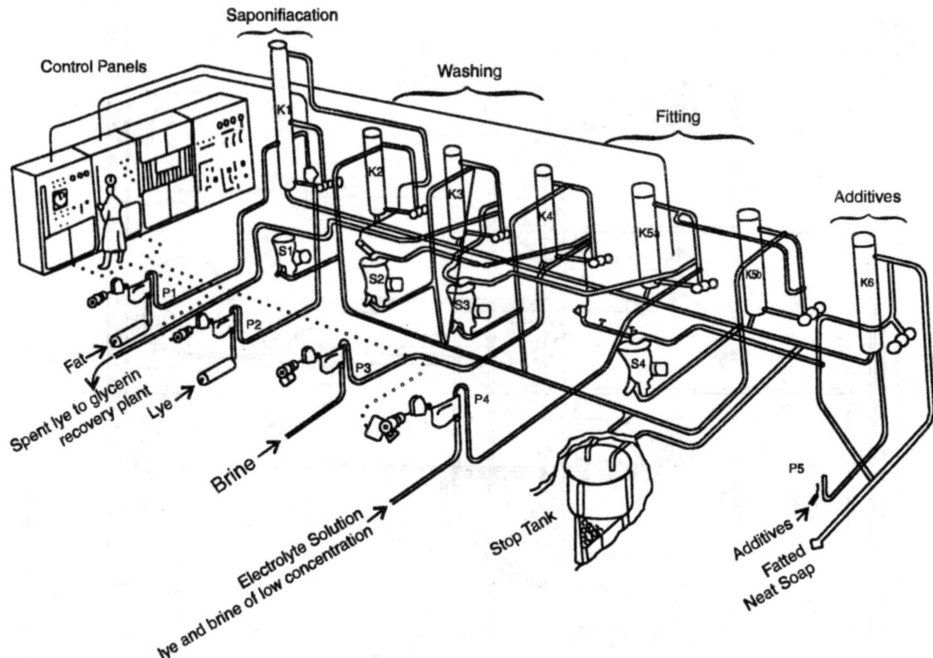

Fig. 18.5: DeLaval Cetripure Process

Fatty acids and fats both are used in. the semi boiled and cold processes.

Continuous saponification: The continuous saponification process has a number of advantages over the cold and semi-boiled processes. The continuous process achieves the saponification at a lower operating cost, stricter process control and the yielding of neat soap. The steam costs are lower. The efficiency of glycerine recovery is higher, the neat soap produced is better coloured.

Continuous saponification is a process which is mainly used in many countries. Continuous processes can be classed into two groups. The first group of processes are based on to continuous saponification of fats. The most important of these processes are the Delaval, the Sharples, Mechaniche Moderne and the Mazzoni SCN-LR processes. The second group is based on the continuous splitting of fats into fatty acids, continuous distillation of the acids followed by the neutralisation. The processes of importance in this group are the Mills, the Mazzoni-SC and the Armour processes. The processes are described in the following sections. The process

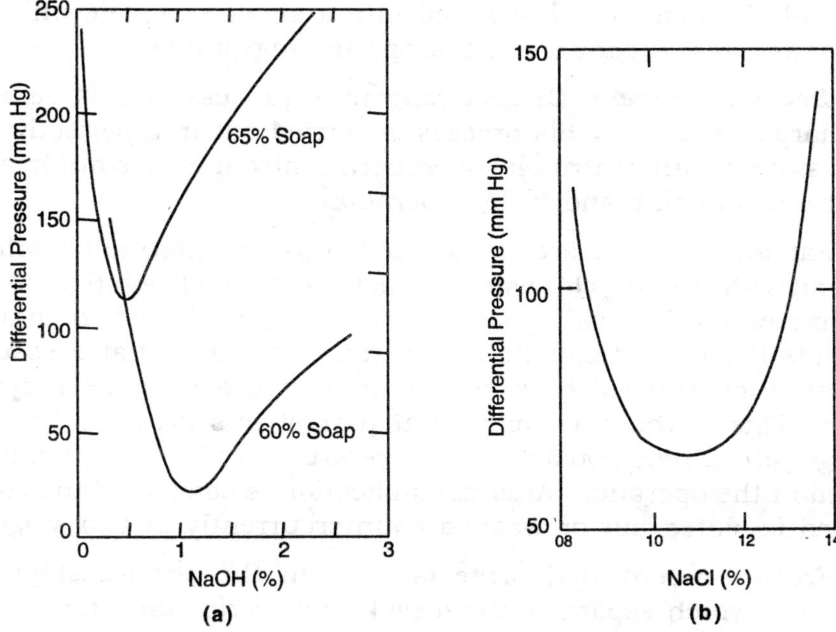

Fig. 18.6: Differential pressure versus concentration of (a) lye and (b) salt.

is economical in respect of steam, labour and maintenance and the spent lye contains two or three times the glycerine that is obtained from the kettle process.

In the Sharples method the fat charge plus alkali and brine solutions of the proper strength flows continuously through a mixer. The resultant soap is then separated by centrifuges into neat soap and lye. The soap is then strong changed and washed several times in succession by means of intermediate mixers and centrifuges and comes out a finished product, having the same specifications as a kettle soap.

A flow sheet of the very interesting process is shown in Fig. 18.4. It is not difficult to trace the various steps back to fundamentals of kettle soap practice. Stage 1 for example is the saponification, stage 2 corresponds to the 1st, 2nd and 3rd wash, stage 3 corresponds to the strong change, and stage 4 to the finish. The principal point of interest to note in this method is the lack of a nigre. This is not necessary as the dirt and colouring matter usually associated with nigres in kettle boiling are accumulated on the inside of the centrifugal bowls and are removed periodically. All equipment is integrated into a single complete unit by means of electric controls and proportioning pumps.

***Delaval process*:** Delaval centripure process (5,6) is similar to Sharples process. This process is carried out in a perfectly air tight system with centrifugal operation. It also includes automatic control of washing and fitting operations .

Fat and caustic are introduced into an internally baffled column with two recycle systems which cause good agitation of the raw materials. Delaval process is more sophisticated in that it controls the rate of caustic feed through a differential pressure meter which senses the pressure difference across the recycle pump. This is the most important part of this process. The fat charge rate is held constant. A temperature of 90-120°C is maintained in the operation. After saponification is completed the soap formed is salted out or washed countercurrently in two stages.

Fresh concentrated brine is used in the second stage of operation which separates the dissolved glycerine and impurities from the neat soap in a highly efficient hermetic centrifugal separator. The spent lye from this stage is sent to the previous stage, where after dissolving more glycerine and impurities, it is separated and sent to the glycerine recovery unit.

The neat soap, washed from the second stage containing a very little amount of glycerine and impurities, is sent to the third section.

Fitting is done in the third section where the neat soap is converted to the high grade toilet soap. In this stage, two phases - neat soap and niger are separated by adding exact amounts of caustic and brine.

Neat soap and niger are separated in a third separator. Niger is returned to the second washing stage and the neat soap, free from niger, is obtained.

A unique viscosity sensing .device is employed for controlling the feed of fat and caustic in this process. The viscosity of the soap mass reduces quickly to a minimum at the lye concentration of about 0.15% NaOH and increases rapidly as the NaOH content is increased (see Fig.18.6(a)). The alkali left is used in complete saponification in the subsequent stage. Salt is used in the fitting stage, and the addition of salt is controlled by the viscosity sensing device. It is seen that the soap mass shows minimum viscosity at the salt content of 1% (see Fig.18.6(b).

The minimum capacity of Delaval plant is about 1000 kg/hr of neat soap. This process is advantageous in that it is economical in steam consumption, space saving and good quality soap.

Monsavon process: The Monsavon system of soap making (U.S. Patents 2,485,204-5) utilizes a colloid mill to emulsify the fat and the aqueous alkali phase. These are discharged to a vertical reaction column. The saponified fat then passes into a washing column and hence to a settling tank for separation into neat soap and nigre.

Fat at about 70°C and caustic soda are added in the colloidal mill to form an emulsion of water-in-oil type. This emulsion is sent through a reaction tube into a soap kettle where a mild agitation is done and saponification is completed in a single stage. The temperature of the saponification mixture is maintained at 95-100 C. This mixture is sent to a washing tower containing 6 vertical washing sections which are controlled at 85°C thermostatically. The crude neat soap is washed countercurrently.

After washing has been completed the washed soap from the last section is sent to the fitting section at the top of the lower where water and brine or caustic are continuously mixed with the soap. Fitted soap is continuously withdrawn and discharged into a settling tank. Complete saponification is ensured by the measure of turbidity of 15% soap solution, and free alkali content is measured by photoelectric examination of the finished product where alcoholic phenolphthalein is added.

Good quality neat soap containing 62-63% fatty acid content can be manufactured within 24 hours in this process. The advantages of this process are the saving of time, space, power and labour. Production of niger can be avoided (as they are continuously recycled) and high glyccrine recovery is achieved.

Mazzoni continuous fatty acid neutralisation process: In Mazzoni process (7,8) neutralisation of fatty acids is carried out with either sodium hydroxide (SC process) or sodium carbonate (SCC process). The schematic diagram of SC process is shown.in Fig. 18.7.

The manufacture of soap from fats by Mazzoni process is shown schematically in the Fig. 18.7a.

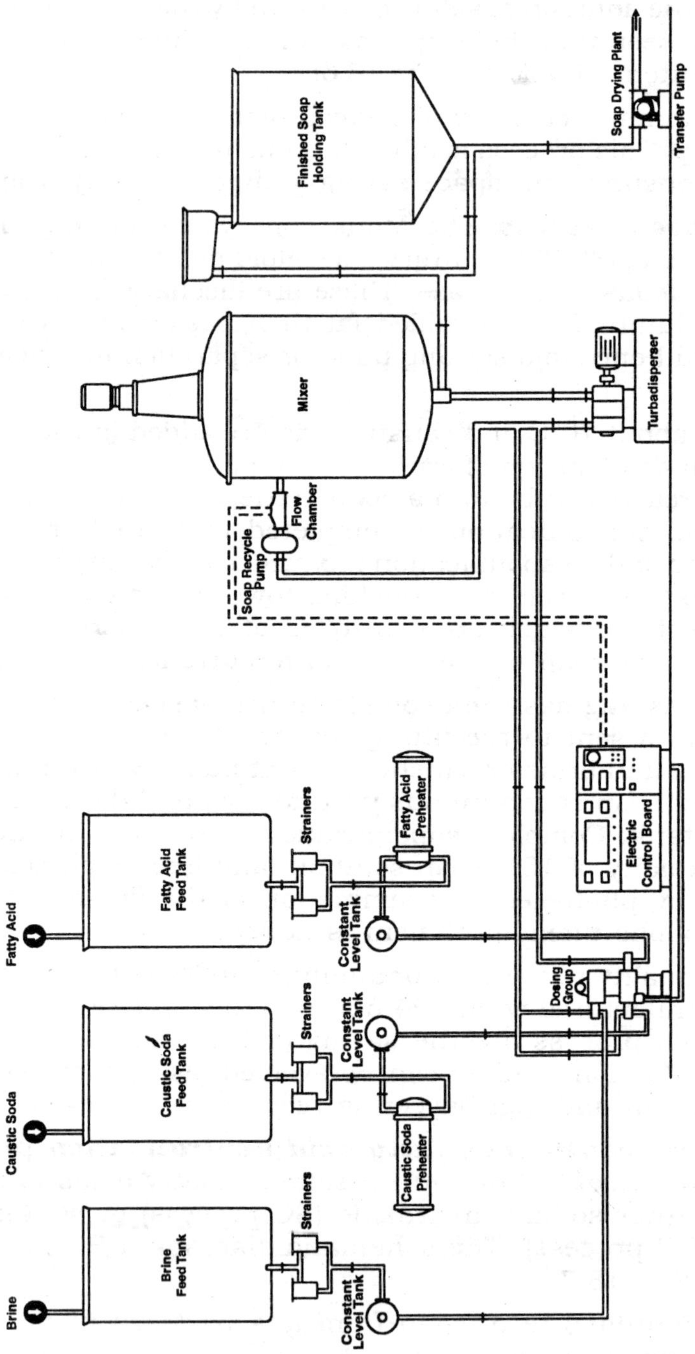

Fig. 18.7: Mazzoni continuous saponification plant (SC)

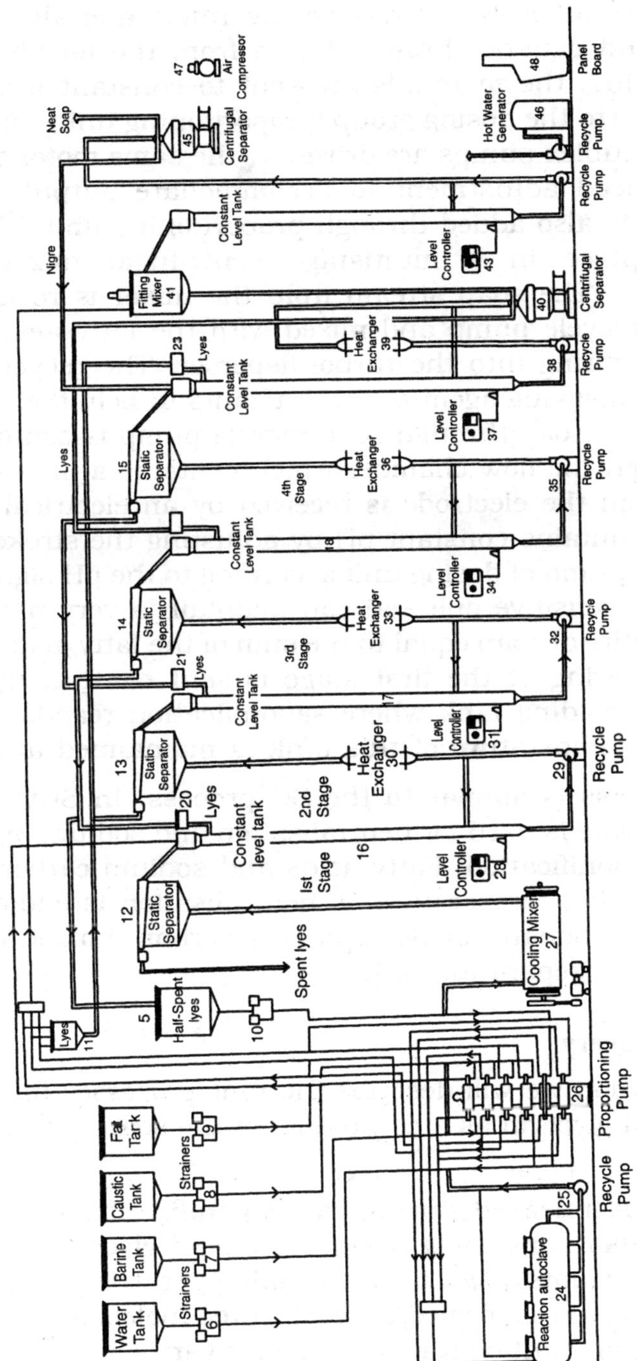

Fig. 18.7a: Mazzoni continuous saponification plant

Preheaters at 50-60° C receive the raw materials, distilled fatty acids and caustic through filters from the feeding tanks. After preheating, the materials are sent to constant level tanks and from there to the dosing group (proportioning unit). In proportioning unit, duplex pumps are driven by the same motor but with a variable speed adjustment to accommodate output requirements. Brine is also added through proportioning unit. The reaction takes place in a multistage centrifugal mixer called turbodisperser. The soap stream from the mixer is recirculated through the recycle pump and mixed with the fatty acid-caustic just before entering into the turbo-disperser. The recycled soap acts as an emulsifying agent of the reactants to help the reaction. The recirculated soap through soap recycle pump is continuously passed through the flow chamber which contains a pH electrode. The signal from the electrode is received by an electrical control unit which maintains constant pH by adjusting the stroke length of one caustic pump of dosing unit according to the pH signal. This unit is a very sensitive one and can maintain a very narrow pH range. A quantity of soap equal to the sum of the fatty acid, caustic and brine entering at the first stage is sent continuously to a finished soap holding tank where saponification reaction is carried out. The temperature of this tank is maintained at 90°C.

SCC process is similar to the SC process. In SCC process sodium carbonate is used for neutralisation and sodium hydroxide is used for saponification. Fatty acids and sodium carbonate are heated at a higher temperature as neutralisation is endothermic with sodium carbonate. In SCC process carbon dioxide is separated through a separating unit.

Glycerine Recovery

In this discussion we will describe the unit processes involved in the recovery of glycerin as a co-product of the soap and fatty acid industries.

Glycerin is released during the saponification of glyceride fats with an alkali or during the hydrolysis of glyceride fats with water. Since the fats used in soapmaking are triglycerides, one molecule of glycerin is formed for each three molecules of soap or fatty acid. The amount of available glycerin varies according to the molecular weight of the fatty acids existing in the fat. For example, neutral beef or hog fats yield 10% glycerin while the laur'ic acid type oils, such as coconut, babassu, and palm kernel, yield 13.5-13.9%. In commercial practice 88-92% of this glycerin is recoverable.

The unit processes involved are:

(a) Purification of the glycerin bearing solutions (commonly called "lye treatment")

(b) Evaporation

(c) Distillation

(d) Bleaching

Purification (Lye Treatment)

Where the fats are saponified with alkali as in the kettle process, the glycerin is separated from the resulting soap by virtue of the insolubility of soap in sodium chloride solutions. Usually a concentration of 10-15% sodium chloride is sufficient to render the soap insoluble. The soap is lighter in specific gravity than the solution and rises to the top while the glycerin remains soluble in the lower layer, which is known as "spent lye" and is withdrawn from the recovery of its glycerin.

Where countercurrent soap-boiling is used the glycerin content of the spent lye will usually be 10-12%, sometimes up to 15%..

Where the fats are hydrolysed with water alone, as in fat splitting, no salt is necessary for the separation since the fatty acids are insoluble and rise to the top of the splitting column. The glycerin-bearing solution in this case is known as "sweet water" and, depending upon the fats used and the water to fat ratio used in splitting, may contain up to 15-20% glycerin.

In the case of spent soap lye the solution will contain impurities such as unused caustic soda, sodium carbonate, soaps of low molecular weight fatty acids which are slightly soluble in salt solutions, mechanically entrained soap, colour bodies, and nitrogeneous impurities. In general, spent soap lye contains more impurities than sweet water since the soaps are not as completely insoluble as are the fatty acids, and there will always be some residual alkali present from kettle saponi-fication. Further, in fat splitting more of the nonglyceride impurities remain in the fatty acid layer.

It is necessary to remove these impurities as far as possible prior to evaporation and distillation, otherwise it would be impossible to conduct the subsequent operations efficiently and a very impure product would result. The treatment of spent soap lye involves two basic factors : (a) neutralisation of the unused alkali (sodium hydroxide and sodium carbonate chiefly); (b) precipitation and coagulation of the soaps and other organic impurities.

The alkalies are neutralised with either sulphuric or hydrochloric acid, and the soaps and organic impurities are precipitated by an aluminium or iron salt such as the sulphate or chloride or ferric alum as freshly prepared *20%* solution to a pH of 3.8-4.4. Soaps are rendered insoluble by conversion from the soluble sodium salt to the highly insoluble metallic salt. The aluminium and iron salts also coagulate nitrogeneous impurities to form insoluble agglomerates.

The choice between hydrochloric or sulphuric acid and between the metallic chloride or sulphate depends not solely on the cost of chemicals but also upon other operating factors. Since the salt is to be recovered during evaporation for re-use in the soap kettles, it is important to avoid the formation of too much sodium sulphate during lye treatment. Sodium sulphate has far less "graining power" than sodium chloride. Further, sodium sulphate is harder to wash free of adhering glycerin. In practice it is desirable to hold the sulphate content of the recovered salt to 10% or less, and seldom over 15%; otherwise an eventual discard of the salt itself would be necessary.

The extent of the impurities is approximately :

Soluble soaps	..	0.25%
Unused alkali	..	0.20% as Na_2O

Other organic impurities - variable according to quality of fats used, probably 0.10-0.20%

The first step is to accumulate the spent soap lyes in a storage tank for a holding period of at least three to four hours to allow entrained soap to rise. In large plants it is customary to provide two tanks for alternate use, one being used to receive new soap lye while the other is resting undisturbed. From the latter the batches for treatment are withdrawn. Soap reclaimed in these tanks is eventually returned to the soap kettles. The treatment is usually done batch-wise in any convenient size, from perhaps 22,727 kgs in a smaller plant to 136,363 kgs or more in a large plant. A sample is analysed for the percentage of unused alkali, and from this the amount of acid necessary to neutralise it is calculated and added to the charge under constant agitation. The amount of precipitant, aluminium sulphate for example, is usually arrived at by experience and varies according to the quality of lye being treated. This is added to the batch and after sufficient agitation a sample is tested.

Several variations in test procedure are used. The one described here is a simple one that has proved adequate. A sample is filtered, and a portion tested by adding to it a few drops of 10% aluminium sulphate solution. If upon heating to near boiling no further precipitate is observed, enough aluminium sulphate has been added. A confirming test is used to prevent wastage of aluminium sulphate (in the event that too much was added) by determining the acidity of a filtered sample in terms of sulphuric acid, which at this point should be between 0.08-0.12%. In some plants the pH is used as a control and at this point should be about 4.5.

When the tests meet specification, the batch is filtered. The temperature maintained during treatment is usually about 80°C, which assists in the coagulation of the precipitate and assists filtration.

Filtration may be done, using recessed plate or plate and frame filterpresses or a continuous rotary vacuum filter. The cake in the filter is later washed with water to reduce its residual glycerin content to 2-3% since this filter cake is discarded and usually amounts to about *2% of* the weight of lye treated.

The foregoing constitutes what is known as the first treatment. The second treatment is merely the removal of the excess aluminum sulphate by adding caustic soda to the filtered lye from the first treatment until a sample is pink to phenol-phthalein, or

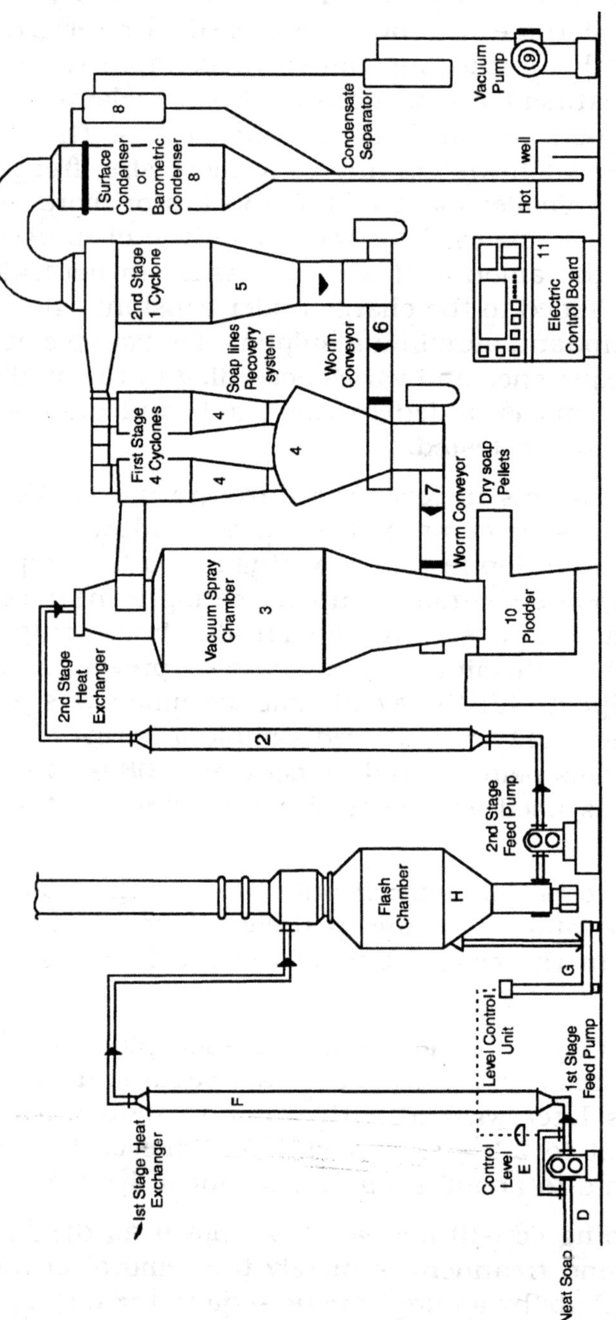

Fig. 18.8: Mazzoni 'C' Multistage Spray-drying Plant.

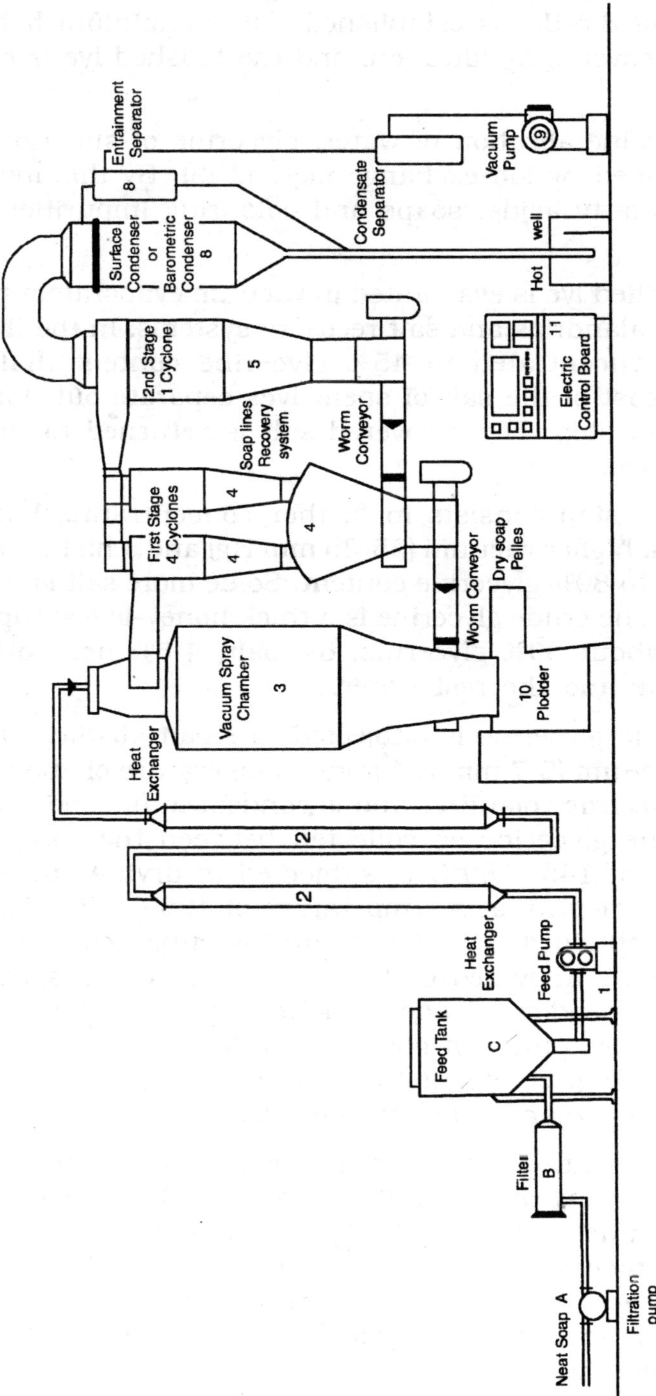

Fig. 18.9: Mazzoni 'C' vacuum spray-drying plant.

until a pH of 8.5-9.0 is established. The aluminium hydrate so formed is recovered by filtration, and the finished lye is ready for evaporation.

Like demineralization of water, glycerine of superior quality can be obtained by ion-exchange method (9). By this method all electrolytes, fatty acids, soaps and colouring impurities are removed.

The purified lye is evaporated in vacuum evaporators provided with steam calandrias and salt recovery systems. In the first step, the lye is concentrated to 45% glycerine content (half crude glycerine). Most of the salt of spent lyes separate out during the concentration step. The recovered salt is returned to the kettle room for reuse.

The next step consists in further concentrating half crude glycerine at a higher vacuum (25-35 mm Hg) and a higher temperature (110°C) to 80% glycerine content. Some more salt is recovered in this step. The crude glycerine is a thick honey-like syrupy liquid containing about 80% glycerine, 6% salt, 4-5% non-volatile organic residue and the rest water.

The crude glycerine is subjected to steam distillation under very high vacuum (5-7 mm Hg) and a temperature of about 158°C. The pure glycerine volatilizes and is condensed in condensers and collected. The glycerine so collected between the condensation temperature of 145°-180°C is subjected to drying and deodorization under vacuum at a temperature of 120°C. The deodorized glycerine so obtained is 98% pure and is subjected to bleaching operation using activated carbon as the bleaching agent. After filtration, the deodorized and bleached glycerine is packed into drums. If higher purity of glycerine is desired, for example, for dynamite grade glycerine, collections made at condensation temperatures lower than 110°C are not included.

Process variations in the distillation plant are possible with continuous feeding of crude glycerine through a pre-heater and drier. Continuous crude feeding necessitates intermittent removal of thickened liquor from the still when the distillation rates fall. These removed liquors can be stripped of the residual glycerine in a separate foots still, yielding glycerine of a poorer colour and quality.

It is important to add some caustic lye to the crude glycerine prior to distillation. This suppresses the formation of esters of glycerine with fatty acids. However, presence of caustic soda in the crude glycerine also catalyses polyglycerol formation at the prevalent temperatures. Polyglycerol formation, apart from representing a loss of value of glycerine, also affects the odour and taste of the product and hence is undesirable. The dosage of caustic soda to the crude glycerine prior to distillation should, therefore, be just sufficient to keep the content of esters and fatty acids in the final product measured by the s aponification equivalent, within acceptable limits.

Transformation of neat soap to finished formCrutching: Crutchers are the machines where blending operationof soaps is carried out. Before framing or drying, soaps are blended with builders or other materials in this equipment. These are small cylindrical vessels of about 150-2500 kg capacity equipped with agitators which, in most cases, are spiral screws. Crutchers are also used for manufacturing soap in cold-made process. These are sometimes insulated and used for remelting and reworking scrap from framing.

Fig. 18.10: Twin worm duplex vacuum-plodder.

Manufacture of framed soaps: The method of solidification of neat soap or other hot liquid soap to form a bar is called framing. The operation is carried out by running the liquid soap at 55-65 C into portable frames and is allowed to solidify spontaneously. The frame can be easily removed from the solidified soap cake. Each soap slab weighs about 0.5 to 0.6 ton and has a rectangular shape which can be cut according to the standard size. 3-7 days are required for solidification of neat soap. The cakes cut from the soap slabs must be dried to some extent before wrapping. Drying is carried out by warm, dry air circulation over the cakes placed in drying rooms of tunnels. After drying, the moisture content of the soap is usually reduced by about 3-5%.

Drying: Neat soap obtained from the saponification process contains about 30-35% water. The soap is dried and the soap chips obtained from the operation contains about 5-15% water. This soap may then be milled or plodded or processed further to produce toilet soap bars.

Drying may be done in different ways. In practice, the older process which is followed for the solidification of soap is accom-

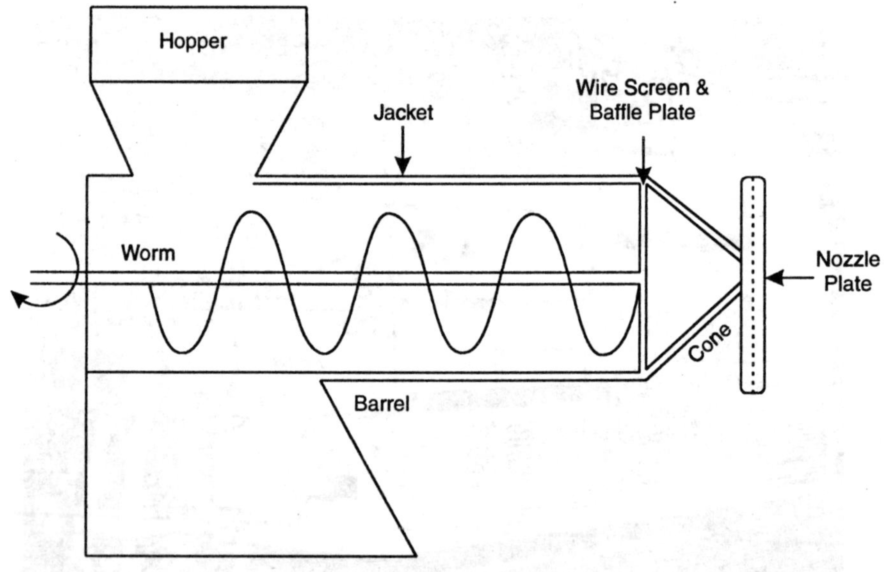

Fig. 18.11: Mazzoni plodder mill.

plished on a chilling roll about 4-5 ft. in diameter and two to three times as long where neat soaps are solidified by the circulation of water. The soap is spread as a thin and uniform layer on the roll by the action of a smaller feed roll set above the chill roll. The soaps are solidified and the roll revolves slowly and come out in the form of thin ribbons which are marked by the needles pointed against the roll and the ribbons are removed by a scraper blade from the roll.

The soap ribbons are then carried by a wire mesh converter which is heated in a tunnel dryer by steam coils by moving back and forth slowly and with the provision of forced circulated air. The

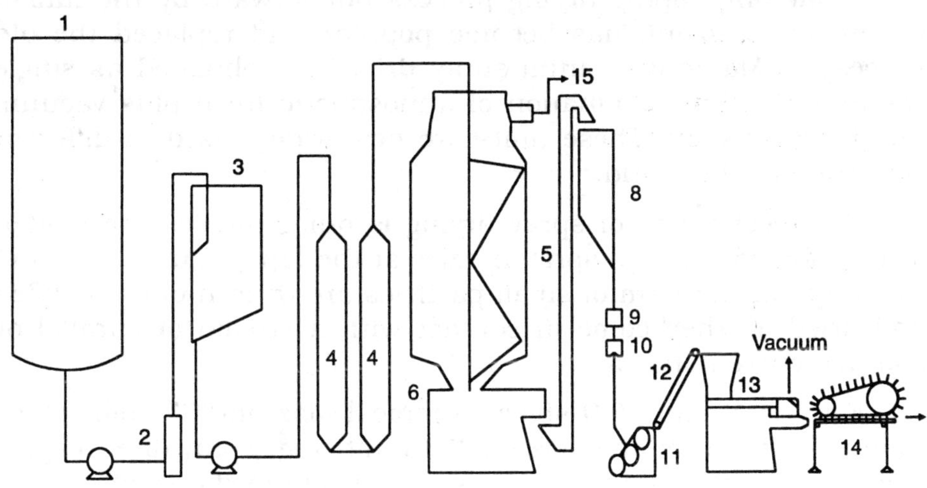

1. Soap Stock Tank	2. Filter	3. Soap Feed Tank
4. Heat Exchanger	5. Spray Chamber	6. Noodler
7. Elevator	8. Storage Bin	9. Weight Hopper
10. Mixer	11. Mill	12. Conveyor
13. Duplex Vac. Plodder	14. T.V. Cutter	15. Vacuum System

Fig. 18.12: Mazzoni soap finishing line.

moisture content of the soap can be regulated by controlling the temperature of the dryer and the speed with which the conveyor of the converter separates. The ribbons are broken into chips after sufficient drying. Care must be taken for the chips to avoid excessive heating so that chips do not melt and stick on the conveyors. The temperature of the chips is reduced by circulating cold air on them. Milled flakes are the best quality and most expensive soaps among all the quick dissolving soaps. These are obtained from only pure soaps without adding any builder. Low moisture (10-11%) dried chips are repeatedly milled in special, close-set water - cooled steel rolls and the drying of these flakes is obtained in the rolls. After milling the flakes are sent for packaging. These flakes have a special glossy appearance and quick dissolving property in the bar form.

The above drying processes were used before World War II. After that time, spray drying process put forward by the Italian Company, Mazzoni, has become popular and replaced the old processes. Mazzoni vacuum spray driers are obtained as single vacuum or as a combination of atmospheric fresh plus vacuum multistage system. These units are now widely used in different countries of the world.

The mechanism of spray drying is quite simple. The molten soap is forced through spray nozzles at the top of a tall tower and the soaps in the form of small particles are thrown out, solidified and dried as they come in contact with a stream of parallel or counter current hot air,

Figs. 18.8 and 18.9 shows a typical Mazzoni 'C' single stage vacuum spray drier and Mazzoni 'C1 multistage spray drier respectively in the single stage drier neat soap is pumped through a filter (B) into a feed tank (C) and then through pump (1) it goes to the heat exchangers (2). The hot soap is passed through a rotating nozzle into a vacuum spray chamber (3) and then deposited on the walls as a thin film. The extent of drying and the soap temperature are dependent on the vacuum applied and the heat supplied at stage (2). The soap film after drying and cooling is scraped from the walls and falls into a plodder (10). The fine soap dust is carried out by the steam in the direction of the condensers (8) and is collected at the bottom of the first and second stage cyclone system (4 and 5). The soap is then passed through the worm conveyors (6 and 7) to the plodder.

After removing most of the soaps the steam from the vacuum chamber passes to a barometric or surface condensor. Vacuum in pump (9) or steam jet is used for achieving the required vacuum.

In the multistage spray drier vacuum and atmospheric flash drying are in the combination. This type of drying process is used when high degree of drying is required. Now several expansion chambers are used in series and vacuum stage is eliminated. Expansion is donc to avoid 'over-drying' of the soap particles which cause 'roughness' in the finished toilet soaps.

Manufacture of toilet soap bars : Standard finishing processes: High quality toilet soap bars are produced by milling and plodding. This quality cannot be achieved by only solidifying the neat soaps in frames. The working on the soaps produces a certain degree of orientation of the soap crystals which causes distinctive physical properties of the soap. In the framed soap ω-phase exists but working converts it to the β-phase. Milling process allows the addition of perfumes, bacteriostats and colour to the soap in cold condition and so the perfumes which are unstable in the presence of hot alkalies or are volatile in the presence of heat can be used easily.

The limit of moisture content of the soap flakes to be milled must be between 13-14% because lower moisture content soaps are difficult to mill and higher moisture content soaps become soft and gummy, hence difficult to handle.

The first stage of soap manufacture consists of mixing the weighed amount of flakes in amalgamators or heavy hori-zontal mixers where perfumes, dyes, titanium dioxide, bacter-iostats, antioxidants etc. are mixed. From this stage the mixture is transferred to the mills. Mill consists of 3 to 5 rolls through which the soap mass passes in the form of a thin sheet. Every roll rotates at different speeds to maintain a shearing action as well as compression on the sheet. From the last roll, soap is obtained as a thin ribbon.

In the older milling process milling stones were usually made up of granite but they were later replaced by hollow iron balls.

These are cooled by circulated water which takes out the frictional heat generated during milling.

Plodder is used to manufacture compressed soap bars from the milled soaps. Plodder consists of a hopper which is attached to .a horizontal circular chamber where a spiral is forced out through an orifice at the end or 'nose' of the device in the form of a continuous bar. The compression cylinder is jacketted and cooling water is circulated. The 'nose' is electrically heated. The soap rod thus formed is cut by a wire cutter or by a continuous automatic cutter.

Modern Soap Finishing Techniques and Equipments

Continuous operations equipped with highly efficient machines are used for the large scale production of toilet soap bars (10,11). The rate of production in the modern process is about 300-400 bars/min. The units in these operations include amalgamators, automatic dosing devices for liquid and solid additives, simplex or duplex refiners (pre-plodders), duplex vacuum plodders, cutters, conditioning tunnels, soap presses, wrapping and cartoning machines.

Amalgamators blend different additives like perfumes and colours with the soap and plodder produces continuous rod like soap by physical working. In refining, the fine screens are used through which soap is forced under high pressure; these screens serve the same purposes as milling. The operating conditions choosen determine the crystallinity and ultimate physical properties of the soap.

Fig. 18.12 shows a modern soap finishing line. The different important units of this process are as follows :

Soap pellet situ	=	1
Scale tank	=	3
Storage unit	=	4
Holding tank	=	7
Automatic amalgamator	=	12
Duplex refiner	=	14

Vacuum plodder	=	16
Cutter	=	17
Conditioning tunnel	=	18
'STU' universal press	=	19
wrapping machine	=	20
Cartuning machine	=	21 and 22

Deodorant and Antibacterial Soaps

Deodorant soaps contain antibacterial agents which prevent or reduce the bacterial decomposition of perspiration', which is odorous to odorless compound.

Deodorant soaps, a major commercial product, contain only two materials, namely, trichlorocarbanilide (TCC) and Irgasan DP 300, while the use of hexachlorophene and tribromo-salicylanilide has been banned.

'Lifebuoy Soap' and 'Dettol Soap' are two well known disinfectant and deodorant soaps.

Different techniques and processes (12, 13) have been developed for evaluating deodorant and anti-bacterial soaps and among them the most extensive are the 'invitro' and 'in vivo' tests. 'In vitro' tests determine the anti-bacterial activity in soap, sustantivity of skin, i.e., whether the bacteriostats are held on the skin and the microbiological availability, which is the determination of the bacteriological activity and skin sensivity of *a* soap. 'In vivo' tests are used to evaluate the usefulness of anti-bacterial agents in soaps for effectiveness in deodorant activity, reduction of cutaneous bacterial populations, performance in different clinical studies, namely diaper rash, erythrasma, secondary skin infections and other dermatological diseases.

After 1970s most of the anti-bacterial agents used in soaps were banned by international organisations like FDA*. Among them hexachlorophene and tribromosalicylanilide are main. It was noticed that use of these anti-bacterial agents at a somewhat

---

* U.S.Federal Drugs Administration

higher percentage, caused harm to the infants and may also harm the adults. Now the only anti-bacterial agents used are TCC, Irgasan CF3 and to a lesser extent Irgasan DP 300. Although these three agents are now used they are also under review.

Transparent Soaps[14]

Transparent or glycerine soap is another type of soap which is much preferred by consumers of some countries.

It seems that the aesthetic nature of the product, i.e. transparency or translucency is the preferred criteria even though detergency properties are of no difference from normal laundry or bath soaps.

In transparent soap making, the following factors have to be observed:

i. Choice of the suitable fats or oils or fatty acids and appropriate proportions of C16-C18 and C12-C14 fatty acids.
ii. Additives which promote optimum transparency,
iii. Rate and temperature of cooling,
iv. Phase changes due to mechanical treatment, v) Purity of the alcoholic soap solution i.e. free from suspended materials.

Raw Materials

The raw materials used in making transparent soap should be selected carefully for colour and purity. It should be free from particles or suspended materials because these materials provide the seed for crystal growth. This is undesirable because transparent soap should consist of a mass of fine ultramicroscopic crystallite.

Fatty materials used in transparent soap making are chiefly coconut oil, refined bleached palm oil, tallow and castor oil or their corresponding fatty acids. These raw materials should be of good colour and have low free fatty acid content. Sodium hydroxide solution is used to saponify the oil and fats or to neutralise the fatty acids. It should not contain any heavy metal contamination.

Castor oil enhances the transparency of the soap but if too much is used will result in a soft soap which rapidly becomes rancid with aging.

Rosin also imparts a good lathering property and provides aromatic odour. It also prevents rancidity but the amount used should be moderate because it may cause the soap to be soft and tacky and to darken with aging.

Other Additives: Glycerine and alcohol are two necessary ingredients to aid transparency. Glycerine also acts as an emolient. Sugar is sometimes used to replace the two materials but excessive usage causes "sweating" of the soap due to its hygroscopic nature. Thermostable polyols such as sorbitol, mannitols, polyglycols are sugar alcohols which can also be used to enhance the transparency in the soap. The choice of perfume and colour is important to ensure that they do not affect the transparency of the soap.

Processing

Transparent soap is manufactured by means of alcohol process, cold process, semi-boiled method and direct saponi-fication of fatty acids.

Alcohol Process: In this process soap is dissolved in alcohol and heated to 50°-70°C to give a soap solution. When all the soap has dissolved, the solution is allowed to stand for the insoluble matter to settle. The clear portion is coloured and scented before being transferred into frames. The soap is then cut into bars.

Cold Process: Liquefied fat and oil are strained into a steam-jacketed kettle. Glycerine and caustic soda solution are crutched in, followed by addition of alcohol. The batch is mixed intimately for 1-5 hours for complete saponification. The liquid soap is poured into frames and cooled rapidly. After cooling the soap is cut into bars.

Semi-boiled Method: Oil or fat in liquid form is poured into a s team-jacketed pan and caustic soda added with occasional stirring. At the same temperature, other ingredients are added slowly in *a* stream, in the following order sodium silicate solution, melted soda, sugar solution and alcohol. Lastly, colouring agent and perfume are added.

Direct neutralisation of fatty acids: The mixture of fatty acids is heated to 60°C and saponified at 100°C with aqueous sodium hydroxide in an autoclave. The other additives are added to the neat soap and mixed thoroughly and dried to moisture content of 16%. The soap mass is vaccuum dried, milled to get noodles, plodded and made into bars.

Detergent bars and Combination Soap Detergent Bars

This kind of product also called 'combo soap' or 'combo bar' is developed for better use of toilet and laundry soaps in hard water - reducing the deposition of calcium and magnetsium soaps. The soaps are provided with additives called lime soap dispersing agents, 'LDAs', which are surface active usually. These include AOS*, α-sulpho methyl esters of fatty acids, acylated acids. In India, toilet soaps containing AOS are available.

Other Commercial Soaps

Castile soap: Castile soap is prepared from pure olive oil and also from a mixture of oils showing the fatty acid composition similar to olive oil. This soap is mainly prepared by cold process. It is supposed to cause almost no irritation in the skin and is used as a soap for babies and for medicinal purpose. Probably, the absence of medium molecular weight fatty acids which are present in other good quality toilet soaps obtained fron coconut oil or palm kernel oil, imparts mildness to the castile soap. Mainly capric and lauric acid cause irritation to the skin of infants; hence these acids are largely replaced by oleic acid.

Bar laundry soaps: Bar laundry soaps are now not popular as detergents but they are still produced and consumed to some extent. There are two main types of laundry soaps, white and yellow. White laundry soaps are prepared from 25-40% coconut oil and 60-75% tallow in full boiled process. Soap builders like sodium silicate are added in sufficient quantity which ' reduce the anhydrous soap content to less than about 50% This soap is never milled.

Yellow laundry soaps contain a high proportions of rosin which gives the yellow colour to the soap. This soap is soft sticky and hence builders are not added in high amount as in white soaps.

---

* AOS= Alpha olefin Sulphonare

In this soap the percentage of coconut oil used is less than the white soap to give greater solubility to yellow soap.

Liquid soaps: Liquid soaps are aqueous solutions of potassium soaps made from coconut oil or mixed with palm kernel oil or from oils with high lauric acid and with soft oils like cottonseed, soyabean etc. The soap solutions contain 15% soap.

Soaps in toiletries and cosmetics: Soaps used in toiletries and cosmetics are made from stearic acid or oleic acid and sodium or potassium hydroxide, amines or alkanolamides. The soaps are used in many products like shaving creams, hand and body lotions, face creams and hair colourants. The main fatty acid used is the so called triple pressed stearic acid containing 45% stearic acid and 55% palmitic acid.

Toilet soaps - Practical characteristics and test methods: The manufacture of toilet soap of desired quality is not easy although many modern techniques and equipments have been developed. This is due to the complex nature of the natural raw materials like fats and oils whose composition may alter with the place of origin, year of harvest, feed etc. Modern analytical techniques like gas chromatography etc. are employed to control the quality.

The tests for the determination of quality include evaluation of lathering ability, cracking tendency of the soaps, 'feel' and the degree of roughness of soaps etc. Other common tests are 'smear test' and penetration tests which use a penetrometer to measure the extent of slushing.

Finishing and Packing[1]

The neat soap produced by any of the foregoing processes, is taken in a crutcher which is a jacketed vessel equipped with a stirrer. The neat soap is mixed here with a chelating agent, an opacifier like TiO_2, fillers like talcum powder and is brought to a free alkali content of 0.03% for a high grade toilet soap or around 0.05% for a washing soap by the addition of requisite amount of sodium bicarbonate. In the case of washing soaps and unmilled bath soaps, the colour and perfume, are also added at this stage.

The crutched soap is sent to a soap free tank through a strainer that retains foreign matter that might have got into the soap accidentally. The soap from the feed tank is pumped continuously through heat exchangers and sprayed into a high vacuum spray chamber maintained at a vacuum of 5-35 mm of Hg as per process requirements. The temperature in the heat exchangers is maintained at 110°C for washing soaps at 44% T.F.M., 125°C for bath soaps at 55% T.F.M. and at 155° C for toilet soap noodles. The temperature and vacuum conditions may, however, be varied to suit specific process needs. The hot soap when sprayed into the vacuum chamber loses moisture through flashing and simultaneously gets cooled and solidified. The solidified soap is scraped off the walls of the spray chamber by means of a scraper moving in the spray chamber.

The solidified and dried soap then falls into a plodder, which is worm screw rotating at close clearance inside a barrel. Double worm plodders are also used wherein two worms rotate in opposite directions. The plodder extrudes the dried soap through a perforated plate. The soap ribbons produced are cut by means of a cutter mounted on the worm shaft. These ribbons then fall into a final plodder which extrudes them either in bar form or in the form of noodles which are further processed for the production of milled toilet soaps. In the case of washing soaps or bath soaps (unmilled), a builder like sodium silicate is sprayed into the spray chamber along with the soap. It is often necessary to introduce a certain percentage (2-3%) of salt in the silicate to prevent soap from sticking to the walls of the plodder barrel or the plodder worm, thereby causing difficulties in plodding.

The plodder barrels are provided with jackets for the circulation of chilled water at 5-6°C to dissipate the heat of extrusion and to help in maintaining the soap in the desirable beta-phase during the operation.

In the case of milled toilet soaps, noodles are produced in the drying and cooling step at 76-80%. T.F.M. Temperature and vacuum conditions are varied to attain the various degrees of drying required. However, temperature in excess of 155°C are seldom employed. The noodles are mixed with colour, perfume, emmolient, anti-oxidant, etc., passed through a graining plodder and then milled twice in cooled triple roller mills, passed through a series of jacket cooled refining plodders followed by extrusion in bar form through a final plodder fitted with a compression cone and an extrusion eye plate. The extrusion mouth of the plodder is kept at 60-75°C by jacket heating. This imparts a surface finish to the soap bars. The heat of extrusion is regularly removed by circulating chilled water through the plodder jackets. The extruded bars are cut into billets, stamped and wrapped on automatic or semi-automatic machines and packed into cartons for onward despatch to sales outlets through the bonded godown.

Modern stamping machines are chilled dies that give a better finish and stamp to the product. The chilling is provided by circulation of a chilled solution of ethylene glycol in water at temperatures as low as —40°C through the die jackets. Movement of unstamped billets and stamped cakes by the use of vacuum suction instead of mechanical pushes further improves the get-up of the product.

With the realisation that it is only the more soluble soaps of fatty acids that actually perform in the cleaning action, with the other less soluble soaps of fatty acids like stearic acid providing structure and body to the soap cake, structured soaps with much less T.F.M. (Total Fatty Matter) containing inorganic structurants like china clay, kaolin, talcum, etc. in appreciable quantities, have appeared on the market, with a view to conserving valuable fatty matter and providing cheap toilet soaps to the consumer. The T.F.M. of these soaps may be as low as 50% against the earlier 76-80%. With toilet soaps now falling under the purview of Drugs & Cosmetics Act, it is important, however, that all ingredients used are of cosmetic grade.

REFERENCES

1. Narula: Treatise on Fats. Fatty Acids & Oleochemicals, Volume I, p. k-1/4, k-1/12 (1995).
2. J. Am. Oil Chemists' Soc., Nov. 1952, p. 501
3. R.H. Ferguson and A.S. Richardson, Ind. & Eng. Chem. 24, 1330 (1932).
4. L. Lascaray, Fette, Seifen, Anstrichm, 46, 531-536 (1939).

5. F.T.E. Palmquist and F.E. Sullivan, J. Am. Oil Chemists' Soc., 36, 173-178 (1959).
6. F.T.E. Palmquist, Soap Chem. Spec. 37, 169-177, 207-208 (Sept. 1961).
7. Trade Publ. Sc. SCC. Mazzoni S.p.A., Bus to Arsizio, Italy.
8. M. Fock, Seifen, Oele, Fette, Wachse, 90, 343 (1964).
9. G.W. Busby and D.E. Grosvenor, J. Am. Oil Chemists' Soc., 29, 318-320 (1952).
10. Trade Publ. C-LTC, Mazzoni, S.p.A. Bus to Arsizio, Italy.
11. A.M.D. Carlton, Manuf. Chem. 33, 176-179 (1972).
12. E. Jungermann and D. Taber J. Am. Oil Chemists' Soc., 48, 318-323. (1971).
13. E. Jungermann, Cosmet Toiletries, 91, 50-58 (1976).
14. H. Kifli and A. Kuntum, Malaysian Oil Science and Technology, July 1992, Vol. 1, No.2, pp. 27-29.

CHAPTER 19

Fat-Based Surface Active Agents

Fat-based surface active agents,also called surfactants, hold a very important position as a commodity in terms of volume production and extent of applications in the detergent industry. Fat-based surfactants are characterised by their diverse molecular structures, which display very distinctive surface active properties like the lowering of surfacetension of water and the interfacial tensions of the aqueous solution of water against a hydrocarbon oil or chloroform.

Soap is still now the most important surface active agent and constitutes 50% of the total production of the surfactant products of the world.

The surfactants other than soap also contain the usual hydrophobic part that is typified by the paraffinic chain of the higher fatty acids (C_{12}-C_{18}) and the hydrophilic part that may be a carboxylic, sulphuric, sulphonic acid group or other functional groups or atoms.

The fat-based surface active agents have emulsification, wetting, spreading, foaming and detergent properties which make them useful. They also show properties related to surface energy and in the case of certain agents bactericidal activity. The naturally occurring fats, and more specifically the fatty acids such as lauric, myristic, palmitic, stearic, oleic, and ricinoleic acids, are particularly suitable as a raw material for preparing surface-active agents. The long straight chain $CH_3(CH_2)_{10-16}$ and the chains $CH_3(CH_2)_7CH = CH(CH_2)_7$ and $CH_3(CH_2)_5CHOHCH_2CH = CH(CH_2)_7$ are natural sources for the hydrophobic part. The carboxyl group lends, itself readily to reactions for the introduction of the necessary hydrophilic part. Hydrophilic groups can also be introduced at the double bond of oleic acid or at the hydroxyl group of ricinoleic acid. A greater variety of surface-active agents has been derived from the fatty acids than from any other single source.

Examples of fat-based surface-active agents are the following: sulphated and sulphonated oils, fatty acids, esters, and amides; sulphated coconut oil alcohols and sulphated tallow alcohols; sulphated monoglycerides, tallow-based N-methyl taurates and related compounds; fatty acid esters of condensation products of anhydrosorbitols and ethylene oxide; fatty acid alkanolamides; and the quaternized and nonionic derivatives of long chain fatty amines and fatty acid amides.

CLASSIFICATION OF SURFACTANTS

The fat-based surfactants, as exemplified above, are classified into four categories : anionic, cationic, nonionic and amphoteric. The differences between them are illustrated below :

1. ***Anionic fat-based surfactant (1)***: The fat-based anionic surfactants are primarily either sulfates or sulphonates, although carboxylates, sulphites, thiosulphates, persulphates and phosphates can also be made. In these surfactants the hydrophobic chain is attached to a hydrophilic, negatively charged group :

   $$R - x^- \text{ where } x^- - COO^-, OSO_3^-, SO_3^-$$

   Products of the general formula

   $$RCON(CH_3)CH_2CO_2H$$

   from the condensation of sarcosine, $CH_3NHCH_2CO_2H$, with a fatty acid chloride have been marketed as the Medialans.

2. ***Cationic fat-based surfactants***: In these surfactants the hydrophobic chain is the cation

$$R - x^{+} \text{ where } x^{+} = \overset{|}{\underset{|}{N}}+-$$

3. ***Nonionic fat-based surfactants***: Here the hydrophobic chain is attached to an uncharged hydrophilic chain or group

$$R - x \text{ where } x = (OCH_2CH_2)n\ OH \text{ or}$$

$$(OCH_2\underset{|\atop CH_3}{C}H)_nOH\ (n =- 6 - 30)$$

$$= - COO\underset{|\atop HOCH \atop | \atop HOCH_2}{C}H_2,\ CONHCH_2CH_2OH$$

4. ***Amphoteric surfactants***: Here the hydrophobic fatty chain is attached to a hydrophilic group containing both positive and negative charges.

$$-x \text{ where } x = - \overset{|}{\underset{|}{N}}{}^{+}\text{- } CH2\text{- } COO^{-}$$

$$x = - \overset{|}{\underset{|}{N}}{}^{+}\text{- } (CH_2)_3 - SO_3^{-}$$

$$x = - \overset{|}{\underset{|}{N}}{}^{+} \rightarrow O^{-}$$

Fat Based Surfactants: *Some Commercial Products. Application and Their Production*

Some commercially important fat-based surfactants, belonging to the above categories are described in brief, dealing with their scopes of applications and the technology of their production.

ANIONIC SURFACTANTS

Sulphated and sulphonated fat-based anionic surfactants are the major products. These include the following : "sulphonated" derivatives of oils, fatty acids, esters, and amides; sulphated alcohols; sulphated monoglycerides; sulphated alkanolamides; sulphated nonionics; α-sulphonated acids; sulphoethyl esters (Igepon A); sulphoethyl amides (Igepon T); and sulphoacetates.

Among the above anionic surfactants, the sulphated and sulphonated materials constitute the largest group of anionic surfactants. In fact, their volume of production is next to soap.

Sulphates: The sulphates are having the following structure:

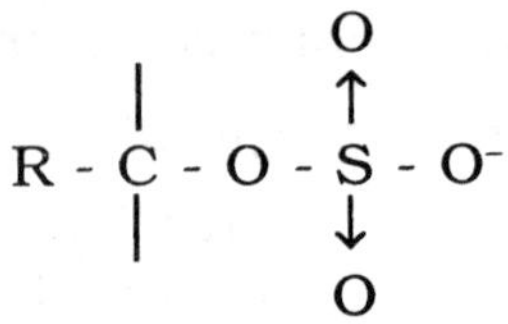

$$\begin{array}{ccccccc} & & & & O & & \\ & & | & & \uparrow & & \\ R & - & C & - O - & S & - & O^- \\ & & | & & \downarrow & & \\ & & & & O & & \end{array}$$

Sulphonates: The sulphonates have the structure:

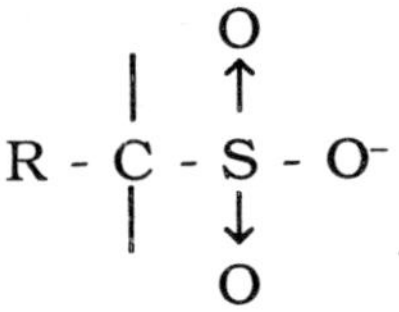

$$\begin{array}{ccccccc} & & & & O & & \\ & & | & & \uparrow & & \\ R & - & C & - & S & - & O^- \\ & & | & & \downarrow & & \\ & & & & O & & \end{array}$$

Many sulphated and sulphonated surface active agents are available commercially.

The general method of preparation of sulphated oils is the addition of concentrated sulphuric acid to the oils under a variety of conditions *at* a temperature below 25°C. The sulpha-tion occurs predominantly at the carbon-carbon double bond. It is mainly at the hydroxyl group of castor oil and the ricinoleates. A number of side reactions occur like the partial hydrolysis of the glycerides, partial hydrolysis of the sulphated oil to hydroxy acids; lactone and lactide and estolide formation; and the formation of highly oxidised or polymerised products of unknown composition (2).

The oil or fat after sulphation is usually allowed to stand, washed with aqueous sodium chloride or sodium sulphate to remove the spent acid, leaving the oil as an upper layer. After neutralisation the sulphated oil is a yellow to dark brown viscous liquid.

Sulphation of ricinoleic acid of castor oil takes place readily at the hydroxyl group rather than at double bond and form turkey red oil.

$$CH_3(CH_2)_xCH = CH(CH_2)_yCOOH + H_2SO_4 \longrightarrow$$

$$CH_3(CH_2)_x\ CH(OSO_3H)\ CH_2(CH_2)_7 - COOH$$

$$(x + y = 14)$$

$$CH_3 - (CH_2)_5\ CH(OH)-CH_2 - CH = CH\ (CH_2)_7 - COOH + H_2SO_4 \longrightarrow$$

$$CH_3\ (CH_2)_5\ CH(SO_3H)-CH_2-CH = CH - (CH_2)_7\ COOH + H_2O$$

In this reaction 98% sulphuric acid is used. Completely saturated fatty oils cannot be sulphated.

The principal uses for the sulphated oils are in the leather and textile industry. Sulphated cod liver oil, neat's foot oil, and tallow are used in the fat-liquoring of leather. Sulphated oils have a number of miscellaneous uses as wetting, dispersing, and emulsifying agents. They are not ordinarily classed as detergents. Their structure, $RCHOSO_3NaR'$, is that of a wetting agent rather than a detergent.

The fatty acids, esters, and amides can be sulphated instead of the glycerides. Sulphated oleic and ricinoleic acids have much greater wetting and foaming power than the sulphated glycerides. Sulphated lower alkyl oleates and ricinoleates are rewetting and softening agents (3). Sulphated amides are stable to use at high temperature, in the presence of acid or alkali, and resistant to hard water.

Sulphated Fatty Alcohols (Alkyl Sulphates) (1)

The sulphates of the long chain primary alcohols, RCH_2OSO_3Na, are important fat-based, surface-active agents.

The fatty alcohols that are generally sulphated are made from palm oil, palm kernel, coconut oil, tallow, or rice bran oil etc.

The saturated alcohols can be sulphated best with sulphur trioxide in commercial practice as it is and is not required in excess. Other reagents used are oleum, concentrated sulphuric acid, chlorosulphonic acid and sulphonic acid.

The individual sodium alkyl and alkenyl sulphates compare as shown in Table 19.1. The surface tension value and interfacial tension values depend on the nature of the sulphate. For sodium

***Table 19.1*: Surface Active Properties of Sodium Salts of Sulphated Alcohols**

	Surface tension 0.1%, 25°, dynes/cm.	*Interfacial tension 0.1%, 25°, dynes/cm.*	*Foam height 0.1% solns., 100 p.p.m 60° mm*	*Sinking time, 0.1% solns., 25° seconds*	*Detergency, 0.1% solns., 100 p.p.m. 60° R*	*Solubility % 25°*
Na dodecyl sulphate	49.0	20.3	240	13	28.6	28.8
Na tetradecyl sulphate	25.3	9.1	246	12	34.2	0.237
Na hexadecyl sulphate	35.0	7.5	178	59	35.9	0.05
Na octadecyl sulphate	40.6	14.2	151	280	38.6	0.02
Na oleyl sulphate	35.0	7.4	226	19	37.1	easily soluble
Na elaidyl-sulphate	36.1	6.3	202	20	37.6	easily soluble

dodecyl sulphate at concentrations near the critical micelle concentration (about 0.25%) the surface tension is 37.4. Sodium dodecyl sulphate is not as good a detergent at 60°. Sodium oleyl sulphate and sodium elaidyl sulphate are good detergents readily soluble at room temperature.

The sulphated fatty alcohols act as detergents, foaming,, wetting, dispersing, and emulsifying agents. They can be used in light and heavy duty laundering, dish-washing, shampoos, brushless shaving creams, tooth-pastes, and metal cleaning and are assistants in the textile and leather industries. Other uses are in the preparation of cosmetic creams, in emulsion polymerisation, in disinfectants, fungicides, germicides herbicides, and insecticides, in flotation compositions, and in the treatment of gastric and duodenal ulcers.

Fat-Based Sulphates with Intermediate Linkages (1)

Fat-based sulphates with an intermediate linkage can be formed from fatty acids, alcohols, and amines. Products from fatty acids are the sulphated mono-glycerides , $RCO_2CH_2CHOHCH_2$ OSO3Na, and sulphated ethanolamides, $RCONHC_2H_4OSO_3Na$, from the reaction of a fatty acid with ethanolamine and subsequent sulphation.Sulphated nonionics, are made from the fatty alcohols. An example of a product from a fatty amine is the sesquisulphate of N- tetradecyl-D-gluconamide

$$[C_{14}H_{29}NHCO\ (CHOH)_4CH_2OH$$

sulphated in an average of 1.5 sulphate groups in the molecule (4)].

The sulphated monoglycerides can be. made by sulphation of the crude monoglyceride or by a simultaneous esterif ication of glycerol with fatty acid and sulphuric acid in a continuous process (5). Sulphated monoglyccridcs are sufficiently stable to hydrolysis to be used in household cleaning and washing and are marketed as a light duty household detergent, a synthetic detergent bar, and as a textile assistant.

The sulphated ethanolamide of coconut oil is made by heating equimolar quantities of the fatty acid and ethanolamine at about 170-180° with *the* removal of water, and then sulphating with sulphuric acid at 30° (6).

Alkylether sulphates: Long chain alcohols are ethoxylated and then sulphated with similar reagents and under similar conditions as the production of alcohol sulphates and produce alkyl ether sulphates. Their properties are also similar to the alkyl sulphates.

$$R\ (OC_2H_4)_n\ OH \longrightarrow R(OC_2H)_n\ OSO_3\ Na.$$

A number of batch and continuous processes for the sulphation of oxyethylated alcohols with sulphur trioxide and chlorosulphonic acid have been reported. Increase in the number of oxyethyl groups ($n = 10$) decrease detergency but improves emulsifying properties and metal-ion stability.

Alkyl ether sulphates have been used widely for different purposes, particularly in shampoos and light duty liquid detergents.

Fat-Based Sulphonates (1)

Surface-active agents containing a sulphonate group, -SO3Na, have the advantage over those containing a sulphate group, -OSO_3Na, in greater stability to hydrolysis, particularly in acid solutions.

Methods for the direct attachment of a sulphonic acid group to a hydrophobic chain include the following : sulphona-tion at the double bond; α-sulphonation of saturated fatty acids; double decomposition of sulphates or halides with inorganic sulphites; and the oxidation of polysulphides or mercaptans.

Saturated fatty acids can be sulphonatsd exclusively in the α-position by the addition of liquid sulphur trioxide to a dispersion of the fatty acid in carbon tetrachloride or tetrachloroethylene. The α-sulphonated acids can be isolated and converted to salts, or to carboxylic esters by heating with an alcohol in the absence of an esterification catalyst. Some properties of α-sulphonated compounds are summarised in Table 19.2 (7).

Esters of α-sulphonated acids are readily soluble and unexpectedly resistant to hydrolysis in hot acid solutions.

The ease with which saturated acids can be monosulphonated exclusively in the α-position by an inexpensive process suggests that α-sulphonated acids may be versatile intermediate in the utilisation of fats.

Table 19.2: **Surface Active Properties of Salts of α–Sulphonated Acids and Esters**

	Solubility 40°g/100 ml	*Surface tension 0.1%, 25° dynes/cm.*	*Sinking time 0.1%, 30° dynes/cm.*	*Foam height, 0.25% 60°, mn*	*Detergency, 0.25% 60° 100 p.p.m. 60° %*
Na α-sulpholauric acid	0.74	32.7	> 1,000	< 10	23.9
Na α–sulphomyristic acid	0.17	30.2	98	160	56.6
Na α-sulphopalmitic acid	0.096	...	176	...	55.6
Na α-sulphostearic acid	0.05	....	...	...	46.4
Na α-sulpholaurate	7.4	69.3	> 1,000	< 10	19.2
Na_2 α-sulphomyristate	1.5	62.8	349	< 10	26.1
Na_2 α-sulphophalmitate	0.45	44.3	45	169	49.5
Na_2 α-sulphostearate	0.14	....	37	...	57.7
NH_4-sulphopalmitic acid	0.5	40.4	100	185	56.1
Triethanolammoniuma α-sulphopalmitic acid	130.0	42.4	77	215	55.4
Na Isopropyl α-sulphostearate	15.00	37.6	43	160	...

Fat-based α-olefin Sulphonates and α-sulpho Methyl Esters of Fatty Acid Production

The process of manufacturing Alpha-olein sulphonate from natural oils and fats consists of the process steps like (i) fat splitting to produce fatty acids from appropriate fats or oils by high temperature and high pressure; (ii) hydrogena-tion of the fatty acids into fatty alcohols by high pressure hydrogenation using copper-chromite catalysts; (iii) dehydration of the fatty alcohol (saturated) into alpha olefin at high temperature 300°C-400°C in the presence of alumina catalyst; (iv) sulphonation cf the alpha-olefin by sulphur trioxide; and finally (v) neutralisation of the alpha-olefin sulphonic acid with sodium hydroxide at high temperature.

Alpha-sulpho Methyl Ester of Fatty Acid production involves the following process steps : (i) Conversion of fat/ oil into Methyl Esters by Methanolysis reaction using caustic soda as a catalyst at atmospheric pressure at about 80°C or at high pressure and high temperature, followed by isolation of the Methyl Ester in high quality; (ii) Sulphonation of the Methyl Ester with two molecules of S03 to form an adduct in a standard sulphonator, followed by rearrangement of the adduct to form Alpha Sulpho Methyl Ester (ASME) by digestion at high temperature.

Fat-Based Sulphonates with an Intermediate Linkage (1)

Surfactants with an intermediate ester or amide linkage between the hydrophobic chain and the sulphonate group are prepared by the reaction of the acid chloride of oleic acid with sodium isethionats or the sodium salt of N-methyltaurine to give

$$CH_3(CH_2)_7CH = CH(CH_2)_7CO_2C_2H_4SO_3Na \text{ (Igepon A) or}$$

$$CH_3(CH_2)_7CH = CH(CH_2)_7CON(CH_3)C_2H_4SO_3Na$$

(Igepon T), respectively

Sodium isethionate, $HOC_2H_4SO_3Na$, is made by the reaction of ethylene oxide with aqueous sodium bisulphite at 70-80° under sufficient pressure to retain the ethylene oxide. The acid chloride is made by means of phosphorus trichloride. A mixture of the acid chloride of oleic acid, dry sodium isethionate, and sodium carbonate is agitated and heated in the molar proportions of 1:1.2:0.1, respectively, to 115°, with the evolution of hydrogen chloride. The reaction mixture is neutralised and ground to a finished product containing 70% active ingredient.

The sodium salt of N-methyltaurine.

$$CH_3NHC_2H_4SO_3Na,$$

is prepared *as* an aqueous solution by the reaction of 42% aqueous $HOC_2H_4SO_3Na$ with excess CH_3NH_2 at 280° under a pressure of 200 atmospheres.

The acyi taurates, are more resistant to hydrolysis than the sulphoethyl esters, particularly in alkaline solutions. They are well established surfactants, principally developed as textile assistants, but useful as detergents, lime soap dispersants, in shampoos, the fat-liquoring of leather, metal plating, and in a number of miscellaneous applications. Variations in the acyl and the alkyl group attached to nitrogen are possible as well as variations in the structure $-CH_2CH_2SO_3Na$. An acyl taurate from the acid chlorides of tallow acids is a recent development.

Sulphoacetates of the general formula

$$RCH_2O_2CCH_2SO_3Na$$

are made by the esterification of oleyl, cetyl, stearyl, or coconut oil alcohols with chloroacetic acid, and replacement of the chlorine atom by reaction with an alkali metal sulphite. Sulphoacetates are useful in toothpastes, shampoos, and cosmetics.

Sulphonated polycarboxylic acid surfactants: These surfactants are sulphonated esters or amides of polycarboxylic acids, the best known are sulphosuccinates and sulphosuccinamates.

The sulphosuccinate diesters are prepared by the reaction of an alcohol and maleic acid or anhydride followed by sulphonation with sodium bisulphite.

$$\begin{matrix} CHCOOH \\ \| \\ CHCOOH \end{matrix} + 2ROH \xrightarrow{H^+} \begin{matrix} CHCOOR \\ \| \\ CHCOOR \end{matrix} \xrightarrow{NaHSO_3} \begin{matrix} SO_3Na \\ | \\ CHCOOR \\ | \\ CH_2COOR \end{matrix}$$

Sulphosuccinamates are prepared by the following steps:

$$\begin{array}{l} HC\text{-}C(=O) \\ \| \quad \;\; O \\ HC\text{-}C(=O) \end{array} + RR'NH \longrightarrow \begin{array}{c} HC\text{-}C - CONNR' \\ \| \\ HC\text{-}COOH \end{array} \xrightarrow{NaSHO_3} \begin{array}{c} SO_3Na \\ | \\ HC\text{-}CONRR' \\ | \\ H_2C\text{-}COONa \end{array}$$

$$\begin{array}{c} HC\text{-}C - CONNR' \\ \| \\ HC\text{-}COOH \end{array} \xrightarrow{R''OH} \begin{array}{c} HC\text{-}CONRR' \\ \| \\ HC\text{-}COOR^{-} \end{array} \xrightarrow{NaSHO_3} \begin{array}{c} SO_3Na \\ | \\ HCCONRR' \\ | \\ H_2CCOOR'' \end{array}$$

These surfactants have many applications in wetting, dispersing and foaming purposes, for example in ink dispersions lubricating oils, hydraulic fluids and antifogging agents for glass.

N-acyated amino acids: These anionic surfactants contain amide group and are prepared by condensing fatty acids with

hydrolysed proteins.

$$RCONH \longrightarrow \text{(Peptide) } COONa$$

Lauroyl, 'cocoyl' and oleoyl derivatives of different protein hydrolysates have been prepared. They are mild and have good foaming properties and are used in cosmetics and shampoos.

Different compounds of N-acylated amino acids serve different purposes. N-lauroyl derivatives of sarcosine (CH_3NHCH_2COOH is used as an important 'antienzyme' agent in toothpaste formulations and in mouthwashes because it can reduce the cavity formation (10). Besides these, acylated amino acids and salts have use in ore floatation and as additives or corrosion inhibit ors in petroleum products or processing.

Phosphorous-containing surfactants (11): The most important phosphorous containing surfactants are the natural phospholipids which occur in a number of vegetable oils.

Lecithin which is actually an amphoteric surfactant and obtained naturally from soyabean oil is a good oil soluble emulsifying and dispersing agents. They are mainly used in food products.

Other uses are as antioxidants, antispattering agents in margarine and viscosity reducing agents in chocolate.

Reaction of lecithin with aikanolamines and alkanolamides gives surfactants whose structures are not well defined.

Synthetic phosphate and phosphonate detergents have high cost and hence have speciality use. Commercially important phosphorous based surfactants are phosphoric and polyphosphoric acid derivatives. These are prepared by the reaction of fatty alcohols and phosphorous oxychloride followed by hydrolysis and neutralisation which yield alky! phosphates.

Cationic Fat-Based Surfactants (1)

Cationic surface-active agents also called "invert soaps" are mostly fat-based as fats are a natural and convenient sourc for the hydrophobic chain. The cationic surfactants are almost exclusively substituted ammonium salts based on amines or amide derived from long chain fatty acids, or on long chain alkyl halides derived from fatty alcohols. The long chain amines, RNH2, are almost insoluble in water but soluble in the form of salts or quaternary ammonium compounds.

The important uses of the surfactants ars as germicides, fungicides, and disinfectants, water-repellents, softening agents, and the fixation of sulphonic and dyestuffs by union with long chain cation.

The bactericidal activity of cationic surfactants has been related to their affinity for wool and to the critical micelle concentration of these colloidal electrolytes. Fat based cationic surfactants include structures such as

$$[RN(CH_3)_3]^+\ Cl^-,\ [RN\langle\equiv\rangle]^+\ Cl^-$$

and

$$[RN\ (CH_3)_2CH_2C_6H_5]+Cl^-$$

Trimethylalkylammonium chlorides and related coupounds can be based on tallow acids, converted to nitrlles, hydrogenated and methylated. Alkylpyridinium halides and related compounds can be based on fatty acids, converted to alcohols, and then to alkyl halides. Maximum bactericidal activity occurs at a long chain alkyl group of 16 carbon atoms.

Cationic surfactants are used for many disinfecting purposes; sanitisation of fabrics in laundries; sterilisation of dishes and glassware in restaurants; and sterilisation of instruments and skin in surgery.

Cationic softening agents are more effective than anionic softening agents, such as sulfated tallow. They are strongly adsorbed on cotton or rayon at low concentrations and impart a soft, rich hand to the fabric, an effect which resists wash in and is semipermanent. The softening effect is essentially a lubricating action between the individual fibers.

Cationic Surfactants with Nitrogen Joined Directly to the Hydrophobic Chain (1)

The largest and most important group of cationic surfactarnt centres on the use of fat-based amines of 8 to 18 carbon atoms. The amines and their salts are widely used for their surface activity, particularly as flotation agents, but principally they are intermediates for the preparation of cationic surfactants .

Amines can be ethoxylated to make the products having the following structure:

$$RN\begin{cases}(C_2H_4O)_xH \\ (C_2H_4O)_yH\end{cases}$$

The products are more soluble and less basic the larger the values of x and y.

Cationic Surfactants with an Intermediate Linkage (1)

Cationic surfactants with intermediate amide, ester, and other linkages have been prepared.

Cationic surfactants with amide and ester intermediate linkages can be made from $ClCH_2CO_2C_2H_5$, $NH_2C_2H_4OH$, a fatty acid chloride, and $(CH^3)_3N$ as follows:

$$ClCH_2CO_2C_2H_5 + NH_2C_2H_4OH \longrightarrow ClCH_2CONHC_2H_4OH$$

$$RCOCl \longrightarrow ClCH_2CONHC_2H_4O_2CR$$

$$(CH_3)_3N \longrightarrow (RCO_2C_2H_4NHCOCH_2N(CH_3)_3]+Cl^-$$

The simple betaine esters,

$$[RO_2CCH_2N(CH_3)_3]^+Cl^-$$

are made from a fatty alcohol, $ClCH_2CO_2H$, and $(CH_3)_3N$. Cationic surfactants such as

$$[ROCH_2N(C_2H_5)_3]^+Cl^-$$

are made from the chloromethyl ether of a fatty alcohol and $(C_2H_5)_3N$.

Nonionic Fat-Based Surfactants (1)

Non ionics are the most versatile in application. It is conceivable that the production of nonionics will continue to grow at *a* rate greater than all surface-active agents. The raw materials for manufacturing nonionics are ethylene oxide, ethylene oxide derivatives such as polyethylene glycol and diethanolamine, glycerol, and sorbitol, for the hydrophilic part. The principal hydrophobe sources are the fat-based long chain alcohols, acids, amines, and amides.

Nonionic surfactants have generally advantage of stability under acid or alkali or hard water or metal ions; and compatibility with anionic or cationic surfactants. They are also versatile in application through control of the balance between hydrophobic and hydrophilic characteristics.

The nonionic surfactants are mostly liquids and they have poor foaming qualities.

Fat-based nonionic surface-active agents are detergents, emulsifying agents, foam stabilizers and thickening agents, dispersing agents and solubilisers; and they are useful in dry cleaning.

Fatty Alcohol Ethoxylation (1)

As it is well known, fatty alcohol ethoxylates are produced by the reaction between an alcohol and ethylene oxide in the presence of catalyst :

$$\underset{\text{Fatty Alcohol}}{R-OH} + x\,\underset{\text{Ethylene Oxide (EO)}}{CH_2\overset{O}{\frown}CH_2} \xrightarrow{\text{Catalyst}} R-O\,(CH_2CH_2O)_xH$$

The catalyst types are strong bases such as NaOH, KOH, Lewis acids such as SbCl5, SnCl4, and other categoires such as Peaking catalysts and the NOVELII catalysts. The NOVEL II catalyst is a Ca and Al alkoxide-type catalyst, partially neutralised and dispensed in an Alfomic A10-12-2.5 EO alcoholethyoxylate. The Ca/Al ratio in the catalyst is 5:1. The alcohol ethoxylate itself makes up about 90% of the catalyst.

Starting with oleyl alcohol and adding increasing amounts of ethylene oxide, a useful degree of water solubility is obtained when n is about 6 to 8. These products are suitable for wool cotton, and rayon. When n is about 20 to 30, excellent emulsifying agents are obtained.

The manufacture of a typical product using sodium hydroxide as a catalyst is described as follows[1]: Powdered sodium hydroxide, 80g, is added to 800 kg of oleyl alcohol in a glass-lined kettle, the mixture is heated to 160°-180° and the kettle is evacuated to 10 mm pressure. Ethylene oxide from a storage tank holding 2,000 kg is allowed to flow in under a nitrogen pressure of about 3 atmospheres. The flow is regulated by this prossure. The reaction is highly exothermic and requires about 12 to 14 hrs. for completion. The product, a soft waxy paste, contains about 15 moles of ethylene oxide per mole of oleyl alcohol. It gives a clear solution of cold water which becomes turbid on heating in about 40°, and has been marketed as a 20% or 30% aqueous solution.

The longer the hydrophobic chain, the more molecules of ethylene oxide are required for adequate solubility. In this respect the coconut oil alcohols, and oleyl alcohol, might be preferred to cetyl or stearyl alcohol. It has been suggested in the case of the saturated alcohols that the number of ethylene oxide molecules required is 3 less than the number of carbon atoms in the fatty alcohol (14).

The phenomena of inverse solubility (greater solubility in water at lower than at high temperature) is characteristic of the nonionics. Decreasing solubility with rise in temperature is due to decreased association of water and solute molecules. The "cloud point", or the temperature at which turbidity first appears, can be correlated with the number of molecules of ethylene oxide, n, condensed with one molecule of the fatty alcohol. The value of n is an average value, and the actual product is never a chemical individual.

Fatty alcohol-ethylene oxide condensates are effective in washing wool and cotton but are poor foamers in comparison with ionic detergents. They have the advantage of being nonsubstantive; the aqueous solutions are not depleted or exhausted by adsorption. Sulphation converts them to anionic detergents but preserves their excellent wool washing characteristics, improves cotton detergency and foaming and economises on the number of ethylene oxide molecules required. A sulphated product from oleyl alcohol and about 5 moles of ethylene oxide is similar in performance to an unsulphated product from oleyl alcohol and 15 moles of ethylene oxide.

Fatty Acid Ethoxylation: Fatty acid-ethylene oxide condensation products, $RCO_2(C_2H_4O)_xC_2H_4OH$, have the same general characteristics as those from fatty alcohols except that they are hydrolysed in hot alkaline solutions. The fatty acids may be esterified with polyethylene glycols or condensed with ethylene oxide.

Fatty acids such as lauric, palmitic, stearic, and oleic acid can be esterified with anhydrosorbitols to give emulsifying agents, marketed under the trade name Span.

Fatty Esters (Methyl Esters and Triglycerides) Ethoxylation

Ester alkoxylation technology is an important development. The ethoxylated ester products are emerging nonionic surfactant: Catalysts like the Peaking ethoxylation catalysts have been successful in efficient conversion of methyl esters of fatty acids into the ethoxylates. The length of the hydrophobic chain as well as the EO content effects detergency. The coco-range MEE (Methyl ester ethoxylates) with an adequate EO level provide good detergency. The tallow-range methyl esters with high EO content show also very high detergency. Shorter chain MEE at lower EO levels provide superior hard surface cleaning performance on graphite, wax and grease soil.

Lion Corporation (Tokyo, Japan) patented a process (14) involving the use of magnesium oxide based catalyst to achieve ethoxylation of methyl ester effectively. The ethoxylates having high conversion of the starting methyl ester, peaked EO distributes, show good surface chemical properties.

KaO Corporation (Tokyo, Japan) (13) disclosed the use of magnesium-aluminium oxide type catalyst for the production peaked MEE.

In 1998, LG Chemical (Tasjon, S. Korea) disclosed the production of methyl esters using a solid metal oxide (13). The EO distributions appear to be peaked. In 1999 BASF Corporation (Germany) disclosed the use of hydrophobiesed hydrotalcoles as catalysts for the production of MEE.

Fatty triglyceride ethoxylate has been produced with the aid of the NOVEL II catalyst. Partial saponification of the triglyceride ethoxylate leads to a blend of surfactants composed of monoglyceride ethoxylate, diglyceride ethoxylate, unreacted triglyceride ethoxylate, diglyceride ethoxylate, unreacted triglyceride ethoxylate, glycerol ethoxylate and fatty acid soap. The entire product blend is very mild to skin.

The ester ethoxylation technology may offer a commercially viable rout toward the production of low-cost surfactants.

Non-ionics from Fatty Acid and Diethanolamine

An important surface-active agent, generally considered in the nonionic class, is the reaction product from one mole of a fatty acid and two moles of diethanolamine, $NH(C_2H_4CH)_2$. From the method of formation the product would be presumed to be $RCON(C_2H_4OH)_2$. The process is extremely simple (14). A mixture of one mole of coconut oil fatty acid and two moles of $NH(C_2H_4OH)_2$ is heated at 150-170° until nearly one mole of water has been recovered and the free fatty acid value has fallen to about 5%. The resulting viscous amber liquid, readily soluble in water with a pH of about 5%, has been considered to be a mixed micelle of alkanolamine, amine, and fatty acid.

The alkanolamides are widely used for their synergism with other surface-active agents. They are foam stabilizers and thickners for detergents, clarifying and emulsifying agents, and they improve the detergency of soap in hard water.

POLYOL SURFACTANTS

Non-ionic surfactants such as the fatty acid esters of polyhydroxy compounds like polyethylene glycols of varying molecular weights, glycerol, sorbitol, mannitol and sugar etc. constitute some examples of polyol surfactants. The polyhydroxy portion, as such or as ethyloxylated serves as hydrophilic moiety while the fatty acid chains act as hydro-phobic moiety.

Glycerol esterifics with fatty acids at high temperature and under vacuum with or without a catalyst to give a mixture of mono, di- and triglycerides. The products mostly called monoglycerides show distinct surface activity.

The fatty acid esters, particularly the monoesters of sorbitol and mannitol, like most of polyhydric alcohol esters, are recommended for emulsification. They have been found to be effective emulsifying agents for cooking fats. The esters of sorbitol and the hexitols, in general, have been claimed and described as dispensing agents, as wetting and penetrating agents, and as detergents. Sorbitan monostearate and its polyoxyalkylene derivative has been used for emulsifiers in ice cream mixes. Commercial sorbitan monolaurate is a recommended wetting agent for cocoa powder and as an antifoaming agent in certain types of aqueous solutions. It is also used alone or admixed with the corresponding polyoxyalkylene derivative as an emulsifier in formulating water emulsion of kerosene, fire oil, linseed oil, resins, softening oils for fat liquoring, crude coal tar, water-thinned paints etc.

The monopalmitate is also used in similar emulsions. The monostearate and oleate, alone or admixed, find use in head creams, vanishing creams, head lotions, liquid cleaners etc.

Among the six hexitols, sorbitol and mannitol are cheap attractive industrial raw materials. Fatty acid esters of hexitols are prepared by heating the reactants under vacuum for 3 to 10 hours without a catalyst. They have also been prepared by transesterification (alcoholysis reaction) of triglycerides and polyols (e.g. sorbitol) at 240°-250°C with an alkali catalyst.

The hexitols undergo dehydration at high temperature and form hexitans and hexitols. These derivatives when subjected to esterification with long chain fatty acids form accordingly mono, di- or triesters predominantly monoesters.

Reaction of carbohydrates with long-chain plant derived lipids yields carbohydrate - alkyl ester surfactants with a polyol structure. Their renewability, low cost, and environmental acceptability make carbohydrate-alkyl ester surfactants an excellent alternative to pertochemically derived product.

BIOSURFACTANTS

Microbial surface-active molecules, also called bio-surfactants, are unique in their character for reducing interfacial surface tension and also for stabilizing emulsions. These surface-active molecules include a wide range of natural surfactants and emulsifiers, such as glycerophospho-lipids, glycolipids, glycosphingolipids, bile acids, saponins proteins, polysaccharides etc.(15). A number of economically viable biosurfactants are obtained from microorganisms, such as Nocardia corynebacterides SMI(16), Rhodococcus erythropolis DSM 43215(17), Pseudomonas rubescons, Gluconabacter cerinus, Agrobacterium tumefacinas, Bacillus subtilis, Bacillus lecheniforrnis, Candida petrophilum(17), Serratia marcescens (18), Rhodococcus sp.(19), Pseudomonas aeruginosa (20) Rodococcus erythropolis(21). A species of Arthrobacter (22) isolated from soil and designated H13 A, produced a variety of glycolipid surfactants when grown on a hydrocarbon substrate and Corynebacteriun lepus (23) produced phospholipid surfactants.

A very recent study investigatad the production of surfactants from the fungal Mortierella strains isolated from soil of a vegetable oil industry of Calcutta, India(24). The surface-active property of the extracted fungal lipids and the amount of glycolipids and phospholipids in the lipids were determined.

Most biosurfactants show low CMC values, reduce the surface tension (around 30 mN/m) and the interfacial tension against n-alkenes to values below 1 mN/m. These values compare favourably with those obtained from commercial synthetic surfactants.

Biosurfactants have remarkable applications as food additives, in cosmetic and pharmaceutical applications and in petroleum recovery. However, they are used in limited quantity because of their higher production costs.

Biosurfactants are easily biodegradable and thus are suited for environmental applications such as bioremediation and dispension of oil spills. In this aspect, biosurfactants have edge over the synthetic surfactants.

World Scenario of Fat-based Surfactants

Total world surfactant consumption is around 13 million tons which includes soap 50%, LAS (linear alkyl benzene sulphonate) (15%), fat-based Ethoxylates, sulphates and Ethoxy-sulphates (12%), APEOs (alkylphenolethoxylates, 4%), and others such as cationics, amphoterics, SAS (Alkane Surfactants), AOS (Alpha Olefin Sulphonates), APG (Alkyl polyglycosides). The world surfactant consumption excluding soap and personal care products, by region indicates that the largest consumer is the Asia-Pacific region with 32% of the total world consumption followed by North America 28% and West Europe, 23% and Latin America and other regions, the rest.

Fat-based surfactants are being increasingly produced as the production of vegetable oils (palm oil, palm kernel oil etc.) have increased quite significantly. In fact, oleo-chemical and petrochemical surfactants share a different portion of the total market depending on the geographical location. In Europe petrochemicals account for over 80% of the market. In the South East Asian countries, there is a more predominant one of oleochemicals (55-65%) because of local availability of basic raw materials.

The surfactant industry represents a major market in which petrochemicals and oleochemicals compete.

The oleochemicals are expected to have increasing use as more vegetable oil based chemical and surfactant plants come on stream. In 1989 40%. of world detergent alcohol capacity was oleochemical based and it is now over 50%, the vast majority of capacity being added in South East Asia.

In fact, the natural fats and fatty acids are suitable materials in producing the fat-based anionic, cationic, and non-ionic surfactants. The molecular nature of fats and fatty acids having the characteristic hydrophilic chain and the hydrophobic carboxyl group that can be utilised to impart further hydrophilic character offer means of making a great variety of surface active agents.

REFERENCES

1. A.J. Stirton, J. Am. Oil Chemists' Soc., **31**, 579-586 (1954).
2. D. Burton and G.F. Robertshaw, Sulfated Oils and Allied Products, Chemical Publishing, New York (1942).
3. Miller, L.M., Am. Dyestuff Reptr., **42**, 435-40 (1953).
4. Mehltretter, C.L., Furry, M.S., Wellies, R.L. and Rankin, J.C., J. Am. Oil Chemists' Soc., **29**, 202-7 (1952).
5. Mancie, F.W., U.S. 2,242,979 (1941).
6. Orelup, J.W., U.S. 1,981,792 (1934).
7. Stirton, A.J., Well, J.K., and Bistline, R.G. Jr., J. Am. Oil Chemists' Soc., **31**, 13-16 (1954).
8. Well, J.K., Bistline, R.G. Jr., and Stirton, A.J., J. Am. Chem. Soc., 75, 4859-60 (1953).
9. Weil, J.K., Witnauer, L.P., and Stirton, A.J., J. Am. Chem. *Soc., 75*, 2526-7 (1953).
10. L.S. Fosdick, J.C. Calandra, R.O. Blackwell and J.H. Burrill, J. Dent. Res., **32**, 486 (1953).
11. D.R. Howton, Science, **119**. 420-421 (1954).
12. U. Weerasooriya, Journal of Surfactants and Detergents, *2*, 373 (1999).
13. Yuji, F., H. Itsuc, and N. Yuichi, U.S. Patent 5,374,750 to Lion Corp. (1994).
14. Tanaka, T., T. Imanaka, T. Kawaguchi, and H. Nagumo, European Patent 0,783,012 Al to Kao Corporation (1997).
15. S. Kudo: Biosurfactants as Food Additives; in: Proceedings of the World Conference on Biotechnology for the Fats and Oils Industry, Ed. Thomas H. Applewhite, American Oil Chemists' Society 1988, pp. 195-201.
16. M. Powalla, S. Long, and V. Wray, Penta- and Disaccharide Lipid Formation by Nocardia corynebacteroides Grown on n-alkane, Appl. Microbiol, Biotechnol. 31 (1989), 473-479.
17. J.S. Kim, M. Powalla, S. Long, F. Wagner, H. Leunderfahd and V. Wray, Microbial Glycolipid Production Under Nitrogen Limitation and Resting Cell Conditions, J. Biotechnol. 13 (1990), 257-266.

18. V. Pruthi and S.S. Camcotra, Production of a Biosurfactant Exhibiting Excellent Emulsifying and Surface Active Properties by Serratia Marscescens, World-J. Microbiol. Biotechnol. 13 (1997), 133-135.

19. M.J. Espuny, S. Egido, I. Rodon, A. Manresa, and M.E. Mercede, Nutritional Requirements of a Bio-surfactant Producing Rhodococcus sp. 51T7, Biotechnol. Lett. 18 (1966), 521-526.

20. U.S. Ochsner, A.K. Koch, A. Feichter, and J. Raiser, Isolation and Characterization of a Regulating Gene Affecting Rhamnolipoid Biosurfactant Synthesis in Pseudomonas aeruginosa, J. Bacteriol. 176 (1994) 2044-2054.

21. F. Wagner: Strategies for Biosurfactant Production, World Conf. Biotechnol. Fats & Oils. Ind. (1988), 189-194.

22. M.J. Brown and V. Moses: Biosurfactants for Petroleum Recovery, in: Proceedings of the World Conference on Biotechnology for the Fats and Oils Industry, Ed. Thomas H. Applewhite, American Oil Chemists' Society 1988, pp. 202-205.

23. S. Bandyopadhyay, S. Chaudhuri, and D.K. Bhattacharyya, Fett/ Lipid **101**, 439-442 (1999).

CHAPTER 20

Alkaline Earth and Heavy Metal Soaps

The constitution of metallic soaps is important along with their manufacture, application and characteristics. The compounds commonly known as 'soaps' in this area include both the 'metallic soaps' and the 'organic nitrogen base soaps'. The latter compounds are formed by the reaction of organic nitrogen base and complex organic acids; triethanolamine oleate is a good example of this type of soaps. Metallic soaps are the alkali, alkaline earth or heavy metal salts of monobasic, complex organic acids. Among metallic soaps there exist two types, water soluble and water insoluble soaps. The water insoluble soaps like alkaline earth metal soaps are highly soluble in non-polar solvents.

The two types of soaps have dissimilar characteristics and uses except in lubricating greases where both types are used. In this chapter mainly the alkaline earth and heavy metal soaps are discussed. These compounds are formed by replacing the cation, the acid hydrogen or its equivalent in a complex monobasic organic acid by a metal. Mono-, di-, tri- or tetravalent metals may also be involved in this reaction. The reaction is shown below :

$$R{-}C\begin{matrix}\diagup OH \\ \diagdown O\end{matrix} + MOH \longrightarrow R{-}C\begin{matrix}\diagup OM \\ \diagdown O\end{matrix} + H_2O.$$

where R is the nucleus of the organic acid and MOH is the hydroxide of a monovalent metal.

Metallic soaps are important in this respect that they are virtually the only compounds of reasonable price which contain considerable metal and are also soluble in organic solvents. In these soaps structural peculiarity results because these are the combination of a non-polar organic radical and a strongly polar carboxyl group in which hydrogen is replaced by a metal. The weak metal to oxygen bonds provides active catalytic property to some soaps.

The raw materials for metallic soaps are classified as saturated fatty acids, unsaturated fatty acids, hydroxy acids, rosin acids, or naphthenic acids. Theoretically there are no limitations as to which of the alkaline earths and heavy metals may form metallic soaps, but the soaps of copper, silver, calcium, strontium, barium, beryllium, magnesium, zinc, cadmium, mercury, cerium, aluminium, thallium, tin, lead, vanadium, antimony, bismuth, chromium, uranium, manganese, iron, cobalt, and nickel etc. are commonly considered. Fats and oils containing fatty acids used as the raw materials for soap manufacture must have desirable properties, reasonable prices and must be obtained regularly. Various fats and oils including castor oil, coconut oil, corn oil, cottonseed oil, linseed oil, soyabean oil, tung oil, lard, tallow and mixed fatty acids are used which serve as the organic part of the constituents. Inorganic part consists of the metal salts like copper sulphate, caustic soda, calcium hydroxide, barium chloride, zinc oxide, aluminium sulphate, lead acetate, litharge, lead nitrate, manganese dioxide, manganese sulphate, cobalt acetate, cobalt sulphate etc.

Solvents are very important in metallic soap manufacture as soaps are desired to be in a dispersed form. Petroleum solvents like high aromatic naphtha, mineral spirits, synthetic petroleum derivatives including cellosolves, carbitols, acetone etc. lubricating oils, turpentine, rosin oils and some vegetable oils such as linseed, soyabean and perilla are used as solvents.

Metallic soaps may be of either the precipitated or the fused type (1). The precipitated type may be formed because of the low solubility of metallic soaps in water. Metathetic reaction of inorganic chemistry is the reaction when solutions of two soluble salts are mixed and the reaction will tend to go to completion if one product is highly insoluble. This is the case in the formation of precipitated soaps. By choosing proper conditions of temperature, concentration and other factors, a solution of the sodium soap or salt may be obtained. When this is reacted with a solution of an appropriate soluble metallic salt, metathesis occurs and a hydrated insoluble metallic soap precipitates while a soluble, inorganic sodium salt is formed. If $\overline{FA}$ represents the organic acid radical, then this reaction could be represented as follows :

$$2Na\overline{FA} + CoSO_4 \longrightarrow Co\overline{FA}_2 + Na_2SO_4$$

Water of hydration is always accompanying these precipitates; it is not shown here for simplicity. Like sodium soaps, metallic soaps may also be formed by adding a suitable solution of a metallic salt to the sodium soap solution under suitable conditions of temperature and concentration. In this process, a metathetic reaction occurs with the formation of a water insoluble hydrated metallic soap and a soluble sodium salt. The reaction is shown below, where $\overline{FA}$ is a fatty acid radical and M is a metal, in this case divalent :

$$MSO_4 + 2Na\overline{FA} \longrightarrow M\overline{FA}_2 + Na_2SO_4.$$

By this reaction mainly two classes of precipitates are obtained having different physical properties. The soaps having low melting points when precipitated are soft, plastic masses which after washing are dehydrated by heating in a kettle. The other class consists of high melting point soaps precipitated as finely divided particles. This granular product may be washed and the washed precipitate is dried on trays or in a continuous drier.

Precipitation process is the most important method of manufacturing metallic soaps. The formation of aluminium stearate is given by the following equations depending on the variation of concentration of caustic soda used.

(a) $C_{17}H_{35}COOH + 3NaOH \longrightarrow C_{17}H_{35}COONa + 2NaOH + H_2O$

$C_{17}H_{35}COONa + 2NaOH + Al^{3+} \longrightarrow Al(OH)_2OOC.C_{17}H_{35}$

$+ 3Na^+[Al(OH)_2OOC.C_{17}H_{35}]$

(b) $2C_{17}H_{35}COOH + 3NaOH \longrightarrow 2C_{17}H_{35}COONa + NaOH + H_2O$

$2C_{17}H_{35}COONa + NaOH + Al^{3+} \longrightarrow Al(OH)(OOC.C_{17}H_{35})_2$

Precipitate

$+ 3Na^+[Al(OH)(OOC.C_{17}H_{35})_2]$

(c) $3C_{17}H_{35}COONa + 3NaOH \longrightarrow 3C_{17}H_{35}COONa + H_2O$

$3C_{17}H_{35}COONa + Al^{3+} \longrightarrow Al(OOC.C_{17}H_{35})_2$

Precipitate

$+ 3Na^+ [Al(OOC.C_{17}H_{35})_3]$

The equations are oversimplified because other intermediate stearates are formed.

The fused type of soap is prepared by reaction which are more or less the same chemically as those used to prepare precipitated soaps. In this procedure, either an oil or a fatty acid is reacted with an oxide, a hydroxide or even an acetate or carbonate by heating the two to a relatively high temperature in the absence of any added water. Under these conditions, saponification or neutralisation takes place with the formation of a metallic soap.

Fusion process has limited application as compared to precipitation process, but it is preferred for the manufacture of metal soaps for use as driers when conditions permit. The preparation of metal soaps of lead oxide or cobalt acetate with oleic acid is illustrated by the following equations :

$2C_{17}H_{35}COOH + PbO \longrightarrow (C_{12}H_{35}COO)_2Pb + H_2O$

$2C_{17}H_{33}COOH + Co(OOC.CH_3)_2 \longrightarrow (C_{17}H_{33}COO)_2Co + 2CH_3COOH$

The reaction materials are heated in the absence of water at a sufficiently high temperature (180-300°C) with vigorous agitation to produce substantially complete reaction. The process may be followed for the preparation of lead, zinc, copper, cadmium and barium octoates and ricinoleates also for the preparation of zinc and copper naphthenates.

Though the production methods of precipitated soaps and fused soaps are somewhat similar, the difference lies in the direct formation of a metal soap instead of its formation by a 2-stage process. Also the reaction of fused soaps occurs in a non-aqueous medium.

Metallic soaps have a vast field of application. They are used as catalysts particularly in organic processes because of their solubility in many organic chemicals. The processes include oxidation, polymerisation, hydrogenation, metathetic and hydrolytic reactions.

Metal soaps used in driers act as positive oxidation catalysts (2). The first products of this type were the lead and manganese soaps of linseed oil fatty acids and of rosin acids, the so called 'linoleates' and 'rosinates'. Other soaps used are naphthenates, octoates, tallates etc. obtained from naphthenic acid, octanoic acid and tall oil fatty acid respectively.

The principal drier metals in use at present can be divided into three groups depending on their primary function, namely (a) polymerisation catalysts—lead, cerium, (b) oxidation catalysts—cobalt, manganese, vanadium, (c) auxiliary driers—iron, zinc, calcium, copper.

Lead is one of the oldest and most universally used metals in the manufacture of driers in the form of soaps of a variety of acids, e.g. resinates, linoleates, tungates, naphthenates, octoates etc. Cerium, especially in the form of naphthenate closely resembles lead drier. Cobalt soaps, in the form of rosinate, naphthenate and linoleate are extremely active oxidation catalysts, but if used in excessive amounts they may cause rapid skinning of oils and wrinkling in air dried films. Manganese soap is also an important drier being intermediate between lead and cobalt in drying properties. Other metal soaps like vanadium, iron, zinc etc. occupy important places in the paint drier field.

Metallic driers used in the manufacture of varnishes are mainly those of lead, cobalt and manganese in the form of naphthenates, tallates ethyl hexanoates (octoates), linoleates and resinates. The principal driers used in printing ink are the oil soluble salts of cobalt, manganese and lead usually in the form of oleates, linoleates, resinates, naphthenates or ethyl hexanoates. Cobalt catalyses surface oxidation, whereas lead and manganese promote polymerisation and general hardening of the ink film.

The largest use of metallic soaps is in the petroleum industry. In this field, metallic soap is used as gelling agents in the manufacture of greases, as lubricant modifiers and oxidation catalysts, in treating oil wells and for other purposes.

The gelling agents most commonly used now are the fatty acid (oleic, palmitic, stearic and other aliphatic acids derived from tallow, hydrogenated tallow, hydrogenated castor oil, soyabean oil, cottonseed oil, hydrogenated fish and sperm oil etc.) soaps of lithium, sodium, calcium and aluminium (di-type). Aluminium soap (e.g. aluminium stearate) has the water proofing property and high melting point. Lead soaps, especially the naphthenates and oleate are used as additives to lubricating oils and give the so called 'extreme pressure' lubricating property to the oil. Cobalt, nickel, zinc and magnesium ethyl hexanoates (octoates) have been proposed for use as sludge deposition inhibitors in mineral oils. Cobalt and manganese soaps are used as catalysts in the oxidation of waxes and other petroleum products.

Metallic soaps are used for different purposes in the rubber and plastic industries. Zinc laurate and stearate and lead oleate are used to soften rubber and improve its processing characteristics. Aluminium, magnesium and zinc stearates act as mold releasing agents in forming rubber objects and are also used to reduce tackiness in uncured rubber. Sodium soaps serve as important emulsifiers in the polymerisation of butadiene or in the co-polymerisation of butadiene and acrylonitrile, styrene, acrylic ester etc. to produce synthetic rubber.

Barium, calcium, lead, magnesium and tin stearates etc. are used as heat stabilisers in polyvinyl chloride type resins. Stearates are said to act as plasticisers and internal lubricants in improving the processing characteristics of some types of plastics.

Due to the insolubility and water-repellency properties of metallic soaps, they act extensively as water-proofing agents of many materials like textiles, paper, fiber board, concrete, magnesite, bricks, mortar etc. For this purpose aluminium stearate is very important. The use of rare-earth soaps as well as zirconium, thorium and uranium soaps have been patented as water repellent agents for textiles. Aluminium oleate and calcium, aluminium or zinc stearates uniformly mixed with cement, magnesite, paper and other fiber pulps impart water-repellency to these products.

Metallic soaps, principally copper and zinc soaps are used for the preparation of fungicides. Copper soaps are more effective than zinc soap. Copper sebacate, octoates, naphehenates etc. are good fungicides. Textiles, especially canvas and tobacco shade cloth are often treated with copper soaps dissolved in mineral spirits to increase their resistance to attack by fungi.

Large quantities of metallic soaps are used in the manufacture of face and body powders. Zinc and magnesium stearates are used as bases for these products. Zinc stearate and oleate and lead oleate have mild antiseptic action and are used in skin creams for treatment of minor burns, wounds, inflammation and irritation of skin etc.

Calcium and magnesium ricinoleates have various pharmaceutical uses. Other salts used are salts of 10-undecenoic acid of mercury, bismuth, aluminium etc.

REFERENCES

1. *Bailey's Industrial Oil and Fat Products*, Vol. I, John Wiley & Sons, New York, p. 105 (1979)
2. *Bailey's Industrial Oil and Fat Products*, Vol. I, John Wiley & Sons, New York, p. 761 - Caption Driers (1979)

CHAPTER 21

Fat-Based Coating Products

Fats and oils of plant, animal and marine sources, find a substantial out let as raw materials in making products for coating applications on surface, principally for protection purpose and for decorating also. Fats and oils are actually used in surface coating applications, as such or as suitably structured modified products to serve as the film-forming materials, depending on the nature of the oils and fats. The materials used in coating industry for protective and decorative purposes are paints, varnishes, lacquers, printing inks. Oils and fats are utilised to a large extent in manufacturing these items.

USAGE PATTERN OF FATS AND OILS IN COATING APPLICATIONS

Oils are used as such or after modification in the coating formulations:

A. Direct use of oils in Coating Formulations

Oils, classified as 'drying', 'semi-drying' or 'non-drying' are used in paints and varnishes either as such or after due modification. Oils rich in the polyunsaturated fatty acids like linolenic (non-conjugated, $C_{18:3}$) and elaeostearic acid (conjugated, $C_{18:3}$) and having high iodine values, are known as 'drying oils' and are directly used in protective and decorative coatings.

The high iodine value oils (due to having high percentage of polyunsaturated fatty acids), when applied to a surface, form a thin film thereby coating the surface with a distinct polymeric material. The surface becomes dry to touch, without any oily feeling.

The drying oils used mostly are linseed oil and Tung oil oiticica, and perilla, stillingia etc. The semi-drying oils like soyabean, tobacco seed, sunflower, safflower etc. are used to a certain extent. Fish oils are also used only after a heat treatment.

B. Use of Drying Oils after Modification:

Drying oils are often modified for enhancing their film formation on surfaces after application. The modifications can be achieved in different ways. The simplest method involves addition of certain metal salts of organic acids and other additives and heat-treating the oils at appropriate temperature to promote the viscosity of the oils slightly by controlling oxidation and polymerisation. The oils thus made are known as 'Boiled Oils'. Another very classical process but an extremely useful process to modify oils particularly the drying oils, consists of the heat-polymerisation of the oils at high temperatures to promote extensively polymerisation and alter the viscosity characteristics and other properties. The heat-polymerised oils are known as 'Bodied' or 'Stand Oils'.

One of the methodologies includes the modification of the oils by promoting oxidation reaction of oils by bringing in contact with air at temperature from 40°C to 150°C. This kind of processed oils is known as 'Blown' oils (also called oxidised oils).

The modification process also include reactions like epoxidation, dehydroxylation, co-polymerisation.

Before each process is described, it is important to know the basic aspects like the mechanisms of the modification of the oils.

MECHANISMS OF MODIFICATIONS OF OILS

Oils, particularly the drying oils, when applied on surfaces form films, which are distinctly polymeric in nature. The polymerisation reaction is mostly oxidative one. On the other hand, when the drying oils are treated at high temperatures, the polymerisation is almost exclusively thermal one.

The unsaturated fatty acids may undergo the ionic polymerisation in the presence of acids catalysts. The free radical polymerisation process is also involved in making drying oils polymer.

OXIDATIVE POLYMERISATION MECHANISM

The polymerisation of oils, involving the unsaturated fatty acids by oxidative means is, relatively complex. The mechanism of the reaction was demonstrated as below :

$$-CH_2 - CH = CH - + O_2 \longrightarrow \underset{\displaystyle OOH}{\underset{|}{CH}} - CH = CH-$$

It was seen that when oils containing linoleate or more highly unsaturated acids are oxidised, conjugated diene concentration increases (measured by UV absorption spectra. at 2340A) at the same rate as oxygen taking and peroxide formation. The modern concept of oxidation of linoleate is as follows (1):

-CH = CH - CH$_2$ - CH = CH-

- H° Abstraction of a hydrogen atom

$-CH = CH - \dot{C}H - CH = CH-$

-CH = CH - CH = CH = CH- | Resonance hybrid free radicals

$-\underset{\bullet}{C}H - CH = CH - CH = \dot{C}H-$

O_2

$$-CH - CH - \underset{\displaystyle 00^{o}}{\underset{|}{CH}} = CH = CH-$$

$$-CH - CH - CH = CH - \underset{\displaystyle 00^{o}}{\underset{|}{CH}}-$$

Three possible peroxy radicals

$$-\underset{\displaystyle 00^{\circ}}{\underset{|}{CH}} - CH = - CH - CH = CH-$$

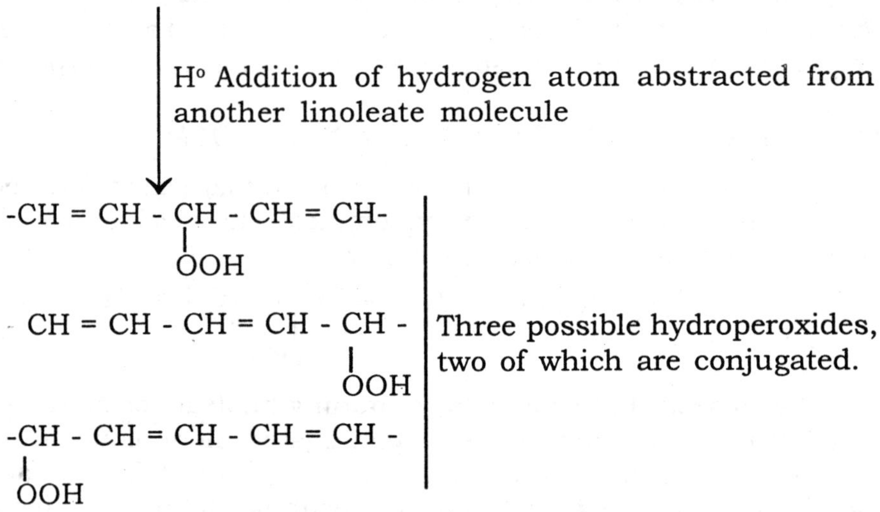

H^{o} Addition of hydrogen atom abstracted from another linoleate molecule

↓

-CH = CH - CH(OOH) - CH = CH-

CH = CH - CH = CH - CH(OOH) -

-CH(OOH) - CH = CH - CH = CH -

Three possible hydroperoxides, two of which are conjugated.

During the oxidation to form hydroperoxides , the natural cis, cis unsaturation of linoleate is converted to cis, trans and trans-trans isomers.

The hydrogen abstraction from α-methylene group involves high energy requirements and proceeds through free radical mechanism.

$$RH + A^* \longrightarrow R^* + AH$$

$$R^* + O_2 \longrightarrow RO_2^*$$

$$\rightarrow RO_2^* + RH \longrightarrow ROOH + R.$$

(R. is returned to the start of the last step, forming a cycle.)

RH is the molecule of linoleate and A* free radical abstracts hydrogen from RH.

This chain reaction involving free radicals goes through autoxidation mechanism where free radicals formed by the decomposition of materials act as catalysts.

Another mechanism suggested by some investigators (3) is that the oxygen directly attacks on the double bond having low thermodynamic probability and initiation of autoxidation is catalysed on the trace metal contaminants like Co etc. Once hydroperoxides are formed they play the role of catalysts. From kinetic viewpoint the reactions are stated as follows :

Initiation: Production of R. and RO2 radicals at the rate r., Propagation :

$$\rightarrow R + O_2 \xrightarrow{K_2} RO_2.$$

$$RO_2. + RH \xrightarrow{K_3} ROOH + R.$$

Termination

$$R. + R. \xrightarrow{K_2} R - R$$

$$R. + ROO. \xrightarrow{K_5} ROOR$$

$$ROO. + ROO. \xrightarrow{K_6} ROOR + O_2$$

Termination step is of prime interest in film forming process where polymers are formed by combination of free radicals. Except the formation of dimers, the higher polymers are formed by vinyl polymerisation.

THERMAL POLYMERISATION MECHANISM

The thermal polymerisation of the fatty acid radicals proceeds by carbon-carbon linkages by Diels-Alder reaction:

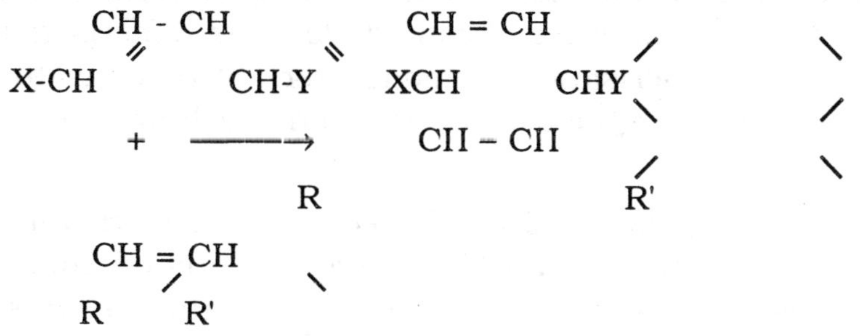

```
      CH - CH                CH = CH
      //      \\                     /         \
X-CH          CH-Y   XCH        CHY
                                        \         /
      +   ———→          CII – CII
                                        /         \
                R                        R'

      CH = CH
          /      \
    R     R'
```

The reaction indicates that the nonconjugated fatty acids isomerise to conjugated forms, which later form the above polymer.

The formation of conjugated acids during heat bodying is evident from the changes in refractive index of the oil. The scheme is as follows :

$$-CH_2 - CH = CH - CH_2 - CH = CH - \text{ non-conjugated (N)}$$

$$\downarrow$$

$$- CH = CH - CH - CH -$$

conjugated (C)

The scheme for uncatalysed thermal polymerisation of methyl linoleate in bulk is shown (2) as follows :

Main reactions :

Non-conjugated fatty acids (N) ⟶ Conjugated isomers (C) (relatively slow) .... (1)

N + C ⟶ Dimer D (relatively fast) .... (2)

Minor reactions :

N ⟶ Cyclic monomer .... (3)

N + D ⟶ Trimer .... (4)

N + N ⟶ Dimer .... (5)

It was reported by the investigators (2) that the reaction 1, i.e. isomerisation reaction is rate-controlling and is of first order, but it was later reported (3) that in the uncatalysed reaction in an inert solvent, the initial rates of conjugation and dimerisation are second order. Polymerisation of monoesters of polyunsaturated acids proceeds through the trimer formation which have polycyclic structures.

When non-conjugated oil like linseed oil, is air dried it takes up approximately 12% by weight of oxygen. Tung oil containing eleostearic acid having conjugated trienoic structure absorbs only 5% oxygen by weight to form a dried coherent film.

OTHER REACTIONS DURING POLYMERISATION: SECONDARY PRODUCTS OF AUTOXIDATION

With the main products of autoxidation a number of secondary products are formed. During air drying of oils different volatile and odourous ma-terials are formed and develop as a result the characteristic odour of paint. Carbon dioxide and water are also formed. The volatile compounds include saturated acids, aromatic acids, unsaturated acids, keto acids, aromatic acids, saturated and unsaturated aldehydes and ketones, alcohols, esters, lactones and heterocyclic compounds. A reaction sequence was proposed (4.) for the autoxidation of ethyl linoleate to produce a diketone which is shown below :

$$\text{-CH - CH - CH}_2\text{ - CH = CH -} \xrightarrow{O_2} \text{-}\underset{\substack{|\\ \text{OOH}}}{\text{CH}}\text{ - CH = CH - CH = CH-}$$

$$\xrightarrow{\quad} \text{-CH - }\underset{\substack{|\\ \text{OH}}}{\text{CH}}\text{ - }\underbrace{\text{CH - CH}}_{\text{O}}\text{ -CH =} \xrightarrow{CH^-} \text{- }\underset{\substack{|\\ \text{OH}}}{\text{CH}}\text{ - }\underset{\substack{\|\\ \text{O}}}{\text{C}}\text{ - CH}_2\text{ - CH - CH-}$$

$$\xrightarrow{O_2} \text{-}\underset{\substack{\|\\ \text{O}}}{\text{C}}\text{ - }\underset{\substack{\|\\ \text{O}}}{\text{C}}\text{ - CH}_2\text{ - CH = CH}^- + H_2O$$

The major initial products of autoxidation are hydroperoxides and almost all autoxidised samples contain non-cyclic and cyclic dimers and other polymeric materials. OTHER KINDS OF MINOR REACTIONS

During the autoxidation process, along with additional polymerisation, some condensation and other reactions also occur when low molecular weight compounds like water, carbon dioxide, volatile acids etc. are formed. Peroxides are also obtained during the air drying of a paint or varnish film. If phosphatides (5) and antioxidants like tocopherols (6) are present in the oil, then drying and polymerisation are inhibited. There are many oxidation accelerators like metallic oxides and salts, sulphur, selenium and their oxides chlorophyll and carotenoid pigments etc.

The acidity of drying oils measured by the 'acid value' increases gradually in heat bodying and markedly in air drying.

'Frosting' or drying with production of a wrinkled, folded surface is characterised by the oils which contain appreciable amount of conjugated acids. This paint is used for decorative purposes. The wrinkling occurs due to the volume changes in the oil during the drying process but the cause is obscure. Most of the applications requires a smooth surface. It was seen by microscopi-c examinations that dried tung oil films show a shrinkage in volume with the formation of small cracks near the beginning of the drying period. Frosting of tung oil is prevented or minimised by heat treatment of oil (7), blending with certain resins or addition of iron driers (8).

Functionality Concept in the Polymerisation of Unsaturated Fatty Acids

The nature of polymer formation depends on the number of double bonds, which is a functional concept. The unsaturated fatty acid chains can be monofunctional, bifunctional etc.

Monofunctional molecules can form only a dimer, but difunctional molecules can form higher polymers. This can be shown by the thermal or oxidative polymerisation of a monoester, ethyllinoleate (Et-L) where reaction stops after the dimer formation because functional groups have disappears in the reaction.

$$\text{Et - L + L - Et} \longrightarrow \text{Et - } \overset{\frown}{\text{L L}} \text{ - Et}$$

The curved line defines the union which may proceed by carbon-carbon, carbon-oxygen or other bonding structures.

In the case of bifunctional compounds the dimer has the same degree of functionality as the unpolymerised compounds. The reaction of a bifunctionai compound dilinoleoglyceride (L_2GS) is shown below :

```
                              ⌒              ⌒
nL - G - L ──→ L - G - L   L - G - L      | L - G - L
       |               |        |     |    | n - 2   |
       S               S              S              S
```

If the compound contains three or more functional groups then the functionality of polymer increases as the reaction proceeds. This growing complexity of polymeric structure can be shown by the cross-linking of trilinolein (GL_3):

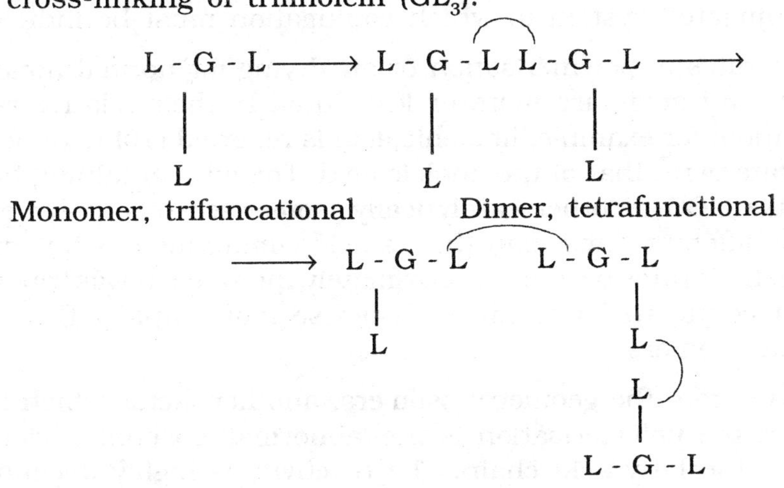

The functionalities of the double bond systems of different drying oil fatty acids are (9) as follows :

Oleic	Greater than 0, less than 1
Normal linoleic	About 1
Conjugated dienoic	About 1
Normal linolenic	Greater than 1, less than 2
Conjugated trienoic	Greater than 1, less than 2

The functionality concept helps in many respects about the variations in air drying characteristics of fats and oils. The classification of glycerides into drying, semi-drying and non-drying oils depends on the degree of unsaturation measured by their iodine value.

Practically, an important aspect of functionality concept is in the application of new techniques for improving drying of oils by separating fatty acids and reesterifying with a polyhydric alcohol having more hydroxyl groups than glycerol. The alcohols may be pentaerythritol, dipenta-erythritol or other polyfunctional alcohols etc.

POLYMERISATION RATES

Oils having conjugated double bond system polymerise thermally through Diels-Alder reaction more rapidly than the oils having non-conjugated system in which conjugation must be induced.

The rates of polymerisation or air drying for normal nonconjugated fatty acids are more or less equal to their relative rates of oxidation; for example, linolenic acid is reported (10) to dimerise at the rate twice that of the linoleic acid. The oils containing trans acids like DCO or other catalytically isomerised oils, polymerise at quite different rates than the cis acid containing oils but upon heating, their rates become approximately the same. It was reported (II) that conjugated linoleate polymerise more rapidly than the cis-trans isomers.

Apart from the geometric isomers, another factor which may influence the polymerisation is the 'abnormal' position of double bonds in the fatty acid chain. The reactivity is highly dependent on the distance of double bonds from the end of the chain.

AFTER-EFFECTS OF OIL-BASED PAINT FILMS

Oil-based paint films exhibit certain after-effects of deterioration like the synthetic polymer-based paints. The deteriorations or defects occur mostly in the case of externally applied coated films. Normally an external surface coated with oil-based paint or varnish shows signs of distinct failure of the film due principally to sunlight that promotes unusual rate of drying thereby causing to the failure of the film. The use of liquid oil in sufficient amount prevents cross-linking to a large extent. In fact, the oil is a plasticiser. When a major part of this liquid becomes solid by polymerisation, the film becomes brittle, cracking and peeling occur readily. This adversed polymerisation is assisted largely by the ultraviolet light of sun (12).

During aging of films acidity develops due to oxidation of unsaturated acids or from hydrolysis of ester linkages. Hence, if basic pigments like Zinc oxide are present then brittle polymeric salts are formed by the reaction of acid and base.

Mostly used exterior paints are of linseed oil and latex paints. Linseed paints have advantages of tint retention, adhesion over chalky surfaces, milder resistance and grain cracking (cracking parallel to the grain of the panels.

EFFECT OF WATER OR ALKALI ON OIL-BASED PAINT FILM

Oil-based paint films undergo degradation from the action of water or alkali. Tung oil and other conjugated acid oils absorb oxygen more slowly than the non-conjugated oils and hence have superior water and alkali resistance. But aged tung oil film absorbs oxygen at the same rate as the non-conjugated oil and more resistant towards water and alkali than ordinary paints. Fresh tung-phenolic varnishes are preferred for floor finishes and floor enamels due to their high resistance towards alkali.

DISCOLOURATION IN OIL-BASED FILMS

Very often the oil-based films exhibit a kind of a yellow or brownish colour in the film. The phenomenon is known as 'yellowing' and it increases with the amount of unsaturation (13) present in the film. In fact, 'yellowing' is a problem of practical significance for the oil film.

It was reported (14) that yellowing occurs due to the presence of chromophoric group such 'as diketone. It is seen that diketone-stearic acid is yellow.

```
R - C - C - R'
    ||  ||
    O   O
```

diketone

According to another suggestion proposed by some investigators the yellow substance is probably a metallic compound formed from the enol form of the diketone of the formula

```
R - C - C = CH-R'
    ||  ||
    O   OM
```

where 'M' may be any basic substance like zinc or lead from basic pigments.

The chromophores resulting from the oxidation of the active methylene groups in linoleic and linolenic acids without disturbing the double bonds have the formula:

$$\text{-CH = CH-CO-CH = CH-CO-CH = CH-}$$

Chromophores having yellow colour may be the quinoid compounds formed by the condensation of two diketone molecules.

```
            O                                       O
            ||                                      ||
            C                                       C
           / \                                     / \
  R  -  C     CH2.  -  R        R  -  C     C   -  R
        ||                                  ||     ||              + 2H2O
        O  +  O                  ——→    R ——C  +  C   -  R'
              ||                            \     /
  R  -  CH2   C     -  R'                    \   /
          \   /                                C
           \ /                                 ||
            O                                  O
            ||
            O
```

Yellowing is inhibited in perfectly dry atmosphere and at low temperatures. Yellowing is favoured by the red portion of the visible light but is prevented by the UV radiation. Hence, this effect is observed particularly in white or light coloured interior finishes and not in the exterior. Yellowing is also assisted by ammonia, amines or oxides of nitrogen, various aldehydes etc.

MATERIALS IN PAINTS AND RELATED PRODUCTS

The materials used for making different types of paints, varnishes, enamels, printing inks and other coatings may be classified as (a) oils, (b) resins, (c) pigments, (d) thinners or solvents, (e) driers and (f) miscellaneous ingredients.

Oils used in the coating industry: The main oils used in coating industry are linseed, soyabean, dehydrated castor, fish oil, and tall oil to a limited extent. Safflower oil having composition of about 75% linoleic acid and low saturated and linolenic acids are good drying oils and have considerable applications in surface coatings. Sunflower oil has similar properties and uses as safflower oil.

Semi-drying or non-drying oils are also in coating industry, such as corn, cottonseed, untreated castor or coconut oils and palm kernel oils. These oils are used in the manufacture of alkyd resins and they act as plasticisers in synthetic coatings. The formulations of synthetic coatings have now been shifted to the formulations using synthetic resins and non-oil latex emulsions.

Unmodified oils: Unmodified oils are those whose glycerides are not essentially altered by processing treatment. These are to be treated for removing non-oil impurities, or adding some essential components.

Raw linseed oil and to some extent soyabean and other oils are used as grinding oils in the wet grinding of pigments. In the past, these oils were treated with acids to make them rich in free fatty acids. But now degummed and alkali refined linseed oils are modified by adding linseed fatty acids or polymerised oils for manufacturing grinding oils.

Oils with variable free fatty acid content are used for grinding purpose. For treating basic pigments such as zinc oxide, the acid value of oils must be about 1-2. On the other hand, for grinding the old white lead pastes the oils must have acid value 12-16. On an average the acid value of grinding oils is in the range of 4-6. Phosphatides and gums must be removed as completely as possible from oils particularly drying oils. Phosphatides can be separated by hydration by injecting water or steam to the oil. In addition, the oil is mixed with a small quantity of diatomaceous earth and then filtered. Drying oils are usually refined with caustic soda for production of high grade varnish oils and bleached by acid-activated earths.

Waxes of the refined oil are removed by refrigeration and then filtration to clear varnishes or oils.

Fish oils are treated with heat to improve their drying properties and to remove their fishy odour. These oils are refined with alkali and also must be refrigerated or so called 'winterised' not for removal of waxes but for the removal of solid glycerides called 'stearin'. After proper winterising, fish oils become suitable for use in paints and varnishes.

Boiled oil: 'Boiled oils' are the drying oils with added metal salts of organic acids. Generally unmodified linseed oil is heated with metal oxides, carbonates or acetates with or without bubbling air through the oil to keep the compounds in suspension until dissolved. The term 'boiled oil' is used as frothing occurs due to the evolution of water, carbon dioxide or acetic acid. Due to the incorporation of metals, viscosity of the oil increases slightly and acid value also increases. Now-a-days driers are prepared in the form of a concentrated solution of metal salts of various organic acids such as naphthenic acid (naphthenates), 2-ethylhexaonic acid (octoates) or tall oil fatty acids (tallates). When these driers are mixed with raw linseed oil the drying time decreases to a marked degreje from some days to some hours.

Catalytic fast bodying oils: Catalytic fast bodying oils have been used for a long time in varnish manufacture, printing ink processing etc. These oils consist of alkali-refined linseed oil with a small amount of polynuclear aromatic catalyst such as β-methyl-anthraquinone or phenanthrene. These materials isomerise nonconjugated polyunsaturated acids to the conjugated structure.

Tall oil: Tall oil obtained as a by-product of sulphate-pulping (kraft) of pine wood contains rosin acids, fatty acids and unsaponifiables. Tall oil fatty acids are particularly used in the production of alkyd resins. This oil has little use in India.

MODIFIED DRYING OILS

Drying oils are modified by heat and oxidation polymerisation processes.

Heat-polymerised oils: Heat-polymerised oils also termed as bodied oils or 'stand oils' are used in the production of varnishes, enamels, printing inks, lithographic varnishes and also in house paints. The heat-polymerised oils reduce the amount of oxygen needed to form a film and improve stability.

In modern methods, the drying oils are heated in a large, mechanically agitated closed copper, monel or more preferably stainless steel vessels with proper temperature control. Oxidation is avoided by making an inert atmosphere or vacuum within the vessel. Conjugated oil require less temperature (225-240°C) than linseed or soyabean oil (300-320°C) to body. By controlling pressure, oils of several acidity ranges can be obtained.

The extent of polymerisation in heat processing can be monitored by means of change in viscosity. Oils can be bodied to any desired level of viscosity upto near gelation according to their use. With change in viscosity, other properties such as iodine value, acidity, refractive index, saponification value, specific gravity, hydroxyl value etc. also change. The characters of oils are not only determined by the viscosity but also by the conditions obeyed at the time of processing. According to these conditions, the oils have different pigment wetting properties and variation in pigment reactivity or tendency to flocculate pigments during paint storage.

Tung oil and other conjugated acid oils must go through the 'gas-checking' operation by heating at high temperature about 300-350°C. The oils must be 'gas-proofed' to resist the access of combustion products of gas-fired ovens to the paints used during the baking operation.

Limed oils: Limed oils are special modification of heat-bodied oils. High acidity bodied oils are thinned to 30-40% non-volatile content with mineral spirits and are reacted with lime to yield calcium soaps of polymerised fatty acids. Limed oils are used in primers, flat paints and semi-gloss paints.

Oxidation-polymerised oils: Blown or oxidised oils are obtained by bubbling air through the oils at temperatures ranging from 40 to 150°C. The viscosity of the oil increases as the oxidation progresses.

Blown oil obtained from low temperature processing contain a variety of peroxides, hydroxy, ether and other oxygen containing compounds. This oil is used as film forming plasticiser for nitrocellulose lacquers. At higher temperatures the peroxides decompose and different polymers are formed. Blown oils have many of the characteristics of heat-polymerised oils but form less resistant films than heat bodied oils due to high content of oxygen containing groups.

Blowing is used to prepare boiled oils used in house paints. Blown oils have high surface activity and hence they can promote the pigments to wet in paints and enamels. These are also used in caulking compounds, oilcloth, patent leather, oiled fabrics, and lithographic varnishes.

Epoxidised oils are, another class of oxidised oils. Partially epoxidised oils (15) such as 10-20% epoxidised linseed oil . is used in house paints. The film formed by these oils show less film embrittlement and less blistering under high moisture condition.

Dehydrated castor oil: Dehydrated castor oil is one of the major drying oils produced in India. Brazil is another country to produce this oil in a large amount. Castor oil produced from castor beans can be directly used as a plasticiser in alkyd and urethane system protective coatings and blown castor oil is an important nitro-cellulose plasticiser. Castor oil contains mainly the glycerides of ricinoleic acid (12-hydroxy-9-octadecenoic acid) and upon hydrogenation it forms 12-hydroxy stearate. Hydrogenated castor oil is used as an additive to introduce thixotropy into solvent thinned paints. Besides these, the most important coating used of castor oil is in production of dehydrated castor oil (DCO). It is used as a drying oil, having excellent drying properties.

Dehydration of castor oil is done chemically by a catalytic method. In this method, the hydroxyl group and a hydrogen from an adjacent carbon atom in the ricinoleic acid portion are removed. This is done by heating castor oil under vacuum at temperature in the range of 230-280°C in the presence of a catalyst until the evolution of water stops. Complete dehydration of castor oil is indicated by the change in the solubility characteristics. The starting castor oil is soluble in methanol and insoluble in petroleum ether. But as dehydration proceeds the oil becomes insoluble in methanol and soluble in petroleum ether. Dehydrated castor oil is less viscous and higher in iodine value and refractive index than the original castor oil. By the changes in these properties the extent of dehydration and polymerisation can be controlled.

Dehydration occurred during mineral acids such as sulphuric acid, phosphoric acid or sodium acid sulphate goes through the mechanism of formation of hydroxonium ion and carbonium ion (16).

Almost no evidence of formation of cis-cis isomer has been obtained, it is usually assumed that the newly formed double bond has trans-structure with the formation of cis-trans conjugated and isolated double bonds.

```
            ↑      ↑
           13 H   12H   11H   10H      9H      O
            |      |     |     |        |      ||
CH3(CH2)4   C  -  C  -  C  -  C   =   C (CH2)7 C - O -
            |      |     |
            H     OH     H    cis
                   | +H+
13 H   12H    11H   10H     9H
 |      |      |     |       |
- C  -  C  -   C  -  C  -   C = C -
 |      |      |
- H    OH2+    H    cis     hydroxonium ion
                            (Protonated alcohol)
               | -H2O
               ↓

13 H   12H    11H   10H     9H
 |      |      |     |       |
- C  -  C  -   C  -  C   =   C  - Carbonium ion
        +      |        cis
               H
             /   \
          -H+     -H+
13 H  12H  11H  10H  9H        13 H   12H   11       10H   9H
 |     |    |    |    |          |     |              |     |
- C =  C  - C -  C =  C -        C -   C =  C  -      C  =  C-
  |trans    |     cis            |     trans|          cis
  |         |                    |          |
  H         H                    H          H
```

9-cis, 12-trans octadecadieonic acid (65% - 75% of total)

9-cis, 11-trans octadecadieonic acid (35% - 25% of total)

Many inorganic salts, acids or oxides and also natural and acid activated clays are used as dehydration catalysts. The most common catalysts are sulphuric acid, phosphoric acid, sodium acid sulphate and potassium acid sulphate (17, 18).

Another method of preparation of dehydrated castor oil is the esterification and pyrolysis of the ester. Polyesters termed 'estolides' are prepared by the condensation reaction between the carboxyl group of one molecule of ricinoleic acid and the hydroxyl group of another. By continuous heating the polyester pyrolyses to give dehydrated castor fatty acids.

The dehydrated castor oil contains about 65 per cent linoleic acid (18:2 9,12) and 22 per cent conjugated linoleic acid (18:2 9,11). About 5 per cent hydroxy acids remain in the product while the proportions of saturated acids and oleic acid remain unaltered as in the raw oil.

DCO based coatings have good flexibility through dry and colour retention property.

ISOMERISED OILS

Isomerised oils can be prepared according to several patents (19,20) which describe the process of reacting fatty acid soaps of an oil with an excess of aqueous alkali at a temperature of 200-230°C and elevated pressure and acidifying the resultant soap to recover the isomerised acids. Linseed oil, soyabean oil, sardine oil and dehydrated castor oil can be converted to isomerised oils.

Alkali isomerisation process is more costly and uneconomical from commercial point of view than other isomerisation processes. The other processes include nickel-carbon catalytic method or platinum or palladium-carbon method though a conjugated safflower oil obtained by the alkali isomerisation process has been marketed commercially. A unique isomerised oil is obtained from dehydration of epoxidised linseed oil (21). The product has been termed 'Delinois' which has excellent drying property. Isomerised oils dry slowly compared to other drying oils and form soft films. Their water resistance is high though alkali resistance is poor. Bioisomerisation process has been developed to make conjugated-fatty acid containing oil. Lipoxygenase enzyme converts linoleic acid into a hydroperoxide that can be reduced by sodium arsenite to an hydroxy acid, which in turn on dehydro-xylation yields conjugated fatty acid. A typical example, involving soyabean oil soap stock, points out the preparation of tung-oil like material by bioisomerisation process (22).

FATTY ACID ESTERS OF HIGHER POLYHYDROXY ALCOHOLS AS DRYING OILS SUBSTITUTE

Many polyhydroxy alcohols or polyols such as pentaerythritol, sorbitol, xylitol, o-methyl-D-glucoside, sucrose, polyallyl alcohol, polyvinyl alcohol, aliphatic diepoxides, epoxy resins etc. yield fatty acid esters having high functionality than triglycerides. Among them pentaerythritol is the most important polyol as it gives economic advantages; its esterification is rapid and esters have good drying properties. Sorbitol esterifies at a slower rate than pentaerythritol because of the presence of secondary hydroxyl groups. The sorbitol esters have drying properties intermediate between glycerol esters and pentaerythritol esters.

To meet desirable combination of melting point, viscosity and solubility, more than one alcohol are to be esterified together, such as the mixtures of sorbitol with glycerol, pentaerythritol or other polyols show some advantages over sorbitol esters.

Sucrose esters can be prepared from methyl linoleate or methyl esters of linseed fatty acids and sucrose by alkali catalysed alcoholysis using dimethylformamide as solvent. One to about seven of the eight available hydroxyl groups of sucrose can be esterified. Higher esters give rapidly dried, hard, durable films with less oxygen consumption than that required for triglycerides. Esters having two to four hydroxyl groups can act as water-thinned vehicles.

Polyvinyl alcohol and fatty acids directly react to form esters in phenol as solvent (23). These are used in adhesives, emulsions and other products.

Epoxy resins can be esterified with linseed, soyabean, DCO, tung etc. fatty acids without requirement of a catalyst. Due to the high functionality of epoxy resins, even oleic acid ester, dry to give films having good alkali resistance.

Epoxy resins, when blended with polyamide resins obtained from fats, give corrosion inhibiting vehicles for paints.

Many polyols have been suggested for preparing esters but small number of them have been used commercially. Among them pentaerythritol (PE), trimethylolethane (TME), trimethyl-olpropane (IMP) and epoxy resins are important.

Fatty Acid Dimers: Monoenoic acid (e.g. oleic acid) and dienoic acid (e.g. linoleic acid) dimers are important materials in coating industry.

Oleic acid dimer is produced by ionic type polymerisation of oleic acid with catalysts such as boron trifluoride or hydrogen fluoride at temperatures of 150°-200°C. The structure of the dimeric acid obtained by polymerisation of oleic acid in the presence of boron trifluoride-phosphoric acid (24) catalyst has been deduced as :

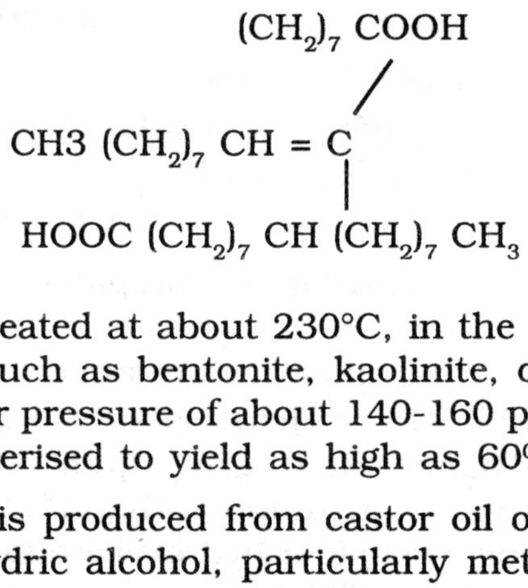

$$
\begin{array}{l}
\qquad\qquad\qquad\quad (CH_2)_7\ COOH \\
\qquad\qquad\qquad\qquad\quad / \\
CH3\ (CH_2)_7\ CH = C \\
\qquad\qquad\qquad\qquad\ \ | \\
HOOC\ (CH_2)_7\ CH\ (CH_2)_7\ CH_3
\end{array}
$$

When oleic acid is heated at about 230°C, in the presence of 1-5% water with clays such as bentonite, kaolinite, or montmorillonite, preferably under pressure of about 140-160 psi for about 2 hours, the acid is dimerised to yield as high as 60%.

Linoleic acid dimer is produced from castor oil or castor oil fatty acid or its monohydric alcohol, particularly methyl alcohol ester by heating at 200°-250°C with a clay catalyst like tonsil earth optimum as such or clay with a dehydration catalyst in presence of 5-10% water at 150-psi for 10-16 hours when the dimer could be obtained in about 35-88% yields. By using a combination of free radical initiator and a dehydration catalyst, dimers of 90-92% could be obtained from castor oil fatty acids at 250°C and 180 psi pressure (25).

WATER-THINNED DRYING OILS

Water-thinned paints based on synthetic emulsion polymers are used for house paint vehicles. There are three types of commercial products - water soluble linseed oil (25), emulsified bodied linseed oil (26) and linseed alkyds, and emulsifiable oils marketed as 100% solids systems with added emulsifier. Non-ionic, anionic and cationic emulsifiers are added to drying oils or alkyd resins to develop various emulsion vehicles. Among them, emulsions based on heavy bodied linseed oil and non-ionic emulsifiers such as ethoxylated allyl phenols are the most commercialised emulsion The water based emulsion paints are especially used to improve adhesion over chalky surfaces and tinting of the paints using water dispersed tinting systems, the so-called 'universal colourants'.

MODIFICATION OF OILS BY FRACTIONATION, INTER-ESTERIFICATION ETC.

Drying properties of oils such as soyabean or fish oils such as menhaden can be improved by liquid-liquid fractionation with furfural or propane or direct interesterification process Solvent crystallisation or fractional distillation (27) are used to improve the unsaturated fatty acids obtained from linseed, soyabean, sardine or cottonseed oils to use in drying oil products.

In the furfural system (28) furfural and oil are subjected to a countercurrent extraction in a packed or bubble cap column. Furfural acts as selective solvent which extracts the more unsaturated glycerides and also free fatty acids, unsaponif iables, chlorophyll and other natural pigments. Hence, if soyabean oil is treated in this process the product from solvent rich phase gives a light coloured oil having excellent drying properties which can replace linseed oil.

In the liquid propane or Solexol process (29), glycerides are completely miscible in the propane at moderate temperature like 60°C and an enuilibrium pressure of 450 lb/in2.But as the temperature increases the solubility of glycerides decrease and at the critical point an abrupt separation occurs into the solvent-rich and oil-rich phases. More unsaturated glycerides and pigments go to the oil-rich phase just opposite to the furfural system. Fractionated soyabean, linseed and fish oils produced by both the two above methods are used commercially in coating vehicles.

Interesterifications of glycerides are done by using alkaline catalysts such as sodium alkoxides, sodium hydride or alkali metals. By this process most of the saturated acids of oils are separated and the drying characteristics are improved.

Saturated acids of vegetable oils or fish oils can also be removed by using crystallisation in wet acetones or aqueous ethanol. Combination of solvent crystallisation and fractional distillation produce fatty acids of any desired purity.

Maleic oils: 'Maleic oil' or 'maleinized oils' are among the most important synthetic drying oils obtained by the two commer cial methods: maleic anhydride may be added directly to the glyceride followed by the esterification of the free acid by glycerol, pentaerythritol or other polyol, or the glyceride may be first alcoholised by a polyol followed by esterification with added maleic anhydride. Treatment with maleic anhydride transforms the slow drying oils such as soyabean oil, tall oil etc. to the rapid drying oils.

OIL-SOLUBLE AND OIL-DERIVED (BASED) RESINS

Resins, natural and synthetic, when added to the drying oils in particular under the influence of heat, form fast drying, hard glossier and durable films compared to drying oils alone. Some synthetic resins do not interact but by becoming soluble in the oleoresinous vehicle give the same effect. Natural resins such as rosin, dammar, kauri, copal, congo, amber etc. are obtained from natural in the form of fossilized exudations of trees. These natural resins are reacted with oils by heating or cooking till dissolved. The reaction produces are later thineed with suitable solvent like a solvent naptha. These products contain, varying proportion of resin and oil. These may be classified as short (20-60% oil), medium (60-74% oil) and long oil (74-96% oil) varieties.

Synthetic resins like phenolic resins, alkyd resins, urethane resins etc. find even today extensive applications in coating formulations for their having specific properties and reactivitiy.

Phenolic resins, which are the condensation products of phenol, particularly substituted phenols and formaldehyde, are obtained by using acid or alkali as catalyst. The acid catalysed phenolic resin is thermoplastic product called Novolac, and most, used for air-drying finishes. Another is obtained by using alkali as catalyst; this resin is thermosetting and insoluble in oil, and cannot be used in varnishes.

Phenolic resins react with drying oils to give a copolymer of oil and resin. It is seen that phenolics give best results in combination with tung oil or other conjugated acid oils. These resins show good water and alkali resistance, generally good solvent and chemical resistance and better exterior durability than most other varnishes.

Alkyd resins: Alkyd resins (30, 31) are the most important class of synthetic resins used in coating industry. The alkyd resins used in coating industry are the products obtained by the reaction of a polyfunctional alcohol with a dibasic acid and a monofunctional acid. Dibasic acids used for alkyd resins are mostly phthalic anhydride, maleic acid, fumaric acid, isophthalic acid, terephthalic acid, tetrachlorophthalic acid, succinic, adipic and sebacic acid.

The polyfunctional alcohol or the polyol is mostly glycerol. In order to obtain almost exclusively a linear polymer, the glycerol is reacted with a monobasic acid fatty acid, to form principally a monoglyceride, which in turn is esterified with a dibasic acid. The resulting resin esters are no longer cross-linking three-dinentional structure but are now soluble in oils and solvents as well.

The products are called 'oil-modified alkyd resins'. The extent of esterification-polymerisation can be varied by varying the amount of oil or fatty acid. The presence of long and flexible fatty acid chains has a plasticising effect.

Degree of oil modification is an important factor which can be expressed in various ways. The common method of expression is based on the phthalic anhydride content. Another useful expression is the so called 'alkyd ratio' which can be defined as the per cent by weight of glyceryl phthalate calculated on the basis of a hypothetical stoichiometric combination between glycerol and phthalic anhydride (32). The alkyd ratios of most of the alkyd resins are between 18 and 65.

Short oil alkyds are those which have 35-45% oil length range (more than 35% phthalic anhydride) with alkyd ratio range 50 to 65. These are usually prepared from non-drying oils and modified with urea or melamine formaldehyde resins. These oils are used mainly for baked finishes on automobiles, refrigerators, washing machines etc.

Long oil alkyds contain oil 55% to 90%, more commonly 70%. These are used predominantly in brushing enamels for interior or exterior architectural finishes where fast drying, durability, good colour and gloss retention are required.

Medium oil alkyds have 46-55% oil and 30-35% phthalic anhydride with alkyd ratios of 39-50. These resins usually modified with soyabean, linseed or tall oil fatty acids. These oil-alkyds have good air-drying properties, gloss retention and durability.

Alkyd resins are prepared by reacting oil *or* fatty acids, polyol and phthalic anhydride together in a single batch. Among the polyol, pentacrythritol and among the dibasic acids, phthalic anhydride are the most important. The properties of alkyd resins can be varied by varying the proportion of either the polyol or the dibasic acid.

Another novel resin having free terminal carboxyl group can be prepared by reacting conventional alkyd resins with an excess of phthalic acid, so that more cross-linking can -occur with the terminal acid group.

Water-dispersible or emulsifiable alkyd for use in water-thinned paints can be obtained by cooking a surfactant into the alkyd so that it will disperse readily when water is added or by introducing propylene glycol into the conventional oil-modified alkyd resin.

Styrenated alkyd resins (33) are prepared by the reaction of styrene with a DCO or tung oil modified alkyd resin using a peroxide catalyst. These are fast drying and used in various industrial finishes but they have poor solvent resistance.

Silicone modified alkyd resins (34.) are prepared by reacting conventional alkyd resins with reactive silicone intermediates. These polymers have excellent durability and are used in exterior maintenance finishes in metals in high rise buildings, bridges, ships etc.

Urethane resins: Urethane resins (35) called urethane oils, polyurethanes, urethane alkyds etc. are prepared by reacting glycerol (or alcohol) with diisocyanate. Here diisocyanate replaces phthalic anhydride. Toluene diisocyanate (TDI) is the most important diisocyanate, prepared by dinitration of toluene, then reduction to diamlne, reaction with phosgene and dehydrochlorination to yield diisocyanate having *80%* 2, 4-isomer and 20% 2, 6-isomer, Isocyanates react with alcohol to yield urethanes:

$$R-N=C=O + R'OH \longrightarrow \left[R-N=C\begin{matrix} OH \\ OR' \end{matrix} \right] \longrightarrow$$

$$\longrightarrow R-NH-C\begin{matrix} OH \\ OR' \end{matrix}$$

The product of TDI and glycerol is shown below:

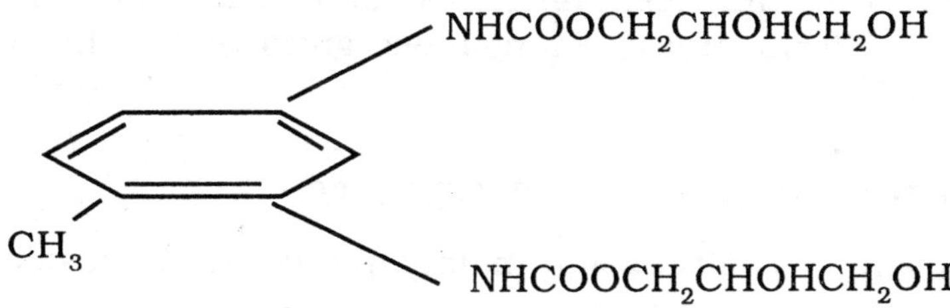

The free primary and secondary hydroxyl groups are available for esterification with drying oil fatty acids.

Urethane products, compared with the conventional alkyd resins, give fast dried harder films with better water, solvent, chemical and abrasion resistance. Other diisocyana such as hexamethylene diisocyanate (HDI) can be used. HDI gives better colour retention but is less reactive than aromatic isocyanates and is toxic.

There are a variety of commercially important drying-oil-free urethane coatings systems including moisture cure urethanes, one-package blocked urethanes, two-package catalysed urethanes, two-package polyurethane systems.

OIL-BASED COPCLYMERS

Copolymers are obtained by reacting drying oils with a variety of simple monomers. Monomers react at the unsaturated centre of the drying oils to yield true copolymers. This reaction can be carried out by thermal polymerisation without any catalyst, emulsion polymerisation with peroxide catalysts or co-polymerisation using basic catalysts.

The important monomers used for modification of drying oils are dicyclopentadiene, styrene or vinyl toluene, α-methylstyrene, divinylbenzene etc.

FATTY ACID-BASED METALLIC SALTS (DRIERS) AS COMPONENT OF COATING MATERIALS

Salts of metals like cobalt, manganese and lead and fatty acids derived from oils including tall oil are used in accelerating the polymeric film formation by promoting carbon-carbon bonding by enhancing the oxygen uptake. Cobalt is the most effective metal for air dried finishes. Manganese is intermediate, lead is slow but gives uniform drying. Manganese gives better through-dry than cobalt.

SOME COMMERCIAL PRODUCTS

Following is a discussion mainly about the oleoresinous products such as paints, varnishes and inks :

Oil paints: Paints are the pigmented compositions made up of a vehicle consisting predominantly of unmodified drying oil or oil of low viscosity. Oleoresinous vernishes or enamels are made up of vehicles fortified with resin. Paints are slow drying and the resultant films are lacking in hardness and chemical resistance but varnishes and enamel films are hard and are quite resistant towards water, alkalies or other reagents.

House paints: Common house paints are the combinations of raw linseed oil and bodied linseed oil with pigments, driers and solvent. The bodied oil increases durability and helps in the elimination of brush marks. Soyabean oil and fish oil are also used but with a sufficient amount of linseed oil to improve their slow drying property. Tung, oiticica and dehydrated castor oil are also used but to a limited extent due to their high cost.

Exterior trim paints: These paints are mainly used for decorative purposes which can give coloured, glossy finishes without chalking and retain their appearence on exposure to weather. These are mainly enamels rather than paints and contain the common vehicle, a long oil alkyd resin. These paints are used in windows, screen frames, shutters, commercial sign painting etc.

Industrial finishes: Industrial finishes are used for both the decorative and protective purposes in all kinds of manufactured products such as toys, automobiles, farm equipments, furniture etc. Industrial substrates may be metals, wood plastics, leather, paper and textiles.

Now the industrial finishes are based on varnishes containing either natural or more generally synthetic resins. The oil-modified aikvd resins are the most important for industrial finishes. These include acrylics, melamine, urea, phenolics and silicones.

Varnishes: Varnishes may be classified into two types—oleoresinous varnish and spirit varnish. Oleoresinous varnishes are the important coating materials obtained by polymerising drying oils at high temperature with various natural and synthe resins and thinning the resultant material by adding volatile solvent. Pigmented varnishes are enamels.

Varnishes and enamels are quick drying and form hard, water resistant, high-glossy films. By varying drying oils, resins and solvents, different types of varnishes and related materials can be prepared and their properties also vary to use them for specific works.

Tung oil and other conjugated acid oils are generally used for making varnish. But relatively poor drying oils can also be used for varnishes if synthetic resins, particularly alkyds, are used with them. The soyabean oil, safflower oil, sunflower oil etc. are used along with synthetic alkyd resins. But the properties of a varnish depend to a considerable extent upon the proper balance between the two ingredients, i.e. oil and resin.

Water dispersible paints: Water-paints or 'calcimides' were cheap, non-washable finishes used for years and consisted of lime, whiting, china clay etc. with glue as a binder. Later, the formulation had been changed to some extent by the incorporation of casein with lime or soya protein in place of glue and also the inclusion of some emulsified drying oil.

For exterior use, long oil alkyd resins (36) are useful for good adhesion, durability, tint acceptance etc. Water-thinned drying oils are also used but to a limited extent.

FAT-BASED PRINTING INK MATERIALS

The vehicles used for printing inks are in general raw or unbodied oils.

Printing ink formulations consists of the so called lithographic varnishes which are actually heat-bodied oils, generally linseed oils of different viscosities. The use of a single varnish in printing ink is rare and combinations of varnishes are used.

Although linseed oil is the main binder for printing ink bodied tung oil and soyabean oil are also used to some extent Dehydrated castor oil and maleic modified oils are also used in printing ink vehicles. Apart from these, castor, groundnu or olive oils and animal oils like lard and neatsfoot oils ar used as adhesion modifiers. Other ingredients of printing inks are blown oils, burnt oils, mineral oils, driers, solvent natural and synthetic resins, pigments and 'compounds'. 'Compounds' are waxy materials such as soap, tallow, vegetable waxes etc. introduced in a trace amount to modify the tackiness of the product.

MANUFACTURE OF COATING MATERIALS

Varnish manufacture: Modern methods of production of different varnishes like alkyd resins, urethane oils, epoxy etc. include the use of closed, mechanically agitated kettles made up of stainless steel and jacketted for Dowtherm heating or cooling. Provisions are made for pumps, steam ejectors for vacuum, condensers, scrubbers etc. with the equipment.

Drying oil and natural resins (except fossil resins) are cooked in the kettle at about 260-35O°C. To use the fossil resin, it must be passed through some preliminary processes like melting, cracking or running at 330-345°C or higher. If tung oil is used, it must be subjected to a short period of high temperature treatment if the product is to be 'gas-proof'. The end point of this cooking operation is determined by checking the viscosity of the batch. The viscosity is actually determined (after dilution with a thinner by double-tube viscometer. Before thinning, the batch must be cooled to 175-225°C. If separate thinning takes with agitators and condensers are provided for cooling then the cooling operation becomes simplified and better control of polymerisation can be achieved.

The ultimate viscosity of varnishes and enamel vehicles is made about 0.5-1.5 st at 25°C by thinning. The non-volatile material content of the thinned,product varies between 40-60%.

Manufacture of alkyd resins (30, 31): There are two basic processes for the preparation of alkyd resins. One is the direct esterification process called 'fusion cooking' and another is solvent esterification process called 'azeotropic esterification or 'azeotropic cooking'. The two methods differ in the procedure used to remove water of esterification.

In the fusion process, water is removed by inert gas, like carbon dioxide, nitrogen or other oxygen-free gas, which is bubbled through the reaction mass and water of esterification is vented out of the kettle together with the exhaust gas.

In the solvent process, a solvent, usually xylene, is used at 5-10% of the total charge to remove water of esterification.

The solvent and water are condensed together and collected in a trap. As water is immiscible in xylene it drops at the bottom of the decanter and xylene flows back into the kettle where it is again vaporized with water by forming an azeotrope.

Solvent method gives higher yield than the fusion method. Solvent method also gives more uniform polymer structure, faster processing, lower inert gas costs and cleaner process equipment In fusion cooking phthalic anhydride sublimes at the esterification temperatures of 225-235°C and collects at the top of the kettle and in the condenser system, but in the solvent process, solvent keeps the phthalic anhydride in solution. The disadvantage of solvent process is the high fuel costs for solvent reflux and the need for cooling water for condenser.

In either process, the progress of reaction is verified by checking the acid values and viscosity of the samples taken at regular intervals. After the completion of esterification, the resin reaction is stopped by chilling quickly.

When alkyds are prepared by using fatty acid as modifier, the process is one-stage esterification reaction. In this case all ingredients are charged to the cold kettle, then heated with agitation by sparging inert gas and esterification is carried out by either fusion or solvent process.

Two-stage process is required when oil is used as alkyd modifier. In this case oil is reacted with a part or all of the polyol to form the corresponding mixed partial esters and the resultant alcoholysis product is reacted with the remaining ingredients to form the final resin using either fusion or azeotropic processing. Alcoholysis catalysts used are lithium hydroxide, naphthenate and ricinoleate. The oil and polyol are heated at about 220-230°C. Excess polyols are taken to preven" the esterification to be too long.

In situ preparation of other resins: *In situ* method is used for the preparation of varnishes other than alkyd resins. In this process resin is formed in the oil during the cooking process. The reaction is carried out by reacting rosin with glycerol, pentaerythritol or another polyhydroxy alcohol and a drying oil or a drying oil fatty acid at a temperature of 285-300°C for several hours. There are different methods of preparation of different types of resins.

The consumption of vegetable oils and derived products in paint industry has already decreased remarkably and it is certain to get further reduced in view of the superiority displayed by a number of synthetic polymer materials whose use in coating formulations exhibit unique benefits from protective point of view of various surfaces.

REFERENCES

1. D. Swern, in Autoxidation and Antioxidants, edited by W.O. Lundberg, Vol. 1, Chapter 1, Wiley -Interscience, New York (1961).
2. R.F. Paschke and D.H. Wheeler, J. Am. Oil Chemists' Soc. 26, 278 (1949).
3. D.F. Rushraan and E.M.G. Simpson, Trans. Faraday Soc. 51, 230 (1955).
   F.W. Heaton and N. Uri, J. Lipid Res., 2, 152 (1961).
4. S.S. Chang and F.A. Kuounerow, J. Am. Oil Chemists' Soc., 30, 403 (1953).
5. P. Desnuelle and R. Massoni, Peintures, Pigments, Vernis, 29, 390 (1953).
6. H.O. Kunkel and W.L. Nelson, J. Biol. Chem. 183, 149 (1950).
7. A.E. Rheineck and S.C.G. Peng, Offic. Dig. Fed. Soc. Paint Technol., 36, 890 (1964).
8. R.F. Schneider, R.H. Coe and E.E. Ware, Canadian Pat. 373,394 (1938).
9. D.H. Wheeler, Ind. Eng. Chem., 41, 252 (1949).
10. I.M. Bernstein, J. Phys. Colloid. Chem. 52, G13 (1948).
11. R.F. Paschke, J.E. Jackson, and D.H. Wheeler, Ind. Eng. Chem., 44, 1113 (1952).
12. M.J. Hendrickson, R.P. Cox and J.C. Konen, J.Am. Oil Chemists' Soc., 25, 73 (1948).
13. H. Rakoff, F.L. Thomas and L.W. Gast, J. Coatings Tech., 48, No.619, 55 (1976).
14. A.C. Elm and G.W. Standen, Ind. Eng. Chem., 24, 1044 (1932).
15. A.E. Rheineck, U.S. Pat. 3242196 (1966).
16. K.T. Achaya, J. Am. Oil Chemists' Soc., 48, 758 (1971).
17. G.W. Priest and J.W. Von Mikusch, Ind. Eng. Chem. 32, 1314 (1940).
18. J. Walton, U.S. Pat. 2429380 (1947).
19. T.F. Bradley, U.S. Pat. 2350583 (1945).
20. A.H. Stevens, British Pat. 558881 (1944).
21. G.S.R. Sastry, B.G.K. Murthy, and J.S. Aggarwal, Paint Manuf., 40(8), 32 (1970).
22. E. Emken, J. Am. Oil Chemists' Soc., 59. 1982.
23. A.E. Rheineck, J. Am. Oil Chemists' Soc., 28, 456 (1951).

24. A.V. Topchier and T.R. Vishnyakova, J. Gen. Chem. (U.S.S.R.), *21,* 1618-1625(1951).

25. B. Chattopadhyay, Ph.D. (Tech.) Thesis, Calcutta University, 1982.

26. J.C. Cowan and L.E. Cast, Am. Paint J., 47(33), 85 (1963).

27. S.W. Gloyer, J. Am. Oil Chemists' Soc., 27, 462 (1950).

28. S.E. Freeman, U.S. Pat. 2200390 and 2200391 (1940); 2313636 and 2316512(1943), 2539661 (1951).

29. A.W. Hixon and R. Miller, U.S. Pat. 2631157 (1953).

30. C.R. Martens, Alkyd Resins, Reinhold, New York (1961).

31. W.M. Kraft, E.G. Jamese, and D.J. Sughrue, in Treatise on Coatings, edited by R.R. Myers and J.S. Long, Vol. 1, Part 1, Chapter 3, Marcel Dekker, New York (1967).

32. E.M. Beavers, Ind. Eng. Chem., 41, 738 (1949).

33. R.H. Whoff, D.C. Davis, and H.A. Hoppens, Cffic. Dig. Fed. Paint Varn. Prod. Clubs, 33, 220 (1961).

34. R.C. Hedlund and C.R. Olson, Paint Varnish Prod., 53(5), 40 (1963).

35. Symposium on Polyurethane Coatings, Office. Dig. red. Soc., Paint Technol., *32,* 196-290 (1960).

36. K.W. Merkel, J. Paint Technol., 58, 543 (1966).

CHAPTER 22

Energy Conservation and Efficiency of Oils and Fats Plant

INTRODUCTION

Oil and fat processing industries are the major consumers of energy (1) as substantial quantities of steam and electrical power are consumed for various commonly used processing operations covering refining, bleaching, hydrogenation, deodourisation etc. Directly or indirectly the energy costs are reflected in the cost of production of the final product. Any savings or conservation of energy effected at the processing stages would therefore lead to reduced operating cost and enhanced profit. With the rising cost of fuel oil and petroleum products the world over, the oil producers have been compelled to design more efficient equipments for heating and cooling apart from saving energy either by the use of supplementary fuels or through application of conservation techniques. As unit cost of energy is expected to increase in future the different ways and means of saving and harnessing energy have assumed great significance. It is usually good to invest in energy recovery equipment as its cost is generally recovered within 2 to 3 years.

ENERGY SAVING MEASURES

Two main forms of energy are of major interest in oil based industry i.e., heat energy (in the form of steam) and electrical energy. Energy saving programmes have to aim at the following :

(1) Recover maximum possible heat from furnace and minimise radiation loss through proper insulation.

(2) Maintain continuity of the process.

(3) Reduce steam consumption by the vacuum equipment and optimise recovery of the condensate.

(4) Minimise cost of producing steam and electricity and maximising boiler efficiency.

(5) Utilise released heat from air-conditioning for production of hot water.

STEAM GENERATION AND ECONOMY

Depending on quantum and pressure of steam required, water-tube or fire-tube boiler may be used. Modern oil plants use water-tube boilers which generate high pressure steam (upto 40-50 bar) and high pressure steam without super-heating. These boilers are expensive, but generate the steam readily. Their efficiency ranges between 80-90%. For low and medium pressure steam (10-20 bar) fire-tube boilers are employed, their efficiency being 70%. The oil industry uses saturated dry steam and injection of condensed water into the steam pipes results in better heat transfer.

The furnace attached to the boiler should be designed to use at least two types of fuel from among coal, fuel oil or husks and shells from seeds. The condensate or other feed water from the boiler should be properly softened and treated economically by ion exchanger to free it of hardness and any other organic impurities besides involved gases i.e. oxygen and nitrogen. Condensate should be collected and re-used for boiler feed water, or for the heat extracted from it by exchange with fresh boiler feed water. Soft water could also bc heated by recovering heat from deodoriser or ejectors. Any wastage of steam or heat loss to air or water should be avoided. Good insulation should be maintained in hot pipes, storage tanks and high temperature process vessels. The insulation should be quickly repaired if and when it is broken. However, if the steam pipes are spread over a long distance there may be a transmissional energy loss of 10% even with good insulation. Slight super-heating of steam (about 30°C) would help to reduce such transport loss. Special attention should be paid to bridge and repair any steam and condensate leak especially on the steam trap system. The steam traps should be regularly surveyed and replaced when necessary.

HEAT EXCHANGE EQUIPMENT

One of the most convenient ways to regulate heat energy is by use of regenerative heat exchangers. In this way heat is recovered by thermal exchange between incoming and outflowing oil based upon gradient between the oil temperatures. These are regularly practised in edible oil industry. There are various types available too. All the more popular types are plate heat exchanger (PHE) and spiral heat exchange (SHE).

The plate heat exchanger consists of a series of embossed heat transfer plates spaced by gaskets to contain pressure and the fluid within. The end plates do not transfer heat. The gasket plates are mounted on guide rails and pressed together in the frame by compression boards. The PHE is designed often according to flexibility as always the capacity of the unit can be easily extended by addition of more plates. Based on needs, this exchanger may be designed for either single or multi-plates fluid flow. The operation of PHE is limited in the range at the designed maximum pressure range of 75-100 psig and temperature range of 130-140°C. Also PHE is imperative when an expensive non-corrosive material of construction is required. Where mild steel is acceptable easily the tube exchanger is usually economical. A plate unit need not be insulated. A spiral heat exchanger (SHE) is the ideal equipment as it has only one channel for each liquid and permits the counter-current flow. SHEs are compact, require less insulation and servicing steps compared to conventional exchangers of equivalent surface. Radiation losses are negligible. SHEs have a somewhat lower heat transfer coefficient compared to PHE and hence are more expensive. PHE's maximum working pressure range is 140-280 psig, and maximum working temperature depending on the gasket material where synthetic rubber or asbestos, range from 150-250°C and show heat transfer coefficient water-to-water in KCal/m²/h°C as 2000-5000. In the case of spiral heat exchanger maximum working pressure is 140-210 psig, maximum working temperature is upto 400°C and heat transfer coefficient in KCal/m²/h°C is 2500. The following are the conservation practised as per the sketch given.

ENERGY CONSERVATION PRACTICES

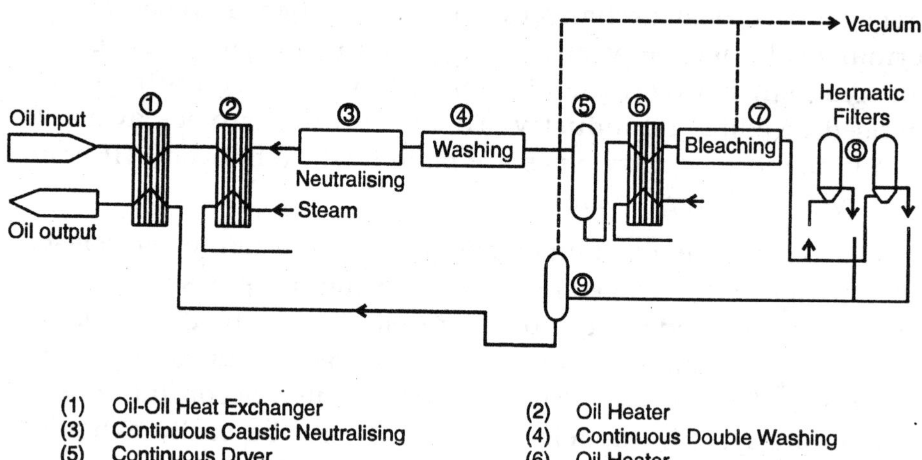

(1)	Oil-Oil Heat Exchanger	(2)	Oil Heater
(3)	Continuous Caustic Neutralising	(4)	Continuous Double Washing
(5)	Continuous Dryer	(6)	Oil Heater
(7)	Continuous Vaccum Bleaching	(8)	Hermatic Leaf Filters
(9)	Buffer tank		

Fig. 22.1 Continuous Neutralising and Bleaching with Heat Recovery (DeSinet).

This represents a system where oil is to be heated during processing and cool down again on being processed. The incoming oil feed at 40°C is first heated, regenerated by hot processed oil in a plate exchanger and then again the oil temperature is raised to 90°C by discharge through a heater which uses steam or hot water as the heating medium. The refined and washed oil is then dried before entering another heat exchanger after being heated to the bleaching temperature of 110-120°C. The bleached oil after filtration is further cooled down by exchanging heat with the raw oil feed before storage. The Fig. 22.1. depicts diagrammetically such an equipment covering conservation of heat energy.

Further economies in heat recovery may be made by re-using the steam after blowing out the filter press cakes in the bleach filters. Only a portion of the steam gets absorbed by the spent-earth cake whereas the remaining steam is practically dried and can be used directly to produce hot water for washing the oil in the neutralising section. Steam consumption both with and without recovery obtained from this system (using 10% bleaching earth) is given in Table 22.1.

***Table 22.1*: Comparative Steam Consumption**

Operation	*Without heat recovery*	*With heat recovery*
(a) Heating the raw oil from 40° to 90°C	75 kg	30 kg
(b) Production of hot water	25 kg	15 kg
(c) Vacuum ejectors	55 kg	35 kg
(d) Steam blowing of filter press	10 kg	10 kg
	165 kg	90 kg

In this way the saving of steam with 75 kg per tonne of processed oil is effected. For a plant producing 100 tonnes per day by refined/ bleached oil, the energy saving per annum (320 days) would aggregate to Rs. 7,20,000 (approx. 25,000 U.S. dollar) reckoned at steam cost at the rate of 300 per tonne based on coal as fuel.

Another source of heat recovery is provided after the degumming-washing operation from where the heat energy is available from the hot wash water discharged from the water washing centrifuge at 85°C. If an additional heat exchanger is included nearly 70% of the heat could be recovered through heat exchanged with soft water fed at 20–25°C yielding further economy. The advantages of recovering heat energy from this waste water lies in the fact that the solubility of the oil in cooled water decreases with the result that more neutral oil could be recovered from washed water in the setting tanks.

CONTINUOUS BLEACHING

In a continuous bleaching system the hot bleached oil is used to partially heat the oil-clay slurry before entering the bleach vessel where it is brought to the bleaching temperature through steam heating. The bleached oil is water cooled through the cooler before filtration. The 65-70% heat recovery is made possible in this way.

HYDROGENATION

More than 1/3 of edible oils and fats processed in the world are hydrogenated, but this step is very expensive. The hydrogenation is carried out as a batch-wise discontinuous operation. At temperatures ranging between 150 and 200°C the reaction is exothermic and therefore considerable quantities of heat are generated. Optimum utilisation of this energy as also its conservation, therefore, deserve attention. The following areas are usually looked into :

(1) heating the material to start the reaction to raise temperature,

(2) energy released due to exothermic nature of the reaction, and

(3) energy required for hydrogen dispersion.

HEAT RECOVERY

In a typical batch hydrogenation plant about 100 kg of steam at 10-12 bar per tonne of refined oil is generally required for initial heating of the oil to start the hydrogenation reaction. By employing the oil-to-oil heat exchanger located below the autoclave and providing coils, the heat requirement can be minimised. The hydrogenated oil is dropped into this exchanger, called drop-tank, and bleached oil at 50-60°C for hydrogenation is pumped through the coils on its way to the autoclave, thereby exchanging the heat with the surrounding hydrogenated oil and raising the temperature of the oil feed to about 190°C. Sufficient heat is recovered by this operation to start the hydrogenation reaction without applying any external heat, besides reducing the cooling water requirement by about 40%. The economies on 100 TPD by this heat exchange may be about Rs. 10,000/- by calculating the cost of steam at about Rs. 300/- per tonne apart from achieving economy in cooling water. As hydrogenation is an exothermic reaction heat energy is released through the process and raises the fat temperature approximately by 1.6 to 1.7°C per unit drop of iodine value per tonne of oil. A portion of this energy is used up to heat the oil from 150-190°C which is the reaction temperature and this is equivalent to iodine value drop to about 25 units. The balance has to be removed by cooling just to maintain the temperature constant at 190°C. The excess heat energy can also be used for heating boiler feed water or heating the fat in the refinery or storage tank. Heat energy can thus be recovered from hydrogenation process. By recovering this heat for a 100 TPD plant would effect a saving upto nearly Rs. 8 to 9 lakhs per annum.

There are newer technologies such as the Buss Loop technology (2). The loop consists of a product pump which circulates the oil with suspended catalysts through an external heat exchanger and the reaction mixture. There is another loop which consists of a gas circulation system where hydrogen gas is drawn out of the gas space of the converter through the self-priming action of the reaction mixer and distributed very evenly in the suspension. Finally, the hydrogen leaving the second part of the mixture get separated from the suspension in the bottom part of the hydrogenation vessel and is drawn again in the mixing nozzle by the self-priming action of the reaction mixture.

HYDROGEN PRODUCTION

Electrolysis of water continues to be the predominant method for the production of gaseous hydrogen for the hydrogenation process whereby oxygen is released as a by-product. The primary consumption being electric energy, it is necessary to have a cost effective equipment capable of sustained production with inexpensive maintenance. Two types of equipments are used for the electrolysis reaction, e.g. (1) the unipolar cells and (2) bi-polar electrolyser. In the case of unipolar cells electrode has a single polarity. The alternating cathodes and anodes of like polarity being connected in parallel while various cells are connected in series comprising the battery. Each cell operates at approximately two fields and the hydrogen and oxygen gases are collected in a separate compartments. The bi-polar electrolyser operates at a relatively high voltage and in this, each electrode has dual polarity. Bi-polar cells employ activated electrodes to operate higher current densities compared to the lower cost uni-polar cells.

Power consumption being the product of current and amperage, the operating cell voltage is a measure of energy efficiency with the production of hydrogen being directly proportional to the flow of current. With an increase in the cell voltage the unit power consumption also increases. A reduction in the cell voltage would eventually lead to energy savings. The optimum cell voltage is 1.9 to 2.0 volts per cell. At normal operating current density of 134 mA per sq. cm or 125 A/ft.sq., voltage of currently available unit polar cells is approximately 2.04 volt. However, the production of hydrogen may be increased through higher current densities. Presently, research is being carried out for operating cells at a current density of 250 mA per sq. cm. with a reduction of cell voltage to about 1.7 volt.

HYDROGEN DISPERSION

In view of the heterogeneous nature of the hydrogenation reaction, an intimate contact of the hydrogen gas with the liquid oil is necessary in order to enhance the reaction rate, this is accomplished by mechanical dispersion of gaseous hydrogen through the medium of stirring by installing motorised drive on the autoclave. Although according to the speed of reaction the energy requirement for the electric motor may differ; in general, an installed power of 2 KW is considered sufficient. However, this can be reduced to half—about 1 KW per tonne of oil batch by improvement of the design of the system for starting a gas liquid dispersion. It is possible to achieve an annual saving of over 30,000 KW hour of a 100 tonne TD hydrogenation plant. This energy saving may add upto over Rs. 70,000/- per year based on the unit cost of the electrical energy at Rs. 2.25 per KW hour.

DEODORISATION AND PHYSICAL REFINING

These operations, which are conducted at rather high temperatures, provide opportunities for saving a lot of energy by arrangement for cooling the oil from high to low storage temperatures. The energy which has been conserved in this way can be utilised leading to lowering of the cost of production, and such savings may amount to 45-85% depending upon the technique employed. Efficient recovery of enormous amount of heat energy available from the hot oil reduces the fuel consumption for steam generation. A deodoriser operation may be semi-continuous or continuous. A very easy way to recover heat energy is to exchange heat of the outgoing hot oil in the incoming deaerated bleached oil. Shell and tube or oil-to-oil heat exchangers are normally employed. Another way would be to cool the oil in a heat exchanger with water for vaporisation of water and production of low pressure steam, which may be used for sparging the deodoriser or as motive steam for production of vacuum through the ejector system or for heating the thermic fluid. The exchanger may also be used for cooling the hot oil to produce hot water which could be utilised as heating medium in other operations. In modern deodoriser heat recovering heat exchanger may be located outside the deodoriser. Deodoriser shell or inbuilt boiler in the semi-continuous deodorisers have exchangers located outside the deodoriser vessel. Generally, Plate Heat Exchanger (PHE) and Spiral Heat Exchanger (SHE) are used for these. The temperature of the refined oil feed is raised from 100°C to 240°C by heat interchange with the deodoriser oil at 260°C which gets cooled down to 120°C. The final operating temperature of 275°C for the deodoriser is attained by passing this oil through a booster heater which circulates a high temperature thermic fluid at 300°C by passage through a cooler. The temperature of the deodorised oil is brought down to 125°C to 40°C for storage. Recently continuous deodorisers have been designed which are provided at the bottom with the inbuilt oil-to-oil heat exchanger. Additional heat exchanger between bleached oil feed and deodoriser is effected through the use of plate heat exchanger (PHE) before the oil is finally cooled with the water.

VACUUM EQUIPMENT

Steam jet ejector system continues to be the preferred method for the production and maintenance of vacuum during various oil processing operations such as deodoriser, physical refining, hydrogenation, bleaching etc. Generation of vacuum during deodorisation is the highest energy consuming operation. As this affects the over all processing cost, an attempt should be made to obtain energy savings by controlling operating variables and good equipment design.

HEAT PUMP CONCEPT

The difference in temperature makes a heat stream flow. Thus, heat may be transferred from a lower temperature to a higher temperature level by means of heat pump which requires energy for driving the pump. These pumps are characterised by the fact that their output is normally higher than energy. Because the amount of efficiently usable energy at higher temperature is greater than the amount of driving energy necessary to keep the process going. The basis of most of the heat pumps is compression and expansion of a gas or vapour-liquid mixture. The ejector heat

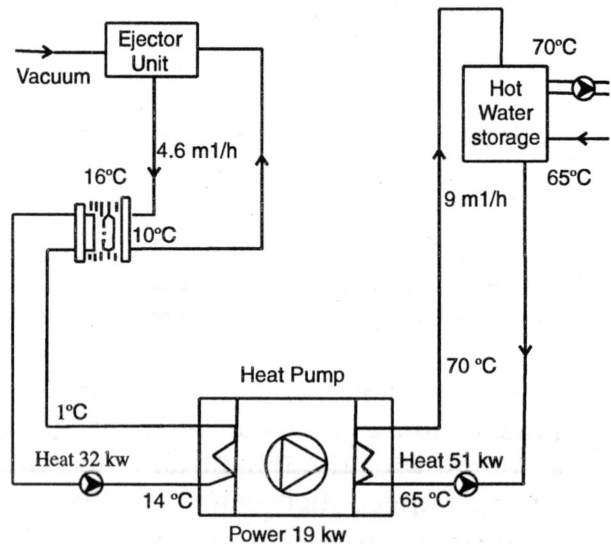

Fig. 22.2: Heat Pump to Cool Barometric Condenser Water.

pump is a special form of compression heat pump where pressurised steam is a driving force. Three basic components which comprise steam ejector include a nozzle, the mixing chamber and a condenser. Passage of steam beyond the moving fluid through the converging diverging nozzle produces high velocity steam. The ejection fluid is entrapped in the steam in the mixing chamber and is directed through the deodoriser which serves to convert the velocity energy into pressure energy. The steam requirement varies thermally with the degree of compression as the ratio is directly related to temperature of the cooling water used. The temperature of cooling water entering the barometric condenser for cooling down the vapour steam mixture gets raised by 5-7°C before leaving the condenser. This water is generally re-circulated directly or indirectly after cooling through a cooling tower. The temperature gradient indicates the desirability of recovering heat from the barometric water for which purpose a heat pump may be incorporated into the deodoriser circuit. This is shown in the Fig. 22.2. This is a new device and it is stated that the net heat energy transfer per tonne of deodorised oil is about 32 KW hour.

FRACTIONATION OF FAT

For fractionation, the solid and liquid portions of the fat are separated. This can be done through plate heat exchanger which has a very high heat transfer efficiency. This reduces the energy consumption to a very great degree.

SOAPS AND DETERGENT

Soap : Soap is a comparatively low priced commodity and any savings on energy will be reflected by reduction of production cost and enhancement of profits. A study has indicated that for production of 50 tonnes of toilet soap per day by the conventional process, which takes about 4 days, with the changes and continuous saponification employing refined, bleached and deodorised fat, which can be done in one day, considerable saving in fuel costs and steam costs is possible. Usually fatty acids are used in a number of continuous processes. So, heat required for steam for long operation is higher compared to operation in a continuous process.

Detergent : Detergents are prepared by reaction between sulphur trioxide with an organic substance such as alkylbenzene, fatty acid ester and fatty alcohols. These reactions are exothermic causing considerable heat generation. Therefore, it is essential that the heat generated be removed as rapidly as possible to avoid thermal decomposition and to produce a high quality product. The regained heat is usually removed by a heat exchanger using oil.

For preparing sulphonic acid, sulphur trioxide, sulphur is burnt with air. This operation is amenable to recovery of heat as the reaction is exothermic. The best way to recover this heat is through generation of steam by transfer of the heat from hot air to water. Recovered heat may also be utilised for spray drying of detergent of for refrigeration systems.

SOLVENT EXTRACTION

Two primary forms of energy are consumed during solvent extraction of oil cakes or oil seeds, e.g. steam and electricity. Steam constitute about 60-70% energy and electricity constitutes about 6-10%. This apart, the expensive solvent usually hexane also has serious energy requirements during manufacturing operations. Any savings from hexane reduce the quantities of steam and electricity. This is sought to be done by recovery of hexane to the utmost degree.

(1) *Steam* : The cost of steam production is a very important factor in an industry. Fuel coal is generally used in India for generation of steam. It is becoming usual to use hulls and shells or use rice husks in producing steam which is comparatively more economical from cost stand point. Various means are adopted : steam used for vacuum ejectors could be used in the first stage of distillation for concentration of the miscella (solvent oil mixture) by more than 50% in the first stage, which indirectly results in the steam conserved in the successive stages. An efficient utilisation of the hot water from the condensate embodies the use of this for degummimg of the oil, the balance being re-circulated to the polar water tank.

(2) *Electricity* : Where maximum vacuum is not required it is advantageous to employ a water ring vacuum pump instead of steam

ejectors. Similarly, electricity used for lighting can save considerable energy if high pressure sodium vapour lamps or mercury vapour lamps are used.

(3) *Hexane* : Losses of hexane occur in 3 ways :

(i) Residual solvent in extracted cake

(ii) Loss in crude oil through miscella

(iii) Hexane loss in decanting water from atmospheric discharge.

It is customary to limit the solvent loss below 500 ppm. Loss of residual hexane in crude oil may be limited to below 300 ppm by smooth functioning of a stripping column. The extent of original moisture content in the set has a marked effect on the residual hexane content in the extracted meal. It has been found that initial drying of the set before extraction of lower moisture content by 7-4.5% resulted in lowering the content of residual hexane in oil cake by half to 1/3. The initial oil content of the cakes which are subjected to solvent extraction has a bearing on the residual oil content in the extracted meal. Therefore, oil cakes having lower oil content are used as it eventually results in lesser solvent loss in final extraction, besides reducing the quantity of miscella to be distilled thereby reducing the quantity of steam to be used.

FATTY ACIDS AND GLYCERINE

The principle of heat conservation and recovery is more or less the same as in other operations, i.e. to recover the heat from the heated material. Judicious employment of preheaters and the condensers to act as heat exchangers for the in-coming and outgoing streams in the fat distillation unit can often lead to a saving and heat recovered upto 80%.

In modern installations an attempt is made to utilise a part of the waste heat from splitting process for the concentration of sweet water upto 80% glycerol or for fatty acids distillation. It is a general practice to conduct the major unit process of sweet water or spent lye evaporation by the use of multiple effect vaporisers whereby only the first stage is heated with 4 bar steam, thus reducing the overall steam consumption to 0.3 to 0.4 kg steam per kg of evaporated water.

GENERAL

As has been stated for earlier operations, the general approach is to re-utilise the heat from a previous operation by exchange through a suitable system for heating substances which are to be subsequently heated.

REFERENCES

1. *Treatise on Fats, Fatty Acids and Oleochemicals*, O.P. Narula, Editor-in-chief, 1995, Chapter S-2 'Heat Recovery and Energy Savings in Oils and Fats Plants' by O.P. Narula and R.K. Chopra. Second Volume, Industrial Consultants (India)
2. Buss, AG, Prathein 1, Switzerland.

CHAPTER 23

Safety Organisation and Pollution Abatement

INTRODUCTION

No industry can operate efficiently without an effective programme of safety, environment and pollution control measures (1). Maintenance of good and harmonious industrial relations and productivity are conditioned among other things by the importance attached by industrial management to the safety and environmental aspects of day to day factory operations.

(a) *Section A : Industrial Safety*

Industrial accidents are caused either due to unsafe acts and conditions and can therefore be avoided by controlling these factors.

(b) *Section B : Minimisation of Un-safety Acts*

Un-safety acts are committed due to lack of safety awareness on the part of the working force. The following steps are known to be effective in minimising un-safety acts:

(i) Induction course in safety for new entrants to industry : This course covers both supervisors and workers which is regularly conducted at the factory with the presence of managers.

In-plant safety training programme covers visit of workers to the shop followed by explanation of the mechanism of typical accidents that have taken place in the past and steps taken to avoid them in future.

At an appropriate stage chosen representatives are also exposed to safety courses conducted by outside agencies for exchange of opinion. Specialised courses are conducted for electricians, fitters, welders etc. since their jobs involve various types of hazards. In addition to training courses the following steps have been found to be very important.

(ii) *Safety awareness of factory personnel* : Safety slogan posters are arranged in which workers are encouraged to participate. Safety kits for demonstration and also safety songs for singing are steps to increase awareness. Similarly, safety films on mechanism of accidents and safety posters dealing with accidents and measures to avoid them are also part of the safety awareness programme along with safety quiz and also the National Safety Week celebration organised by various production and engineering departments, which put up stalls at chosen spots of their operating area to display safety equipments, accident statistics, mechanism of accidents and safety targets for the forthcoming year. Wide publicity is given to display all safety equipment and the use of safety equipment depending on the nature of the work involved. For example, when handling liquid caustic soda, it is necessary to use goggles, gloves and shoes. Similarly, due precaution is necessary during welding operations. Gas masks are necessary when working in polluting atmosphere or dusty atmosphere. The display of various types of safety equipments increases the awareness about observing safety precautions from the point of view of health and accident prevention.

(iii) *Fire-fighting equipment* : Demonstration of fire-fighting equipment and fire-fighting techniques is very necessary so that skills achieved during training sessions may be used in the case of eventualities. It is useful to display sign boards such as 'No Smoking', 'No Naked Lights' and 'Restricted Area' etc.; it would go a long way to improve safety consciousness.

However, accidents can still happen. It is necessary, therefore, to provide built-in safeguards for plant and machinery so that the operator has no access to machine parts in running conditions for preventing the fingers getting in an area where they can get cut in moving parts. In most factories safety conditions are framed to over-view good maintenance conditions. It is also mandatory, to appointment of a Safety Officer in a factory, who is responsible for accident statistics and investigating all accidents and formulating a safety programme. It is also his duty to motivate the workers and the supervisors for active participation in all safety controls and programmes. The record is maintained in each factory in statistical terms regarding the frequency rate, severity rate and incidents, rate of failure of safety measure and untoward occurrences. There are prescribed norms for expressing these.

Accident statistics may reveal that a number of accidents take place due to power plant layout. A good layout ensures adequate moving space, safety access, safe maintenance, space for future expansion, ventilation, lighting and noise control.

GOVERNMENT EFFORTS

There are prescribed factory acts and rules—the pollution acts and regulations, electricity act and rules on petroleum, which have been enacted by the Government. Compliance with these acts helps to ensure basic safety measures. The factory inspectors play the role of guardians of the law and as expert officers of the management.

VOLUNTARY ORGANISATION

The National Safety Council, the Indian Association for Occupational Health, the Safety First Association of India and the Council of Industrial Safety play and important role by conducting adequate programmes such as seminars, film shows, exhibitions etc.

SAFETY IN OIL PROCESSING INDUSTRY

In general oil processing industry does not pose serious problems of safety if basic precautions of good house-keeping are followed and provisions exist to ensure plant safety. One common unsafe condition is oily floor which may cause accidents due to slipping if the floor is not kept free from oil by adequate cleaning from time to time.

The chemicals handled are solvents in solvent-extraction, phosphoric acid for degumming and bleaching chemicals or earth, caustic soda for neutralisation of free fatty acids in oil, nickel catalysts for hydrogenation, sulphuric acid used for bleaching soap stock and hydrogen for hydrogenation. Due caution is to be exercised in handling these items.

Safe storage and handling of solvents as hexane requires knowledge of their properties. Storage tanks must be vented through seals. Bunds around storage tanks should be provided with capacities equal to those of storage tanks. In case of accidental spillage or leakage, the entire plant can be contained within the bund. An adequate fire-fighting facility should be available near at hand for taking care of accidental fire hazards. Regular maintenance of tanks, pipes, valves and pumps ensures absence of major leaks.

There is risk of fire hazard for spent bleaching earth if too much oil is left behind in the earth and the earth is discharged hot and dry from the filter press. The normal procedure is to recover as much oil as possible from the earth by pumping hot water through the filter press on completion of filtration of the bleached oil.

Smoking should be strictly prohibited in oil processing section, especially where hydrogenation is carried out. Hydrogen gas is generally stored in the fluid-sealed gas holders using water as the fluid. Autoclaves for hydrogenation need to be inspected annually or as per schedule laid down by the appropriate Government agencies.

Repairs to some coils located within the vessels should be carried out by observing adequate safety precautions such as purging of the vessel by positive removal of gases prior to repair work and by provision of venting arrangement during repair.

Fire extinguishers are to be provided at strategic locations to take care of human failures and to combat autoignition, e.g. spent bleaching earth. Emergency showers and eye wash solutions must be provided in all areas for storage, handling and transfer of hazardous chemicals, especially acids and alkalies. Valves, pumps and pressure gauges should be readily accessible and where necessary, platform should be provided.

All places should be protected by hand rail. Belt drives of motor should be protected by suitable guards. Guards should also be provided for the moving parts of the projecting machinery to prevent fingers getting cut by moving parts. Insulation of pipes and valves for reducing surface temperatures is a common precaution to be scrupulously followed.

ENVIRONMENT

Introduction : The quality of environment prevailing in the industrial establishment has a direct bearing on industrial relations. Safety of working personnel, productivity and degree of efficiency go hand in hand. In this respect all efforts should be made from the point of view of ventilation, comfortable postures, lighting and acoustics.

In oil based industries the plants are generally located in large rooms with adequate space between vessels for movement of personnel. Fatty acid plants are generally located in open space. Where plants are located in close areas, provision of windows should be adequate to take care of ventilation, suitable exhaust fans must be provided to improve ventilation.

Lagging of vessels, steam and hot oil pipes is necessary for keeping ambient temperature under control. The thickness of insulation will depend on cost effectiveness and it is considered beneficial to aim for a more complete and liberal insulation in view of the increasing cost of fuels.

Postures : The question of posture is applicable to a greater extent when dealing with a large number of workers on a job such as a packing line. Chairs with back rest are necessary for those who perform jobs in sitting posture. A large number of designs of chairs are available to suit particular needs and the choice is made with the help of the factory medical officer. Absence of back rests can lead to problems associated with back pains and can affect productivity.

Lighting : The well lit factory is a pleasant work place and is conducive to good productivity. By introducing fluorescent tube lighting simulating day light is an accepted form of lighting. Good lighting minimises safety hazards too and enables the working force to be alert at work to avoid fire hazards. Covered lighting may be provided particularly in the hydrogenation section.

Acoustics : The noise level in the processing factories should also be minimised, particularly in the packing section using tin containers. However, with the use of plastic containers and flexible pouches the noise level is minimised. The acceptable noise level is in the range of 85-90 decibels. Continuous exposure beyond this range can lead to deafness.

Noise can be controlled by the provision of adequate sound insulation between source of noise and desired environment. Simplest form of sound insulation is prepared by a barrier and the degree of protection is dependent upon the frequency content of noise source, size of the barrier and its location with respect to the sound to the listener. Mineral wool and glass fibre is a good sound absorber; other methods are also used.

POLLUTION CONTROL

Pollution control refers to the part played by the industry is effectively controlling air and water pollution and ensuring clean surroundings. Air pollution is caused by the release of noxious gases, sometimes also containing particulate matter, into the atmos-phere. These should be controlled by appropriate measures.

Water pollution is caused by the discharge of harmful and untreated effluents into nearby rivers or running streams. Absence of pollution control leads to health hazards of the community at large apart from the workers themselves. Pollution control is a social responsibility of all industrial establishments. There are legislations which have been provided by the Central and State Governments which make pollution control measures mandatory. Limits have been set for the discharge of gases into the atmosphere and for the discharge of effluents into municipal and surface drains. Before any industry is set up it is necessary to obtain prior clearance from Pollution Control Boards who scrutinise the plants and ensure that adequate facilities have been provided for controlling atmospheric and water pollution before the operation is started.

SOURCE OF POLLUTION IN OIL FACTORIES

1. *Oil Spills* : Spillage and leakage of oil take place at the time of their receipt in drums, road and rail tankers, during transfer to storage tank and during withdrawal from tanks. Some suspended oils also escape to the drain whenever water separating from storage tanks is drained. The oil unloading areas are generally enclosed so that spillages may be received and re-used. If any oil leak enters the drain located within the factory it can still be recovered by connecting the drain to a fat trap. The conventional fat traps have provision for trapping free oil which can be recovered either by skimming or by pumping. Design of the fat trap is made by study of the settling rate. Fat traps are made of cement concrete. But if the water has stagnated it will be necessary to provide a protective linking such as FRP (Fibre Glass Reinforced Plastics) to prevent direct and prolonged contact of concrete with acidic water. Care should be taken that effluent a streams containing oils are not allowed to mix with alkaline streams as the alkali can emulsify the oil. Roads or other areas where oil spills have occurred can be washed only with powerful jets of water for cleaning these areas of oil.

2. *Chemical spills* : As in the case of oils, chemical spills take place at the time of receipt and handling. Chemicals being reactive it is very important to achieve the right type of materials for construction, for receipt, storage and handling of chemicals. In the oil processing industries the main inorganic chemicals used are caustic soda, phosphoric acid and sulphuric acid. The built-in facilities of recovery of spillages and leakages are usually bunds for tanks and pumps. These should be adequately designed. In addition to these, steps are also necessary to organise a schedule of plant maintenance so that leaks etc. can be repaired in time.

Bleaching with activated earth : The two sources of pollution that are involved in earth bleaching of oil are dust from bleaching earth and spent bleaching earth. Dust from earth can be controlled by water and a separate room for bleaching earth with dust extraction facilities. Spent bleaching earth is a solid waste which can be physically removed from the place of work and can be used for making cheap soaps or for landfill depending on residual oil content. Oils from spent bleaching earth can be usually removed from circulating bleaching earth through the filter press towards the end of filtration.

Chemical bleaching: Chemical bleaching involves the use of chemicals such as sodium chlorate, sulphuric acid or other chemicals and salts. As a result the aqueous drainings are acidic. In hydrogenation factories, the effluent may contain nickel soap, nickel having escaped through the filter press as such or as nickel soap. Effluents from chemical bleaching are passed through a separate fat trap for separating the oil from the aqueous draining. The water from this filter has to be neutralised first before releasing.

Spent catalyst: After repeated nickel catalysation for hydrogenation, the spent catalyst is removed from the filter press and stored in drums for sale as solid waste. Nickel from the spent catalyst is usually recovered for rectification and in the process a part of the oil is also recovered.

Soap stock : During alkali refining of oils, soap stock is produced. This is either used as such in a nearby special factory or applied to seed oil by bleaching with sulphuric acid in FRP or lead-lined acid. The soap stock splitting section can also contribute to pollution and losses through leakages if full precautions by way of the use of corrosion-proof materials of construction and planned maintenance are not employed. Splitting vessels are lined with FRP or lead or acid resistant tiles which are used for flooring. Personnel handling acid are provided with protective equipment besides an emergency shower and first-aid kit.

FATTY ACID PLANT

Fat splitting and distillation leads to the production of residues comprising gummy matter, polymerised oil and other organic impurities associated with the oil. These are either disposed off as solid waste or re-used for manufacturing inferior grade soaps or for making grease.

Barometric Condenser water in oil processing units : Barometric condensers are extensively used in the vacuum systems. There are two types of Barometric Water—Once-through system and recirculation system. The once-through system is generally valid where sea water is available freely. The discharge from the barometric condensers returns to the sea via a fat trap. In the recirculation system, the water used is generally fresh water from rivers of ponds. It is distributed from an overhead tank through the overhead vessel. The discharge from this vessel is passed through a cooling tower for reducing the temperature of the water to the original level. Since the water is circulated through a close circuit system the question of pollution of source of water does not arise.

Boiler House : In the Boiler House, which is an essential section of an oil processing factory, the sources of pollution are :

(a) Coal spillage where coal is used as fuel

(b) Water treatment section

(c) Boiler blow down

(d) Chimney exhaust and boiler ash

Coal spillage can be largely avoided by locating the boiler away from the manufacturing units so that the coal dusts do not spread to the manufacturing area. Adequate storage and handling facilities are provided.

Water treatment section : During water treatment of boiler water the chemicals used are lime and soda which may cause the formation of calcium and magnesium carbonate. This goes down as a slurry which has to be passed through filter press for separating the solids, and water can then be led into the drain. There are standard procedures for the method of treatment where ion exchange method is used.

Boiler blow down : Boiler blow down results in an effluent steam containing inorganic and organic matters. However, this is very small.

Chimney exhaust : To prevent generation of bulk smoke from the boiler chimney careful attention is to be given to the quality and flow rate of coal being burnt and efficiency of combustion. If the combustion is efficient, i.e. with adequate supply of air or oxygen, bulk smoke can be avoided. The gases from chimney exhausts are analysed from time to time to ensure this.

Boiler ash : Ash from the boiler is first cleansed with water and the water is then passed through a filter bed or press to free it from suspended matter before it is led to enter the drain and mix with other effluents.

ENVIRONMENTAL PROBLEMS AND POLLUTION CONTROL IN OIL BASED INDUSTRIES

Oil based industries have a lot of pollution problems which have to be taken care of for the safety of the workers associated with the factory and the community at large (2). There are laws in every country of the world to minimise or prevent the problems caused by pollution. All countries have a set of laws under their statutes. Indian concerns have started more or less in earnestness with the participation at the UN Conference of Human Environment held in Stockholm in June 1972.

There are laws concerning hazardous substances such as solvents which need fire protection. Environmental Pollutance and the Bureau of Indian Standards, formerly Indian Standard Institution, have prescribed specific guidelines on which the Central/State Pollution Control Boards are to act. Generally the governmental authorities can further modify conditions to some extent according to the situation. There is a Minimum National Standard (MINAS) for a particular industry which are to be observed by every industrial establishment by installing pollution controlling measures. General medium standard for discharge of effluents which are applicable have been given by authorities. For example, the general standard for discharge of effluents for in-land surface water, public sewage meant for irrigation and marine coastal areas have been specifically given.

The standard of such effluents discharged from any factory should be such that they do not pose danger to the environment where they are discharged. For example, there are standards for colour and odour, suspended solids, particle size of suspended solids, pH value, temperature, oil and grease, total residual chlorine, nitrogen, free ammonia, Biochemical Oxygen Demand (BOD),

Chemical Oxygen Demand (COD) and such harmful metals like arsenic, mercury, lead, cadmium, chromium, copper, zinc, selenium, nickel, iron, vanadium etc. and chemicals like cyanide, fluoride, phosphates, phosphides, sulphide, phenolic compounds, radio active materials etc.

There are also ambient area quality standards such as the presence in a concentration not above prescribed limits. These have been prescribed for silver oxide, oxides of nitrogen, suspended particulate matter, respirable particulate matter, lead, carbon monoxide.

For area quality they are divided into industrial area, residential and rural and other sensitive areas for which different standards are prescribed. The Government of India has also enacted for protection against noise pollution for various areas such as industrial area, commercial area, residential area, silence zone etc.

EFFLUENT TREATMENT IN OIL PRODUCING INDUSTRY

The pertinent parameters and their levels in the effluent have been specified with a view for protecting and improving environment (3). The authorities do not specify the discharge of absolutely pure water as this would have been impractical. The standards allow a certain degree of pollution, but this must not be extended. Some of the significant parameters are given as under :

(i) *Colour and odour* : The standards are prescribed because these indicate the presence of pollutant substances and may also become offensive, such as hydrogen and methane gas, organic impurities etc. These are offensive and may cause nausea and vomiting.

(ii) *Suspended solids* : There are either inorganic or organic in nature. There is a test which measures the sizes, e.g. those which are of sizes above 1μ are filterable. Those with diameters in the range 1 mu to 1μ are non-filterable. These may also be colloidal solids. These do not settle by themselves.

(iii) *Size of the suspended solids* : Generally coagulation or biological oxidation process followed by sedimentation are adopted to remove these particles. The suspended solids are also classified into volatiles and fixed. The former are mainly due to organic fractions which exist at 600°C. Their quantification is important in passing a biological treatment system.

(iv) *pH* : This characterises the nature of the waste water whether acidic or basic. Values outside the prescribed range are harmful to aquatic life. Low values would also damage the linings of the stream convenience systems.

(v) *Temperature* : This is also important for the well being of the aquatic life which helps to purify the water. Moreover, oxygen is less soluble in warm water. The rate of biological reaction also increase with temperature. When accompanied by low dissolved oxygen concentration effect of warm temperature is lethal to aquatic life.

(vi) *Oil and grease* : These are also contaminants in oil processing water. Water in mineral oil or kerosene are also counted as oily mix. These may be in the free state or in the emulsion. They may float on the surface in the free state, or may appear as film. Oil and grease tend to coat suspended solids and ducting surfaces. They also slowly react with biological oxygen of waste water.

(vii) *Biological Oxygen Demand (BOD)* : This is the measure of organic pollution due to presence of protein, carbohydrates, fats and other biodegradable organic matters. It is expressed in terms of oxygen consumed by microorganisms in biochemical oxidation. Organic matter in the sample in a 5-day period under conditions of the test is known as BOD. This reaction is conducted at 20°C. It has to be noted that it is not the total oxygen required for complete oxidation of all the organic matter. Biochemical oxidation is a very slow process, complete oxidation may take indefinite number of days. Reaction is 60-70% complete within 5 days and 95% in 20 days. BOD values are very useful for practical purposes. It helps to determine the quantity of oxygen required for the treatment system as well as to design the reactor. Discharge of untreated waste water streams with high BOD into water streams can lead to depletion of oxygen in the water stream and to the development of septic conditions. Aquatic life in the water stream will be in danger.

(viii) *Chemical Oxygen Demand (COD)* : This also is a measure of organics in waste water. The oxygen equivalent of the organic matter that can be oxidised by strong oxidising chemicals in a static medium is determined. Potassium dichromate is the oxidising medium chosen, but the test is carried out in boiling temperature and usually in the presence of catalysts such as silver sulphate. The COD of waste water is usually higher than BOD. Most organic contents and even inorganic contents are chemically oxidised. Under uniform conditions it is possible to correlate BOD with COD. In oil refinery waste water, the ratio of BOD to COD is 2 : 3. COD is a good tool for plant control purpose. It can be determined in 3 hours against 5 days required for a BOD test.

There are a number of tests prior to the treatment in any establishment. Every effort is made to prevent the discharge of polluting substances by using processes which will eliminate production of pollutants. For example, substitution of caustic refining by physical refining will eliminate production of pollutants like soap stock and waters. A conscious effort is made that some of the operations may be modified to reduce the pollution load either in limits or level of concentration. Steps such as introducing a fatty acid scrubber between the deodoriser and bigger contact condenser. In solvent extraction plant the waste water from the water solvent separator is taken to a stripper to recover the residual solvent left in the water. Special attention should be paid to leakages, spillages etc. A good house-keeping practice is a very good measure of prevention. The plant of waste water can be reduced by recycling and re-use. Direct cooling water are recycled immediately. Hot water is used for floor washing. Steam condensates are collected and re-used either for hot water wash or for raising steam. The clarified water after an effluent treatment can be used for washing toilet or garden.

(xi) *Segregation* : To further reduce the cost of processing, the various waste streams should be segregated according to the treatment required for the stream. The waste water streams in an oil refinery may be listed as follows :

(a) Domestic waste water

(b) Stream water

(c) Industrial waste water

(i) soap stock
(ii) washed water
(iii) floor washing water
(iv) direct contact condenser water
(v) indirect contact condenser water and
(vi) boiler blow down

Domestic effluent can be directly run into sewer lines. Stream water should be directly led into nullahs or specific drains. The industrial waste water has to be treated. The recoverable oil bearing streams, such as soap stock, wash water, floor washing should be kept separate from non-recoverable oil bearing streams such as direct condenser water. They can be mixed only after oil has been recovered. Boiler blow down etc. which do not require chemical treatment can be united with other streams after chemical treatment.

(x) *Equalisation* : Similar streams but with varying degrees of pollutant load should be allowed to mix and average out before being run into the treatment section. It is usually the practice to neutralise the acidity of acidulated water along with acidulant of soap to recover the fatty matter. The operations are as follows:

The acidic water is mixed with alkaline water, i.e. floor washing by agitation. To separate acid oil, the washed water is allowed to settle. This water is then treated. The treatment consists of three types of operation e.g. physical, chemical and biological. The physical operations are the following : screening, sedimentation, floatation and filtration. The water may be allowed to pass through a screen fixed vibratory or rotary from which solids such as rags, paper bags and polymeric bags, wood pieces, metal cans and other small objects as oil seeds or meal particles are separated. For smaller pieces a screen made of wire mesh of perforated place may be used. The size of the screen is so selected that the head loss is minimum.

(xi) *Sedimentation* : The process employed is for the separation of filterable solids or liquids. Sedimentation may be conducted in bigger tanks or continuous clarifier or separators. But continuous plants are preferred as mechanical working improves both the rate of settling and conpaction of sludge as well as its removal.

(xii) *Floatation* : The process is the same as that of sedimentation. The separation is aided by introducing air bubbles into the liquid phase. The bubbles attached to the particulate matter make it buoyant and raise it to the surface. The particles or oil which separates can be skimmed off.

(xiii) *Filtration* : In filtration the flow of slurry is blocked by a porous medium. The filtration medium may be a bed of sand (sandy filters), textile material or wire mesh. In sand filter the system is reconditioned by a wash. In the other the sludge layer is removed mechanically. Filters may be gravity filters or pressure filters. The sand filters are usually gravity filters.

(xiv) *Chemical Operations* : This operation is essential for oil refinery waste waters using caustic refining process. The pH has to be corrected since this is a statutory requirement. It is also important to have pH around 7 for efficient BOD treatment. Generally, the refinery waste water reaching the treatment section is acidic. The acidity is usually neutralised with lime such as lime slurry or milk of lime. Sufficient lime must be added to neutralise the acidity and to react with acidic materials to form calcium carbonate.

Coagulants : Use of coagulants such as alum, ferric chloride, ferrous sulphate etc. induces floc formation and improves clarification. Ferric hydroxide is oxidised to the very desirable ferric form by the oxygen adsorbed with water. Thus bulky gelatinised flocs are formed. These enmesh the suspended matter in mechanical way and carry it down into the sludge. The adsorption of colloidal organic matter also takes place on the large surface afforded by the precipitate.

Poly-electrolytes (PE) : These are long chain organic molecules (polymers), whose sub-units are ionisable depending on whether their charge, when placed in water, is negative, positive or neutral. The PEs are classified as anionic, cationic or non-ionic. The PEs neutralise the charge of the solid particles and help them to floc together. The solid particles may get absorbed at surface of PE. The bridge may be formed when two or more particles may get absorbed along with the length of the polymer. The settling character of the precipitate is vastly important. The separated sludge may be floated to the surface using air and then skimmed off. More often it is allowed to settle in sedimentation tanks or clarification.

Since flocculation precedes settling, combined flocculators-clarifiers are also used. In a typical flocculator-clarifier the influent enters the flocculation section at the centre where it is gently agitated to help it to form flocs. It flows vertically down and rises in the clarification section. Here it finds conditions for vibrating settlement. Sludge settled at the conical bottom may be scraped towards the outlet at the centre. Clear effluent overflows the circular wier at the top.

Typically the whole chemical treatment scheme may consist of a flash mixer wherein the influent and chemical solutions are dozed in and mixed rapidly. The mixed liquor is pumped into the flocculation part. Polyelectrolyte slurry or gel is dozed into the exit line of the mixed slurry pumped by another dozing pump. The sludge formed is removed continuously. A telescopic pipe with adjustable outlet pipe is used. The sludge overflows into the collecting tank from where it may be led into the dry beds or filter press for dewatering. Clear effluent overflows from the top of the circular wier. It is mixed with filtrate and conveyed to the next operation. In this operation the total suspended salts may be reduced by 80-90%. The reduction in BOD by 40-70% and COD by 30-60% can be expected. This greatly eases the subsequent biological treatment.

Biological Operations : The objective is to reduce the BOD. This is achieved by removing organic compounds by their biochemical oxidation with the help of living microorganism. A part of the organic matter is used up for creating new cells. The remainder is used up for producing energy and stable oxidised products. These metabolise the colloidal and dissolved organics into gases and/or assimilate them as food to form more cells and tissues. These tissues being heavier, settle down and are removed.

Type of operations : The biological process may take place under varying conditions with respect to oxygen; that which occurs in the presence of oxygen is known as aerobic and that which takes place in the absence of oxygen is anaerobic. The bacteria which live in the presence of dissolved oxygen are known as obligated aerobes, those which survive in the absence of oxygen are known as obligate anaerobes. There are bacteria which are indifferent to the presence or absence of oxygen. These are known as facultative microorganisms and the biological treatment process carried out with their help is known as facultative. The biological process may take place as a suspended growth process, i.e. the microorganisms are maintained in suspension within the liquid. In contrast, the process may be an attached growth process. Here the micro-organisms are attached to some inert medium such as rocks, sludge or specially designed ceramic or plastic materials.

Bacteria : Bacteria are the most widely occurring microorganisms in biological waste treatment. These treatments have therefore been used interchangeably. Bacteria are unicellular in structure; other microorganisms may be multi-cellular. They multiply by binary fission.

Nutrients : In their nutritional requirement most living cells are quite similar. To continue to live, the organisms need a source of energy and nutrient material, principally carbon for the synthesis of new cellular material. Inorganic material such as nitrogen, phosphorus and some trace elements like sulphur, potassium, calcium and magnesium are also important for cell synthesis. In case the system is deficient in any of these, they are to be provided. Energy may be obtained from the sun or by an oxidation reduction reaction such as oxidation or fermentation of the organic matter.

Process : The most frequently used biological process in oil refinery is the activated sludge process. Essential to this process are two pieces of equipment : (1) reactor and (2) settler of sludge separator. The influent containing the organic matter to be removed flows into the reactor. Here the aerobic bacteria are kept in suspension. Oxygen is supplied by the use of either defuser or mechanical aerator. Constant agitation keeps the mixed liquor in suspension. Nutrients may be added if necessary. After allowing the required residence time the mixed liquor is allowed into the settler. The cell mass is slightly heavier than the treated water. The mix of few old cells separates as sludge. A portion of the settled sludge is returned to the reactor to maintain the desired concentration of mixed liquor suspended solids (MLSS), a measure of the concent-ration of the organisms in the reactor. The balance is dewatered and disposed off. The residence time 't' is determined by considering

L = initial BOD of the waste water

r = ratio of waste water flow to return sludge flow

p = proportionate increase in the number of active cells per day

s = MLSS in aerator section

T = temperature of the mixture in the aerator such that

$$t \alpha \frac{L}{r.p.s.T}$$

Higher inlet BOD would require a larger aeration tank to provide higher residence time. Since temperature speeds up the reaction a lower MLSS would be sufficient in summer as compared to winter. Again, during colder nights the reaction will be slower thereby needing less of air. It is to be noted that sufficient air should be provided to keep the dissolved oxygen level between 1 and 2 ppm. It should not fall below 0.5 ppm otherwise anaerobic reaction will take place. The efficiency of BOD removal is generally between 85 and 90%. A 2-stage system may be used if initially BOD is above 600 ppm.

Disposal of the excess sludge : The sludge has to be dewatered. This may be done over drying beds or by means of a filter press or a centrifuge. Centrifugal decanters have also been used. The dewatered sludge is also used as landfilling. Ideally it should be stabilised. It should be biologically degraded by aerobic or non-aerobic digestion. In the latter case methane gas is produced as a by-product and the residue is usable as manure.

REFERENCES

1. *Treatise on Fats, Fatty Acids and Oleochemicals*, O.P. Narula, Editor-in-Chief, 2nd Volume, 1995, Chapter T-1 — 'Safety Organisation and Pollution Abatement' by K.S. Parasuram
2. *Treatise on Fats, Fatty Acids and Oleochemicals*, O.P. Narula, Editor-in-Chief, 2nd Volume, 1995, Chapter T-3, 'Environmental Protection and Pollution Control in Oil Industry' by R.K. Suri & A Mehrotra
3. *Treatise on Fats, Fatty Acids and Oleochemicals*, O.P. Narula, Editor-in-Chief, 2nd Volume, 1995, Chapter T-4 'Effluent Treatment in Oil Processing Industry' by K.C. Mathur and Surinder Kumar

CHAPTER 24

Supply of Edible and Industrial Oils

INTRODUCTION

Oils and fats are the essential components of human diet. The importance of these products has been increasing day by day. We consume about 75 lakh tonnes of edible oils and fats of vegetable origin besides about 10 lakh MT of dairy fat. Our current per capita consumption is a little more than 8 kg per annum excluding invisible fats. This is around 50 per cent of the world average not to speak of the per capita consumption level of developed countries which are much higher.

No doubt, there has been very significant increase in the production of oilseeds and, hence, of oils over the years. However, the growth in production has not been able to keep pace with the growth in demand consequent upon growth in population and increase in the income. The consequence is that there continues to be an imbalance in the demand and availability of edible oils. The shortfall is currently estimated at around 1 million tonnes annually. As a supplemental measure, to tide over this situation, Government has had to resort to import of edible oils so as to ensure adequate availability of edible oils and hold the price line within reasonable limits.

Our production increase so far has been more because of area increase rather than the increase in productivity. Horizontal expansion depends upon substitution of oilseeds for less profitable crops particularly the coarse grains, which are invariably grown in unirrigated conditions. There is a limitation of land availability for crops other than foodgrains in the country primarily due to national priority to achieve self-sufficiency in the production of foodgrains.

The key oilseeds/oil bearing materials, where major opportunities for increasing the production exist, are groundnut, rapeseed/mustard, soyabean, sunflower and rice bran in the short run and oil palm/coconut, particularly oil palm, in the long run. In fact, having regard to the high productivity of oil palm (yield of palm oil per hectare is 10 times or even higher than that from cultivated oilseeds), it has been promoted in agro-climatically favourable regions of the country (Andhra Pradesh, Karnataka, Kerala etc.) both by private and government agencies. The total potential area so far identified as suitable for oil palm cultivation is nearly 8 lakh hectares. The progress so far made is not encouraging. There are constraints to be overcome. But even if we are able to achieve 25 % success, we can reasonably hope to have an additional availability of 6 lakh MT of oil on the assumption of a modest average yield of 3 MT of oil per hectare. Apart from contributing to edible oil resources, promoting red palm oil as a very important source of beta-carotene can also go a long way in combating vitamin A deficiency which is a major nutritional problem in the country.

In order to help improvement in the process technology so as to increase production of quality oils from oil bearing materials, as also optimum utilisation of these materials, the Directorate of Vanaspati, Vegetable Oils & Fats has been funding various research projects. A number of institutes such as Central Food Technological Research Institute (CFTRI), Mysore, Department of Applied Chemistry, Calcutta University, Ganesh Scientific Research Foundation, New Delhi, Mechanical Engineering Research & Development Organisation (MERADO), Ludhiana, Harcourt Butler Technological Institute, Kanpur, who have availed if this facility, have already made important contributions.

INDIA'S OILSEED/OIL PRODUCTION OUTLOOK

The production target of oilseed for 1996-97 as set by the Union Ministry of Agriculture is 23 million tonnes. The oilseeds production target for the year 2000 AD has been fixed at 26 million tonnes. The productivity of oilseeds in India is 944 Kg per hectare against the world average of 1340 Kg per hectare during 1994-95. The total production of nine cultivated oilseeds during the year 1995-96 has been around 224 lakh tonnes (Table 24.1). The trade estimate is 209.8 lakh MT which was declared at 34th All India Convention of Oilseeds and Oils Trade and Industry held at Bangalore.

The normal area under the rabi oilseeds is around 108 lakh hectares. The sowing of rabi oilseeds especially mustard, sunflower and linseed is completed. So far the coverage of 98.0 lakh hectares has been reported by Directorate of Oilseeds Development, Hyderabad, against the coverage of 101.0 lakh hectares during the corresponding period last year.

The condition of the standing rabi crops is reported to be satisfactory. The sowing of summer groundnut is in progress while for sunflower in the northern states, this is yet to start. The coverage under summer oilseeds may be a little more than last year's coverage. But so far, no report of area coverage for summer has been received.

The Agriculture Ministry's production target of oilseeds at 23 million tonnes for 1996-97 is likely to be achieved. In addition to this, oil is also available from non-conventional secondary sources such as copra, cottonseed, oilcake, oilseeds of tree and forest origin. The net domestic availability of oils from all these sources excluding the oil which goes in the form of export of oilseed/oil equivalent as also for industrial uses is estimated to be of the order of 65 lakh MTs. This is against the demand for edible oils which is estimated at 78.5 lakh MTs for the year 1996 97. Thus, there is likely to be a shortfall to the extent of 13.5 lakh MTs of edible oils for the year 1996-97. The shortfall on industrial oil front is estimated at around 2.5 lakh MTs.

FACTORS RESPONSIBLE FOR LOWER PRODUCTION

One of the major reasons for this situation has been lower productivity of oilseeds compared to the situation in other countries. Though our average yield of oilseed has increased from about 700 Kg per hectare in 1986-87 to around 944 Kg per hectare in 1995-96, this compares very poorly with the world average of around 1350 Kg per hectare and 2000-3000 Kg per hectare in developed countries. A number of factors seem responsible for this situation :

(i) Only about 20 % of the oilseed crops are being irrigated. Thus, the extent and spread of rainfall has a critical role in production.

(ii) Horizontal expansion depends upon substitution of oilseeds for less profitable crops particularly coarse grains, which are invariably grown in unirrigated conditions. There is the limitation of land availability for crops other than foodgrains in the country primarily due to national priority to achieve self-sufficiency in production of foodgrains.

(iii) Lack of HYBRID/HYV seeds. As per the available information, the availability of this quality of seeds is hardly 10 % of the requirement.

(iv) Susceptibility of oilseeds to pests and diseases.

(v) Oilseeds production is yet to receive the desired priority in the extension set-up in the country.

MEASURES NECESSARY FOR HIGHER PRODUCTION

The key (primary) oil bearing sources , where major opportunities for increasing the production exist, are groundnut, rapeseed/mustard, soyabean, sunflower and rice bran in the short run and oil palm coconut, particularly oil palm, in the long run. There is considerable scope for increasing the production of oils from these sources. This is because the yield per hectare of these cultivated oilseeds are much lower than those achieved elsewhere.

It is possible to achieve an additional production of around 10 lakh MT of oils on assumption of 20 % increase in the productivity of cultivated oilseeds. Some of the measures to be taken/being taken for higher production are :

(i) Need to concentrate efforts on developing HYBRID/HYV seed industry as well as promoting import of technologies for HYBRID/HYV seed production without any sectoral bias so as to sustain growth.

(ii) Timely availability of HYBRID/HYV seeds, particularly short duration soyabean varieties, HYBRID/HYV mustard and short duration sunflower varieties in adequate quantities.

(iii) Watershed management to provide protective irrigation particularly during long dry spells.

iv) Application of improved agricultural practices including fertilisers etc.

(v) Oil palm cultivation having regard to the high productivity of oil palm (yield of palm oil per hectare is 10 times or even higher than that from cultivated oilseeds). Oil palm cultivation has been promoted in agroclimatically favourable regions of the country (Andhra Pradesh, Karnataka, Kerala etc.) both by private and Government agencies. The total potential area so far identified as suitable for oil palm cultivation is about 8 lakh hectare. The progress so far made is not encouraging. There are constraints to be overcome. But even if we are able to achieve 25% success, we can reasonably hope to have an additional availability of 6 lakh MT of oil, on the assumption of a modest average yield of 3 MT of oil per hectare.

WORLD OILSEED PRODUCTION OUTLOOK

In the nine years ended 1995-96, the area harvested for the 10 major oilseeds increased almost uninterruptedly by an average of 3.8 million hectares per year. The growth was particularly sharp during the last three years at an average of 5.8 million hectares per year, bringing the total area harvested in the year 1995-96 season to a record 189.2 million hectares. Most of the acreage boost during the past few years was in soyabean, sunflower seed and rapeseed.

The current world production of 10 major oilseeds is 258.6 million tonnes which is an increase of 6 million tonnes from the last season and assumes a recovery in the average yield by 3%. The world production of 10 major oilseeds is given in Table 24.2.

The world production of oils and fats during the year 1996-97 (October to September) is expected to reach a record 97.2 million tonnes. The prospective annual increase of 2.1 million tonnes will be better than this season's 1.7 million tonnes. The production of oils and fats from oilseeds is projected to rise only about 0.5 million tonne based on the present oilseed crop prospects and assuming a reduction of oilseed stocks to minimum levels. The world soyabean oil production is likely to show a healthy growth of 3.0% to a record 20.3 million tonnes in the next season. The two major oils, namely rapeseed and sunflower oils, are projected to show a combined decline of almost 0.6 million tonnes as they are suffering most of the acreage shifts to grains. Palm oil is estimated to be top performer among the other oils and fats. An excellent growth of over 0.6 million tonnes is expected also for olive oil. The world production of oils and fats is given in table 24.3.

Malaysia currently accounts for 51% of the world's palm oil production and 62 % of world exports. It also accounts for 8 % and 22 % of the world total production and exports, respectively, of oils and fats. The average productivity of palm oil vis-a-vis other major oilseeds are as follows:

Oilseeds	*1994-95 Oilseed yield (tonnes/ha)**	*Oil Conversion Factor(%)*	*Oil Equivalent (Kg/ha)*
Soyabean	2.19	18-19	405.2
Cottonseed	1.02	18-20	193.8
Groundnut	0.93	45-50	441.8
Sunflowerseed	1.24	40-45	527.0
Rapeseed	1.30	40-45	552.5
Sesameseed	0.41	45-50	194.8
Palm oil	—	20	3230.0 **
Palm kernel oil	0.97	45-50	460.8
Coconut	0.57	65-68	379.1

* *Source :* Oil World Statistics Update, Jan. 1996.
** Malaysia's yield is 3495 kg/hectare.

As already indicated, having ragard to the high productivity of oil palm, in an effort to achieve self-reliance in edible oils, Government of India has already launched a programme for cultivation and promotion of red oil palm in agro climatically favourable regions of the country. Apart from contributing to the edible oil resources, promoting red palm oil as a very important source of beta-carotene will go a long way in combating Vitamin A deficiency which is a major nutritional problem in India.

***Table 24.1*: Production of Nine Cultivated Oilseeds in India**

(In lakh tonnes)

Oilseeds	*1994-95**			*1995-96**		
	K	*R*	*T*	*K*	*R*	*T*
Groundnut	60.6	20.0	80.6	58.2	19.9	78.1
Soyabean	39.3	—	39.3	49.9	—	49.9
Rapeseed/Mustard	—	57.6	57.6	—	60.7	60.7
Sunflower	3.3	8.9	12.2	3.9	9.3	13.2
Sesamum	3.9	2.0	5.9	3.5	2.0	5.5
Safflower	—	4.2	4.2	—	4.0	4.0
Niger	2.0	—	2.0	1.9	—	1.9
Linseed	—	3.2	3.2	—	3.1	3.1
Castor	8.5	—	8.5	7.8	—	7.8
	119.5	93.9	213.4	127.4	97.0	224.2

K - Kharif
R - Rabi
T - Total
(**Source** : Ministry of Agriculture)

A detailed statistical data has been prepared recently (April, 1997) by the Directorate of Vanaspati, Vegetable Oils & Fats, Ministry of Civil Supplies, Government of India. This data is presented in Tables 24.4 to 24.12 in addition to the statistics already given in earlier tables.

As has been stated already, the total production of oilseeds and oils in India can go up much higher as productivity per hectre in India is one of the lowest in the world partly because oilseeds are grown in unirrigated land and agricultural inputs are very low. With better irrigation and proper agricultural inputs like fertilisers, growth promoters and also appropriate post-harvest technology there is no doubt that at least 3-4 times of the present volume of production of oilseeds can be achieved with the next 5 years. It is hoped that the awareness of this possibility would motivate those who are concerned to the various aspects of the implementation of programmes for betterment of oilseeds sector.

Table 24.2 : World Production of Oilseeds (Year : October - September)

In million tonnes

Oilseeds	1996-97	1995-96	1994-95	1993-94	1992-93	1991-92	1990-91	1989-90	1988-89	1987-88	1986-87	1985-86
Soyabean	132.70	123.16	135.85	116.98	117.61	108.05	104.21	107.48	96.05	104.47	98.03	97.41
Cottonseed	34.80	34.29	33.03	29.79	31.90	37.59	33.82	31.23	32.70	31.76	27.37	31.04
Groundnut Shelled	19.70	19.42	19.62	18.13	17.18	16.71	16.56	16.38	17.72	15.19	14.88	14.44
Sunflower seed	24.29	25.24	24.31	21.39	21.63	22.61	22.86	21.78	20.71	21.10	18.75	19.57
Rapeseed/ Canola	30.76	34.62	29.98	27.04	25.59	28.49	25.32	22.07	22.65	23.38	19.64	18.77
Sesame seed	2.50	2.47	2.48	2.46	2.39	2.19	2.35	2.07	2.14	2.21	2.23	2.27
Palm kernels	4.98	4.71	4.49	4.24	4.15	3.57	3.35	3.30	3.15	2.68	2.44	2.38
Copra	5.17	4.86	5.20	4.81	4.73	4.62	4.95	5.50	4.45	4.43	4.95	5.28
Linseed	2.37	2.57	2.45	2.18	1.97	2.46	2.88	2.45	2.30	2.79	3.29	3.01
Castor seed	1.33	1.33	1.32	1.14	1.14	1.17	1.35	1.11.	1.01	0.83	1.05	1.18
Total	258.60	252.66	258.72	228.17	228.30	227.47	217.64	213.37	202.88	208.85	192.65	195.33
Total Supplies	289.65	288.62	286.14	259.22	258.35	257.51	247.06	242.21	231.47	234.89	223.26	220.70

Source : 'Oil World', December, 1996.

Table 24.3: World Production of Oils & Fats

(In 1,000 MT)

Grand Total	*Oct - Sept 1996-97*	*Oct-Sept 1995-96*	*Oct-Sept 1994-95*	*Oct-Sept 1993-94*
Opening Stocks	12154	11337	9589	10663
Production				
EV - 15	13737	13683	13840	12591
Other W. Europe	388	397	364	372
East Europe	2684	2599	2420	2329
Ex-USSR	3771	4028	3698	4187
Nigeria	1024	1028	1041	1042
South Africa, Rep.	463	447	335	348
North Africa(Z)	653	413	476	633
Canada	1792	1833	1696	1506
U.S.A.	14183	14010	14635	13122
Mexico	1037	1062	981	987
Argentina	4160	4285	3772	3182
Brazil	4892	4922	4633	4275
Columbia	502	512	506	456
China, PR	10082	10198	9447	8774
India	7639	7359	7207	7155
Indonesia	6161	5653	5201	4849
Japan	1902	1932	1907	1941
Korea, South	410	401	393	349
Malaysia	9544	9449	8896	8183
Pakistan	994	981	854	786
Philippines	1445	1275	1760	1133
Taiwan	594	596	582	567
Thailand	578	560	523	478
Turkey	999	868	906	737
Australia	808	761	736	718
Newzealand	404	395	376	383
Other Countries	6606	6382	6268	6068
Total	97454	96028	93451	87153

Source : 'Oil World', December, 1996.

It also needs to be recognised that apart from oilseeds *per se* the de-oiled oilseeds residue, i.e. the oil cake is also a valuable item which is exported in large quantities being utilised for the production of edible protein, animal feed, poultry feed and also as bio-fertiliser. India has already embarked on marketing this commodity and the total quantity of all cakes was around 2,038,358 MT in 1995-96.

In the context of what has been stated, one can be optimistic about the future development in India of the oilseed sector with its great potential in terms of economic factors, which has a great bearing on the industrial sector as a whole. The total financial involvement may be between 60,000 crores to 101,000 crores if all aspects of the oilseed industry, including industries that are connected with it, are considered. Future development of industrial economy in India will depend to a very large extent on its agriculture and the oilseed sector is a major part of this agricultural sector, second only in terms of importance to the textile and sugar industry. With this view of things much more attention needs to be given to this sector by the agricultural agencies, industries and also research and development institutions. It is hoped that within the first 10 years of the 21st century this would happen.

***Table 24.4*: Indian Production of Vegetable Oils 1985-86 to 1994-95 (Nov.-Oct.)**

(000' Tonnes)

Oils	85-86	86-87	87-88	88-89	89-90	90-91	91-92	92-93	93-94	94-95
Groundnut	1188	1364	1357	2241	1879	1742	1644	2052	1799	1936
Rapeseed & Mustard	891	861	1150	1456	1370	1739	1952	1619	1792	1835
Seasmum	150	135	174	204	225	225	216	258	173	228
Soyabean	157	136	138	240	279	403	386	482	718	574
Sunflower	90	136	207	120	204	282	386	383	454	483
Safflower	70	70	92	86	96	63	39	66	113	80
Niger	73	49	68	68	72	72	68	68	68	76
Linseed	126	106	130	130	116	116	102	95	116	105
Castor	117	87	76	160	203	281	225	242	250	273
Total for 9 Oilseeds	**2862**	**2944**	**3392**	**4705**	**4444**	**4953**	**5019**	**5265**	**5483**	**5590**

Oils	85-86	86-87	87-88	88-89	89-90	90-91	91-92	92-93	93-94	94-95
Coconut	200	188	215	252	277	288	298	335	335	350
Washed Cottonseed	309	244	226	313	409	349	346	405	379	400
Ricebran	203	279	291	316	350	357	402	428	448	425
Solvent Extracted Oils from Oil Cakes	77	113	115	163	230	242	330	270	280	409
Oils from Tree and Forest Origin	130	102	154	150	130	125	135	125	125	135
Minor Oils like Watermelon, Corn etc.	25	35	35	40	45	70	70	70	60	60
Total for other Sources	**944**	**961**	**1036**	**1234**	**1441**	**1431**	**1581**	**1633**	**1627**	**1779**
Grand Total	**3806**	**3905**	**4428**	**5939**	**5885**	**6384**	**6600**	**6898**	**7110**	**7369**

***Table 24.5*: India - Oilseeds Area, Production and Yield**

Area : '000 Hectares Production : '000 Tonnes Yield : Kg. Per Hectre

Oilseeds	89-90	90-91	91-92	92-93	93-94	94-95
GROUNDNUT						
Area	8529	8309	8672	8166	8322	7922
Production	9658	7515	7090	8565	7829	8255
Yield	1132	904	817	1049	941	1042
SESAMUM						
Area	2448	2516	2608	2129	2217	2034
Production	682	835	710	758	564	618
Yield	279	332	272	356	254	304
RAPESEED						
Area	4832	5982	6470	6193	6289	6230
Production	4377	5229	5870	4803	5328	5884
Yield	906	904	907	776	847	944

Table 24.5 (Contd.)

Table 24.5 (Contd.)

Oilseeds	89-90	90-91	91-92	92-93	93-94	94-95
SUNFLOWER SEED						
Area	1104	1633	2097	2085	2668	1970
Production	370	873	1190	1182	1348	1204
Yield	335	535	567	567	505	611
SOYABEAN						
Area	1734	2564	2821	3789	4371	3993
Production	1547	2602	2490	3387	4745	3666
Yield	892	1015	882	894	1086	918
SAFFLOWER SEED						
Area	315	823	491	734	793	771
Production	445	320	200	348	521	419
Yield	546	389	407	474	658	543
LINSEED						
Area	1199	1099	913	895	954	964
Production	361	332	290	278	330	325
Yield	301	302	317	311	346	337
CASTOR SEED						
Area	631	810	714	660	684	787
Production	415	716	580	623	635	850
Yield	658	884	812	945	927	1079
NIGER SEED						
Area	604	611	642	585	599	589
Production	178	186	180	162	195	196
Yield	295	304	280	277	327	333
TOTAL NINE						
Area	21896	24347	25428	25236	26897	25260
Production	18033	18608	18600	.20106	21496	21417
Yield	824	764	731	797	799	848

Source : Directorate of Economics & Statistics, Ministry of Agriculture.

Note : Trade Estimate 1993-94 20.40 Million Tonnes & 1994-95 19.97 Million Tonnes.

Table 24.6 : State-wise Area, Production and Yield of Nine Oilseeds (Area : Lakh Hect., Production : Lakh Tonnes)

	1994-95					1993-94				
State	*Area (Lakh Hectare)*	*% of Total Area*	*Production (Lakh Tonnes)*	*% of Total Production*	*Yield (Kgs/ Hect)*	*Area Lakh Hectare)*	*% of Total Area*	*Production (Lakh Tonnes)*	*% of Total Production*	*Yield (Kgs) Hect)*
Gujarat	30.4	12.1	37.1	17.3	1219	30.2	11.2	15.7	7.3	521
Andhra Pradesh	30.0	11.9	21.5	10.0	714	32.5	12.1	29.0	13.5	893
Madhya Pradesh	48.1	19.1	35.9	16.8	746	53.2	19.8	47.4	22.0	890
Rajasthan	34.9	13.8	28.3	13.2	812	36.1	13.4	24.1	11.2	666
Maharashtra	26.6	10.5	18.1	8.5	682	28.4	10.5	23.5	10.9	825
Karnataka	25.6	10.1	15.9	7.4	621	31.3	11.6	18.9	8.8	604
Tamil Nadu	14.2	5.6	19.9	9.3	1404	13.6	5.1	19.7	9.2	1446
Uttar Pradesh	17.2	6.8	14.4	6.7	839	17.2	6.4	14.2	6.6	828
Orissa	4.5	1.8	3.0	1.4	651	5.5	2.0	2.9	1.3	532
Haryana	6.3	2.5	8.7	4.1	1393	6.2	2.3	8.7	4.0	1392
West Bengal	5.3	2.1	4.2	2.0	797	5.3	2.0	4.2	2.0	781
Punjab	2.1	0.8	2.6	1.2	1261	1.8	0.7	2.4	1.1	1329
Assam	3.1	1.2	1.6	0.7	530	3.1	1.2	1.4	0.7	474
Bihar	2.3	0.9	1.5	0.7	682	2.4	0.9	1.5	0.7	623
Others	2.0	0.8	1.5	0.7	—	2.2	0.8	1.4	0.7	—
All-India	252.6	100.0	214.2	100.0	848	269.0	100.0	215.0	100.0	799

1 Million = 10 Lakh Tonne means Metric Tonne.

Table 24.7: Production of Vanaspati, Refined Oil, Industrial Hard Oil, Margarine and Bakery Shortening by Vanaspati Factories

'000 T)

Oil Year (Nov.-Oct.)	Vanaspati	Refined Oil	Industrial Hard Oil	Margarine	Bakery Shortening
1981-82	903.2	87.2	64.1	1.7	NA
1982-83	893.4	83.3	67.9	1.1	NA
1983-84	881.6	42.4	80.0	1.2	3.2
1984-85	916.8	64.0	80.3	6.4	2.6
1985-86	917.3	64.6	59.2	6.3	3.2
1986-87	929.3	86.6	62.3	5.5	2.3
1987-88	985.6	85.9	38.7	0.6	0.8
1988-89	970.1	56.9	42.5	0.7	0.9
1989-90	880.8	56.1	33.2	1.1	1.6
1990-91	816.6	41.4	38.0	1.4	2.0
1991-92	885.7	56.2	22.5	1.7	5.3
1992-93	890.7	47.8	17.3	1.8	4.9
1993-94	937.4	50.2	20.4	2.8	8.3
1994-95	945.1	103.9	25.4	3.8	24.1
1995-96	991.9				

Source : Directorate of Vanaspati Vegetable Oils & Fats N.A. Not Available

Table 24.8 : World Production (MN.T), Yields (T. HA) and Harvested area (MN. HA) of Major Oilseeds

Oilseed	Production			Yield			Area		
	95/96 (P)	94/95	93/94	95/96 (P)	94/95	93/94	95/96 (P)	94/95	93/94
Soyabean	123.24	135.51	117.06	1.99	2.19	1.95	61.96	61.88	60.15
Cottonseed	34.32	32.94	29.69	0.98	1.02	0.96	35.15	32.40	30.97
Groundnut	19.31	19.60	18.13	0.94	0.93	0.88	20.59	21.01	20.53
Sunflower Seed	25.18	24.19	21.39	1.24	1.25	1.18	20.23	19.36	18.20
Rapeseed	34.72	29.87	27.04	1.42	1.31	1.35	24.52	22.87	20.02
Sesame Seed	2.47	2.48	2.46	0.40	0.41	0.40	6.10	6.06	6.09
Palm kernels	4.77	4.56	4.23	0.96	0.97	0.95	4.96	4.69	4.45
Copra	4.91	5.22	4.80	0.53	0.57	0.52	9.19	9.21	9.19
Linseed	2.56	2.45	2.18	0.67	0.67	0.62	3.85	3.68	3.52
Castor Seed	1.37	1.31	1.14	0.88	0.88	0.83	1.56	1.50	1.38
Total	252.85	258.13	228.12	1.34	1.41	1.31	188.11	182.66	174.50

Source: Oil World Annual 1996 P Provisional

***Table 24.9*: Imports of Edible Oils (Oils Year 1976-77 to 1995-96)**

(Qty in Lakh M.T)

Oil Year	*Total Imports**	*Vanaspati*	*PDS*
1976-77	4.05	4.27	0.28
1977-78	5.23	5.23	0.50
1978-79	8.21	4.47	0.93
1979-80	11.49	6.04	3.54
1980-81	10.74	6.41	4.37
1981-82	9.98	5.73	4.34
1982-83	11.50	6.00	5.36
1983-84	16.34	6.95	7.04
1984-85	13.69	6.20	7.32
1985-86	12.76	4.90	6.72
1986-87	16.45	4.80	8.89
1987-88	18.19	6.17	10.75
1988-89	3.73	0.58	3.58
1989-90	6.07	S	3.58#
1990-91	1.02	S	1.02#
1991-92	2.15	S	2.15
1992-93	0.52	S	0.25
1993-94	2.00	S	1.46
1994-95	8.32	S	1.50
1995-96(E)	12.50	S	2.00

* Includes imports under OGL which tapered off after 1980-81. Includes open market distribution by STC.

\# Includes open market distribution by NDDB/STC.

S System of allocation of imported oils for Vanaspati Manufacture discontinued from 1989-90.

& Edible oil import placed under OGL w.e.f. 1.3.94.

Sources: Ministry of Civil Supplies, Consumer Affairs and Public Distribution Approximately.

Table 24.10: India's Share In World Production of Oilseeds

(Qty. in 000 Tonnes)

Oilseeds	1995-96 (P)	1994-95	1993-94	1992-93	1991-92	1990-91
SOYABEAN						
World	123240	135506	117058	117508	107847	104203
India	4450	3100	3700	3106	2492	2419
Percentage	3.61	2.29	3.16	2.64	2.31	2.32
COTTON SEED						
World	34325	32940	29688	32196	37691	33891
India	4700	4590	4120	4740	4090	3980
Percentage	13.69	13.93	13.88	14.72	10.85	11.74
GROUNDNUT						
World	19312	19602	18134	17187	16621	16511
India	5100	5600	5390	6198	4967	5336
Percentage	26.41	28.57	29.72	36.06	29.88	32.32
SUNFLOWER SEED						
World	25180	24188	21390	21642	22733	23107
India	1380	1350	1400	1185	1194	889
Percentage	5.48	5.58	6.55	5.48	5.25	3.86
RAPESEED						
World	34724	29869	27039	25601	28534	25310
India	6100	5300	5390	4872	5863	5152
Percentage	17.57	17.74	19.93	19.03	20.55	20.36
SESAME SEED						
World	2467	2479	2464	2398	2125	2320
India	880	842	790	853	706	810
Percentage	35.67	33.97	32.06	35.57	33.22	34.90
PALM KERNELS						
World	4767	4559	4233	4140	3552	3344
India	3	2	2	2	3	2
Percentage	0.06	0.07	0.05	0.05	0.08	0.05
COPRA						
World	4910	5220	4801	4648	4612	4893
India	490	470	455	445	440	420
Percentage	9.98	9.00	9.48	9.57	9.54	8.58

Table 24.10 (Contd.)

Table 24.10 (Contd.)

Oilseeds	1995-96 (P)	1994-95	1993-94	1992-93	1991-92	1990-91
LINSEED						
World	2555	2448	2177	1969	2458	2993
India	340	345	330	268	410	339
Percentage	13.30	14.09	15.16	13.61	16.68	11.60
CASTOR SEED						
World	1368	1318	1135	1140	1163	1326
India	930	900	700	617	575	716
Percentage	67.98	68.29	61.67	54.12	49.44	54.00
TOTAL						
World	252847	258128	228120	228429	227335	217737
India	24373	22500	22277	22285	20621	20061
Percentage	9.64	8.72	9.77	9.76	9.07	9.21

Source : Oil World Annual 1996.

Table 24.11: Per Capita Availability of Edible Oils & Vanaspati

Financial Year	*Liquid Oil (Kg.)*	*Vanaspati (Kg.)*	*Total Edible Oils & Fats*
1979-80	3.7	1.0	4.7
1980-81	3.8	1.2	5.0
1981-82	5.1	1.3	6.4
1982-83	4.5	1.3	5.8
1983-84	5.8	1.2	7.0
1984-85	5.5	1.3	6.8
1985-86	5.0	1.3	6.3
1986-87	5.0	1.2	6.2
1987-88	5.8	1.2	7.0
1988-89	5.3	1.2	6.5
1989-90	5.3	1.1	6.4
1990-91	5.5	1.0	6.5
1991-92	5.6	1.0	6.6
1992-93	5.9	1.0	6.9
1993-94	6.0	1.0	7.0
1994-95 (P)	6.5	1.0	7.5

Source : Economic Survey 1994-95 (P) Provisional.

Table 24.12: Estimated Availability of Vegetable Oils & Demand During 1995-96 (Nov.-Oct.)

(Qty. in Million Tonnes)

Supply Source		Oil Recovery %	1995-96 Marketable Surplus for crushing	Oils availability
OILSEEDS				
1.	Groundnut (Kernel)	40	3.80	1.52
2.	Soya	17	3.50	0.60
3.	Rape	33	6.00	1.98
4.	Sunflower	35	1.47	0.51
5.	Seasame	45	0.55	0.25
6.	Castor	42	0.80	0.12
7.	Niger	30	0.20	0.06
8.	Safflower seed	30	0.34	0.10
9.	Linseed	43	0.30	0.13
10.	Cottonseed	11	3.83	0.42
11.	Copra	66	0.70	0.46
12.	Rice Bran Oil	—	—	0.45
13.	S E Oils	—	—	0.30
14.	Minor Oils	—	—	0.12
	Sub Total	—	21.49	7.02
IMPORTED OILS (Est.)				
15.	Edible Oil			@ 1.00
16.	Non-edible oil			@ 0.20
Sub Total		—	—	1.20
Total Supply/Demand		—	—	8.22

Note @ Estimated import requirement

	Castor Oil Production	Export	Net availability
1995-96	0.34	0.22	0.12
1994-95	0.32	0.16	0.16

REFERENCE

1. *News Bulletin* VOP, Vol.2, No.1, March 1997, Published by the Directorate of Vanaspati, Vegetable Oils & Fats, Ministry of Civil Supplies, Government of India

CHAPTER 25

Intellectual Property Rights and Patents Materials in India

Intellectual Property (IP) is the product of human mind resulting in ideas, inventions, designs, computers etc. and the legal right arising out of such IP is called Intellectual Property Right (IPR). This is however, the colloquial description of IPR, but the legal description is slightly different from the colloquial because it focusses upon the rights which are enjoined in the produce of the mind rather than upon the produce itself. IPR cannot reside in an idea unless such idea is translated generally into some form of expression in nature. A writer may conceive a new plot for his novel but there is no IPR until such plot is reduced to the form of a novel or a fiction. Another author may use a similar idea independently conceived to create another novel by a different form of expression and be the subject matter of IPR. Likewise inventions can be protected as IPR when a new or old idea is translated or reduced to a new product and/or process. The principal facets of IPR are patents, designs, trade-marks and copyrights.

Inventions are protected as patents under the Patent Law. Industrial designs of aesthetic appeal such as design of a ball pen, tooth brush or a textile design may be protected as industrial designs under the Designs Act 1911. Artistic work, musical work, literary work are protected under the Copyright Act. Software and other computer programmes come under the purview of Copyright Act 1957 as literary work.

Trade-marks, trade-names, logo, monograms which are applied to products of commerce are protected under the Trade and Merchandise Marks Act 1958. The various IPR laws were essentially adopted from the British laws. IPR related to scientific work is protected by the Indian Patents Act of 1970. The patent is a privilege granted by the State to a person or persons for his/their invention. The patent is a contract between the inventor and the State under which the State guarantees monopoly under certain conditions to the inventor for his invention, and in return the inventor is required to make a full disclosure of his invention to the State so that on the expiration of the monopoly the public can benefit by using such invention. By such arrangement the inventor is protected by the State for a substantial period and the public is benefited by way of access as well as the liberty to make use of such invention after the expiry of the patent. A patent granted by the State is a property, more particularly having the characteristics of a moveable property and can therefore be sold, licensed, mortgaged and any other interest created therein.

The creation of the monopoly by a patent does not take place automatically but is an intricate exercise which is a blend of technical knowledge in a legal descriptive form which provides the inventor the protection. By patenting his invention the inventor can stop others from using the invention by the strength of his patent. A patent is a useful tool for transfer of technology and transmission and propagation of useful technical knowledge for the elevation of the State of technology.

Patents enable the creation of a competitive atmosphere forcing innovative work by others. If one pharmaceutical industry invents and patents a new drug molecule of immense use it will have a tremendous response and the new product will fetch handsome profits to the company. The demand for the new product will automatically bring down the demand for the competing and less effective products, and the price and profitability of the conventional products will crash. Thus if the competitors, i.e. other pharmaceutical companies have to suvive they will also have to engage themselves in high quality research to find new and useful products.

Such constant research will lead to new products and consequently generate immense investment opportunities which will eventually take care of all the financing problems for R&D. Such R&D will eventually give rise to products which will also find decent demands not only nationally but also in the international markets, and this will create global openings for Indian companies and provide opportunities for Indian companies to become MNCs. There is no substitute to scientific research and investment in the era of globalization. In the intellectual world the most effective weapon of immense economic value to individual as also to its nation is technology and one can counter such weapon with technology only.

Transfer of technology in India has so far been one sided and India has always been the recipient. The import of technology of whatever worth eats up substantial part of the profits of the Indian enterprises and therefore hinders growth of the Indian enterprises. The Indian enterprises are looking to foreign collaborators to solve every technical problem and thereby wholly relying and depending on foreign collaborators. Most of the Indian enterprises who import technology basically involve themselves in trading and screw driver technology rather than meaningfully developing the technology. The trend amongst the Indian entrepreneurs is to run abroad for technology without really exploring what technology is indigenously available or may be generated indigenously. This is essentially due to lack of coordination between the R&D at university and other levels and the industry. One effective means of establishing or narrowing the gap is patents. If one makes a search in the Indian Patent Office he would hardly find any Indian patent on most of the subjects and therefore finds it appropriate to look to the West for the technology. On the other hand, if the search reveals relevant patents by Indian industries or R&D such patents could be useful bridges between the industry and the R&Ds.

In India the level of patent consciousness is very low particularly amongst the scientific community as the concept of protection of ones original work is lacking and the primary priority is to publish the work. Lot of original and useful work made in India has been published without protection although many of such works could have been subject matters of strong patents and attracted substantial investment and benefit and could have also been exploited outside India. Even today the foreigners are major patent holders in India and the holding is due to the fact that the Indian universities and the industries are not doing co-ordinated work with the result that good and useful work is not finding its way into the industry. The industry may be purchasing inexpensive foreign technology while our own scientists may be well versed with the science and the related technology. With the advent of globalisation, GAAT and other treaties which are all aimed at common markets are threatening the countries like India which still operate with traditional and out-dated technology. Unless India comes up to international standards in terms of R&D and technology it will find it difficult to compete on equal terms and will soon dwindle in the face of advanced technology. Technology can originate from industries, institutes and laboratories and therefore it is the research workers who have to contribute in their respective fields to upgrade technology. Of course, this will require a lot of investment, and such investment can come from private sector as well as from the Government provided the output of the research promises good results.

It is not enough to make inventions and create technologies, it is also important that the inventions should be adequately protected under patents in India and abroad, where necessary, and it is extremely important that researchers should make themselves acquainted with the patent laws and procedures.

The most basic requirement of patenting is that one should not publish, use or otherwise disclose his invention to anyone before applying for patent. This is an essential criterion for patenting whether in India or abroad. Patent is of immense importance in today's scenario, and quality and quantity of patenting are today considered as indicators as to the state of technology of a given country or industry. Besides, patent information is a phenomenal source of technical information—about 80 % of the latest scientific information is available in patent literature. If one has to have an insight into the state of science in a particular field the best source is patent literature. Before starting R&D on a particular project it is essential that one should conduct a patent literature search otherwise there could be duplication and overlapping of work. It is therefore high time that we learn to make use of patents and simultaneously learn to protect our valuable findings through the process of patents.

Scientists and technocrats should learn how to design their research for obtaining patents and making invention patent worthy. Patent specifications are available at the Indian Patent Office, and a study of these specifications of the respective discipline would give a fair idea to an inventor as to how to design the research and how to and what to disclose in a patent. A patent search of foreign patents may also be done sitting in India for which the government agencies are offering free services to the scientific community, especially to the universities. There are certain criteria for patenting. The Patent Act defines both patentable and non-patentable inventions.

WHAT MAY BE PATENTED ?

Any invention which satisfies the definition of 'invention' given in the Act may be patented.

(a) As defined in Section 2 (i) (j) of the Act an invention means any new and useful:

  (i) art, process, method or manner of manufacture;

  (ii) machines, apparatus or other article;

  (iii) substances produced by manufacture, and includes any new and useful improvement of any of them and alleged invention.

(b) The following are not "inventions" whithin the meaning of the Act :

  (i) an invention which is frivolous or which claims anything obviously contrary to well established natural laws;

  (ii) an invention the primary or intended use of which would be contrary to law or morality or injurious to public health;

  (iii) the mere discovery of a scientific principle or the formulation of an abstract theory;

  (iv) the mere discovery of any new property or new use for a known substance or of the mere use of a known process, machine or apparatus unless such known process results in a new product or employs at least one new reactant;

(v) a substance obtained by a mere admixture resulting only in the aggregation of the properties of the components thereof or a process for producing such substances;

(vi) the mere arrangement or re-arrangement or duplication of known devices each functioning independently of one another in a known way;

(vii) a method or process of testing applicable during the process of manufacture for rendering the machine, apparatus or make the equipment more efficient or for the improvement or restoration of the existing machine, apparatus or other equipment or for the improvement or control of manufacture;

(viii) a method of agriculture or horticulture;

(ix) any process for the medicinal, surgical, curative, prophylactic, or other treatment of human beings or any process for a similar treatment of animals or plants to render them free of disease or to increase their economic value or that of their products.

(c) Certain inventions not patentable

As provided in section 4 of the Act, no patent shall be granted in respect of an invention relating to atomic energy falling within Sub-section (1) of Section 20 of the Atomic Energy Act, 1962.

(d) Inventions in respect of which only the process of manufacture is patentable

In respect of inventions :

(i) claiming substances intended for use or capable of being used as food or as medicine or drug, or

(ii) relating to substances prepared or produced by chemical processes (including alloys, optical glass, semi-conductors and intermetallic compounds), no patent shall be granted in respect of claims for the substances themselves, but claims for the methods or processes of manufacture shall be patentable.

The term "medicine or drug" includes insecticides, germicides, fungicides and the like.

For obtaining an invention the inventor has to show that his process is novel and inventive and useful as well. In his patent document the inventor should also disclose and discuss all relevant prior art and distinguish such prior art with his invention. Every inventor should try to assess the potentiality of his invention in the global scenario and try to work out a programme for protecting the invention also in other countries. Patent is a national right and does not extend to other countries. An Indian patent protects the inventor only in India and not at other places outside the territories of India. If a person uses the invention patented in India and Pakistan there will be no infringement unless he has obtained a patent in Pakistan under the Pakistani laws. It is therefore important that we should first learn to recognise what is protectable and how best to exploit the intellectual property.

SCIENTIST'S RESPONSIBILITY TO PROTECT IPRS

Most scientists are not sure about their intellectual property rights and, most often, find legal language to be confusing. However, they have an equal responsibility to protect the research work being done by them or else the opportunities underlying their research could be unduly lost to others. They must learn to manage protection of their IPRs and dovetail their learning with day-to-day R&D activities and practices.

PATENTING IN RESEARCH AND DEVELOPMENT

Patent information has not been frequently used in India as an input for deciding research topics. This would become increasingly important in the days to come as India also catches the attention of foreign companies for patenting. Hence, a thorough patent search and study of available patents becomes essential. The notion that no new patent can be obtained in respect of already patented area is not correct. Further, patents provide knowledge about the state of art in technology and thus could determine R&D priorities. Two examples of this kind of exercise which was undertaken a few months ago are presented here.

ANALYSIS OF THE PATTERN OF PATENTING IN CHEMICALS IN INDIA IN RECENT YEARS IN RELATION TO OTHER COUNTRIES

The chemical industry globally is experiencing a gradual shift in emphasis to value-added, specially performance-oriented, products and cleaner/safer technologies. Staying ahead in this evolving competitive scenario will necessarily require well focussed R&D activities, translating innovations faster to viable technologies/products, ensuring their protection through IPR, quality manufacturing management practices and reaching the market place first. Likewise, the Indian industry is now engulfed in this whirlwind of change. It will have to refine its strategies for survival and sustained future growth.

Before moving to the subject of Patents, it would be useful to take a bird's eyeview of the chemical industry. Tables 25.1 and 25.2 are snapshots of the global and Indian chemical industry. Though India's output compared to other nations is fairly small, it grew at an average rate of 10-15 % during the 1980s. It is expected to cross the US $ 60 billion mark in the next 15 years. Growth will continue to be in fertilizers, consumer products and specially chemicals.

***Table 25.1*: Global Chemical Industry**

Region	*Output (US $ billion)*
Western Europe	340
North America	275
Japan	190
Eastern Europe	170
India	15
Rest	140

Table 25.3 shows a comparison of the granted patents in USA, Japan, Europe and India in terms of their break-up (in percentage) into chemicals, pharmaceuticals, electrical, general and mechanical. It is clear that pharmaceuticals occupy a small share (2-7 %) as compared to other areas. In the four regions, patents in the chemical category vary from 23-36 % of the granted patents.

Table 25.2: Indian Chemical Industry : Composition

	Percentage
Fertilizers	23
Cosmetics and toiletries	22
Drugs and pharmaceuticals	19
Inorganic chemicals	10
Organic chemicals	9
Paints	7
Dyes	6
Pesticides	2
Others	2
	100

Table 25.3: Comparison of Granted Patents in USA, Japan, Europe and India in the Year 1992-93.

Country	Total granted patents (%)			
	Chemical	Pharmaceutical	Electrical	General & Mechanical
USA	25	7	30	38
Japan	36	2	43	20
Europe	23	4	26	47
India	25	7	15	53

Note : General also includes food patents. In India the contribution from patents in foods is only 1 %.

Source : Data on USA, Japan & Europe from Patent World, February 1994. Data on India from Patents : 21st Annual Report of the Controller General of PDTM under section 155 of Patent Act, 1970 (1992-93).

THE INDIAN SCENE

Figure 25.1 shows the total number of patents sealed in India from 1972 to 1990. The data in Tables 25.4 and 25.5 are the patents filed in various fields. It will be noted that chemicals stand out with the maximum number of patents filed every year. On an average, only

42 % of the patents filed in the areas of chemicals actually proceed to the granting stage compared with those in the areas of drugs, food, electrical, mechanical and general where the granting rate is above 60 %. It should be noted that most patents in India are filed by foreign inventors.

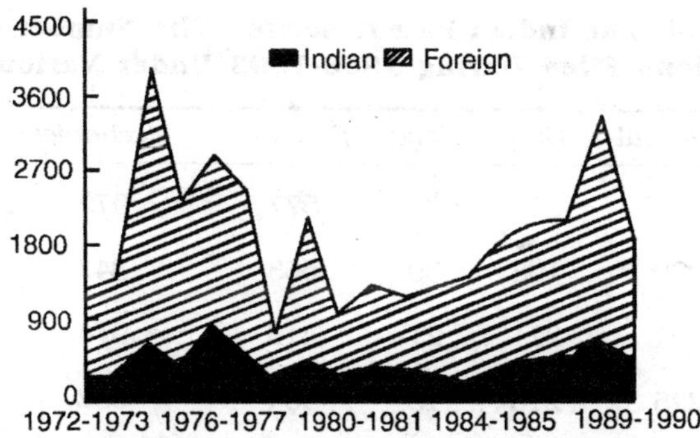

Patents sealed in India (1972-1990) *Source : Pocket Data Book – Science and Technology, Department of Science and Technology,* Government of India, New Delhi. 1992

Fig. 25.1

Most of the filing in chemicals are inventions in the field of organic chemicals with maximum contribution in polymer chemistry. In 1992-93, the largest number of filings are by foreign inventors as in the previous years. Contributions from Indian inventors have been mostly in the application of surface-active agents, hydrocarbons, surface coatings, cosmetic preparations, lubricants, synthetic paper and dyes. Organic chemicals like amino anthraquinone, refrigerants like halo carbons, hydrocarbons etc. also found magical presence. In drugs and pharmaceuticals, slow or controlled release formulations are gaining significance. Patenting in inorganic chemicals have been mostly in the processes for manufacture of diverse catalysts, silicates, and recovery of industrial wastes etc.

When the patentings trends in India are compared to those elsewhere, it is evident that the patenting regime is extremely restricted to processes and formulations. However, elsewhere, practically all areas which find useful applications in industry are patentable. The number of patents worked in India is also insignificant, as seen from Table 25.4. Such trends can result from protection of inventions restricted by the prevailing Patents Act and the nature of projects carried out by industrial R&D centres in India.

***Table 25.4*: The Indian Patent Scene : The Number of Patent Applications Filed During 1986-1993 Under Various Fields.**

Year	*Chemical*	*Drug*	*Food*	*Electrical*	*Mechanical*	*General*	*Total*
1986-1987	1112	214	34	577	972	579	3489
1987-1988	1020	198	39	563	847	772	3598
1988-1989	1191	221	21	419	974	772	3598
1989-1990	1225	216	13	454	932	821	3661
1990-1991	1297	258	41	492	1173	502	3763
1991-1992	1185	323	38	468	994	544	3552
1992-1993	1138	234	29	461	946	659	3467

***Table 25.5*: The Number of Patents Granted During 1986-1993 Under Various Fields**

Year	*Chemical*	*Drug*	*Food*	*Electrical*	*Mechanical*	*General*	*Total*
1986-87	474	185	21	293	705	448	2125
1987-88	731	124	15	228	489	517	2104
1988-89	704	300	44	608	1183	541	3380
1989-90	389	300	44	322	692	321	1890
1990-91	339	87	10	285	535	235	1491
1991-92	474	118	10	167	181	726	1676
1992-93	318	94	12	194	372	282	1272

Source : Patents - 21st Annual Report of the Controller General of PDTM under section 155 of Indian Patent Act, 1970 (1992-93)

Table 25.6: Number of Patents Worked During 1987 - 1991 with Classification.

Year	Chemical	Mechanical & General	Electrical	Total
1987	151	318	89	558
1988	155	301	118	574
1989	64	172	53	289
1990	80	134	38	282
1991	93	210	40	343

Source : Patents - 21st Annual Report of the Controller General PDTM under section 155 of the Indian Patent Act, 1970 (1992-93).

BIOTECHNOLOGY AND PATENT PROTECTION WITH PARTICULAR RELEVANCE TO INDIAN PATENT LAW NEEDS TO BE DISCUSSED

Introduction

The exciting developments in biotechnology have resulted in intensive R&D activities all over the world including India. The technologies emerging by the application of biotechnology provide vast potential for industries based on new inventions in the field. Most of the inventions developed in biotechnology fall mainly into the following broad categories:

(i) Preparation of chemical substances utilising genetically engineered organisms. The substances prepared may be new or may be known but can be prepared by a new improved process by the application of biotechnology which may be simple and economical.

(ii) Genetically engineered organisms and altered genes, DNA probes, vectors etc.

(iii) The process techniques that are employed in the production of abovesaid substances which fall in the area of genetic engineering, protein engineering, cell fusion and hybridoma technology, tissue culture, gene therapy and fermentation technology.

(iv) Basic studies dealing with the phenomena of various biochemical, physiological processes in living cells so as to know the signals which are given for expression, secretion of certain substances, reproduction etc.

(v) studies on the structure and role of the molecule such as chromosomes, cytoplasm, DNA, RNA, specific hormones, neurons and inter and intra communications amongst them.

Majority of the abovesaid inventions relating to biotechnology fall within the field of chemical inventions, more particularly the fields of food, agriculture, drug and medicines. It is now common knowledge that there are specific problems in patenting inventions relating to biotechnology. The problems arise mainly due to the fact that inventions in the area of biotechnology involve living materials which are very difficult to describe. It is difficult to differentiate such inventions from discovery and there is difficulty in repeatability of the inventions. The inventions in the fields of agriculture, plant and animal breeding etc. were not patentable. Later, with the development of fermentation industry the meaning of technical character was altered so as to include the processes of fermentation utilising bacteria/microorganism for the production of industrial products like alcohol, antibiotics etc.

These developments can be considered as exceptions from the earlier view that inventions belonging to the field of technology encompassed only objects and phenomena of *inanimate nature* and their utilisation and control by techniques of physics and chemistry. In 1980 after the famous Chakravorty case, it was decided in USA that relevant distinction for an invention to be patentable and not patentable was not between living and non-living things, but the consideration as to whether the invention is developed by human intervention or not.

These developments are milestones in the history of patent protection, since they opened up avenues for the protection of various categories of biotechnology inventions, which earlier were not considered patentable. In USA, the concept of technical character no longer constitutes any obstacle for the protection of biotechnological inventions. They have accordingly been granted patents for a mouse (a mammal genetically engineered to contain a cancer causing gene), US Patent No. 4,736,866 to Harvard University, designed to radically improve the process of detecting cancer producing substances.

PATENTABILITY OF INVENTIONS IN THE FIELD OF BIOTECHNOLOGY IN INDIA UNDER THE PATENTS ACT, 70

In the abovesaid background, the position of protection of inventions in the field of biotechnology by patents in India under the Patents Act, 1970 is examined below:

The Patents Act, 1970 was enacted after a very careful consideration keeping in view the pace in which the development of science & technology is taking place with the main objective of encouraging inventions and to secure that the inventions are worked in India within a reasonable period of time.

A scrutiny of the nature of patents granted under the Act in the field of biotechnological inventions included in the publication titled "Indian Patents in Biotechnology" published during the period 1972-88' and the update covering the period 1989-90, it can be observed that majority of the inventions falling within this field are patentable under the Patents Act, 1970. It may be observed that patents are being granted under the Act for inventions relating to a process for the preparation of DNA, RNA, peptides by recombinant DNA techniques, a process for preparing inter-ferongamma. The inventions falling in this category which are not allowed patent protection are those falling in the area of genetic engineering involving and producing live forms, plant variety etc.

There is a wide spectrum of opinion on the question of providing patent protection for inventions in the field of biotechnology. One school of thought considers that inadequate patent protection is the most important constraint on further developments in this field, while there are others who consider this constraint to be irrelevant believing that the secrecy backed up by aggresive marketing for the final product can build a quick lead over potential rivals and achieve market domain. It seems the answer to this problem lies between these two extreme positions.

The progress being made in the field of biotechnology indicates that in the very near future the existing pharmaceutical industry is likely to be superseded by unique developments in this field. Similarly, in the field of agriculture, biotechnology can be applied for the production of biocides, useful for exterminating unwanted organisms, bio-fertilizers, growth promoting hormones, and the like. In the case of industrial applications, it may relate to fermentation processes and also for the treatment of wastes for the production of useful products. Many areas of biotechnology are relatively expensive at R&D stage, but are relatively cheap for commercial applications due to low investment in plants and equipments. There is also an inherent danger of copying and repeating the invention, thereby affecting the interests of initial inventors and investors. The need of the hour is, therefore, to give a thorough and immediate attention to the subject of providing patent protection for the inventions involving and producing living materials, taking into account the advantages and disadvantages of providing such protection in the fast developing area of biotechnology and arrive at an appropriate decision which will be most beneficial to our people and for the development of our country.

Index

C

T

U

V

W

Z